Irrigation Water Pricing

The Gap Between Theory and Practice

Irrigation Water Pricing

The Gap Between Theory and Practice

Edited by

F. Molle

and

J. Berkoff

www.cabi.org

CABI is a trading name of CAB International

CABI Head Office
Nosworthy Way
Wallingford
Oxfordshire OX10 8DE
UK

Tel: +44 (0)1491 832111
Fax: +44 (0)1491 833508
E-mail: cabi@cabi.org
Website: www.cabi.org

CABI North American Office
875 Massachusetts Avenue
7th Floor
Cambridge, MA 02139
USA

Tel: +1 617 395 4056
Fax: +1 617 354 6875
E-mail: cabi-nao@cabi.org

A catalogue record for this book is available from the British Library, London, UK.

Library of Congress Cataloging-in-Publication Data

Irrigation water pricing : the gap between theory and practice / edited by François Molle and Jeremy Berkoff.
p. cm. -- (Comprehensive assessment of water management in agriculture series)
Includes bibliographical references and index.
ISBN 978-1-84593-292-3 (alk. paper)
1. Irrigation water--prices. 2. Water-supply, Agricultural--Economic aspects I. Molle, François. II. Berkoff, Jeremy, 1943- III. Title. IV. Series.

HD1714.1774 2008
333.91'3--dc22

2007021697

Typeset by SPi, Pondicherry, India.
Printed and bound in the UK by Biddles Ltd, King's Lynn.

Contents

Contributors

J. Berbel is an Assistant Professor, Agricultural Economics Universidad de Cordoba. His main research interest is operational research applications to environmental economics and water management. He is a member of the Strategic Committee for Common Implementation of European Water Framework Directive and he has been also involved in the management of public and private environmental firms.

J. Berkoff is an independent Water Resources Economist and consultant with more than 40 years' work experience. Before becoming independent, he worked on water resources, irrigation and development issues for the World Bank, the British Government and consultancy firms. His work has primarily been in South, Central and East Asia but he has also worked in the Middle East, Africa and elsewhere.

J. Calatrava is an Associate Professor of Agricultural Economics at the Universidad Politécnica de Cartagena, Spain. He has a PhD in Agricultural Economics (Universidad Politécnica de Madrid, Spain, 2002). He is an agricultural engineer and the author of 20 papers in scientific journals and books on the issues of Water Economics and Policy and Agricultural Resource Economics. He has participated in nine Spanish and European research projects on water economics, agricultural and environmental policy and the economics of soil erosion.

J.-P. Fontenelle is an agronomist specialized on social water management. He has been working for more than 10 years with the Group of Research and Technological Exchanges (GRET), a French NGO. He spent 5 years in the Red River delta in Vietnam from 1994 to 1999, where he conducted research activities on agriculture and water management within the Red River Program. He completed his PhD on 'Agrarian Dynamics, Irrigation and Institutions in the Red River Delta (Vietnam)' with the University of Louvain-la-Neuve in Belgium.

C. de Fraiture is a Senior Researcher at the International Water Management Institute. Her background is in water engineering and economics. She is currently working on the development of global scenarios on future water and food demand.

Z. Gao is Director General and Professor for the Department of Irrigation and Drainage in the China Institute of Water Resources and Hydropower Research and the past Secretary for the International Committee on Irrigation and Drainage. He leads a department that has extensive research experience into irrigation policy issues and water allocation in China.

A. Garrido is an Associate Professor of Agricultural and Resource Economics at the Universidad Politécnica de Madrid, on sabbatical at the University of California, Berkeley, California, during 2005/06. His work has focused on natural resource and water economics and policy. He has conducted consultant work for OECD, IADB, European Parliament; European Commission, Ministry of Environment, Ministry of Agriculture and Spanish Parliament, Autonomous Communities of Aragon and Navarre, Spain; IFPRI. He is the author of more than 65 papers.

Y. Hasan is Planning Director at the Ministry of Water and Irrigation/Jordan Valley Authority in Jordan. He has worked in many projects and programmes dealing with the integrated development of the Jordan Valley, and has presented a number of papers in irrigation water management in local and international workshops and conferences.

N. Hatibu is the founding Regional Coordinator of the Soil and Water Management Research Network (SWMnet) of ASARECA, since September 2003. He is a Professor at the Sokoine University of Agriculture in Tanzania where for more than 12 years he was Leader of the soil and water management research group, which is credited with research and promotion of rainwater harvesting, making it a central water management strategy in Tanzania. An agricultural engineer, he has experience in project development and management, knowledge management and strategy formulation.

P.J.G.J. Hellegers has more than 12 years' experience as a Water Economist with the Agricultural Economics Research Institute (LEI) in the Netherlands and has currently a joint appointment with the International Water Management Institute (IWMI). She has particular interest in the costs, price and value of water and the role of policy instruments in irrigation water management.

Q. Huang is an Assistant Professor in the Department of Applied Economics at the University of Minnesota. Her research addresses water and irrigation issues primarily in China as well as estimation methodologies for these issues. She has conducted extensive fieldwork into irrigation practices in China and has researched the effects of water policy on rural incomes, crop production and environmental outcomes, with a focus on groundwater extraction policies.

A. Kishore is a student of public policy at Woodrow Wilson School of Public and International Affairs at Princeton University, Princeton, New Jersey, USA. He has worked for 4.5 years at IWMI-Tata Water Policy Program as a researcher. He has a postgraduate degree in rural management from the Institute of Rural Management (IRMA), Gujarat, India.

B.A. Lankford is a Senior Lecturer in Natural Resources at the School of Development Studies at the University of East Anglia, UK. He has worked in irrigation and water resources for over 20 years, mainly in sub-Saharan Africa. He has just completed managing a collaborative research project based in Tanzania looking at river basin management.

B. Lei is an Assistant Researcher in the Department of Irrigation and Drainage at the China Institute of Water Resources and Hydropower Research. He has researched several aspects of China's rural economy and has conducted numerous fieldwork projects. His current research looks at how water pricing and irrigation management reform affect water use and rural livelihoods. He has a PhD from the Chinese Academy of Agricultural Sciences.

B. Lohmar is an Economist and a Specialist of China at USDA's Economic Research Service. He has researched several aspects of China's agricultural economy including labour allocation, the land tenure system, reform of commodity marketing and trade policies, in addition to the research in irrigation and water policies on agricultural production. He has a

PhD in Agricultural Economics from the University of California at Davis and a bachelor's in Mathematics and Chinese from the University of Minnesota.

H. Mahoo is the Team Leader of the Soil-Water Management Research Group (SWMRG) and Associate Professor, Sokoine University of Agriculture, Tanzania.

F. Molle is a Senior Researcher at the Institut de Recherche pour le Développement, France, with a joint appointment with the International Water Management Institute, Colombo. He has experience from South America, Africa and Asia on irrigation and river basin management and now focuses his research on water policy and governance issues.

C.J. Perry has 20 years' experience in irrigation and water resources project design and implementation with the World Bank followed by 5 years at the International Water Management Institute (IWMI) as head of the Performance Assessment Program and Deputy Director-General. He has particular interests in the overall framework within which successful water management is practised, the economics of water and water management, and performance assessment. He now works with WaterWatch on potential of remote sensing to contribute to improved understanding of water systems.

T. Petitguyot is a former PhD student at Cemagref (Montpellier, France). His work looked at the potential of economic instruments in preventing aquifer depletion in a context of conjunctive water use in the irrigated scheme of Tadla, Morocco. He is now in charge of water resource management in a French department.

I. Ray is an Assistant Professor at the Energy and Resources Group, University of California at Berkeley. She has a bachelor's from Oxford University and a PhD from Stanford University. Before joining the ERG faculty, she worked at the Turkey office of the International Water Management Institute. Her research interests are water and development, technology and development, and common property resource management.

C. Scott is a Hydrologist who headed the IWMI Hyderabad office from 2001 to 2005. From 2005 to 2006 he worked on river forecasting collaboration with India and Mexico, based at the US National Oceanic and Atmospheric Administration (NOAA). He is currently Assistant Professor at the Udall Center for Studies in Public Policy & Department of Geography and Regional Development, University of Arizona, Tucson, USA.

T. Shah is currently a Principal Scientist at International Water Management Institute, in their India office. He was trained as an Economist and Management Specialist and his research interest lies in institutions and policies for water resources management, a subject on which he has published extensively.

A. Sharma is currently working as a Consultant with PriceWaterhouseCoopers. He co-authored Chapter 9 when he was an employee with IWMI-Tata Water Policy Program. He has a postgraduate degree in rural management from the Institute of Rural Management (IRMA), Gujarat, India, and a master's in Public Policy and Management from the Carnegie Mellon University.

C.S. Sokile is an expert in Water Management Institutions. He has 6 years' experience in water management research. His research interests include informal water use associations, customary water rights and informal arrangements for water tariffs. He is currently a Programme Officer responsible for Social and Human Sciences at the UNESCO, Dar es Salaam Cluster Office.

H. Turral is an Irrigation and Water Resources Engineer and works as Theme Leader for 'Basin Water Management' at International Water Management Institute (IWMI). Prior to this he worked in development projects in Asia and as a researcher at ODI and the

University of Melbourne on water management for agriculture and has experience mainly in South, South-east, Central and East Asia, and Australia.

B. van Koppen is Principal Researcher, Poverty, Gender, and Water at the International Water Management Institute, Southern Africa Regional Program, Pretoria, South Africa. She specializes in the institutional and legal conditions to improve poor women's and men's access to water for multiple uses.

C. Varela-Ortega is a Professor of Agricultural Economics at the Polytechnic University of Madrid, Spain. She has an extended experience in research in the EU and international networks in the fields of water economics, agricultural policy and institutions. She has been the scientific coordinator of numerous EU research projects related to water management, a consultant for the FAO, IADB, WB, CIHEAM, EU, and has worked in a number of countries. She has published extensively in scientific journals and books, and is the Spanish representative in the IAAE, a member of the scientific advisory committee in various international institutions and a member of the advisory group of DG Research of the European Commission.

J.-P. Venot was formerly a master's student at the Institut National Agronomique de Paris-Grignon, working with the French Agricultural Mission for Water and Agriculture (MREA) in the French Embassy in Amman, Jordan. He is now a PhD scholar with the International Water Management Institute, Hyderabad, India, and the GECKO Laboratory at Paris X-Nanterre, and studies river basin water management in the Krishna basin.

P.Z. Yanda is an Associate Professor at the University of Dar es Salaam. He is currently the Acting Director of the Institute of Resource Assessment, University of Dar es Salaam. He is a physical geographer with research experience in natural resources and environment.

1 Water Pricing in Irrigation: The Lifetime of an Idea

F. Molle and J. Berkoff

Irrigation Financing and Cost Recovery

Providing irrigation always entails a measure of human labour and capital investment. In traditional small-scale systems investments were made by the communities themselves and the initial commitment generally defined rights to access water (Coward, 1980). Such undertakings were often limited (e.g. tapping a spring or a run-of-the-river diversion using a few stones or logs laid across a small stream) but could also be quite costly (as in the case of *qanats*, underground drainage galleries commonly dug over several kilometres). Larger-scale ventures were financed directly by rulers (e.g. river diversions in Mesopotamia or large tanks in South Asia) who derived economic surpluses from the increased production.

The view of irrigated agriculture as a means of ensuring both population needs and generating returns to capital was made explicit during colonial times. Investments in irrigation by the British in Sudan, Egypt, India and Sri Lanka, for example, are all well documented, and income generation and profitability were central concerns. Farmer (1976) observed that in Sri Lanka 'the English government was always concerned, and sometimes obsessed, by the protection and the increase of its income, as was the case in other colonial territory'. Colonial administrators sought both to protect and to uplift the poor masses, when considered to be in a state of misery, and involve them in productive capitalistic investments that would yield net revenues to the Crown (Bastiampillai, 1967).[1] Stone (1984) also documented the endless debates between supporters of irrigation and the guardians of the royal purse.

In contrast to narratives which assume that a focus on the economic value of water was characteristic of a late phase of water resources development, British colonial documents clearly show that most questions currently debated on the economics – perhaps more accurately the financing – of irrigation were already centre stage. The questions of who was to finance the infrastructure (local revenue, the Crown, or private interests), whether and how a water fee should be levied, what its impact on different categories of people would be, whether it should be increased, whether it could influence crop choice or water use behaviour, to cite a few examples, were fiercely debated. Opinions

[1] For example, arguing for investments in the south of Sri Lanka, a British administrator referred to the 'magnificent and really noble and philanthropic, enterprise [to be] accomplished. Nor will it be a barren philanthropy, I mean, in point of pecuniary profit even' (Steele, 1867).

diverged between the British Government, the Government of India and other colonial authorities, local governments, canal engineers, etc., and alternatives such as private investments, bulk volumetric pricing and crop-based differential rates were all tested (Bolding *et al.*, 1995).

The financial (or economic) view of irrigation lost its prominence in the four decades following World War II. Irrigation and dams became pivotal investment options for developing countries, notably newly independent states, to deliver on the promise of feeding the masses, providing income opportunities to rural populations, balancing regional development and alleviating poverty, and hence building self-sufficiency and state legitimacy. Development was seen largely as a matter of infrastructure and technical transfer, and large dams, irrigation schemes, flood control structures and other water projects received massive capital outlays (see Molle and Berkoff, Chapter 2, this volume, and Molden *et al.*, 2007). The national, as well as geopolitical, interests vested in such investments and in the increase in lending by development banks contributed to an outburst of projects, frequently undertaken on political rather than on sound economic grounds (Barker and Molle, 2004). Cost–benefit analyses often remained shoddy and there was limited scrutiny on the assumptions and projections made. All parties involved (governments, local politicians, consultants, construction firms, lending agencies, etc.) had incentives to go ahead (Repetto, 1986; Molle and Renwick, 2005), while the concerned populations were most of the time considered mere recipients of projects rather than partners in their own development. Whether politicians and engineers were infected by the 'desert bloom' syndrome (Carruthers and Clarck, 1981), fulfilled a 'hydraulic mission' through politically rewarding iconic mega-projects or aimed to revitalize an impoverished countryside, free land and water resources were seen as the basic material of agricultural development.

These investments yielded mixed results. Although much was achieved, land productivity, distribution efficiency and management often remained suboptimal, economic returns were often disappointing and environmental externalities (salinization, waterlogging) became more evident with time. Technology alone proved unfit to deal with these growing challenges and attention shifted to organizational aspects, including farmers' participation, turnover and capacity-building. Initially, the World Bank funded only new projects, but poor performance led to a policy shift towards rehabilitation in the late 1960s (Jones, 1995). A first operational policy memorandum (OPM 2.61), issued in 1971, stated that the recovery of all project costs was a normal aim but offered a loophole by adding that 'as a minimum, operation and maintenance costs should be recovered completely' (Jones, 1995). During the 1970s, the questions of why charge, and whom and how much to charge, for water stirred much debate at the World Bank. Proponents of irrigation lending and engineers perceived policy instructions as interference in their job. The prevailing philosophy remained that of 1971, though it was recognized that investment costs might be too high for beneficiaries to pay back and that a 'reasonable' share would be acceptable. Covenant language was accordingly often vague ('. . . to the extent practicable' or '. . . as much as possible') and there was virtually no capital cost recovery (Duane, 1986). An earlier study (W.A. Wapenhans, IBRD, 1969, unpublished data) had shown that 17 projects completed in the 1960s had estimated levels of charge collection that exceeded operation and maintenance (O&M) but only amounted to 29% of full costs.

In 1976, an 'informal discussion paper to assist staff in developing satisfactory approaches to cost recovery' (Ray *et al.*, 1976), followed by Central Projects Memorandum No. 8.4 (World Bank, 1976), defined new overall policy principles and guidelines, stressing three objectives as the basis for cost recovery: public savings, income distribution and economic efficiency. The objective of public savings was to 'enable governments to undertake additional rural development projects that would reach a larger number of the rural poor'. It was also recognized that recovery of all costs might not be possible and that the poor should be identified and exempted.[2] 'Efficiency pricing of irrigation water is usually not possible' but 'even a nominal price for water would offer users some incentive to eliminate at least some of the conspicuous waste and overwatering . . . which occurs when water is treated as a free good' (Ray *et al.*, 1976). Volumetric pricing was desirable but, if not practical, a benefit tax (linked to the land tax), 'although constrained by various administrative and political factors', should be considered a second-best option.

In 1981, the Operations Evaluation Department (OED) released an analysis of 26 irrigation projects completed in the 1970s (World Bank, 1981). Aside from severe problems with water management and maintenance, the survey found that cost recovery covenants had been breached in 11 cases, with no or limited water charges. Reasons included reluctance by government to reduce farm income, cultural or religious resistance, the political clout of farmers and a common 'operational' constraint: 'If project management cannot guarantee continuous and adequate water deliveries to most, or all, project beneficiaries, the Government becomes liable.' While, on the one hand, insufficient attention had been given to differing local conditions, on the other, large discrepancies in the way the Bank handled negotiations with different countries could not be explained by the policy guidelines. Lastly, no relation was found between charges and irrigation efficiency and 'factors, other than water charges, always proved to be much more important in explaining farmer behaviour than the presence, absence or absolute cost of water charges' (World Bank, 1981).

Application of the guidelines[3] in different countries proved difficult. In Indonesia, reinvestment of charges in O&M was hindered by a fiscal problem of flow of funds between central, provincial and local governments, and the willingness to pay was affected by quality of service and by a taxation on rice amounting to 37% of the world price (D. Thompson, World Bank, 1982, unpublished data); in Bangladesh irrigation remained heavily subsidized with benefits accruing to the 'better off' (World Bank, 1978); in some countries studies on farmers' ability to pay were made at the Bank's insistence but their conclusions were disregarded (World Bank, 1981).

The 1976 policy was broadened and simplified in a Policy Note (World Bank, 1984), informed by yet another survey on cost recovery performance. This note distinguished between resource mobilization and allocation and emphasized again the failure to fund O&M, regardless of how much was recovered. It was proposed that assurances should be sought of adequate funds for O&M as a substitute for demanding cost recovery but this was edited out of the final text (Jones, 1995). The lack of incentive for non-autonomous agencies to collect fees or improve management, inadequate collection mechanisms and transaction costs of collecting fees (especially if they were to be volumetric) were listed as constraints. Although the 'longer term objective to have a system of resource mobilization that will recover capital costs so permitting replicability of investments' (World Bank, 1984) remained, most Bank economists were incensed by the weakening of the principle of long-term marginal cost pricing (Jones, 1995).

A further review of conditionality and cost recovery in 1986 confirmed that in only about 15% of irrigation projects were loan covenants fully met and that recovery rates ranged from 0% to 100% of O&M costs, with most in the range of 15–45% (World Bank, 1986). Limited adherence to covenants was ascribed to: (i) the lack of government commitment; (ii) unreliable water supply due to poor O&M of irrigation systems; and (iii) the often heavy burden of direct and indirect taxes already imposed on the farming sector (World Bank, 1986).[4] The lack of relation

[2]It was proposed that an 'indicator of benefits' taken as the incremental gross value minus all incremental costs (irrigation service fees or their equivalent not considered) should be used. Farmers below a critical consumption level (CCL) to be defined would not be taxed.

[3]Reissued with minor changes in 1980 under Central Project Note No. 2.10 (World Bank, 1980).

[4]Preliminary results of the study of the political economy of agricultural policy by Krueger *et al.* (1988, 1991), as well as the review by Small *et al.* (1986), seem to have been influential in bringing this issue to the fore.

between recovery and O&M effectiveness questioned the Bank's emphasis on cost recovery, with Duane (1986) considering the Bank's approach as 'heavily influenced by its thinking about authorities supplying public utilities such as electricity, water for domestic use, etc. which were expected to be self-sustained by commercial revenues'.

The Bank policy had to come to terms with the fact that countries such as India or Thailand were clearly opposed to direct charges, either because irrigation was targeted towards the rural poor and was not expected to be self-sustaining or generate revenue, or because price distortions already siphoned off much of the agricultural surplus (Mexico, Thailand, Sri Lanka, Indonesia, Egypt, etc.) (Duane, 1986; Krueger *et al.*, 1988, 1991; Small, 1990). In 1986, the Asian Development Bank (ADB) also carried out an evaluation of its irrigation projects and came to conclusions similar to those of the World Bank's 1981 review (ADB, 1986a). In most cases, executing agencies had remained in complete or partial default of irrigation service fee covenants.

Management and Cost Recovery

Despite these disappointing reviews, 1986 was notable for a growing consensus that coalesced in a number of converging analyses of the role of irrigation service fees and their relationship to other mechanisms for improving irrigation performance. A World Bank study, for instance, condensed ideas collected from a few country-level analyses and concluded that 'it is time to take a more pragmatic and comprehensive approach to this issue' (World Bank, 1986); the ADB held a regional seminar (ADB, 1986b) and commissioned the International Irrigation Management Institute to carry out a regional study (Small *et al.*, 1986). Concurrently, US Agency for International Development (USAID) commissioned a report on 'Irrigation pricing and management' (Carruthers *et al.*, 1985), and FAO and USAID (1986) conducted an expert consultation on irrigation water charges. Several subsequent papers and reports were consonant with these views (e.g. Moore, 1989; Sampath, 1992; Vaidyanathan,[5] 1992), which were eventually summed up in a remarkable book on irrigation financing by Small and Carruthers (1991).

Although emphasis differed, there was general agreement that water charges alone were an inadequate mechanism for improving irrigation performance and that primacy needed to be given to water distribution and control. Staff members of development banks acknowledged that 'an element of subsidy in irrigation projects is not necessarily sub-optimal' (Ghate, 1985) and that 'bidding for water should not be promoted' (Frederiksen, 1986). The following list by and large summarizes this consensus:

1. *The primacy of management.* Irrigation water charges influence individual farmer behaviour in only a very few on-demand systems. By far the most important mechanism for achieving rational water use is by careful control of distribution and by allocations that broadly meet crop requirements. Fee policies have little or no impact on irrigation system performance (Svendsen, 1986).
2. *Control of supply a prerequisite.* 'Many of the frequently cited inefficiencies of water use in irrigation projects stem more from inadequate control over the distribution of the supply of water than from failure to regulated demand through prices. Supply control can reduce wastage of water associated with excess amounts of water flowing through uncontrolled canals and ungated turnouts onto fields and into drainage channels. It may also encourage more efficient use of water at the farm level by imposing a degree of water scarcity on the farmers. A substantial portion of the large efficiency gains which are sometimes expected from a demand-based pricing system would thus most probably be realized by implementation of the prerequisite supply control' (Small *et al.*, 1986).

[5]In 1992, a Committee on Pricing of Irrigation Water headed by Professor Vaidyanathan (1992) issued a report to the Planning Commission of the Government of India with recommendations regarding the pricing of irrigation water in India.

3. *Financial autonomy.* 'The way in which fees are assessed, collected and expended is more important than the actual level of fees in improving system efficiency and effectiveness. The most critical factor is the level of fiscal autonomy of the irrigation agency, i.e. the extent to which the level of its operating budget is tied to the amount of revenue generated by irrigation systems operations. This provides an incentive for cost-effective goal-oriented performance that is otherwise often weak or lacking' (FAO and USAID, 1986).

4. *Contextualized cost recovery.* The principle of charging for water should be contextualized to consider ability to pay and the overall taxation of agriculture, indirect charges often providing an indirect (but straightforward) means to recover investment costs. Cost of collection needs to be evaluated carefully, price structures tailored to the particular situation and prices indexed. The evaluation of what should be the ideal level of O&M activities should receive more attention.

5. *A contribution principle.* Subsidized water users should repay some of the investments but they should not be asked to repay the cost of 'over-elaborate gold-plated designs, incompetent, expensive construction, cost overruns for reasons of corruption, bad scheduling of construction activities or the like, nor overmanning of the public sector'.[6] While making farmers pay for O&M costs is achievable in most cases, in very few projects (if any) would farm revenues be enough to repay investment costs.

The exception to this consensus was Repetto's (1986) discordant but influential paper on rent-seeking and the performance of public irrigation schemes, which heralded the coming critiques of the 1986 consensus. Repetto convincingly showed how the design and development of irrigation projects were influenced by rent-seeking strategies. From this, he concluded that there was little virtue in objectives other than economic viability, advocating that irrigation projects should be considered as normal investments requiring recovery of full costs, without considering secondary benefits. His analysis of pricing as a means to improve management, however, proved to be weaker: it shrugged off the constraints pointed to by the other studies and extrapolated particular cases, such as private irrigation schemes, to support the generalization of full volumetric pricing and the trading of water rights. Repetto endorsed the model of financial autonomy but in the narrow sense of the utility model, without flagging the difficulties inherent in water allocation and distribution in large-scale surface hydraulic systems.

Repetto's analysis coincided with a growing awareness in the 1980s and early 1990s, in the wake of financial crises and structural adjustment programmes, of the burden on government finances inherited from ever-expanding schemes of dubious profitability. Several countries including the Philippines, Mexico, Morocco, China and Turkey, opted for reforms primarily aimed at shifting part of the O&M burden to the farmers, blended with varying degrees of transfer of management responsibility (see Molle and Berkoff, Chapter 2, this volume). These experiences were sometimes influential but failed to launch a wider dynamic that would have embodied and imposed the principles identified.

At the Bank, the debate was not interrupted by the series of documents issued in the 1980s. The decade ended with a renewed attempt to clarify issues and break away from past confusion; several mistakes from the past were acknowledged (e.g. 'zeal for the fiscal autonomy model' has been insensitive to borrowers' policies and the 'single-minded application [of the model][7] to a second-best world' might not be adequate; establishing boundaries between poor and other farmers to

[6]Rao (1984) estimated that in India only about half of the officially estimated costs should be taken as real costs.

[7]According to Small (1990) the banks' constant concern for cost recovery (despite the fact that payment of loans is guaranteed by governments) is linked to 'a misplaced concern stemming from the importance of cost recovery in private investments, where the inflow of funds to the investor represents the return on the investment. But it is inappropriate to place the same meaning on cost recovery in the case of public investments.'

be charged is 'unworkable') (O'Mara, 1990). On the other hand, emphasis was put again on the priority to be given to physical sustainability, on accepting 'the diversity of cultures and institutional arrangements in borrowing countries' and on basing cost recovery policy on a full analysis of government interventions (O'Mara, 1990).[8]

Water Pricing and Economic Incentives

Although the ideas can be traced back to earlier periods, 1992 marks a convenient turning point in the debate on water pricing: in 1992, the Dublin International Conference on Water and the Environment proposed a set of four principles, the fourth[9] of which underscored that 'managing water as an economic good is an important way of achieving efficient and equitable use, and of encouraging conservation and protection of water resources'. Although, as seen above, there was nothing novel in the concern with financial profitability, the fourth Dublin principle can be considered a landmark shift in emphasis to the economic dimensions of water use in general and irrigation development in particular. Economic instruments and the economic value of natural resources further found legitimacy in the Rio Declaration on Environment and Development of the United Nations in 1992 (EU, 2000) and its Agenda 21 (United Nations, 1992),[10] which supported the 'implementation of allocation decisions through demand management, pricing mechanisms and regulatory measures'.

More generally, the early 1990s saw the rise of the concept of demand management (which can be defined by 'doing better with what we have' as opposed to continuous supply augmentation), mostly under the influence of resource economists stressing both the economic nonsense of privileging costly and environmentally unfriendly water resources development, and the role and potential of economic incentives in managing demand and reducing the need for additional supplies. The emphasis put on economic efficiency and on the 'user-pay' and 'polluter-pay' principles struck sensitive cords and ushered in heated debates on the right to water, the respective roles of the private sector and local communities, and how to interpret and reconcile the economic and sociocultural dimensions of water.

Conceptually, this period distinguishes itself from the preceding one by a shift in emphasis (Maestu, 2001): earlier justifications of charging for water centred on the financial need for cost recovery to fund further projects (equity), relieve state finances and ensure the physical integrity of, and continued benefits from, irrigation schemes. In the 1990s, water prices, and more generally economic incentives, came to be seen as key policy tools endowed with the potential

[8]O'Mara, Principal Economist at the Agricultural Policies Division (ARD Department), offered his paper as a 'modest effort to clear away the confusion surrounding irrigation policy both inside and outside of the Bank. That there is a need for a policy dialogue within the institution on this topic is increasingly apparent. In its present form, the paper reflects the comments and criticism of many Bank staff concerned with irrigation.'

[9]The full principle reads: Principle No. 4: Water has an economic value in all its competing uses and should be recognized as an economic good. Within this principle, it is vital to recognize first the basic right of all human beings to have access to clean water and sanitation at an affordable price. Past failure to recognize the economic value of water has led to wasteful and environmentally damaging uses of the resource. Managing water as an economic good is an important way of achieving efficient and equitable use, and of encouraging conservation and protection of water resources.

[10]Principle 16 of the declaration reads: 'National authorities should endeavour to promote the internalization of environmental costs and the use of economic instruments, taking into account the approach that the polluter should, in principle, bear the cost of pollution, with due regard to the public interest and without distorting international trade and investment.' More importantly, Chapter 18 of Agenda 21 stresses: 'Implementation of allocation decisions through demand management, pricing mechanisms and regulatory measures . . . [p]romotion of schemes for rational water use through public awareness-raising, educational programmes and levying of water tariffs and other economic instruments.'

to achieve multiple objectives. With demand management-oriented approaches making conservation a critical issue, the conventional role of prices in managing demand moved from the back seat to centre stage. Likewise, increasing intersectoral competition for water and associated environmental externalities made pricing mechanisms appear as a potential and desirable means to arbitrate water allocation[11] and promote desirable environmental objectives, while maximizing water productivity and aggregate economic welfare. Assigning all these roles to pricing could be seen as the embodiment of the Dublin principle stressing the economic nature of water.

Given this anticipated potential for ensuring financial autonomy of the irrigation sector, cutting state expenditures, eliciting water savings and maximizing the economic efficiency of water use across society, water pricing understandingly attracted increasing attention from policy makers, academics, development agencies and banks (OECD, 1999b). With so much frustration generated by the need for repeated rehabilitation (in Indonesia, for example, one-third of the 3 million ha of government-designed irrigation schemes has been rehabilitated twice in the last 25 years; World Bank, 2005a), by failed attempts to improve water management or efficiency substantially and by incomplete turnover of management to farmers, price instruments appeared to hold the promise of promoting several desired policy goals. In addition, they would provide an elegant solution to long-standing problems, changing behaviour directly through incentives, thus seemingly avoiding the painstaking intricacies of irrigation management, and its technical, social and political ramifications.

This economic rationale soon percolated to water policies. The World Bank's *Water Resource Management Policy Paper* of 1993[12] observed that 'waste and inefficiencies have resulted from the frequent failure to use prices and other instruments to manage demand and guide allocation', and established a powerful narrative around the overarching causal link between water crises, water waste and underpricing. Subsequently, the Bank's policy paper remarked that the value of water differed greatly between agriculture and other sectors, 'often indicating gross misallocations if judged by economic criteria'. It followed that 'setting prices at the right level is not enough; prices need to be paid if they are to enhance the efficient allocation of resources' (World Bank, 1993).[13] Besides continuing to ensure basic cost recovery, price mechanisms were thus assigned the further objectives of reducing water waste, minimizing environmental damage and reallocating water towards higher uses.

The 1990s saw a flourishing literature on the theoretical principles and potential impacts of pricing and water markets, with a leading contribution from the World Bank.[14] During a press conference in Washington on 12 April 2000, James D. Wolfensohn (2000), President of the World Bank, reiterated the view that 'the biggest problem with water is the waste of water through lack of charging'. Johansson (2000) saw water pricing as a 'primary means . . . to improve water allocations

[11]In 1985, concern was only expressed for 'the efficient level of use of scarce water and to its allocation to crops where returns to irrigation are higher', not for sectoral allocation (see Ghate (1985) for ADB's point of view). In the EU 'it is only in the early 1990s that attention started switching to the economic value of water' (EU-WATECO, 2003).

[12]Jones (1995) reports that the elaboration of the paper saw a renewed conflict between economic orthodoxy bent on the long-term marginal-value pricing principle and the view defended by operating divisions, Agriculture Department staff and consultants, who advocated more flexibility.

[13]Identification of an 'allocation stress' became commonplace. For instance, Dinar (1998) held that 'the potential for economic benefits from allocation-oriented institutional change are not only substantial but also increasing with each increase in water scarcity'. Rosegrant and Cline (2002) posited that 'there is considerable scope for water savings and economic gains through water reallocation to higher-value uses'.

[14]See, for example, Teerink and Nakashima (1993); Le Moigne *et al.* (1994); Tsur and Dinar (1995); Bhatia *et al.* (1995); Thobani (1997); Dinar and Subramanian (1997); Easter *et al.* (1998, 1999); Dinar (2000); Johansson (2000); and AMAECO and ANAFID (2002).

and to encourage conservation'. The Economic and Social Commission for Asia-Pacific (ESCAP, 1996a,b)[15] saw pricing as an 'essential component of water demand management', which could in particular 'significantly reduce the wastage of resources'. ADB, in its 2000 water policy, reaffirmed that it 'needs to promote efficiencies in water use by supporting demand management, including water pricing'. Jones (2003) stated that 'anything scarce and in demand commands a price', and that consequently 'water pricing is increasingly seen as an acceptable instrument of public policy'. Finally, the World Water Commission's (2000) report proclaimed that 'the single most immediate and important measure that we can recommend is the systematic adoption of full-cost pricing for water services', although acknowledging that full-cost pricing, long advocated in the irrigation sector, 'has seldom happened'. Other UN organizations and development banks, such as ESCWA (1997, 2005), ESCAP (1981), and AfDB and ADF (2000),[16] usually reproduced these principles and objectives, most of them underscoring cost recovery, but some – including the IADB (1998) and CEPAL (1995) – putting their emphasis on decentralization, water rights and water markets.

These views were consonant with, and perhaps partly derived from, policy shifts in developed countries. The late 1990s saw the gradual elaboration of the European Water Framework Directive which put economic incentives in general and pricing policies in particular at the heart of its objectives of financial and environmental sustainability[17] (see OECD, 1999a, 2002; European Commission, 2000a,b). Interestingly, the use of pricing in the EU policy is advocated primarily as a conservational means to manage demand so as to curb excessive abstraction of water from ecosystems, and incorporates the polluter-pay principle, with water charges being instrumental in internalizing environmental costs. This reflects the weight of environmentalism in promoting economic incentives as key tools for water policy (de Moor and Calami, 1997; Avis *et al.*, 2000; Kaika, 2003; Khanna and Sheng, 2000). In contrast, official references to the sectoral allocation and to charging opportunity costs are rare, although some environmentalists regard full-cost pricing as a way of decreasing demand and environmental damage, since 'the price [of water] could be raised until the level of demand was consistent with the environmental constraints on supply' (Hodge and Adams, 1997), and since 'full cost recovery for water services (should) include the costs of damages to the environment' (Avis *et al.*, 2000).

Numerous analysts have embraced the concept of demand management (Frederick, 1993; Hamdy *et al.*, 1995; Brooks, 1997; Winpenny, 1997; Ahmad, 2000; Louw and Kassier, 2002), seeing its application as a primary means to solve the current water crisis. In turn, central ideas such as the persistence of massive water losses in the agriculture sector, poor management and misallocation of water resources, and the crucial role of economic incentives made their way into the mainstream media including *The Economist* (2003), *Scientific American* (Gleick, 2001), *Science* (Gleick, 2003) and *National Geographic* (Frank,

[15]If properly set and implemented, water pricing for agricultural water could significantly reduce the wastage of resources (ESCAP 1996a). 'Water pricing is an essential component of water demand management which is instrumental in achieving two important goals: to generate revenue for capital recovery, operation and maintenance, extension of the system; to promote efficiency in use; and to protect the quality of water resources by reducing the wastewater discharge' (ESCAP, 1996b).

[16]AfDB and ADF (2000), for example, reads like a textbook of ideal principles, peppered with realism, such as: 'Ultimately, the aim of water pricing should be economic cost recovery, taking into account social equity and capacity to pay by the rural and urban poor. Initially, however, RMCs should target the recovery of full financial costs.'

[17]The 'proposed Water Framework Directive promotes the use of water charging to act as an incentive for the sustainable use of water resources and to recover the costs of water services by economic sector. This will contribute to meeting the environmental objectives of this directive in a cost-effective way' (European Commission, 2000b).

2001). Spurred by the Second and Third World Water Forums, newspapers and analysts also echoed prophecies of the 'coming' (Lavelle and Kurlantzick, 2002), 'creeping' (Falkenmark, 2001), 'impending' (Rosegrant *et al.*, 2002) or 'looming' (IRRI, 1995; UNESCO, 2000) water crises.

These ideas trickled down to policy and law-making in many countries. The 1998 South African Water Act specifies that 'water use charges are to be used to fund the direct and related costs of water resource management, development and use, and may also be used to achieve an equitable and efficient allocation of water' (Republic of South Africa, 1998).[18] Article 19 of the 1997 Brazilian Water law recognizes water as an economic good and introduces water fees with the triple objective of indicating the value of water, rationalizing the use of water and levying funds for the further development of water resources (Government of Brazil, 1997). The 1999 National Water Policy of Bangladesh states that '[a] system of cost recovery, pricing, and economic incentives/disincentives is necessary to balance the demand and supply of water' and that 'water will be considered an economic resource and priced to convey its scarcity value to all users and provide motivation for its conservation' (Government of Bangladesh, 1999; Chakravorty, 2004). Many other state policies or legal acts[19] include similar general principles, or focus on particular ones, such as cost recovery in the case of Vietnam (1998) (users have a 'financial duty and the duty to contribute manpower and budget'), or of the 1988 Law of China (as well as succeeding draft versions of its revision).[20]

The apparent overwhelming[21] adoption of pricing principles created an intellectual environment which made it somewhat difficult for alternative or nuanced voices to be heard. Several papers looking critically at the issue were published[22] and several reviews were carried out though they did not significantly alter the debate.[23] An OED study (Jones,

[18]In addition, they may also be used to ensure compliance with prescribed standards and water management practices according to the user-pay and polluter-pay principles. Water use charges will be used as a means of encouraging reduction in waste, and provision is made for incentives for effective and efficient water use.

[19]This is not the case, however, for all national laws and policies. India (GOI, 2002), Pakistan (GOP, 2002) and Malaysia (FAO, 1996a), for example, do not see irrigation pricing as a water management and policy instrument.

[20]Article 42 stipulates: 'Those who use water provided by water supply projects shall pay water charge to the supplying unit in accordance with stipulations. Water price shall be defined as per the principles of cost recovery, reasonable profit, and good price for good quality and fair shares. The system of accumulative pricing shall be conducted to the water use over than the planned amount.'

[21]Many papers emphasized the emergence of a consensus and the alleged growing application of such principles, contributing to create a 'policy bubble'. See, for example, Johansson *et al.* (2002): 'In addressing water scarcity and increased population pressures many countries are adopting water-pricing mechanisms as their primary means to regulate irrigation water consumption'; Saleth (2001): 'Although water continues to be subsidized in most sectors and countries, there is growing recognition of water pricing as a key policy instrument for cost recovery and demand management'; Jones (2003): 'Water pricing is increasingly seen as an acceptable instrument of public policy.' While these statements are correct in the narrow sense that economic and financial concerns have become more salient and incorporated in policies, they tend to convey an overly optimistic view that economic instruments will be both paramount and effective in achieving multiple long-sought goals.

[22]See, for example, Carruthers and Morrisson (1996), Morris (1996), Perry (1996, 2001a,b), Chaudhry *et al.* (1993) and Perry *et al.* (1997).

[23]For a number of economists, the question was no longer the desirability or possibility of using price regulation but a mere technical debate on how to determine the 'optimal price', for example: 'Despite the pervasiveness of water pricing as a means to allocate water, there is still disagreement regarding the appropriate means by which to derive the price' (Johansson *et al.*, 2002; see Kim and Schaible, 2000; Louw and Kassier, 2002). That prices based on concepts of marginal costs or opportunity costs are invariably found to be incompatible with maintaining farm revenues does not seem to have triggered much theoretical debate.

1995) on 'the World Bank and irrigation' questioned the 'Bank's enthusiasm for irrigation cost recovery . . . [based on] a presumed link between cost recovery and better operation and maintenance', because it confirmed earlier findings by OED that 'there is normally no link between higher water charges and better operation and maintenance. Revenue from water charges generally goes to the general treasury and is not earmarked for O&M'.

'Principled Pragmatism': The Idea Comes Full Circle

Despite the hopes vested in pricing policies during the 1990s, a number of elements have gradually made a reassessment of these expectations necessary. This readjustment has been driven not only by the recognition of a host of technical, socio-economic, legal and political difficulties, which will be analysed at length in Chapter 2, but also by the emergence of severe conflicts caused by raised water charges (or curtailed subsidies) in several countries. The question of charging for water has also suffered from an unfortunate lack of distinction between agriculture and the domestic sector, and many of the conflicts that have bedevilled the latter were mistakenly extended to the former. This may have been partly due to insufficient attention given to crucial differences between the two sectors (see Molle and Berkoff, Chapter 2, this volume), apparent in many policy and academic documents that tend to assume that the two sectors are similar.

The empirical literature on water pricing in irrigated agriculture also yields a paucity of cases in which pricing policies have successfully achieved the objectives assigned to them. First, it has been excessively difficult to raise and stabilize cost recovery from users and in most cases even O&M expenditures are not recovered. There are, however, exceptions. Morocco and Tunisia have, for instance, been successful in covering O&M; Mexico has turned over most of its public schemes (and their related costs) to water user associations; water charges were increased by three times in the 1997 reform of Andhra Pradesh, India, though from a very low level (Samal and Kolanu, 2004); the National Irrigation Agency in Philippines has cut its staff by 75% in the last 25 years (Oorthuizen, 2003); China is experimenting with several ways of delegating water management and strengthening incentives (see Lohmar *et al.*, Chapter 12, this volume), etc. Not all these cases have been unmitigated successes, but they perhaps signal a trend towards better cost recovery, with financial autonomy of irrigation units or projects as a major objective.

The impact of water charges on efficiency has, in contrast, remained almost entirely elusive, as revealed by Bosworth *et al.*'s (2002) recent review of the literature. An analysis of the use of economic tools for demand management in Mediterranean countries also showed that their use in agriculture was far more limited than in the urban sector, and that prices alone did not suffice to elicit significant changes in behaviour (Chohin-Kuper *et al.*, 2002). Compilations of cases such as Bhatia *et al.* (1995), Dinar and Subramanian (1997), Dinar (2000) and Johansson (2000) provide some evidence to the contrary but they are drawn almost exclusively from the urban water sector or from modelling exercises. Examples of changes in cropping patterns and technology are more numerous but these changes are typically caused by a host of interacting factors of which water pricing is seldom of more than marginal significance. Finally, Dinar and Saleth (2005) admit that 'efficient water pricing schemes are rare, if not completely absent, even in economically advanced regions with extreme water scarcity levels, [which] provides sufficient evidence for the persistence of a vast gap between the development of pricing theory and its practical application'; and there also appears to be no example of a country having resorted to administered price setting in order to allocate water among sectors (Bosworth *et al.*, 2002).

A review of OECD countries (Garrido, 2002) concluded that progress in the implementation of water pricing policies had been slow and uneven, and that farmers typically paid only a fraction of O&M costs (and nothing for rehabilitation and amortization of investments, let alone environmental or resource costs). 'Irrigation pricing reforms should not expect significant reductions in farmers' water consumption', and quotas[24] are likely to be required, though prices are expected to contribute to the EU's environmental objective based on the polluter-pay principle (Garrido, 2002). A review of the use of economic incentives (EIs) in Canada (PRI, 2005) noted that 'there has been a tendency to promote EIs as being capable of delivering the best of all worlds: environmental protection, economic and technological development, and revenue generation, while maintaining equity, and all in one convenient box' but 'careful examination of real-life experiences' is needed before these objectives can be assumed to be achieved.

It is thus becoming apparent that on-the-ground evidence of the impact of economic tools remains well short of expectations and promises. Since 2000, several official documents and academic papers have scaled down the earlier enthusiasm for water pricing, reflecting not only the widening gap between theory and practice but also the wish to avoid the violent controversies around this issue (mostly it is true relating to the domestic sector). The Ministerial Declaration of the Second World Water Forum (World Water Commission, 2000) advocates a prudent 'move towards pricing water services to reflect the cost of their provision', but adds that 'this approach should take account of the need for equity and the basic needs of the poor and the vulnerable'.[25] Tellingly, the word 'pricing' is absent from the Bonn Conference 27 recommendations for action, issued in December 2001. Similarly, the 2002 Stockholm statement that, under the title 'Urgent action needed for water security', synthesizes the lessons from the five previous symposia lists four principles for action that do not refer to the use of economic instruments in managing water. Recently, the World Water Assessment Program (UNESCO-WWAP, 2006) stressed the importance of non-economic goals in irrigation, the potential limitations to volumetric pricing and the goal of recovering O&M costs only.

More significantly, perhaps, a recent OED assessment of the 1993 World Bank water strategy concluded: 'Globally, most Bank projects pay lip-service to cost recovery,[26] . . . [and] too frequently, Bank water staff promote reform when the enabling conditions are absent due to the programmatic nature of projects.' In sum: 'Pricing promotes efficiency and conservation . . . but there are few successful examples because of the economic and cultural difficulties of putting a value on a natural resource' (Pitman, 2002). In 2003, the Bank issued a new water resources sector strategy (World Bank, 2003), aimed at updating the document issued 10 years earlier. It acknowledged the 'yawning gap between simple economic principles . . . and on-the-ground reality'.

> It has often been stated that having users pay 'the full cost of water' would solve these problems. Experience has shown that the situation is considerably more complex and nuanced, and that it is not enough to just extol the virtues of pricing. This section outlines a different approach – one of 'principled pragmatism.' 'Principled' because economic principles such as ensuring that users take financial and resource costs into account when using water, are very important. And 'pragmatism' because solutions need to be tailored to specific, widely varying natural, cultural, economic and political circumstances, in which the art of reform is the art of the possible.
>
> (World Bank, 2003)

[24]But 'the use of quotas or allotments suggests that efficient allocation can be made without prices, and that the combination of quotas and cost-recovery charges – not including the opportunity cost of water as the European Union foresees in its Water Framework Directive – may be a viable mix of instruments' (Garrido, 2002).

[25]Interestingly, this political statement appears much more prudent than the World Water Council's two parallel reports prepared for the same forum: 'Making Water Everybody's Business' 'recommends that consumers be charged the full cost of providing water services' (Cosgrove and Rijsberman, 2000); see *supra* for quote from the report 'A water secure World' (World Water Commission, 2000).

[26]Among sectors of the water strategy whose implementation was rated as 'ineffective' were 'allocation issues and opportunity cost of water' and 'transparency and full cost accounting of water delivery service', while 'increasing user charges' was rated 'moderately effective' (Pitman, 2002).

Yet, the soundness of the theoretical background is constantly reaffirmed (World Bank, 2003).[27] Difficulties in implementing water pricing, however, are often ascribed to technical or cultural difficulties, and to political resistance of entrenched sectoral interests (Saleth, 2001; Dinar and Saleth, 2005), and there is a continued hankering for a more ambitious role for pricing. The most recent World Bank initiative for 'Reengaging in agricultural water management' (World Bank, 2005b), however, adopts a more balanced position and states that management of large-scale irrigation has 'been plagued by problems of irrigation service charges, both low levels of charge and low levels of collection'. Where demand is not responsive to price increases and where there is a water shortage, a case admittedly quite frequent, 'rationing (in the short term) or the allocation of quotas (for the long term) should be considered as an effective way to reduce demand and encourage efficiency' (World Bank, 2005b).

It is becoming clear that arguments have often been presented in a very broad manner, with general principles repeated without the necessary qualifications. The literature bears frequent confusion across the board between the different possible justifications for water pricing, and the theoretical arguments that may apply to a particular context are often implicitly or explicitly extended to other situations where they cease to be valid. It is evident, in particular, that there are crucial differences between domestic and irrigation water, classical large-scale surface irrigation and pump irrigation, government and farmer-managed schemes, low- and high-tech distribution systems, staple and cash-crop production, and developed and developing countries. Similarly, parallels with land rights provide limited guidance for addressing water rights (Hanemann, 2006), and comparisons between the water and the power sector can also be misleading.

On a more philosophical plan, the principle of 'water as an economic good' has triggered a heated debate, with the emergence of a concurrent paradigm underscoring water as a social good and/or a human right. This confrontation of world views has introduced a main fault line across the debate (ODI, 2002; Hanemann, 2006). All parties agree that water is the 'stuff of life' and, to some extent, that extravagant consumption is to blame. Those supporting 'water as an economic good', however, see waste as the result of underpricing and, consequently, pricing or markets as a way out of the crisis. They see perfect markets as an optimal means to achieve economic efficiency, as a desirable objective for the society as a whole, and alternatives as second-best options. The rationale for cost recovery, linked to the need to fund maintenance and further expand water services, is opposed by supporters of the 'water as a basic human right' paradigm, who consider that domestic supply is a right that warrants subsidized public investments. They view pricing or market instruments with suspicion, stressing that water is foremost a social good and that its allocation cannot be left to mechanisms that will eventually favour the wealthy and powerful. In their view, prices should be controlled by the government to avoid the commodification of water and the exclusion of the poorest, and only volumes beyond vital requirements should be charged (The Water Manifesto, 1999; Shiva, 2002). Here again, the debate has been obscured by an indiscriminate mix of situations, from little to very water-short regions, from domestic use to irrigation and from individual use to large public schemes.

Controversies and debates along this fault line have increased in recent years. At both extremes, rather uncompromising viewpoints have been expressed, which have not been helpful in building bridges across the two world views. They have stuck, on the one hand to market fundamentalism that seems to be impervious to the lessons of reality on the ground and, on the other, to a romantic posture where water is seen as god-given and should not be sullied by mundane issues of cash. Some, however, seek to adopt more nuanced and conciliatory stances. Despite

[27] The neo-classical principles of pricing and allocation are axiomatic. If at fault, it is because of contextual factors that should be removed, not because the theory should better conform to the real world.

such attempts to bridge conflicting viewpoints, the debate remains fairly polarized.

In the 1990s, the academic literature was dominated by theoretical considerations and promotion of economic incentives as key policy instruments to instil economic rationality and regulate the water sector. Recent publications have focused on the practical constraints faced, besides the inadequacy of some of their theoretical tenets. Without going into the details analysed by Molle and Berkoff in Chapter 2 (and illustrated in the subsequent chapters) mention should be made of the evidence provided by the case studies and literature reviews carried out by Bosworth *et al.* (2002), Cornish and Perry (2003), Hellegers and Perry (2004) and Cornish *et al.* (2004). They stress the importance of distinguishing between objectives and the design of charging systems to meet these objectives according to the context. Volumetric pricing is rare and 'the response in demand to volumetric pricing is widely shown to be minimal'. Water markets have been established in a few locations but bureaucratic allocation of water through price setting is nowhere to be observed; the debate on sectoral allocation may have been misconstrued (Savenije and van der Zaag, 2002) and the degree of misallocation overstated (Molle and Berkoff, 2006).

A balanced assessment has also been issued by ICID (2004) which does not consider recovery of the full financial costs of irrigation but emphasizes the need to define negotiated contractual relationships between providers (of any kind) and users, and to charge the latter the cost of O&M plus renewal costs ('the sustainability costs'). 'Opportunity pricing' has no application in pricing services but the determination of all costs helps in assessing values before allocating resources. Defining quotas may hinder flexibility in reallocation but quotas are equitable and effective in managing scarcity. Dinar and Mody (2004) also observe that financial cost recovery, though becoming more common, is hard to implement. In most cases, they note, pricing does not elicit more efficient on-farm water use, and when it does (often through crop shift or technological change), it does not automatically translate into total water savings. Easter and Liu (2005) focus on cost recovery objectives, ponder on why cost recovery rates are low and acknowledge that water demand may be elastic only at levels of charge that are politically unacceptable. Emphasis is put on participatory and transparent definition of charges and on keeping them within the system, ensuring financial autonomy and enhancing accountability of managers.

In other words, a new consensus is emerging which is by and large replicating the conclusions established 20 years earlier. Charging for water is primarily a fiscal issue on which no general statement can be made as long as it is not part and parcel of a wider financing mechanism, whereby users are effectively empowered and managers made accountable through their dependency on fee collection. Other conservation and allocation objectives remain important but the effectiveness of pricing is limited to some specific 'niches', which can be made to grow but which are likely to remain limited, or marginal, in the foreseeable future. Pricing will generally have limited impact alone but is an instrument that can contribute to a package of incentives. *Principled pragmatism* is needed to apprehend the constraints on the ground, and sound management of supply – at all scales, from the farm to the basin – remains the unglamorous yet fundamental prerequisite to improving the performance of the water sector.

This storyline raises intriguing questions on why the debate has gone full circle in a 20-year period, going through different conflicting views,[28] detours and

[28]As suggested along this historical review, the debate showed considerable wavering between opposite viewpoints and statements: as a rule, cost had to be recovered from users but it was proposed that this could be alternatively done by the government; only direct irrigation benefits should be considered but consideration of induced economic activities was also proposed; subsidies were acceptable and optimum might differ from long-term marginal pricing but strict endorsement of the latter principle proved persistent; the utility model was seen adequate for irrigation service but its clear limitations sometimes recognized; irrigation should be seen as any other economic activity but its other social objectives acknowledged; pricing instruments can target several goals at one time but it is not the case in most instances; etc.

dead ends and finally 'rediscovering' both the limits imposed by the real world to policy instruments and the particular conditions needed for their effectiveness. Although it is not the central objective of this chapter to address this question, one may wonder whether economic thinking, coming to prominence in the late 1980s to early 1990s, has not been subjected to the excessive self-confidence that other disciplines (e.g. agronomy, water engineering, rural sociology and planning) have shown earlier, before being confronted with difficulties in raising yields, improving irrigation efficiency, setting up user groups or implementing integrated development projects or policies. Overconfidence leads to excessive faith in theoretical frameworks, and lack of attention to on-the-ground and political economic factors (Dinar, 2000; Green,[29] 2000). Systematic stigmatization of irrigation as a wasteful sector has frequently been based on a lack of understanding of irrigation management and basin hydrology, just as the domestic and irrigation sectors have been confused, despite crucial differences. Likewise, anti-state ideological rhetoric has often supported the idea that bureaucratic water allocation is insensitive to economic rationality[30] (Moore, 1990; Carruthers, 1997), even where evidence suggests otherwise (Molle and Berkoff, 2006). The issue of sectoral reallocation may have been inflated because of its salience in the USA and also because some economists advocate[31] markets out of ideological inclination rather than sound examination of local contexts (Gaffney, 1997; Bauer, 2004). It is also apparent that the constitution of a massive body of literature, largely fed by a few mainstream institutions and overly self-referential, has contributed to mainstreaming ideas that have often been indiscriminately picked up in national universities or policies, without the necessary caveats and contextualization.

Chapter 2 is devoted to giving flesh to this narrative. It starts with some general considerations on pricing and irrigated agriculture before examining the different policy objectives that can be attained through pricing instruments. For each of these, we attempt to confront the theoretical background with field evidence and assess the scope for achieving these objectives. Getting price incentives in irrigation 'down to earth' by no means negates the importance of prices, or the crucial need for economic insight in the development of water resources. It does, however, assert that – as for all other policy instruments – we should neither entertain unreasonable expectations nor justify or propose policies based on general principles that may not hold in a particular context. When there are good reasons to design financial mechanisms, it does not help to confuse objectives by bringing in arguments of limited validity. Through abundant references to the literature, we will also point to discursive and conceptual shifts and finally identify a range of conclusions which might, hopefully, be contemplated as firm ground for future policy making.

[29]Green (2000) contrasts a Panglossian (optimistic) approach with a 'Pragmatic approach, generally characterised by a concern for institutional design, for increasing public participation and a search for ways of supporting decisions with appraisal tools such as benefit–cost analysis . . . [which] lacks the self-confidence of the Panglossian approach and lacks the glorious heroism of economists riding to the rescue of water management. It is more hesitant in claiming success, hoping instead that instances offer lessons which will improve future decisions'. See also Albiac *et al.*'s (2006) remark that 'water pricing advocated by some government advisors and environmentalists starts to look like "armchair economics"', and Embid-Irujo (2005) on Spain in the 1990s: 'For certain economists or the intellectual colleagues of certain economists, this policy [the setting of a "real" price for water] was a sort of "magic wand" that would solve all the current problems at a stroke, while other experts were more realistic.'

[30]See, for example, Anderson and Snyder (1997): 'Because [water] is so precious, we cannot afford misallocation that comes from political control.'

[31]'Faith in market mechanisms for resource allocation has been "politically correct"—often approaching dogma—for more than a decade. Although attractive in principle, the complexity of establishing markets for tradable water rights should not be underestimated' (Siamwalla and Roche, 2001).

References

ADB (Asian Development Bank) (1986a) *Review of Bank Operations in the Irrigation Sector 1966–1985*. Asian Development Bank, Manila.

ADB (1986b) Irrigation service fees. *Proceedings of the Regional Seminar on Irrigation Fees*, Manila, 21–25 July 1986. Asian Development Bank, Manila.

ADB (2000) Water for all: the water policy of the Asian Development Bank. Available at: http://www.adb.org/documents/policies/water/water.pdf

AfDB (African Development Bank) and ADF (African Development Fund) (2000) *Policy for Integrated Water Resources Management*. African Development Bank, Tunis.

Ahmad, M. (2000) Water pricing and markets in the Near East: policy issues and options. *Water Policy* 2(3), 229–242.

Albiac, J., Martínez, Y. and Tapia, X. (2006) Water quantity and quality issues in Mediterranean agriculture. In: Chapter 5, *Water and Agriculture: Sustainability, Markets and Policies*. OECD, Paris, pp. 137–156.

AMAECO and ANAFID (2002) *Les Politiques D'irrigation: Considérations Micro et Macro-économiques*, Vol. 4. Association Marocaine d'Agro-Economie et Association Nationale de l'Amélioration Foncière, d'Irrigation et de Drainage. Agadir, Morocco.

Anderson, T.L. and Snyder, P.S. (1997) *Priming the Invisible Pump*. Cato Institute, Washington, DC.

Avis, C., Tydeman, C. and Gelabert, E.R. (2000) *What Role for Water Pricing? Ten Actions for Internalising Sustainability*. WWF International, European Freshwater Programme, Brussels.

Barker, R. and Molle, F. (2004) *Evolution of Irrigation in South and Southeast Asia*. IWMI Comprehensive Assessment Research Report 5. IWMI Comprehensive Assessment Secretariat, Colombo, Sri Lanka, pp. vi, 45.

Bastiampillai, B. (1967) The revival of irrigation enterprise in Ceylon, 1870 to 1890. *Ceylon Journal of Historical and Social Studies* 10, 1–26.

Bauer, C.J. (2004) *Siren Song: Chilean Water Law as a Model for International Reform*. Resources for the Future, Washington, DC.

Bhatia, R., Cestti, R. and Winpenny, J. (1995) *Water Conservation and Reallocation: Best Practice Cases in Improving Economic Efficiency and Environmental Quality*. Water & Sanitation Currents, World Bank, Washington, DC, p. v.

Bolding, A., Mollinga, P.P. and van Straaten, K. (1995) Modules for modernization: colonial irrigation in India and the technological dimension of agrarian change. *Journal of Development Studies* 31(6), 805–844.

Bosworth, B., Cornish, G., Perry, C. and van Steenbergen, F. (2002) *Water Charging in Irrigated Agriculture: Lessons from the Literature*. Report OD 145. HR Wallingford, Wallingford, UK.

Brooks, D.B. (1997) Water demand management: conceptual framework and policy implementation. In: Brooks, D.B., Rached, E. and Saade, M. (eds) *Management of Water Demand in Africa and the Middle East: Current Practices and Future Needs*. International Development Research Centre, Ottawa, Canada.

Carruthers, I. (1997) The water paradox: scarce and valuable but inefficient use and poor political decisions: who should we blame? In: Dupuy, B. (ed.) *Aspects Economiques de la Gestion de L'eau dans le Bassin Méditerranéen*. CIHEAM-IAMM, Bari, Italy, pp. 125–135.

Carruthers, I. and Clarck, C. (1981) *The Economics of Irrigation*. Liverpool University Press, Liverpool, UK.

Carruthers, I.D. and Morrisson, J.A. (1996) Institutions in water resources management: insights from new institutional economics. In: Howsam, P. and Carter, R.C. (eds) *Water Policy: Allocation and Management in Practice*. E & FN Spon, London, pp. 205–212.

Carruthers, I.D., Peabody, N.S.III, Bishop, A.A., LeBaron, A.D., Mehra, R., Ramchand, O., Peterson, D. and Wood, D.H. (1985) *Irrigation Pricing and Management*. Report to USAID. DEVRES, Washington, DC.

CEPAL (Comisión Económica para América Latina y el Caribe) (1995) Mercados de Derechos de Agua: entorno Legal. CEPAL, Santiago de Chile.

Chakravorty, N. (2004) Water pricing in Bangladesh: a pro-poor framework based on two surface water irrigation projects in Bangladesh. Draft.

Chaudhry, M.G., Majid, S.A. and Chaudhry, G.M. (1993) The policy of irrigation water pricing in Pakistan: aims, assessment and needed redirections. *The Pakistan Development Review* 32(4) Part II, 809–821.

Chohin-Kuper, A., Rieu, T. and Montginoul, M. (2002) Economic tools for water demand management in the Mediterranean. Forum on 'Progress in Water Demand Management in the Mediterranean'. *Fiuggi*, 3–5 October 2002.

Cornish, G., Bosworth, B., Perry, C. and Burke, J. (2004) *Water Charging in Irrigated Agriculture: An Analysis of International Experience*. FAO Waters Reports 28. FAO, Rome, Italy.

Cornish, G.A. and Perry, C.J. (2003) *Water Charging in Irrigated Agriculture: Lessons from the Field*. Report OD 150. HR Wallingford, Wallingford, UK.

Cosgrove, W.J. and Rijsberman, F.R. (2000) *World Water Vision: Making Water Everybody's Business*. Earthscan, London.

Coward, E.W. (1980) *Irrigation and Agricultural Development in Asia*. Cornell University Press, Ithaca, New York.

de Moor, A. and Calami, P. (1997) *Subsidizing Unsustainable Development: Undermining the Earth with Public Funds*. Earth Council, San Jose, Costa Rica.

Dinar, A. (1998) Water policy reforms: informational needs and implementation obstacles. *Water Policy*1(4), 367–382.

Dinar, A. (ed.) (2000) *The Political Economy of Water Pricing Reforms*. OUP, for the World Bank, New York.

Dinar, A. and Mody, J. (2004) Irrigation water management policies: allocation and pricing principles and implementation experiences. *Natural Resources Forum* 28, 112–122.

Dinar, A. and Saleth, R.M. (2005) Issues in water pricing reforms: from getting correct prices to setting appropriate institutions. In: Folmer, H. and Tietenberg, T. (eds) *The International Yearbook of Environmental and Resource Economics 2005/2006*. Edward Elgar, Cheltenham, UK.

Dinar, A. and Subramanian, A. (1997) Water *Pricing Experiences: An International Perspective*. World Bank Technical Paper No. 386. World Bank, Washington, DC.

Duane, P. (1986) Cost recovery in irrigation projects: perceptions from World Bank operation evaluation. In: FAO, Land and Water Development Division and USAID, *Water Management Synthesis II Project*. Technical papers from the Expert Consultation on Irrigation Water Charges, Vol. I, Rome, 22–26 September 1986. FAO, Rome, Italy, pp.122–157.

Easter, K.W. and Liu, Y. (2005) *Cost Recovery and Water Pricing for Irrigation and Drainage Projects*. Agriculture and Rural Development Discussion Paper No. 20. World Bank, Washington, DC.

Easter, K.W., Rosegrant, M.W. and Dinar, A. (eds) (1998) *Markets for Water: Potential and Performances*. Kluwer Academic Publishers, Boston, Massachusetts.

Easter, K.W., Rosegrant, M.W. and Dinar, A. (1999) Formal and informal markets for water: institutions, performance and constraints. *The World Bank Research Observer* 14(1), 99–116.

Embid-Irujo, A. (2005) Water pricing in Spain. *International Journal of Water Resources Development* 21(1), 31–41.

ESCAP (Economic and Social Commission for Western Asia) (1981) Proceedings of the Expert Group Meeting on Water Pricing, held at Bangkok, Thailand, 13–19 May 1980. ESCAP Water Resources Series No. 55. UN, New York.

ESCAP (1996a) *Guidelines for the Establishment of Irrigation Water Pricing Policies and Structures*. ST/ESCAP/1733. Economic and Social Commission for Asia-Pacific, Bangkok.

ESCAP (1996b) *Overview of Water Pricing Policies and Structures in the ESCAP Region*. ST/ESCAP/1646. Economic and Social Commission for Asia-Pacific, Bangkok.

ESCWA (Economic and Social Commission for Western Asia) (1997) *Review of the Impact of Pricing Policy on Water Demand in the ESCWA Region, with a Case Study on Jordan*. (E/ESCWA/ENR/1997/6), Beirut.

ESCWA (2005) *Module Ten: Valuing Water Resources*. Workshop on 'Training of Trainers on the Application of IWRM Guidelines in the Arab Region' Kuwait, 14–18 May 2005. ESCWA, Beirut.

EU (European Union) (2000) *Communication from the Commission to the Council, The European Parliament and The Economic and Social Committee: Pricing Policies for Enhancing the Sustainability of Water Resources*. COM(2000) 477 final, European Union, Brussels.

European Commission (2000a) *The Environmental Impacts of Irrigation in the European Union*. A Report to the Environmental Directorate of the European Commission, Brussels.

European Commission (2000b) *Pricing Policies for Enhancing the Sustainability of Water Resources*. Communication from the Commission to the Council, The European Parliament, and The Economic and Social Committee. COM(2000) 477 final, Brussels.

EU-WATECO (2003) *Common Implementation Strategy for the Water Framework Directive (2000/60/EC)*. Guidance Document No 1: Economics and the Environment – The Implementation Challenge of the Water Framework Directive. Produced by Working Group 2.6 – WATECO. European Union, Brussels.

Falkenmark, M. (2001) The greatest water problem: the inability to link environmental security, water security and food security. *International Journal of Water Resources Development* 17(4), 539–554.

FAO (1996a) Malaysia. Available at: http://www.fao.org/ag/agl/aglw/aquastat/countries/malaysia/index.stm

FAO and USAID, Water Management Synthesis II Project. (1986) *Report on The Expert Consultation on Irrigation Water Charges*, Rome 22–26 September 1986. FAO, Rome, Italy.
Farmer, B.H. (1976) *Pioneer Peasant Colonization in Ceylon: A Study in Asian Agrarian Problems*. Greenwood Press Publishers, Connecticut.
Frank, A. (2001) World has enough water for all, experts say: but only if people pay. *National Geographic News*, 1 May 2001.
Frederick, K.D. (1993) *Balancing Water Demands with Supplies: The Role of Management in a World of Increasing Scarcity*. World Bank Technical Paper 189. World Bank, Washington, DC.
Frederiksen, H.D. (1986) *Organizing for O&M and Charges for Service*. Office Memorandum. World Bank, Washington, DC.
Gaffney, G. (1997) What price water marketing? California's new frontier – Special Issue: commemorating the 100th Anniversary of the Death of Henry George. *American Journal of Economics and Sociology*.
Garrido, A. (2002) *Transition to Full-Cost Pricing of Irrigation Water for Agriculture in OECD Countries*. OECD, Environment Directorate, Paris.
Ghate, P. (1985) Some principles relevant to fixing the level of irrigation charges in bank projects. Asian Development Bank, Economic Office, Manila. Inter-office Memorandum. Draft.
Gleick, P. (2001) Making every drop count. *Scientific American*, February 2001.
Gleick, P.H. (2003) Global freshwater resources: soft-path solutions for the 21st century. *Science* 302, 1524–1528.
GOI (Government of India, Ministry of Water Resources) (2002) *National Water Policy*. New Delhi, April 2002.
GOP (Government of Pakistan) (2002) *National Water Policy*. Islamabad, Pakistan.
Government of Bangladesh (1999) National water policy. Available at: http://www.comnet.mt/bangladesh/NWtrPol.htm
Government of Brazil (1997) Política e sistema nacional de gerenciamento de recursos hídricos Lei Federal N° 9.433, 8 January 1997.
Green, C. (2000) If only life were that simple: optimism and pessimism in economics. *Physics and Chemistry of the Earth* 25(3), 205–212.
Hamdy, A., Abu-Zeid, M. and Lacirignola, C. (1995) Water crisis in the Mediterranean: agricultural water demand management. *Water International* 20(4), 176–187.
Hanemann, W.H. (2006) The economic conception of water. In: Rogers, P.P., Llamas, M.R. and Martinez-Cortina, L. (eds) *Water Crisis: Myth or Reality?* Taylor & Francis, London.
Hellegers, P.J.G.J. and Perry, C.J. (2004) *Water as an Economic Good in Irrigated Agriculture: Theory and Practice*. Agricultural Economics Research Institute, The Hague, The Netherlands.
Hodge, I.D. and Adams, W.M. (1997) Allocating water for production and rural conservation: choices and institutions. In: Kay, M., Franks, T. and Smith, L. (eds) *Water: Economics, Management and Demand*. E & FN Spon, London, pp. 117–127.
IADB (Inter-American Development Bank) (1998) *Strategy for Integrated Water Resources Management*. IADB, Washington, DC.
ICID (International Commission on Irrigation and Drainage) (2004) *Irrigation and Drainage Services: Some Principles and Issues towards Sustainability*. An ICID Position Paper. ICID, New Delhi.
IRRI (1995) *Water: A Looming Crisis*. IRRI, Los Baños, Philippines.
Johansson, R.C. (2000) *Pricing Irrigation Water: A Literature Survey*. World Bank, Washington, DC.
Johansson, R.C., Tsur, Y., Roe, T.L., Doukkali, R. and Dinar, A. (2002) Pricing irrigation water: a review of theory and practice. *Water Policy* 4(2), 173–199.
Jones, T. (2003) Pricing water. OECD, Environment Directorate, Paris.
Jones, W.I. (1995) *The World Bank and Irrigation*. Operations Evaluation Department, a World Bank Operation Evaluation Study. World Bank, Washington, DC.
Kaika, M. (2003) The water framework directive: a new directive for a changing social, political and economic European Framework. *European Planning Studies* 1(3), 299–313.
Khanna, G. and Sheng, F. (2000) Towards full-cost pricing of water. WWF Conservation Economics Unit.
Kim, C.S. and Schaible, G.D. (2000) Economic benefits resulting from irrigation water use: theory and an application to groundwater use. *Environmental and Resource Economics* 17, 73–87.
Krueger, A.O., Schiff, M. and Valdés, A. (1988) Agricultural incentives in developing countries: measuring the effect of sectoral and economywide policies. *World Bank Economic Review* 2(3), 255–271.
Krueger, A.O., Schiff, M. and Valdés, A. (eds) (1991) *The Political Economy of Agricultural Pricing Policy*. A World Bank Comparative Study. Johns Hopkins University Press, Baltimore, Maryland.

Lavelle, M. and Kurlantzick, J. (2002) *The Coming Water Crisis*. Resources for the Future, Washington, DC.

Le Moigne, G., Easter, K.W., Ochs, W.J. and Giltner, S. (1994) *Water Policy and Water Markets*. World Bank, Technical Paper No. 249. World Bank, Washington, DC.

Louw, D.B. and Kassier, W.E. (2002) *The Costs and Benefits of Water Demand Management*. Centre for International Agricultural Marketing and Development, Paarl, South Africa.

Maestu, J. (2001) The political economy of the implementation of changes in pricing practices in Spain. What can we learn? In: *Pricing Water, Economics, Environment and Society*. Conference Proceedings, Sintra, European Commission, Brussels, pp. 247–267.

Molden, D., Frenken, K., Barker, R., de Fraiture, C., Mati, B., Svendsen, M., Sadoff, C. and Finlayson, M. (2007) Trends in water and agricultural development. In: Molden, D. (ed.) *Water for Food – Water for Life*, Chapter 4. *Comprehensive Assessment of Water Management in Agriculture*. EarthScan, London.

Molle, F. and Berkoff, J. (2006) *Cities versus Agriculture: Revisiting Intersectoral Water Transfers, Potential Gains and Conflicts*. IWMI Comprehensive Assessment Research Report 10. IWMI Comprehensive Assessment Secretariat, Colombo, Sri Lanka, p. vi.

Molle, F. and Renwick, M. (2005) *Economics and Politics of Water Resources Development: Uda Walawe Irrigation Project, Sri Lanka*. IWMI Research Report 87. IWMI, Colombo, Sri Lanka, p. vi.

Moore, M. (1989) The fruits and fallacies of neoliberalism: the case of irrigation policy. *World Development* 17(11), 1733–1750.

Moore, M. (1990) The rational choice paradigm and the allocation of agricultural development resources. *Development and Change* 21, 225–246.

Morris, J. (1996) Water policy: Economic theory and political reality. In: Howsam, P. and Carter, R.C. (eds) *Water Policy: Allocation and Management in Practice*. E & FN Spon, London, pp. 228–234.

ODI (Overseas Development Institute) 2002. *The 'Water Crisis': Faultlines in Global Debates*. ODI Briefing Paper. ODI, London. Available at: http://www.odi.org.uk/

OECD (1999a) Future water resources: towards a demand-side solution? Highlight of the OECD Information Base. Available at: www.oecd.org/sge/au/highlight16.html

OECD (1999b) *The price of water. Trends in OECD countries*. OECD, Paris, p.173.

O'Mara, G.T. (1990) *Making Bank Irrigation Investments More Sustainable* (It is time to rationalize policy guidelines on Bank irrigation projects). Agriculture and Rural Development Department. Working Paper. World Bank, Washington, DC.

Oorthuizen, J. (2003) *Water, Works, and Wages: The Everyday Politics of Irrigation Management Reform in the Philippines*. Wageningen University Water Resources Series. Orient Longman, Hyderabad, India.

Perry, C.J. (1996) *Alternative to Cost Sharing for Water Service to Agriculture in Egypt*. IIMI Research Report No. 2. IIMI, Colombo, Sri Lanka.

Perry, C.J. (2001a) Water at any price? Issues and options in charging for irrigation water. *Irrigation and Drainage* 50(1), 1–7.

Perry, C.J. (2001b) *Charging for Irrigation Water: The Issues and Options, with a Case Study from Iran*. Research Report No. 52. IWMI, Colombo, Sri Lanka.

Perry, C.J., Rock, M. and Seckler, D. (1997) *Water as an Economic Good: A Solution, or a Problem?* Research Report No. 14. IWMI, Colombo, Sri Lanka.

Pitman, G.K. (2002) *Bridging Troubled Water: Assessing the World Bank Water Resource Strategy*. World Bank Operations Evaluation Department. World Bank, Washington, DC.

PRI (Policy Research Initiative) (2005) *Economic Instruments for Water Demand Management in an Integrated Water Resources Management Framework*. Government of Canada, Ottawa.

Rao, P.K. (1984) Comment on cost recovery and irrigation water pricing. In: *ODI Irrigation Management Network Paper*. Overseas Development Institute, London.

Ray, A., Bruce, C. and Hotes, F.L. (1976) Cost recovery policies for irrigation projects: informal guidelines. Paper presented for informal guidance of persons involved in cost recovery policies for irrigation projects. World Bank, Washington, DC.

Repetto, R. (1986) *Skimming the Water: Rent Seeking and the Performance of Public Irrigation Systems*. Research Report 4. World Resource Institute, Washington, DC.

Republic of South Africa, Department of Water Affairs and Forestry (1998) National water resource strategy. Available at: http://www.dwaf.gov.za/Documents/Policies/NWRS/Default.htm

Rosegrant, M.W. and Cline, S. (2002) The politics and economics of water pricing in developing countries. *Water Resources Impact* 4(1), 6–8.

Rosegrant, M.W., Cai, X. and Cline, S.A. (2002) *Global Water Outlook to 2005: Averting an Impending Crisis*. IFPRI, Washington, DC; IWMI, Colombo, Sri Lanka, p. V.

Saleth, R.M. (2001) *Water Pricing: Potential and Problems*. Policy Brief Focus 9. International Food Policy Research Institute, Washington, DC.
Samal, C.K. and Kolanu, T. (2004) Water pricing and decentralized irrigation management in Andhra Pradesh: schism between objectives and realities. Paper presented at the 'Deutscher Tropentag', Berlin.
Sampath, R.K. (1992) Issues in irrigation pricing in developing countries. *World Development* 20(7), 967–977.
Savenije, H.H.G. and van der Zaag, P. (2002) Water as an economic good and demand management: paradigms and pitfalls. *Water International* 27(1), 98–104.
Shiva, V. (2002) *Water Wars: Privatization, Pollution, and Profit*. Pluto Press, London.
Siamwalla, A. and Roche, F. (2001) Irrigation management under resource scarcity. In: Siamwalla, A. (ed.) *The Evolving Roles of the State, Private, and Local Actors in Rural Asia: Study of Rural Asia*. Oxford University Press, Hongkong, pp. 183–212.
Small, L.E. (1990) Financial tools for improving irrigation performance. In: Sampath, R.K. and Young, Robert A. (eds) *Social, Economic, and Institutional Issues in Third World Irrigation Management*. Westview Press, Boulder, Colorado, pp. 147–268.
Small, L.E. and Carruthers, I. (1991) *Farmer-financed Irrigation: The Economics of Reform*. Published in association with the International Irrigation Management Institute. Cambridge University Press, Cambridge.
Small, L.E., Adriano, M.S. and Martin, E.D. (1986) Regional study on irrigation service fees: final report. Report submitted to the Asian Development Bank. IIMI, Colombo, Sri Lanka.
Steele, T. (1867) *Report on the Hambantota District*. Administration Reports, 1867. Ceylon Government Press, Colombo, Sri Lanka.
Stone, I. (1984) *Canal Irrigation in British India: Perspective on Technological Change in a Peasant Economy*. Cambridge University Press, Cambridge.
Svendsen, M. (1986) Irrigation system recurrent cost recovery: a pragmatic approach. In: FAO. Land and Water Development Division; USAID. *Water Management Synthesis II*. Project Technical papers from the Expert Consultation on Irrigation Water Charges, Vol. I, Rome, 22–26 September 1986. FAO, Rome, Italy, pp. 122–157.
Teerink, J.R. and Nakashima, M. (1993) *Water Allocation, Rights, and Pricing: Examples from Japan and the United States*. World Bank Technical Paper No. 198. World Bank, Washington, DC.
The Economist (2003) Priceless. Available at: http://www.economist.com/displaystory.cfm?story_id = 1906846
The Water Manifesto (1999) Available at: www.f1boat.com/99/watermanifesto.html
Thobani, M. (1997) Formal water markets: why, when and how to introduce tradable water rights. *The World Bank Research Observer* 12(2).
Tsur, Y. and Dinar, A. (1995) *Efficiency and Equity Considerations in Pricing and Allocating Irrigation Water*. Policy Research Working Paper 1460. World Bank, Washington, DC.
UNESCO (2000) Director-general sounds the alarm over looming water crisis. Available at: www.waternunc.com/gb/ungb04.thm [accessed August 2002].
UNESCO-WWAP (2006) *Water: A Shared Responsibility*. The United Nations World Water Development Report 2. UNESCO, Paris.
United Nations (Department of Economic and Social Affairs) (1992) Agenda 21. Available at: http://www.un.org/esa/sustdev/documents/agenda21/english/agenda21toc.htm
Vaidyanathan, A. (Chair) (1992) Report of the Committee on Pricing of Irrigation Water. Planning Commission, Government of India, New Delhi. Reproduced in *Indian Journal of Agricultural Economics* 49(1), 107–133, January–March 1994.
Winpenny, J.T. (1997) Demand management for efficient and equitable use. In: Kay, M., Franks, T. and Smith, L. (eds) *Water: Economics, Management and Demand*. E & FN Spon, London, pp. 296–303.
Wolfensohn, J.D. (2000) Remarks at the Second World Water Forum: 'From Vision to Action'. The Hague, The Netherlands, 22 March 2000.
World Bank (1976) *Irrigation Water Charges, Benefit Taxes and Cost Recovery Policies*. Central Projects Memorandum (CYM) No. 8.4. World Bank, Washington, DC.
World Bank (1980) *Irrigation Water Charges, Benefit Taxes and Cost Recovery Policies*. Central Projects Memorandum (CYM) No. 2.10. World Bank, Washington, DC. (reedition of No. 8.4 with little modification)
World Bank (1981) *Water Management in Bank-supported Irrigation Project Systems: Analysis of Past Experiences*. Report No. 3421. Operations Evaluation Department. World Bank, Washington, DC.
World Bank (1984) *Financing Operation and Maintenance in Irrigation*. Agriculture and Rural Development Department. World Bank, Washington, DC.
World Bank (1986) *World Bank Lending Conditionality: A Review of Cost Recovery in Irrigation Projects*. Report No. 6283. Operations Evaluation Department, World Bank, Washington, DC.
World Bank (1993) *Water Resources Management: A World Bank Policy Paper*. World Bank, Washington, DC.

World Bank (2003) *World Bank Water Resources Sector Strategy: Strategic Directions for World Bank Engagement*. World Bank, Washington, DC.

World Bank (2005a) Priority issues for Indonesian agriculture. Policy Briefs. Available at: http://www.worldbank.or.id-January 2005

World Bank (2005b) *Reengaging in Agricultural Water Management: Challenges, Opportunities, and Trade-Offs*. Water for Food Team, Agriculture and Rural Development Department (ARD). World Bank, Washington, DC.

World Bank (1978) *Bangladesh: Irrigation Water Charges*. A discussion paper with proposals for action to develop an irrigation water charge policy. World Bank, Washington, DC, p. 56.

World Water Commission (2000) *A Water Secure World: Vision for Water, Life, and the Environment*. World Water Vision Commission Report. World Water Commission, The Hague, The Netherlands.

2 Water Pricing in Irrigation: Mapping the Debate in the Light of Experience

F. Molle and J. Berkoff

Introduction

This chapter provides a broad discussion of water pricing in agriculture, scrutinizes arguments sequentially, gives examples from the literature and indicates links to other chapters. It suggests the conditions under which water pricing is likely (or not) to bear fruit, and assesses its potential for alleviating the global and local water crises. The focus is on public large-scale gravity schemes although groundwater and communal systems are also referred to, albeit in less detail.

Charging for water use or disposal is not an end in itself, but an instrument for achieving one or more policy objectives (Fig. 2.1). A water charge may be *a financial tool* aiming to recover all or part of capital and recurrent costs, recurrent cost recovery being particularly critical to preserve the physical integrity of the system when public funds are not forthcoming. A water charge may also be *an economic tool* designed to conserve water and raise water productivity by promoting: (i) careful management and water conservation; (ii) cultivation of less water-demanding crops and investments in water-saving technologies; and (iii) reallocation of water to high-value agriculture and/or other sectors. Finally, a charge can be *an environmental tool* to counter water pollution and enhance water quality.

Water pricing issues lie at the confluence of two complex 'spheres': on the one hand, the microeconomy of the farm and its linkages to the wider economic system and agricultural policies and, on the other, the hydrology of the plot and its interconnectedness with the irrigation system, the river basin of which it is a part, and the overarching water policy framework (Fig. 2.2).

These nested levels of interaction result in a complex set of dynamics. Economic interactions reflect the multiplicity of factors that govern economic behaviour and the heterogeneity of the different economic actors. Hydrological interactions between upstream and downstream, surface water and groundwater and quantity and quality are compounded by seasonal and interannual variability that creates unstable and unpredictable systems. Economic and hydrological interactions are further embedded within cultural and social contexts that eventually define the distribution of costs and benefits within the society, and are thus highly political in character (Johansson, 2000; Dinar and Saleth, 2005).

In the past, emphasis has typically been placed on influencing the performance of farmers and irrigated agriculture (right sphere) by the manipulation of the

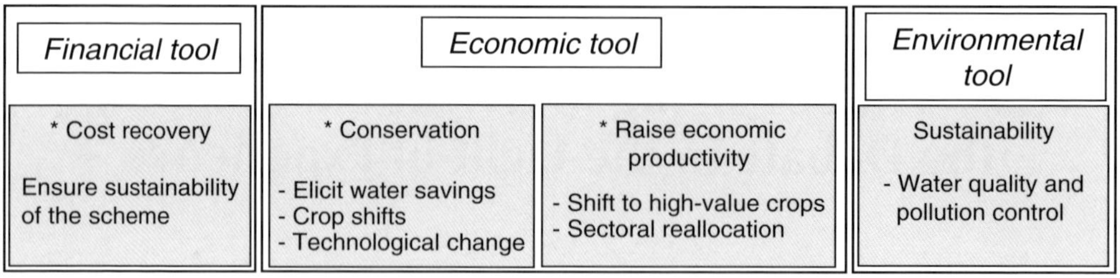

Fig. 2.1. What to charge for?

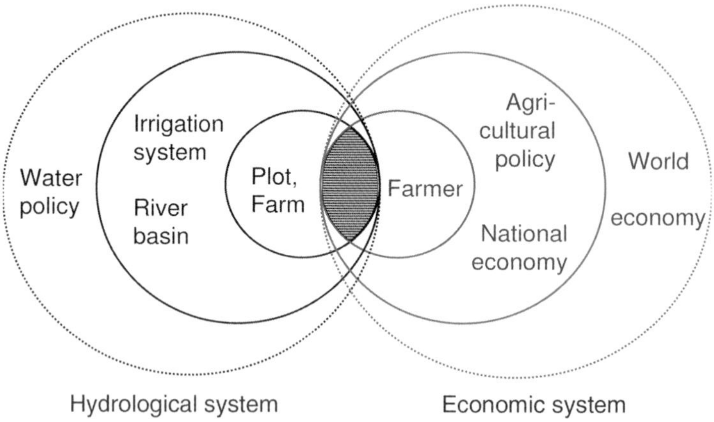

Fig. 2.2. Water pricing issues at the intersection of two spheres of complexity.

hydrologic cycle and the design of canal and pipe networks (left sphere). Increasingly, however, emphasis has shifted to influencing performance of the water system (left sphere) by the adoption of economic and related incentives (right sphere). This chapter reviews the potential and the effectiveness of the latter approach, focusing in particular on the contribution of water pricing. It will argue that water pricing is strongly related to the institutional setting, that is, to the combination of community, government and market regulation, and to the attendant rules that define water governance and management in a particular context. More specialized issues, such as irrigation management transfer, characteristics of water markets, environmental protection, irrigation modernization and politics of water development, though important in their own right and relevant to the issues under consideration, receive less attention in this synthesis chapter, as do related theoretical considerations.

The following section expands on the economic and hydrological systems summarized in Fig. 2.2, and discusses the broad context within which the subsequent discussion is set. Within this framework, we move to examining the practicalities and effectiveness of current water charging practices. The following five sections successively review the main roles commonly attributed to irrigation water pricing: (i) cost recovery; (ii) water conservation; (iii) enhanced water productivity; (iv) intersector reallocation; and (v) control of water quality. The concluding section offers a synthesis of the assessment and corresponding conclusions. While the various sections have been defined for analytical purposes, it will become clear that they are strongly interrelated.

Table 2.1. Evolving priorities of the EU Common Agricultural Policy. (From Gómez *et al.*, 2005.)

	Issues and concerns	Objectives	Agricultural water pricing
Past	Poverty in rural areas	Equity and rural development	Lower prices
	Increasing food demand	Food self-sufficiency	
Future	Water and soil pollution	Sustainable development	Higher prices
	Budgetary constraints	Economic efficiency	

CONTEXT MATTERS

The Economic Context

The rationale for irrigation

For millennia, subsistence and financial self-interest have driven communities to construct village schemes, rulers to develop major projects and farmers to exploit groundwater and make other on-farm investments. During the colonial period, there were those who hoped the self-interest of private investors would drive large-scale irrigation investment, but few such projects proved commercially viable and major irrigation has remained predominantly in the public sector.

Cost recovery has always been a major concern. Communities internalized costs, historic rulers recruited corvée labour mainly from the farming population and colonial governments constantly debated the optimum balance between profitability and income generation. As described by Molle and Berkoff (Chapter 1, this volume), the balance shifted following World War II. Governments and donor agencies continued to pay regard to profitability, re-expressed in economic rather than financial terms (in cost–benefit studies), and also began to raise environmental concerns. But other objectives were often dominant, notably:

- Poverty alleviation, equity and employment generation;
- Regional development and the urban/rural balance;
- Food self-sufficiency and/or food security;
- State building and the search for political support and legitimacy.

These objectives can, of course, be mutually consistent with one another and with economic optimization and environmental sustainability, and such consistency is often claimed. But where they are inconsistent, choices must be made. Despite lip service to economic optimization and sustainable development, large-scale expansion of the irrigated area has, in practice, been driven largely by political interests reflecting these other objectives. Recently, the balance has shifted back in favour of the environment, at least in the USA and Europe, with implications for irrigation water prices (Table 2.1).

Whatever the rationale given for the initial construction of an irrigation scheme, subsequent cost recovery remains a widely accepted policy. In practice, cost recovery is normally limited to the recovery of operation and maintenance (O&M) costs and at most to a (small) share of capital costs. The main driver for cost recovery has been containment of government costs, though recouping at least some of the costs from direct beneficiaries is also advocated on equity grounds. In addition, it is claimed that charging for water can promote favourable economic and financial outcomes, especially if combined with irrigation management autonomy. Some commentators have gone further, arguing that irrigation pricing can lead to economically efficient outcomes. Although such claims are now largely discounted (Molle and Berkoff, Chapter 1, this volume), the idea remains important and is explored later in this chapter.

Cost–benefit analysis

Cost–benefit analysis ostensibly provides the basis for taking decisions on public investments. Standard approaches allow for the adjustment of financial prices as a basis for choosing economically viable projects, with additional studies throwing light on possible economic distortions.[1] The main direct costs are the initial capital costs, which typically account for 80–85% of discounted total costs in surface irrigation. Recurrent costs comprise a higher share in pump schemes though capital costs still largely determine viability. Once built, capital costs are '*sunk*' and the direct marginal costs comprise regular O&M together with the costs of replacement, rehabilitation and modernization. Indirect costs include negative environmental and social externalities and opportunity costs – if any – reflecting an appropriate share of the value of output forgone in alternative uses (see below). The main direct benefits comprise the incremental value of agricultural output *with* relative to that *without* the project. There may also be benefits from domestic supply and other uses, and from positive externalities. If discounted benefits exceed discounted costs, the project is viable.

Although cost–benefit analysis is, in principle, straightforward, its application in irrigation and other water projects has been problematic. Although some claim that ex post evaluation studies show that irrigation projects have performed satisfactorily (Jones, 1995), others suggest that there has been a systematic bias in favour of new construction (Repetto, 1986; Berkoff, 2002; Molle, 2007). Three types of argument support the latter case:

- First, as suggested above, political objectives rather than economic priorities often drive irrigation expansion. Moreover, the political dynamics almost always favour going ahead given the combined self-interest of beneficiary farmers, politicians, contractors, consultants and staff in irrigation, and lending agencies (Repetto, 1986; Merrett, 1997). Finance and other entities serving a broader national interest may restrain irrigation expansion, but can seldom prevent it, even if that is their preference.
- Second, the economic analysis of irrigation is more than usually uncertain. Unwitting optimism is widespread and over-optimistic assumptions are difficult to refute, both with regard to costs and to benefits. 'Costs tend to be high because of: inappropriate design, stemming in part from poor studies done prior to start-up; long gestation periods resulting from funding shortfalls due to changing government priorities and poor capital programming and budgeting; few managerial incentives to control costs; and reported corruption that typically involves kickbacks from construction companies' (Holden and Thobani, 1996). Benefits comprise the difference between two large hypothetical future flows (the values of production *with* and *without* the project). Estimating these flows is based on a host of assumptions that cannot be readily validated (Carruthers and Clarck, 1981; Merrett, 1997; Green, 2003). If prices, yields, irrigation efficiency or cropping patterns are adjusted even modestly, the impact can be surprisingly large. Who is to say the assumptions are wrong?
- Third, the retention of surface irrigation in the public sector and the funding of surface irrigation from the government budget limit financial accountability and help explain why inadequate cost–benefit studies generate such little concern. Canals and related facilities are often classified as infrastructure comparable to roads or power supply, and governments feel responsible for infrastructure. But irrigation is also a productive activity in many ways analogous to industry. Few governments still feel competent to *pick winners* in the industry, yet this is rarely questioned in irrigation.

Cost–benefit analysis is thus malleable, and analysts are invariably under pressure to produce positive results. Feasibility studies that appear competent at the time often prove very over-optimistic in retrospect (Pitman, 2002). Re-estimated rates of return are thus typically much lower at completion of project works than at the feasibility stage, and lower still at impact assessment when actual performance outcomes are available. Moreover, long-term price trends, system deterioration and failure

[1] For example, nominal and effective protection studies.

to account adequately for the *without* case suggest that – even at impact assessment – over-optimism is rife (Berkoff, 2002).

Overriding national priorities

The use of social weights and an opportunity cost for labour are techniques that can, in theory, help address issues of poverty alleviation, equity and employment in cost–benefit analysis (Squire and van der Tak, 1976). These partial equilibrium approaches are, however, controversial, given also the inherent uncertainties described above. Moreover, it is arguable that they do not account adequately for broader issues. Irrigation has both backward and forward linkages, while enhanced incomes have further multiplier impacts. Large-scale irrigation is thus often promoted as the engine that drives rural development as a means to both alleviate poverty and provide job opportunities so as to limit outmigration to cities. Such regional development issues are, in theory, best addressed in a general, rather than a partial, equilibrium context. General equilibrium models are, however, complex and expensive, and well beyond the scope of most project studies. Some advocate a simpler approach, that of increasing benefits by some factor representing multiplier impacts. But, for this to be valid, multiplier benefits should be confined to *incremental* impacts relative to those of the next best alternative, allowing also for opportunity costs and the avoidance of double-counting (Carruthers and Clark, 1981; Gittinger, 1982). It is arguable that such conditions occurred in densely populated Asia at the early stages of development (say, 1950–1980) when other viable regional projects were scarce and labour and water were abundant relative to land. Whether such conditions prevail today, notably in land-abundant Africa and Latin America, is much more questionable. Farmers in these regions often have access to rain-fed lands, population densities are much lower and conventional returns to irrigation have declined drastically.

Even if the case for new irrigation based on multiplier effects is questionable, they may still provide a rationale for *preserving* irrigation that has already been built. If investments in transport, marketing and social infrastructure depend on irrigation for their continued profitability, the case for preserving irrigation as a form of social overhead capital comes into its own (Small, 1990). On the North China Plain, for instance, irrigation is affected by severe water constraints. Water transfers from the Yangtze will help maintain farm incomes and slow rural depopulation. Although new irrigation cannot be justified on economic grounds, the economic returns to the transfer to sustain existing irrigation are strengthened by the costs sunk in existing assets not only in irrigation facilities, but also in rural economic and social infrastructure (Berkoff, 2003a).

Irrespective of these economic arguments, history shows that many schemes have also, in practice, been designed with wider geopolitical motives in mind. The western USA, for instance, illustrates a long history of engagement by the state in support of colonization (Reisner, 1986). The Gezira scheme in Sudan (Gaitskell, 1959), Israeli settlements in Palestine (Lipchin, 2003) and the GAP project in south-eastern Anatolia (Harris, 2002) are other well-known examples of projects promoted to achieve geopolitical goals (Molle *et al.*, 2007). Likewise, the context of the Cold War and the food shortages and fears of rural disintegration that followed the El Niño-related climatic perturbation of 1972 did much to justify the huge investments in dams and irrigation infrastructures that were to follow (Barker and Molle, 2004). Food self-sufficiency or food security has often been a top strategic concern to be addressed at any cost. In such situations, economic or hydrologic rationality is in effect neither here nor there and overriding political decisions dictate public investments.

Shifting subsidies and taxation

Moreover, the public subsidies incurred under such rural development policies need to be placed in a general economic context. In the decades after World War II, many countries adopted a policy of taxation of agriculture, notably by export duties (Harris, 1994) and public procurement programmes that maintained farm-gate prices often well below their world price equivalents. The magnitude of this taxation amounted – to borrow from Schiff and Valdés (1992) – to a 'plunder' of agriculture during 1960–1985. In Mexico, the price distortion amounted to an implicit tax of 20–50% of the value of the project commodities (Duane, 1986) and similar state extractive policies were carried out in most developing countries, including Egypt (Barakat, 2002), Thailand (Molle, Chapter 5, this volume), Malaysia (World Bank, 1986), Pakistan (Chaudhry *et al.*, 1993), Côte d'Ivoire, Ghana and Sri Lanka (Krueger *et al.*, 1991; Schiff and Valdés, 1992). Low food prices benefited the urban poor and landless, and taxes on output generated public savings for investment in industrial and urban development, only partially offset by irrigation and other rural subsidies (Lipton, 1977). Low food prices also had adverse impacts on crop output so that rationing was often required to manage consumption, limit imports and maintain food self-sufficiency.

Over time, the arithmetic of relative taxes and subsidies changed drastically as world prices declined and incomes rose. This and the widespread adoption of liberalization policies led to the abolition of most export duties and food-rationing programmes. Reforms initially boosted farm output and incomes as farmers responded to liberalized markets and exploited the agricultural technologies open to them. But as prices declined further, and as economic growth and diversification took place, urban/rural income differentials were reaccentuated, often provoking farmer unrest. Fearing also adverse impacts on domestic output,[2] some governments (e.g. China and India) have begun to support (rather than – as in the past – tax) farmers by limiting imports and adopting other trade-distorting measures. In this they have followed the lead of developed countries (the EU, the USA and Japan) that have long protected agriculture. This situation helps explain the reluctance of governments to raise water charges or other input prices for fear of losing their competitive edge (Tiwari and Dinar, 2001), since many farmers have to compete with exporters from the North who benefit from lavish subsidies.[3]

These trade distortions (market access, tariffs and export subsidies) are the major concern of the WTO Agricultural Agreement (WTO, 2000). Their removal would raise farm-gate prices significantly by reducing developed country exports, thus moderating the need for interventions by developing country governments in support of their farmers, besides facilitating attainment of food self-sufficiency objectives and promoting developing country food exports and inter-south trade (USDA, 2001). The WTO agreement also aims to reduce direct food and fertilizer as well as other input subsidies that have a direct impact on trade. In contrast, irrigation expenditures are amongst those that can be used freely since it is argued that they have minimal impact on trade (WTO, 2000). This is perhaps debatable. It is true that viable irrigation projects do not distort trade but if – as suggested above – much irrigation has been uneconomic, cumulative worldwide irrigation subsidies have contributed to declining world prices in a manner comparable to that of other trade distortions. Moreover, although irrigated output has risen enormously, rain-

[2]Taxation of agriculture and the resulting 'urban bias' are also seen as reflecting the shifting influence and political clout of interest groups and coalitions (whether defined by sector or income groupings) (Lipton, 1977; Bates, 1981; Sarker *et al.*, 1993), linked to their income, information and education, potential for collective action and political representation (Binswanger and Deininger, 1997). According to Bates (1993) this transformed the agriculture sector from 'an embattled majority that is taxed into a minority powerful enough to be subsidised'.

[3]Yang *et al.* (2003) show how decreasing profitability could put further pressure on domestic food production in China, challenged by international markets since the late 1990s, and even more since China's recent accession to the World Trade Organization (WTO) (Huang and Rozelle, 2002). After adhesion to the WTO, Jordan had to face 'unfair market intrusions by countries with less stringent WTO membership conditions' (WTO, 2001) and realized that abolishing subsidies altogether would be detrimental to its own farmers.

fed yields and output may well have been suppressed (Berkoff, 2003b). If so, food self-sufficiency based on irrigation may have been achieved at the expense of the rain-fed farmer.

Ultimately, all tax and subsidy polices are conditioned by politics, and reflect the cultural, economic and political milieu in each country concerned. Although the WTO negotiations aim to moderate economic distortions, and thus benefit those that are discriminated against, especially by developed country interventions, all such interventions must be understood within the wider political and policy context if they are to be analysed and possibly changed (Sampath, 1992; Speck and Strosser, 2000).

The Hydrological Context

The characteristics of water and water use

The physical characteristics of surface water are well known and include site-specificity, mobility, stochastic variability and uncertainty, bulkiness and solvent properties. Accompanying these are its relatively low value as a commodity, the economies of scale that often make supply a natural monopoly and the pervasive interdependence of water users (Young, 1986; Livingston, 1995; Morris, 1996; Savenije, 2001; Green, 2003). Groundwater shares some of these attributes but has other attributes that set it apart, including its relative immobility, security and divisibility.

Water has numerous human uses, some of which are consumptive (agriculture, industry and domestic) and others non-consumptive (fisheries hydropower, navigation, etc.). Water also has environmental values that are appreciated by humanity. The characteristics of water use in agriculture set it apart in many ways from its use in municipal and industrial use.

Diversions for consumptive use are invariably larger than the fraction that is actually consumed, with the balance returning to the water system. Agricultural withdrawals (predominantly for irrigation) account worldwide for 70% of the water withdrawn for consumptive use (Aquastat, 2004). Its share is typically higher in developing than in developed countries. Evapotranspiration accounts for 40–60% of agricultural withdrawals (rising to above 70% due to repeated reuse, modern irrigation techniques, etc.). In contrast, domestic water withdrawals are largely used for washing and cooking, and domestic diversions largely return – often in a polluted form – to the water system. Similarly, industrial diversions are mainly for cooling and dilution of wastes rather than for chemical incorporation in products. Consumptive use as a proportion of withdrawals is thus much higher in agriculture (70%) than in domestic (14%) or industrial (11%) use, and agriculture accounts for as much as 85–90% of total consumptive use worldwide (Shiklomanov, 2000).

Uses in the municipal and industrial (M&I) as well as the irrigation sectors are not always fully interchangeable. M&I use is usually far more valuable than in irrigation, and logic implies that water should move wherever possible from irrigation to M&I in the event of conflict. But transfers are only feasible if the infrastructure is, or can be, integrated at acceptable cost. Moreover, M&I have much higher quality and security-of-supply requirements than irrigation, which may limit transfer opportunities.

Consumptive use impacts on non-consumptive uses through its effect on flow regimes, water quality and flood risk. Given that irrigation use is so much greater than M&I use, the major quantity conflicts are generally between irrigation on the one hand and in-stream and environmental uses on the other (though M&I can have large quality impacts). Irrigation diversion capacity often exceeds dry season flows and, as use rises, irrigation may be able to divert flows year-round. In-stream uses suffer, rivers and wetlands dry up, affordable groundwater is exhausted and pollution loads rise (though flood risks may moderate). Action to safeguard in-stream and environmental uses may then become desirable and, in effect, irrigation rather than the environment becomes the *user of last resort* (Elston, 1999).

Irrigation efficiency

The concept of irrigation efficiency is often misstated (Willardson *et al.*, 1994; Frederiksen, 1996; Keller *et al.*, 1996; Huffaker *et al.*, 1998; Perry, 1999; Huffaker and Whittlesey, 2000; Loeve *et al.*, 2004; Molle and Turral, 2004) with significant implications for water pricing. If water is abundant, scheme-level efficiency is of limited concern other than for system capacity and capital cost reasons. If basin water is scarce, raising scheme efficiency can be elusive since return flows are fully utilized and the only *additional* source of water lies in reducing unproductive losses.[4] In north China, for instance, apart from uncontrollable floods and releases for silt and pollution control, little water reaches the sea from a vast area containing up to 7.5% of world population. Drainage and wastewater reuse are pervasive, losses recharge groundwater, farmers underirrigate, tail-end areas are abandoned and basin efficiency is high by any standards. Existing irrigation can essentially absorb *all* the water available and shortages relative to theoretical crop water requirements have little meaning (Berkoff, 2003b).

It is not only basin efficiency that is misstated. Scheme and on-farm efficiencies are also often (much) higher than assumed. That water is 'wasted' when it is abundant (e.g. after it rains) is inconsequential – low physical efficiency may even correspond to high economic efficiency since management is eased and labour reduced (Gaffney, 1997). In contrast, farmers fight for water and return flows if it is scarce (and overpump groundwater). The struggle for water *when it is scarce* means that little water is wasted *when it has value* and average estimates of efficiency can be very misleading. Case studies from Thailand (Molle, 2004), California (Zilberman *et al.*, 1992) and China (Loeve *et al.*, 2003) have shown the multifarious efforts deployed by farmers to adjust to water scarcity and make the best use of water. These changes go often unnoticed but statements such as 'farmers waste water just because [they] are not aware of the fact that water has a value' (Roth, 2001) are both unfair and mistaken. Moreover, even if there is potential for increased scheme-level and on-farm efficiency, this can require expensive investments in drip or sprinkler systems that may not be justified either financially or economically.

[4] That there is little water – if any – to be saved in closed basins must, however, be qualified since there are notable exceptions. If return flows from irrigation are degraded in terms of quality (salinity, contamination), they may incur yield losses when reused (Morocco: see Hellegers *et al.*, Chapter 11, this volume; Pakistan) or be unfit for agriculture (e.g. Jordan Valley: Fontenelle *et al.*, Chapter 7, this volume), and therefore losses should be minimized. If the time taken by water to become available again is very long (e.g. percolation to deep aquifers), these volumes are not available for short-term use. Water wasted in the wet season in cities or irrigation schemes could also sometimes be kept in reservoirs for later use in the dry season. Another caveat concerns the costs incurred by possible successive pumping operations associated with reuse.

Irrigation design

Opinions on irrigation design range from those that advocate modern systems of control (Plusquellec, 2002) to those that advocate simple technologies that respond to human and institutional limitations (Horst, 1998; Albinson and Perry, 2002). The critical factor is stochastic water variability: from day to day, week to week and year to year. *Supply* is variable because runoff is variable; *demand* is variable because rainfall and crop water requirements are variable. Reservoirs and groundwater improve predictability, and on-demand systems help farmers obtain water when it is needed. But in practice, most surface water systems are designed to meet peak water requirements for a specified cropping pattern, say, 3 years in 4 (i.e. the 75% year) (the full area being irrigated in the wet season and a restricted area in the dry season). This is a compromise. If greater security is guaranteed to a smaller area, in most years the available resource is underutilized. If canal capacity is increased to expand the area in good years, unit costs rise, security declines and capacity in most years is excessive. In contrast to fully on-demand systems, therefore, it is *by design* that the full area cannot be irrigated in dry periods, in dry years and during the dry season.

As economies develop, shortages increase, water tables fall, other users get priority and variability is increasingly concentrated on irrigation as the residual user. Both the value of water and the costs of insecurity rise. Reservoirs are built, farmers install wells and on-farm ponds and modernization and volumetric measurement become affordable. Operator salaries and skills also rise in line with general living standards. In other words, irrigation responds to the external context. Ultimately, the issue in irrigation design is not that it is innately different to M&I design, but that there is a continuum from simple surface systems suited to low-return agriculture in poor countries, through conjunctive use and partially modernized systems appropriate to countries moving through the rural transition, to advanced technologies appropriate to high-return agriculture in richer countries that are completing the transition. At the limit, design approximates to that for M&I, and volumetric measurement at the level of the individual farmer becomes feasible. Until this point is reached, physical characteristics of irrigation severely constrain the possibility of using efficiency (marginal cost) pricing, and the debate on how economic pricing can be introduced has, in general, been a distraction.

Irrigation performance

Irrigation performance also ranges through a continuum. Traditional systems can be stable, but crop yields and farm incomes often remain low. Productivity and income in public systems are normally higher and manageability improves as an economy develops, agriculture becomes more entrepreneurial and market-driven, farm sizes and incomes rise, O&M agencies are better-funded and accountable and storage and modern control become affordable, manageable and justified.

Nevertheless, despite these trends, in the view of most observers, irrigation performance in developing countries remains generally poor. Water variability is again the main reason why so many schemes are so difficult to manage. *Ex post*, management must respond to conditions that deviate continuously from the average conditions implied by a design cropping pattern that means little to the farmer. Irrespective of design intentions, the farmer typically *wants* more water than he is allowed in the dry season and in dry periods; after rainfall, he may reject his allocation even if this causes problems elsewhere in the system. Differing objectives set up a continuing tension between scheme managers and farmers. Farmers interfere in outlets and water levels contributing to head-end and tail-end problems, while poorly paid system operators living close to the farmers fail to enforce – perhaps cannot or do not want to enforce – the rules. On the one hand, water-use efficiency is enhanced as farmers struggle for water and, on the other, damage is pervasive, inequities emerge and there is a broad failure to operate the system in line with design.

A Typology of Irrigation Systems

Figure 2.3 suggests a simplified typology of irrigation systems that reflects the above discussion. It classifies systems in relation to an index of relative water supply (RWS)[5] and suggests two broad types of management response: *pragmatic management* and *volumetric management* (that are linked not only to the degree of development, but also to the climatic context). With respect to Fig. 2.3:

- Situation W1 is typical of wet regions with abundant water supply. Water tends to be supplied continuously – often for paddy – at, or close to, full supply level, though rotations can be necessary if main canal capacity is a constraint. Occasional shortages may occur due to ill-discipline and farmer

[5]RWS is defined as the ratio of the water delivered to gross irrigation requirements (net of the effective rainfall) after accounting for losses. It provides a broad indication of the amount supplied relative to demand.

intervention. Minimal data on flow, rainfall and land use are typically collected.

- Situation W0 typifies non-arid countries as water is increasingly exploited. Operations reflect experience rather than active management, with water often released in response to farmers' complaints. Head-end and tail-end problems are limited while temporary supply reductions can lead to short-term crises as discipline breaks down. Data are collected haphazardly and seldom analysed. As RWS falls to 1, conflicts intensify and rotations are increasingly adopted.
- As RWS drops below 1 (D0), rotation becomes the rule. Farmers respond by deficit irrigation and conjunctive use (tapping drains, ponds or aquifers) and use water more carefully. Head-end and tail-end problems become pervasive. Data are collected more systematically and basic parameters (efficiency and water applied) are calculated. Supply-driven management predominates with scheduling planned, based on target allotments, and bulk allocations may be negotiated.
- Under situation D1, potential demand cannot be met and supply limits allocations. If the system is uncontrolled, water distribution may be chaotic. Groundwater replaces surface water and conjunctive use is ubiquitous, with land left fallow or abandoned. In systems that are better controlled – depending on design – water is confined to part of the scheme, supplied in turn or allocated proportionally (as under warabandi).[6] In fully controlled systems, volumetric rights are clearly defined and water may be supplied on demand, subject to availability.

When RWS falls below 1, the crucial step is the shift from 'pragmatic' to 'volumetric' management (Fig. 2.3). *Pragmatic management* is weak, reactive and ad hoc, with managers responding to complaints from below and farmers responding as best they can, e.g. by investing in wells and on-farm storage. As scarcity develops, water distri-

[6]All systems have to cope with hydrological variability (i.e. varying values of RWS) but both demand and supply are more predictable in arid climates since rainfall is a less significant factor and reservoirs are the norm. In humid climates, rainfall is a much more complicating factor since it strongly influences not only supplies at the source, but also requirements in the fields.

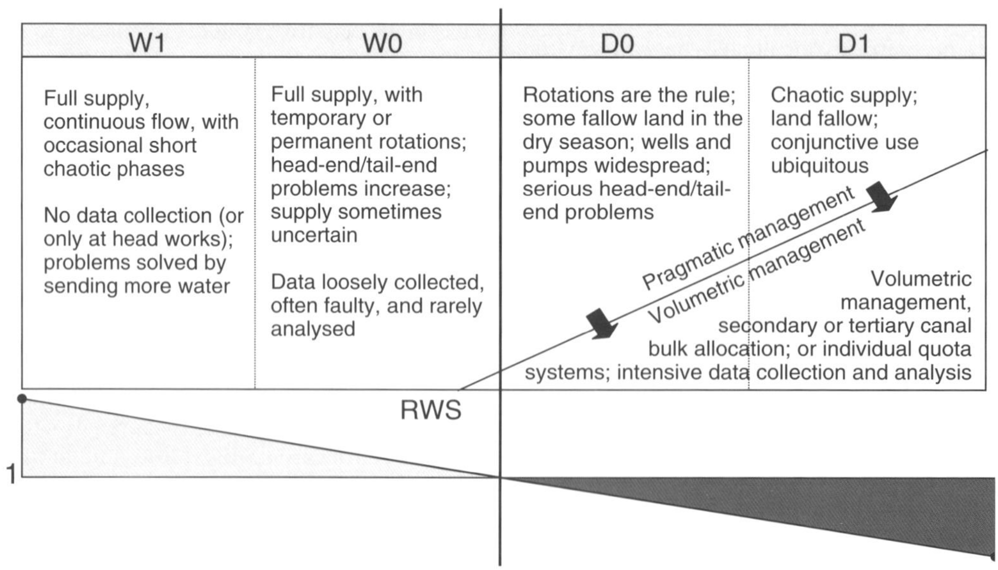

Fig. 2.3. A typology of irrigation systems.

bution becomes increasingly chaotic. Such conditions are common in developing countries, especially when schemes are large, farmers are numerous and poor and surface irrigation is dominated by cereals and low-return crops. Under these conditions, head-enders tend to divert what they want and tail-enders often fail to obtain even minimal supplies. With *volumetric management*, in contrast, a stronger degree of control is maintained. Water may be allocated in bulk or by individual quotas, rotational rules are clear and roughly predictable and risks are defined. At the limit, water may be provided approaching on-demand supply. This situation tends to occur in developed and/or arid countries, especially when farms are large, irrigated agriculture is for high-return crops and farmers incur large on-farm costs and financial risks (see above). Security in supply invites complementary on-farm investments and tends to make farmers willing to pay for water since even high charges comprise a small share of farm costs and service standards are critical.

This classification simplifies real-world diversity and variability. Even so, it can provide guidance in assessing the potential of water pricing policies. The difference between pragmatic and volumetric management corresponds to a 'quantum leap', and efficiency pricing is only possible if the scheme is under volumetric management and control is maintained. Many reforms fail because they assume very lightly that shifting from the former to the latter is simply a question of goodwill or capacity building, whereas it is linked in complex ways not only to RWS, but also to irrigation design and hydraulic control, manager-incentive and farmer-incentive structures and the wider institutional context.

Implications for Irrigation Pricing

Full marginal cost pricing

By analogy with domestic water supply and other infrastructural services, some analysts recommend long-run marginal cost (LRMC) pricing in irrigation (Arriens *et al.*, 1996). But there are important differences between the sectors. One issue is that volumetric pricing is far more problematic in irrigation than in reticulated urban systems, and this greatly restricts the adoption of efficiency pricing in irrigation. Basically, LRMC pricing in the urban sector simulates a competitive market price for a final good and, besides funding recurrent, replacement and related costs, it aims to generate the investment funds needed to match rising demand as a city expands and its population becomes richer (Munasinghe, 1990). If consumers are willing to pay the LRMC price, system expansion is economically justified; if not, effective demand can be met by existing capacity.

In contrast, irrigation water is an intermediate, not a final, good, and canals are sized to serve a specific command area at defined levels of probability (see earlier section). Possibilities for system expansion are thus restricted. Since charging existing farmers for a new scheme is no more justified than charging City A's inhabitants for expansion of City B's system, initial capital costs should usually be treated as *sunk*, in which case marginal direct costs comprise O&M and replacement costs.[7] Of course, if the scheme is inherently profitable, farmers should, in theory, be able to repay full costs (including initial capital costs), and charging them less than full cost gives them a windfall gain. But if expansion of irrigation has been driven by other public objectives (see above) and is uneconomic, charging full capital costs is neither feasible nor equitable (Carruthers and Clarck, 1981). Moreover, over time, capital subsidies are incorporated in land values and, though the initial beneficiaries may receive a windfall gain, inequities arise if charges are imposed on those that subsequently buy irrigated land.

Irrespective of any theoretical rationale for marginal cost pricing, there may still be a case for charging farmers a share of initial capital costs on financial and equity grounds, given

[7]They should also, in theory, cover modernization and system expansion costs if the water saved by the modernization investments is justified specifically in terms of the expansion of the scheme. The analogy with LRMC in expanding urban systems is then valid.

the needs of the economy and adverse impacts on rain-fed farmers. There is also the quite separate issue of whether opportunity values in alternative uses and externality costs should be reflected in some way in the irrigation charge. But competition between irrigation and cities is limited to specific periods and locations and, once urban demands are satisfied, opportunity cost falls drastically. Beyond compensating farmers on a case-by-case basis, water pricing to promote reallocation is generally impracticable (Molle and Berkoff, 2006; more on this later). Once M&I use is met, most conflicts lie between irrigation and the environment. But valuing environmental *externalities* (third-party impacts, soil salinization, water contamination, health hazards) is also a contentious issue, and willingness-to-pay for moderating such costs varies greatly at differing locations and stages of development. In most cases, there is no agreement on how pricing can mitigate negative impacts, and reflecting environmental use and valuing externalities are again impracticable (see section *Pricing as an environmental tool*).

The relevance of marginal cost pricing

Moreover, the need for strict marginal cost (efficiency) pricing in practice is often questionable. As argued above, irrigation performance typically reflects a rational response by farmers and operators to the evolving context and associated incentives. Water is used much more efficiently than is commonly supposed, and the scope for enhanced water-use efficiency and the potential role of water pricing can be greatly overstated. Furthermore, the massive expansion of private groundwater, much of it within surface schemes, has further strengthened irrigation performance. Groundwater is, in effect, available on demand and provides a security of supply that can offset variability of rainfall and canal supplies. Groundwater use, or conjunctive management, has thus accounted for most of the high-return diversified agriculture that has developed in response to economic growth, urbanization and external markets, and groundwater's pervasiveness limits the need for surface irrigation to meet these diversified demands.

In addition, no administered price can reflect short-term stochastic variability and, though at the margin water charges may impact on farmer behaviour and promote favourable economic and financial outcomes (Fig. 2.1), this is far short of true economic efficiency pricing. Modern control systems may be justified and, at the limit, a pressurized on-demand irrigation system approximates to a reticulated urban network. But, while urban systems are, in principle, *designed* to operate on demand, the vast majority of surface irrigation projects *by design* cannot supply water on demand since they cannot meet potential farmer uses when water is scarce (e.g. in the dry season or a drought). Comparing benefits and costs at the margin is therefore meaningless because farmers cannot, like urban users, access as much water as they wish and are willing to pay for. These considerations suggest that efficiency pricing is usually impracticable even in fully reticulated systems; supply management and rationing will inevitably remain the preferred mechanisms for controlling surface distribution in most irrigation in developing countries.

Potential price effects

As empirical evidence will confirm, the economic and hydrological characteristics reviewed above impact on irrigation water pricing in such a way that water charges are eventually, first and foremost, a cost-recovery mechanism. Even confining water charges to this one objective is far from straightforward since, as discussed above, what is meant by cost can vary depending on whether costs are limited to financial costs or extend to the full economic costs to society (Rogers *et al.*, 1998) and what is to be recovered may be limited to recurrent and replacement costs or include some or all of the capital costs invested. Financial O&M costs are invariably a priority since, once a scheme is constructed, production is contingent on continued O&M of the infrastructure.

In addition to financial cost recovery, economists argue that opportunity and

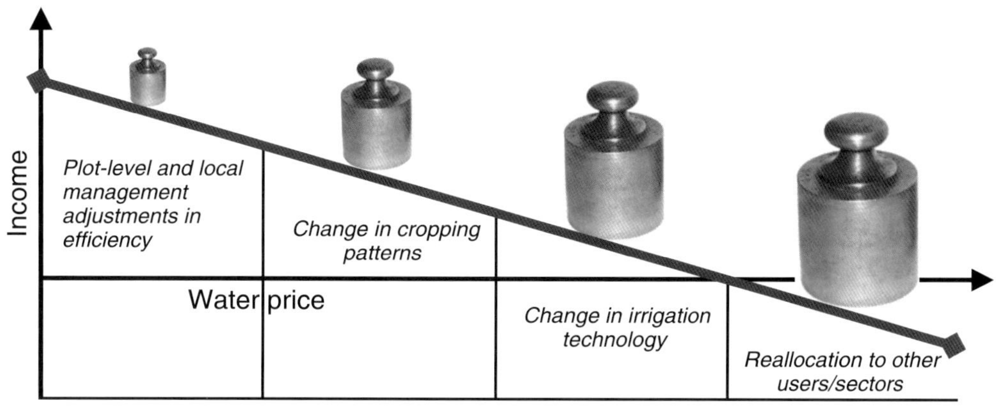

Fig. 2.4. Effects of water pricing as an economic tool.

externality costs are equally valid in societal terms (Rogers *et al.*, 1998; Tsur, 2004). Although their definition and estimation vary, the level of water charges may impact on farmer behaviour and bring economic benefits. Figure 2.4 proposes a tentative hierarchy of responses to increasing water prices, while recognizing that the order of these effects may sometimes be altered by relative factor prices and other aspects. Moderate water prices may trigger low-cost adjustments in water management, while higher prices may successively elicit changes in cropping patterns, in irrigation technology and, finally, release water to other higher-value activities. These effects imply a role for pricing as an economic tool and the likelihood of achieving such outcomes is examined in the following sections.

A Note on Terminology

A *water charge* can be defined as an actual (financial) payment by users to access water and is the term generally adopted in this chapter. It is equivalent to a *tariff*, a term commonly used in the domestic sector when differential rates are set. *Charge* is a term disliked by some decision makers, who fear that it suggests that water – perceived as a gift of nature or god – is taxed. In 1979, several Asian countries agreed to replace it with the term *irrigation service fee* (ISF) (ADB, 1986a). This is now often adopted, though it conflicts with the definition of a *fee* as an administrative payment (e.g. for the registration of a water right). Another term commonly used is *water price*. This is preferably confined to the (economic) price that emerges in a market as the result of the actions of willing buyers and willing sellers, with no connotation of (financial) cost recovery. Since such markets are rare in the water sector, *price* is often used as a synonym for *charge* to indicate the administrative rate set by an agency to a user. Most of the discussion in this chapter uses the term *water charge*, focusing on how water charges are reasoned, justified, determined, enforced, recovered and eventually expended.

A word is also necessary on the terms *ability-to-pay* and *willingness-to-pay*. Many studies conclude that farmers have an *ability-to-pay* much higher water charges than are charged in practice. This is sometimes supported by evidence that they are *willing-to-pay* much higher amounts for private irrigation and by the fact that consumers in the domestic sector are *willing-to-pay* much higher prices to street vendors than the tariffs charged by the utility. The use of these terms can, however, be confusing.

Willingness-to-pay is best used as an economic term to describe consumer behaviour. The poor may be *willing-to-pay* the high unit price charged by a private tube well owner or a vendor but buy little at this price, the amount being determined by profitability (in irrigation) or subsistence needs (in domestic use). As prices and incomes shift, demand also shifts reflecting the price, income and cross-price elasticities described by standard demand curves (Young, 1996). Similarly, private investments (such as wells) and their subsequent operation reflect investor assessment of profitability, that is, by farmers' *willingness-to-pay* (*or to invest*). Purchases from a private tube well owner or vendor and private investment in irrigation are determined in markets governed by the actions of willing buyers and willing sellers.[8]

If *willingness-to-pay* describes behaviour, *ability-to-pay* relates to farmer incomes and public subsidies. If irrigation investment is economically justified, and prices are undistorted, farmers should in principle be *willing-to-pay* all costs including capital cost. But irrigation is driven by non-economic objectives and in most cases farmers *should not* repay full capital costs. If they are unable to pay for marginal (future) costs, then – leaving aside distortions in other costs and prices – continued irrigation is itself uneconomic. In extreme cases, farmers may be unable to pay even recurrent costs since the resulting farm incomes are inadequate to sustain life (Cornish *et al.*, 2004) or the rainfed option is more profitable. But the issue in irrigation is seldom, if ever, an absolute inability-to-pay (although this may, of course, typify extreme cases in respect of domestic water). It is one of fairness, incentive and acceptability, and *ability-to-pay* is best thought of as that level of payment thought reasonable and practical, given the general context and government priorities and objectives. The level of subsidies given to construct a new scheme or sustain an existing scheme is thus ultimately a political decision.

[8]Such markets may, of course, be distorted as a result of monopoly practices, distorted input and output prices, changeable public policies, etc., and there may be a case for interventions by government or a regulator to correct for these distortions. They are also shaped by social relationships and values.

CHARGING FOR WATER IN PRACTICE

This section addresses the practicalities and modes of charging for water, as well as the current situation regarding cost recovery by irrigation schemes.

Main Types of Water Charge

The following are the most common ways of defining charges and their differentiation according to uses and users (Sampath, 1992; Tsur and Dinar, 1997; Garrido, 1999; Bosworth *et al.*, 2002; Easter and Liu, 2005):

1. *Uniform user charge* – users are taken to have similar access and are charged evenly. Even if the level of use varies, differences cannot, or are too costly to, be assessed.
2. *Area-based charge* – the irrigator is charged according to the area irrigated, based either on: (i) the area owned; or (ii) the area cropped (declared by the farmer or assessed by the agency).
3. *Crop-based charge* – the charge is based on area *and* type of crop. Differentials may be justified by crop priority (e.g. cereals for food security) or water diverted or consumed by crop or its value.
4. *Volumetric charge* – water is charged, based on actual diversions to a user or group of users (bulk water pricing). Metering is necessary but volume may be represented by time or the number of 'turns', provided discharges are more or less stable and predictable.
5. *Volumetric block tariffs* – when metered, charges can be fixed for different levels of consumption. Increasing block tariffs discourages excessive use. Decreasing block tariffs promotes sales and rewards economies of scale, being appropriate only if water is abundant.
6. *Mixed tariffs* – charges combine a flat rate (usually area-based) with a volumetric

charge. This provides both a stable minimum revenue to the operator and a variable charge according to use.

7. *Quotas at fixed charges* – quotas may be uniform (e.g. based on area) or vary by crop. Charges can be proportional to nominal volumes or vary with crop type (as in the Jordan valley).

8. *Quotas and marginal volumetric pricing* – users can access more than their quota (subject to availability and within limits), but additional use is charged at higher rates (as in Israel).

9. *Market-based price* – the price of water is determined in a market where allotments can be traded (within season, seasonally or permanently). If the market is regulated, the regulator may set the price, set price limits, serve as broker, etc. (as in the California Drought Bank).

Each method has its advantages and disadvantages, notably the ease with which charges can be calculated, justified and implemented. Additional modalities may also vary: for instance, charges may vary by season, be paid before or after cropping, in one or more instalments, in cash or in kind, etc.

Besides direct charges, farmers may also be charged implicitly via the tax system or in the level of output prices. Land taxes, for instance, often vary to reflect the higher productivity of irrigated land, and betterment levies may be imposed when irrigation is brought to an area for the first time. Similarly, procurement programmes and/or export duties can depress crop prices and can be thought of as an indirect charge. But this is not specific to irrigation and may be offset by other subsidies (e.g. on fertilizer). Moreover, farmers may be protected rather than taxed. These and related issues are thus best considered in relation to the general context rather than to irrigation charges per se (see earlier section).

Who Collects and Uses the Water Charge?

Water charges may be assessed and collected by the state, by a revenue or irrigation department, or by a combination of the two (as in much of India); by an autonomous irrigation entity at the national level (as in the case of the National Irrigation Administration (NIA) in the Philippines) or at the scheme level (as in China and other countries where schemes are managed autonomously or quasi-autonomously); or by a communal organization (such as a Water User Organization) collecting charges directly from its members. Numerous options exist. The state may assess and collect charges at farm level, and consider this levy as revenue. Alternatively, assessment and spending of this revenue can be shared with other levels. Again, a Water User Association (WUA) or some other agent may collect the fees and retain a pre-assigned share for its own requirements (e.g. O&M of the tertiary command), transferring the balance to the irrigation agency, the basin agency or the state, in return for irrigation supply. This can be paralleled by contractual arrangements made for bulk allocations and schedules at each level (e.g. between the river basin agency and irrigation entities, between the irrigation entity and pump/canal organizations and between the canal organization and the WUAs).

In other cases, a state or provincial government may regulate the different rates applied by various entities (including the charge paid by farmers), or each entity or organization may be free to establish its own rates subject to agreement between the different levels and approval under the rules of the organization. Where the state is responsible, payment may be reduced or forgiven in a drought or for some other reason.

There are also options relating to incentives and farmers' involvement in decision making. For instance, incentives may be provided to encourage collection either being paid to officials of the relevant organizations or to private subcontractors. The corresponding levels of farmers' involvement in decision making are equally important (e.g. in allocation decisions or possibility of hiring their own staff). The nature of the arrangements impacts on the rate of collection and on the potential for water conservation and enhanced water productivity, as discussed further below in the appropriate sections.

Who Pays What and How Much?

Types of charge

The most common form is area-based or area plus crop-based, as in Pakistan (Bazza and Ahmad, 2002), Nigeria (Olubode-Awosola *et al.*, 2006), Kazakhstan (Burger, 1998), Vietnam (Fontenelle *et al.*, Chapter 7, this volume), Turkey (Yercan, 2003), Argentina, Greece, Japan, Philippines and Sudan (Cornish *et al.*, 2004), with occasional distinctions by season (as in India, Saleth, 1997; or Nepal). This type of charge accounted for 60% of the sample studied by Bos and Wolters (1990).

Volumetric pricing is usual in the Middle East or North Africa, e.g. Tunisia (Hamdane, 2002a), Iran (Perry, 2001a,b), Jordan (Venot *et al.*, Chapter 10, this volume) and in countries such as the USA, Australia, Southern Europe and Mexico. Volumetric pricing is often associated with a quota, and defined at a bulk rather than at an individual level. Two-part tariffs are also common (e.g. Spain: Maestu, 2001; Colombia: Garcés-Restrepo, 2001; Lebanon: Richard, 2001; Morocco: Ait Kadi, 2002). Volumetric charges are widespread in lift irrigation given the ease of measurement (though not in Vietnam; see Fontenelle *et al.*, Chapter 7, this volume).

Numerous variations occur: in Indonesia charges may be differentiated by head, middle and tail, and be lower in unproductive areas (Hussain and Wijerathna, 2004), and in India they sometimes reflect water dependability (Sur and Umali-Deininger, 2003). In Bangladesh, at one time charges were set as 3% of gross incremental benefit but this proved impracticable (ADB, 1986b). In contrast, simpler approaches may be negated by considerations of equity: a flat per acre rate was, for instance, adopted in Sind in 1972 to reduce irregularities only to be abolished in 1980 since charges based on actual crop areas were thought fairer. Some countries once collected charges in kind (e.g. the Office du Niger, Mali: Aw and Diemer, 2005; Philippines: Oorthuizen, 2003), and in Tanzania this is still an option (Tarimo *et al.*, 1998). Elsewhere, rates are expressed in terms of a paddy quantity (e.g. in Vietnam and Philippines), though rates must be updated if productivity or prices vary (Carruthers *et al.*, 1985).

Some countries impose a resource charge in addition to an irrigation charge. This may simply be an administrative fee, e.g. for registering a water right, but can be a contribution to basin management costs South Africa (Spain, France: Berbel, Chapter 13, this volume; Tanzania: van Koppen *et al.*, Chapter 6, this volume; Colombia: Garcés-Restrepo, 2001). Resource charges are seldom significant to the farmer (e.g. 13% of O&M costs in Peru: Vos, 2002).

Despite occasional claims that models can assist in determining technically optimal prices (Tarimo *et al.*, 1998; Louw and Kassier, 2002; Garrido, 2005), there is little evidence that this has ever occurred: charges are invariably based on historical practice, microeconomic data on crop income or the level of O&M/investment costs (Lee, 2000) and are the result of negotiations or bureaucratic arbitration (Lanna, 2003). In general, a balance is struck between supply costs and what farmers can pay or, maybe more to the point, between tax collection costs and higher charges that would not be politically possible.

Charging mechanisms are not necessarily established once and for all and may evolve with circumstances and objectives (Rieu, 2005). Changes may be triggered by climatic circumstances (volumetric pricing will perform badly in dry years, as experienced in Mexico: Kloezen, 2002), level of state subsidies, O&M costs (which may vary with age of the system), type of incentives needed, etc. (see Plantey *et al.*, 1996; Nicol, 2001 for two French examples).

Rates of recovery

Collection problems have plagued many systems (World Bank, 2005c). Collection is low in Pakistan (30–60%: Bazza and Ahmad, 2002; less than 30% in Sindh: Cornish *et al.*, 2004; and 5–15% in schemes studied by Hussain and Wijerathna, 2004), Kenya (20% in West Kano: Onjala, 2001), Nepal (5%: World Bank, 1997), Bangladesh (less than 10%: World Bank, 2005c) and India (8% in 1989: Saleth, 1997), though 66% and 85% in Andhra Pradesh and Uttar Pradesh, respectively, in 1998 (Sur and Umali-Deininger, 2003).

Recovery rates tend to be higher: (i) under authoritarian governments; (ii) if supply is cut off for non-payment; (iii) if charges are low, recovered with other taxes and/or collected before the crop season; (iv) where users decide on the use of the charges; and (v) when supply is reliable. Thus, it is 98% in Mali (Office du Niger: Aw and Diemer, 2005), 95% in Turkey (Özlü, 2004), 90% in Syria (Bazza and Ahmad, 2002) and Tunisia (Hamdane, 2002a), 80% in Mexico (OECD, 2003) and the Jordan Valley (Venot *et al.*, Chapter 10, this volume) and 50% in Kyrgyzstan (Sehring, 2005). The overall rate of recovery for a sample of 82 irrigation providers was 77% (Lee, 2000).

Water charges come with both administrative and compliance costs that can be quite substantial (Nickum, 1998; Tiwari and Dinar, 2001; Johansson *et al.*, 2002) and differ depending on the type of charge (Tsur and Dinar, 1997). In Bihar, collection costs are said to sometimes exceed the income derived, being estimated at between 52% and 117% of the amount collected (Prasad and Rao, 1991). For Bhatia (1991), collection keeps '5,000 persons busy and unproductive in the fields'. Transaction costs make volumetric charging impractical in Egypt (Bowen and Young, 1986) and similar settings.

The burden of irrigation charges

This burden varies widely. Bos and Wolters (1990) reviewed 150 systems and, in all but one, water charges were less than 10% of the net farm income excluding water costs. The share ranges from zero if water is supplied free (as in Albania, Poland, Croatia: Cornish *et al.*, 2004, Saudi Arabia: Ahmad, 2000, Thailand: Molle, Chapter 5, this volume and Taiwan) to above 30% in pump schemes (e.g. 31% in Niger: Abernethy *et al.*, 2000; 34% in Gujarat: Cornish *et al.*, 2004; and even 65–76% in the Jordan highlands: Venot *et al.*, Chapter 10, this volume). Figure 2.5 shows the ratio for a number of schemes and scheme averages.

Two qualifications should be added here. First, formal charges do not capture in

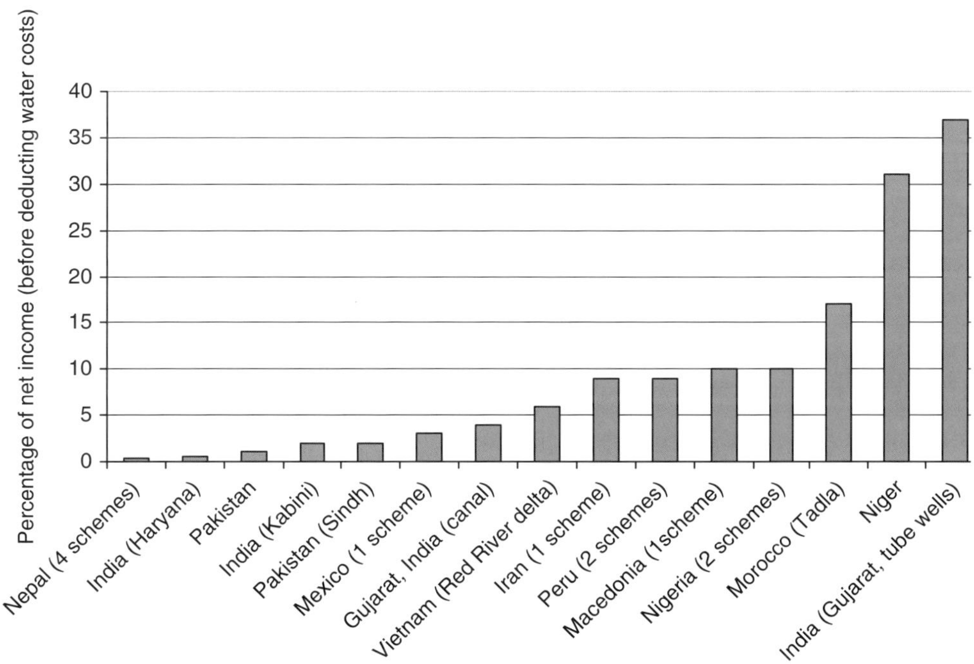

Fig. 2.5. Water costs as percentage of net income.

full the water payments made by farmers. Extralegal payments to local officials are widespread, especially if water is scarce (India: Wade, 1982; Indonesia: Rodgers and Hellegers, 2005; Vietnam: Fontenelle *et al.*, Chapter 7, this volume; Pakistan: Rinaudo, 2002). Farmers are also usually responsible for O&M costs within the tertiary – watercourse – command (in Egypt, India, Pakistan, Indonesia, etc.). Finally, farmers incur major on-farm costs including investments made to augment and/or offset insecurity in main system supplies (not only in private tube wells, but also in hand pumps, reuse systems, on-farm reservoirs, etc.). Second, averages disguise high variability. Low-yielding and tail-end farmers typically pay a higher proportion of net income in water charges (Carruthers *et al.*, 1985). Figure 2.6 shows, for a sample of 101 rice farmers in Sri Lanka studied by Hussain (2005), that water charges would greatly decrease income for the 25–30% of poorer farmers even if, on average, they are only 10–15% of the average net income (Rs 11,000/acre).

In some countries, charges are limited by law in terms of either a maximum share of net income or another measure (e.g. Vietnam); in Iran, regulated surface water charges are limited to 1–3% of the gross value of crop output (Keshavarz *et al.*, 2005); in Cyprus, the charge is limited to no more than 40% of the weighted average unit cost (65% in exceptional cases) (Tsiourtis, 2002); in India, a 1972 policy review recommended that water rates should lie within the range of 5–12% of gross farm revenue (Prasad and Rao, 1991; Vaidyanathan, 1992). Elsewhere, minimum values are sometimes (ineffectively) decreed as in Korea (Sarker and Itoh, 2001) and Peru (Vos, 2002). Block tariffs have been proposed to protect the poor though others conclude that water pricing mechanisms are ineffective in redistributing income, besides having perverse subsidy effects (Tsur and Dinar, 1995; Dinar *et al.*, 1997).

PRICING AS A FINANCIAL INSTRUMENT: COST RECOVERY

Arguments for Cost Recovery

Funds for physical sustainability

The least controversial – and most compelling – argument in favour of cost recovery in irrigation is to ensure the availability of funds needed to sustain physical sustainability of

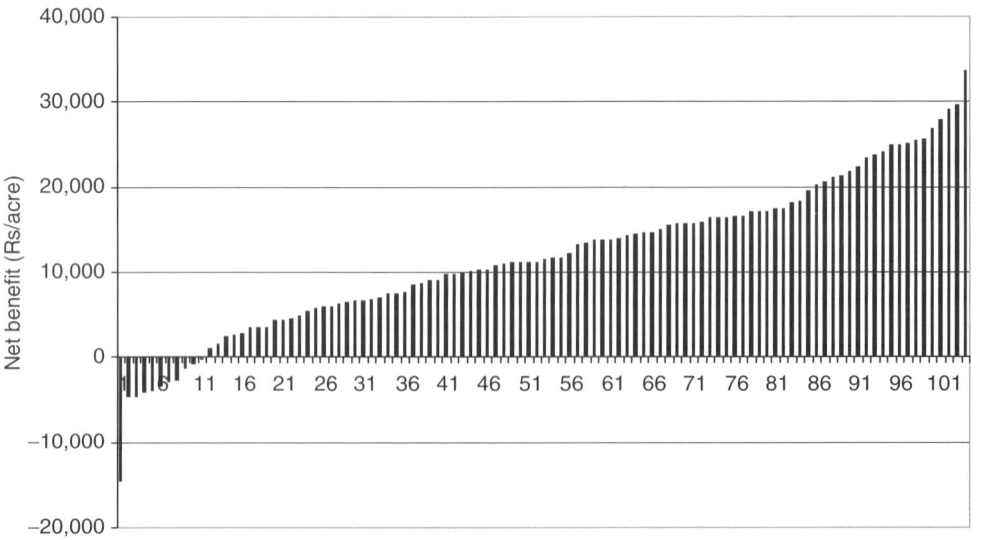

Fig. 2.6. Distribution of net income from rice cultivation (southern Sri Lanka).

the infrastructure. Concerns relating to physical sustainability have a long provenance, but rose to particular prominence in the 1980s when many governments and lending agencies faced the necessity of rehabilitating schemes that had sometimes been constructed only a few years back, but were already in a dilapidated state.[9] In Indonesia, for example, one-third of the 3 million ha of public sector irrigation schemes has been rehabilitated twice in the last 25 years (World Bank, 2005b). In the Philippines, successive projects funded by the World Bank and ADB have similarly returned repeatedly to the same national irrigation systems (World Bank, 1992) and, no doubt, other examples could be quoted. The decay of irrigation infrastructure leads to poor water delivery and is thought to lower agricultural production and decrease farmer income (Tiwari and Dinar, 2001; Hussain, 2005).

Degradation of facilities can be linked to many causes, including faulty design, shoddy construction, lack of incentives to respect covenants, pressures on public finances and a tendency by politicians to adopt a 'build-and-forget' approach to politically motivated projects. Widespread reliance on government for financing O&M has, in practice, led to underinvestment, deferred maintenance and degradation of facilities. This can also be related to 'public goods' and 'freerider' issues, as farmers intervene in low-level public infrastructure to secure their individual interests and as the incentives facing ill-paid operators and farmers have proved unsuited to the effective maintenance of both public and communal facilities. In many countries, tertiary maintenance is the responsibility of the farmers, yet even this is often poorly undertaken, in part due to the inability of the main system to guarantee predictable supplies, and in part due to lack of cooperation, freeriding and incentive issues at farmer level.

Underinvestment in maintenance is believed to be very considerable. For instance, total O&M requirements for public systems in India have been assessed at about Rs. 25–30 billion per year, yet less than a quarter of this amount is actually provided, with wide variation across states (Thakkar, 2000) and revenue receipts covering only 10% of expenditures in 2000 (Sur and Umali-Deininger, 2003). In Egypt, a desirable level of expenditures on O&M/rehabilitation has been put at US$234 million, yet only US$164 million is provided (Bazza and Ahmad, 2002). Comparable situations are found in numerous other countries, contributing to the perceived need for repeated rehabilitation as in Indonesia and the Philippines. The conclusion is that states have been de facto major defaulters and that sustainability depends on users taking over responsibility for maintenance.

Performance incentives

But paying for water does not by itself ensure good maintenance and service. When the receipt from water charges is channelled to state coffers, farmers come to regard charges as a tax rather than a direct benefit to themselves and pressurize politicians to reduce – even abolish – them. The assumption that paying for water in itself creates a sense of ownership has thus no doubt been overstated (e.g. Onjala, 2001, for Kenya).

---

[9] The literature provides uncontroversial evidence that these financial difficulties have been the driving force – or at least the chief justification – behind the revision of pricing policies, and also of many programmes of participatory irrigation management and varied degrees of turnover of management to farmer collectives (Frederiksen, 2005): see Burger, 1998 on Kazakhstan; Çakmak *et al.*, 2004 on Turkey; USAID, 2002 on Egypt; and Rap, 2004 on Mexico. Yet, the rhetorical argument that O&M costs are a 'huge drain' on state coffers appears frequently at fault. In 1997/98, canal irrigation subsidies were equivalent to 2.6% of the fiscal deficit in Karnataka and 7% of the fiscal deficit in Andhra Pradesh, with the same order of magnitude for Maharashtra, Rajasthan and Uttar Pradesh (Sur and Umali-Deininger, 2003). This seems significant, but only amounts to 0.1–0.3% of the respective state expenditures, a limited subsidy if redistribution to farming populations is considered a state policy (Molle, Chapter 5, this volume, and Venot *et al.*, Chapter 10, this volume, provide other examples for Thailand and Jordan).

When incentives are provided to the officials of the relevant organizations or to private subcontractors (these incentives may or may not be passed to users) to encourage collection or improve water management within the area they control, a link is established between payment and benefits to users. In order to close a virtuous circle of incentives, managers should ideally depend financially on farmers' contribution. Another fraction of the charges can be managed internally by a local group – e.g. farmers along a distributary or minor – for local repairs and maintenance or to pay ditch riders, thus ensuring that user payments are used to maintain the infrastructure and improve operations in direct sight of the farmers concerned. The focus here is not on paying benefit taxes to the state, but on ensuring both financial and physical sustainability through direct farmer involvement.

In sum, there are numerous variations of incentive mechanisms, depending on the degree of farmers' involvement in planning, allocation and hiring of staff, the level at which the boundaries are drawn between farmers' and agencies' responsibilities, and the inbuilt accountability mechanisms and incentives for financial contribution. Cost recovery makes full sense when arrangements are centred on financial autonomy, a clear definition of the responsibilities of managers and users and inbuilt accountability mechanisms (Small *et al.*, 1986; Small and Carruthers, 1991; Vaidyanathan, 1992; ICID, 2004; see Molle and Berkoff, Chapter 1, this volume, for a historical perspective). A reassessment of this model of financial autonomy will be attempted in a later section.

Equity considerations

Another important argument for recovering costs from farmers is that, having benefited from exceptional public investments, farmers should repay at least a part to the national budget on equity grounds (World Bank, 1984; Perry, 2001a,b). One mechanism for achieving this is a betterment levy (e.g. by increasing the land tax); another is by levying water charges. The equity argument is often supported by pointing to differences between investment in irrigated and rain-fed agriculture, and by the fact that water charges are seldom more than 5–15% of the incremental value of production relative to that of rain-fed output (Easter and Liu, 2005). Ministries of agriculture and irrigation typically spend much of their budget on irrigation (60% in the case of Thailand) and annual irrigation subsidies are often massive (Rosegrant, 1997; Sur and Umali-Deininger, 2003). Investment opportunities in rain-fed areas are no doubt more limited than in irrigated areas and it is perhaps understandable that governments start by developing regions that lend themselves to irrigation. Nevertheless, as argued earlier, irrigation subsidies have probably discriminated against the rain-fed farmer (ICID, 2004).

A related equity argument is that cost recovery can contribute funds for irrigation expansion in currently deprived regions, an argument notably employed by politicians in advocating investments in their constituencies[10] (World Bank, 1984) and by those who advocate irrigation as the driving force for regional development. However, if income from water charges or betterment levies is accrued to the general public budget, there is no assurance that it will be used to expand irrigation since Ministries of Finance typically allocate resources in line with general political priorities.

Objections to Cost Recovery

Identification of beneficiaries

At first sight, it is obvious that farmers are the beneficiaries of irrigation and the large majority welcome irrigation projects. Even so, they are neither consulted on construction nor are their obligations always clearly defined. Some may have to relinquish land while others may have invested earlier in private or communal

[10]This may unfortunately lead very often to uneconomic projects which are granted against political support to the ruling party, or to other MPs ('pork barrel' in the USA). A perverse outcome can be the 'overbuilding' of river basins (Molle, 2007).

irrigation and gain little by being included in the new scheme (e.g. in Iran, Thailand or Argentina). Demanding repayment of costs decided by the state in these cases seems inequitable. Moreover, irrigation is often provided in the context of multi-purpose projects and irrigation itself may benefit non-farmers (e.g. domestic users or those in the flood plain). Since cost allocation is seldom applied systematically, irrigators may be asked to pay more than a fair share of joint costs (though hydropower rather than irrigation is more typically overcharged). Moreover, as argued earlier, if much irrigation is underpinned by strategic objectives and is inherently uneconomic, recovery of full costs is neither fair nor practicable: 'Is it fair to charge the full cost (including the capital cost) for projects designed without the farmers' say or designed on the basis of higher world grain prices?' (ICID, 2004).

Cost recovery is sometimes taken to imply that all costs should be recouped from direct beneficiaries. However, some argue that the 'joint private/public nature of benefits that result from such projects' and the long-term nature of economic returns may warrant subsidization by the state (Kulshreshtha, 2002). Others assert that irrigation facilities are a form of social overhead capital with farmers being just one category of beneficiaries amongst many (Small, 1996). If so, it is arguable that other beneficiaries – traders, processors and transporters – should be charged a share or irrigation costs. More broadly, a whole region may benefit from the stimulus of irrigation and consumers everywhere benefit from rising farm output in the form of lower prices (Sampath, 1992; Small, 1996; Bhattarai *et al.*, 2003). Thus, it is sometimes argued that 'indirect beneficiaries of irrigation, (notably) consumers of cheap food, should be happy to subsidize irrigation development through taxes' (Perry, 2001a,b).

Care must be taken in disentangling these arguments. If multiplier benefits are limited to *incremental* impacts relative to those of the alternative project (which also, invariably, exhibit such multiplier effects), then – for this and other reasons – the conditions under which they can be included in total benefits are restrictive (see first section). Moreover, food marketing is often amongst the most competitive sectors in developing countries. If so, participants, by definition, pay almost full economic costs so that charging specific indirect beneficiaries for a share in irrigation costs risks double-counting. The justification given for indirect benefits is thus less convincing than sometimes implied.

As Abu-Zeid (2001) recognizes, governments may 'continue to subsidize [new] projects for several reasons, e.g. enhancing national security, maintaining political stability, decreasing population density in certain sensitive geographical regions and conserving water'. Given these national objectives, the level of capital cost recovery that is desirable is ultimately a political judgement given the context concerned, reflecting judgements on the weights given by society to national objectives other than economic optimization.

Cost estimation

Cost estimation – and hence the level of cost recovery implied – is seldom straightforward. For schemes constructed in part with unpaid labour (whether voluntary or otherwise) – as in China, Vietnam, Burma and at the tertiary level in many countries – implicit farmer contributions should be excluded. FAO and USAID (1986) have also suggested that 'farmers should not be asked to repay the cost of over-elaborate gold-plated designs, incompetent, expensive construction, costs overruns for reasons of corruption, bad scheduling of construction activities or the like'. Similarly, farmers should not be asked to pay for overstaffing,[11] poor management and corruption (Rao, 1984; FAO and USAID, 1986; Bhatia, 1991; Gulati and Narayanan, 2002 – Rao has estimated that in India only about half of officially estimated costs represent real costs). Moreover, with regard to maintenance, should actual costs or ideal costs be

[11]Lee's (2000) review of 82 irrigation providers found an average of 38% of O&M costs spent on salaries, with a maximum of 82%; it is 80% in Sindh, Pakistan (SIDA, 2003), but only 10% in northern Vietnam (see Fontenelle *et al.*, Chapter 7, this volume).

considered and how should the ideal be defined? Systematic maintenance may lengthen a project's life, but what is the economic optimum? Finally, convincing farmers that opportunity and externality costs are real, let alone charging them for these costs, is extraordinarily difficult (see later section).

Irrespective of whether actual O&M and related costs are justified, they must be financed either by government or by farmers if irrigation is to be sustained. As noted earlier, scheme autonomy strengthens incentives for containing costs to those justified by prevailing conditions. In the state of Victoria, Australia, for example, when farmers were required to pay the full costs of O&M, increased scrutiny of the supply agency led to a 40% reduction (World Bank, 2003a,b). While farmers tend to take a short-term view of what is required, often in the hope that government will, in due course, rehabilitate the scheme, they also usually have a much better idea than unaccountable public agencies of what is truly required (sometimes less than external experts commonly suppose).

Cost Recovery: Empirical Evidence

The literature suggests that no more than a portion of O&M costs is typically recovered (Dinar and Subramanian, 1997; Cornish *et al.*, 2004; Easter and Liu, 2005), a conclusion that probably holds despite inconsistencies in the definition of these costs. OECD countries often recover full O&M costs (Garrido, 2002; Berbel *et al.* Chapter 13, this volume), while Latin America (notably after management transfer) and the Mediterranean basin (e.g. southern Europe, Tunisia, and Morocco) have fared better than Asia and Africa, and East Asia better than South Asia (ESCWA, 1999 for Western Asia; Ringler *et al.*, 2000 for Latin America; Chohin-Kuper *et al.*, 2002 and Bazza and Ahmad, 2002 for Mediterranean countries; Cornish *et al.*, 2004 for a review). Figure 2.7

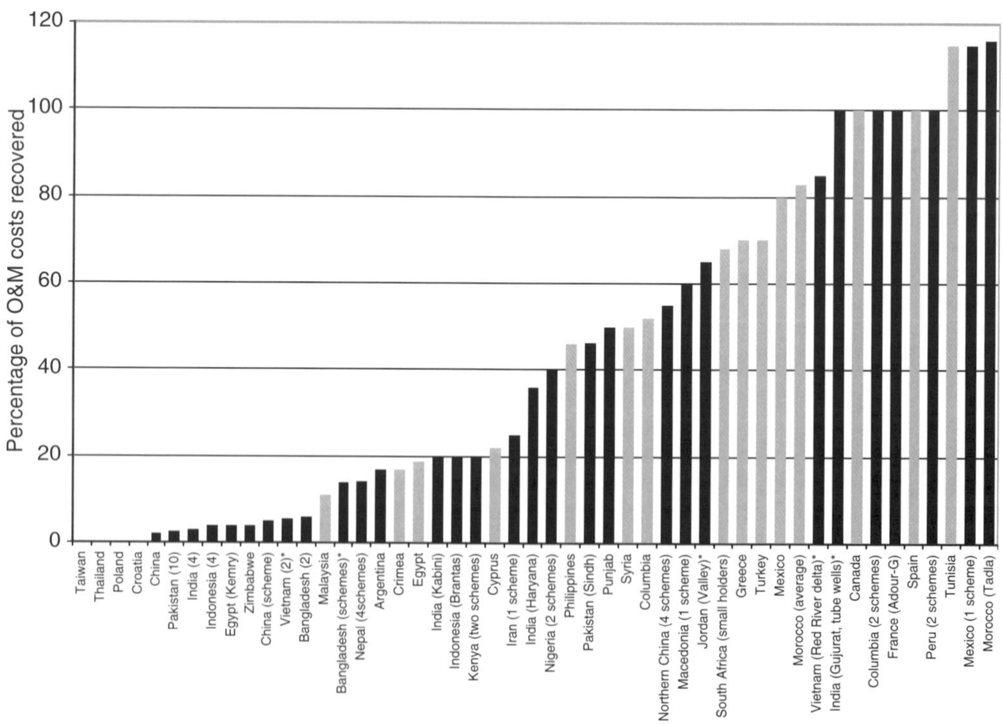

Fig. 2.7. Water charges relative to O&M costs in selected schemes and countries.

plots average levels of cost recovery for a number of cases, distinguishing between particular schemes (both gravity and pressurized marked with*) and country averages (in grey).

Beyond these average estimates drawn from the literature, in practice both O&M costs and cost recovery levels vary over time depending on water use patterns and the age of systems, government policies and organizational arrangements (Carruthers *et al.*, 1985). For instance, the real irrigation charge in Tunisia was raised by 2.4 times between 1990 and 2000 and collections rose from 57% to 90% so that they now cover, on average, 115% of O&M costs (Hamdane, 2002a,b). In Morocco, charges in the Tadla scheme cover both O&M and depreciation (Hellegers *et al.*, Chapter 11, this volume), although they cover no more than O&M costs in three other gravity schemes, and 66% in three major pumping schemes (values for 2001; Belghiti, 2005b).

Historical evidence suggests that in no country have the beneficiaries shouldered a significant share of the initial capital costs of large-scale irrigation, let alone the costs of subsequent irrigation expansion. Many schemes date back to when irrigation expansion was a national policy and are targeted for cost recovery mainly to contain current public expenditures. Even in richer countries, it is difficult to justify the recovery of capital costs of past public projects, given that irrigation benefits have usually been capitalized in land values and, given that relative price shifts often make it financially impossible (see Pigram, 1999 on Australia; Musgrave, 1997). Postel (1992), for instance, reports that 4 million ha in the west USA are supplied 'at greatly subsidized prices' by the Federal Bureau of Reclamation (see also Anderson and Snyder, 1997), reflecting the fact that the 1902 legislation emphasized western settlement rather than full market returns for Federal water projects (Gollehon *et al.*, 2003). Irrigators in the Central Valley Project have repaid only 4% of the capital cost. Currently, repayment of capital costs averages about 15% in real terms (Howe, 2003; Hanemann, 2006).

In South Korea, financially autonomous Farmland Improvement Associations (FLIAs) have repaid part of initial capital costs, in addition to shouldering full O&M costs (ADB, 1986b) and in Japan corporate Land Improvement Districts shoulder 10–15% of the costs of large-scale state irrigation projects and 25% of medium-scale projects initiated by prefecture governments (Sarker and Itoh, 2001).[12] The principle of capital cost recovery has been incorporated in European directives and has the clear potential to ensure that projects are cost-effective and to crowd off marginal and politically motivated water resource development (Garrido, 2002). Yet, perhaps for this very reason, obstacles still prove pervasive and fiscal discipline elusive (Hill *et al.*, 2003).

Morocco is a rare example in the developing world in having an Agricultural Investment Code that specifies 'with the objective to alleviate the [financial] burden on farmers, (irrigation rates) will be called upon to contribute to investment costs *only* to the level of 40% of these costs' (Belghiti, 2005a; emphasis added). Although this level has yet to be attained Morocco has taken bold steps towards financial autonomy. In Egypt, new irrigation areas (New Lands) for commercial entrepreneurs are also being granted with a degree of cost sharing (Perry, 1996), while expansion of the irrigated area in the Office du Niger (Mali) included 20% of contribution by farmers (Aw and Diemer, 2005). In contrast, in Bihar and Haryana, where irrigation remains firmly in the public sector, if capital costs were charged in full, payments would amount to 40–90% of net incremental farm income (Bhatia, 1991).

Development agencies have long been reluctant to recognize that few countries will recover more than a nominal share of initial costs, and that irrigators' 'debt' to the state will be eventually written off, even in developed countries (Garrido, 2002). For example, ADB's 1985 review (ADB, 1986a) calls for 'benefit-conscious project preparation' and notes that the disregard for loan covenants

[12]It is perhaps no coincidence that South Korea and Japan simultaneously subsidize their rice-farming sector through import duties and controls that lead to very high internal prices and promote domestic production.

(in particular on ISFs) by governments is not being addressed. Pitman (2002) observes that 'Globally, most [World] Bank projects pay lip-service to (capital cost) cost recovery', but that those which addressed this issue in practice were largely water supply projects. Recognition of the case against full capital cost in irrigation and greater realism in practice would clearly be desirable (World Bank, 2003a,b).

Empirical evidence also shows that very seldom are incentives linked to charges. Bos and Wolters' (1990) survey of 159 schemes covering 8 million ha showed that there is no relation whatsoever between the level of charge and efficiency. This was confirmed by later findings by Jones (1995) which showed that revenue from water charges generally goes to the general treasury and is not earmarked for O&M. A typical example is Pakistan where revenues from water charges go to the provincial or state treasury, losing the link between payment and O&M and quality of service (Bazza and Ahmad, 2002) (see also Jordan: Venot *et al.*, Chapter 10, this volume; and India: Samal and Kolanu, 2004). Conversely, the failure to ensure reliable supply is one of the major reasons for widespread defaulting (Carruthers *et al.*, 1985; ADB, 1995; Spencer and Subramanian, 1997). Samal and Kolanu (2004) note the 'categorical and explicit refusal of [Indian] farmers to pay the water tax till the irrigation service was improved'. In Sindh, Pakistan, 'farmers are not willing to pay since the financial system is not transparent and they do not see that the charges paid are used to deliver a good service'. The farmers said that they were willing to pay for services, but not for 'someone's wife's jewellery' (Cornish and Perry, 2003).

Even where progress has been made in transferring responsibilities at the tertiary or secondary level to farmer organizations under irrigation transfer and similar programmes, supply has often remained unpredictable. Whether due to suboptimal management, to real constraints in controlling stochastic water variability and uncertainty or to what happens upstream, insecure main system supplies have undermined efforts by farmers to organize at secondary or block level. For example, Parthasarathy (1999) has shown that, in Gujarat, India, WUA members failed to pay higher rates when they appreciated that managing an isolated or terminal portion of the canal system failed to contribute to any real improvement in the reliability of water supplies. As Freeman and Lowdermilk (1991) put it: 'To disconnect farmer payments of assessment for maintenance, whether in cash or kind, from water delivery is virtually to invite organizational decay.'[13]

In most countries, governments continue to be responsible for the funding of main-system O&M, together with replacement, rehabilitation and modernization works, quite independently of charge collection itself. In other countries, notably in East Asia, Latin America and much of North Africa (as well as in most developed countries), irrigation water charges are collected and retained by scheme management (irrigation district). But even in these situations, O&M expenditures can be deficient. In China or Vietnam, for instance, the level of water charges is regulated by national, provincial and local price commissions, and, though in principle authorized charges are based on estimated requirements, in practice increases have been limited with a view to reducing burdens on farmers (Hydrosult, 1999; Lohmar *et al.*, Chapter 12, this volume). Similarly, the Government of the Philippines has repeatedly failed to authorize the NIA to effectuate needed increases in water charges (World Bank, 1992). Financial autonomy – total or partial – has been practised widely in developed countries,

[13]In addition to farmers' reluctance to contribute, low rates of recovery are compounded by agencies' reluctance to enforce collection (Carruthers *et al.*, 1985), due to drudgery avoidance, unwillingness to antagonize farmers and desire to keep good relations, sympathy for their economic situation, or fear to give farmers reasons to question the quality of service.

including the USA, Spain, France, Italy, Mexico, Japan and Korea.[14]

PRICING AS AN ECONOMIC INSTRUMENT: WATER CONSERVATION

Introduction

That water is wasted due to underpricing is a widely held view, from the former President of the World Bank ('the biggest problem with water is the waste of water through lack of charging': Wolfensohn, 2000) to the World Water Vision ('users do not value water provided free or almost free and so waste it': Cosgrove and Rijsberman, 2000), to detached analysts ('water is consistently undervalued, and as a result is chronically overused': Postel, 1992) and environmentalists who favour 'developing a pricing system that prevents excessive use of water' (WWF, 2002). For the EU (2000b): '[E]fficient water pricing policies have a demonstrable impact on the water demand of different uses. As a result of changes in water demand, efficient water pricing reduces the pressure on water resources. This is particularly true for the agricultural sector.'[15]

Seemingly corroborating the assumption of waste is the fact that irrigation accounts for approximately 70% of withdrawals on average. Agriculture 'gobbles up at least 75% and sometimes as much as 90% of the available water', while 60% of water deliveries fail to reach the fields (The Economist, 2003). Profligacy combined with agriculture's dominant share suggests an easy solution: if raising irrigation charges can reduce losses even by a small percentage, sufficient water can be freed to meet the much smaller demands of other expanding sectors (World Bank, 1993; Winpenny, 1997; Gleick, 2001; Louw and Kassier, 2002; Davis and Hirji, 2003; IRN, 2003).

This section evaluates whether low water charges lead to waste and higher charges promote conservation. It first examines the received wisdom that 'water is wasted because it is underpriced'. Then it examines the conditions under which pricing water can be a 'key to saving water' and assesses the empirical evidence. It concludes by evaluating the potential of pricing for promoting conservation.

Is Water Wasted Because It Is Underpriced?

Is water wasted?

The first section showed that the concept of irrigation efficiency is often misstated. If water is abundant – in surplus basins, or during the rainy season, after it rains – excess diversions matter little since they return to the hydrological cycle (though, of course, they can impact adversely on water control, waterlogging and flooding). If water is scarce, farmers compete for the limited flows available: the struggle for water *when it is scarce* means that little water is wasted *when it has value*, and this is shown by observation of shortage situations. Moreover, losses may be used – after a delay – downstream or from aquifer recharge and only if water flows to the sea or another terminal sink is it no longer available for human use.[16] The central issue is thus one of *basin* efficiency and focusing on farm-level or scheme efficiency can be very misleading.

[14]Although this autonomy is partly paralleled with, or allowed by, massive subsidies granted through output prices or direct payments.

[15]See also 'Inefficient pricing and management of irrigation water supply leads to massive wastage' (Hansen and Bhatia, 2004) and similar statements in Holden and Thobani (1996), FAO (1998), ESCWA (1997), UNESCAP (1996), Ringler *et al.* (2002), TDRI (1990), Siamwalla and Roche (2001), Roth (2001), Bate (2002), etc.

[16]Flows to the sea may still, of course, have important environmental functions, including: flushing out sediments, diluting polluted water, controlling salinity intrusion and assuring the sustainability of estuary and coastal ecosystems.

There might be cases of a water-abundant scheme located within a water-short basin. Such a situation may be due to locational reasons, specific water rights or political influence that insulates that particular scheme from overall scarcity. This is a problem of (basin-wide) allocation and equity, which has other roots and will not be solved by pricing policies.

Is wastage due to low prices?

The above explanation implies that much less water is 'wasted' than is commonly supposed. Residual 'real' losses (evaporation from open surfaces, transpiration via unproductive growth, etc.) may be identified on a case-by-case basis but can 'real' losses be attributed to low water prices? A first issue is that shifts in farmer behaviour (induced by prices or otherwise) only impact on the share of diversions they receive. Ray (Chapter 4, this volume), for instance, estimates that farmers in the Mula scheme receive no more than 30–35% of the water released from the reservoir, the remainder being 'lost' from the canal system. Typical losses of 50% imply that raising the water charge to farmers can at best impact on about one-half of the water diverted. A second issue is that scheme-level deficiencies primarily relate to inequities (head-end and tail-end problems) and socio-economic costs rather than physical losses. Whenever wastage (or shortage) occurs, it is because the supply made available at the farm inlet is not in line with needs, and the causes of this mismatch remain largely independent of the users themselves (Grimble, 1999; Rodgers and Hellegers, 2005). Resolving such problems is primarily an issue in design and management, and remedies lie at the system level rather than with changing the behaviour of farmers (Chambers, 1988): effective control of supply is needed but, as Small (1987) aptly observed: '[I]t is likely that once this prerequisite exists, the amount of "wastage" will be greatly reduced, thus lowering the potential efficiency gains from any subsequent attempt to introduce water pricing.'

Conditions for Water Pricing to Elicit Water Savings

Although the causal relationships between low water-use efficiency and low prices are weak, and the fundamental objective is to optimize agricultural returns rather than minimize physical losses for their own sake, there is nevertheless a case for adopting pricing policies whenever they can contribute to this fundamental objective. Although the opportunities may be very limited, there is a continuum from conditions where price has no impact on water use and solutions lie entirely in management, to conditions where water is on demand and farmers can adjust volumes to reflect marginal returns (Fig. 2.3). This subsection addresses the prerequisites for the latter (see also Ray, Chapter 4, this volume). Associated issues related to externality and third-party impacts are considered in a later section.

Is pricing volumetric?

It is sometimes argued that, by making farmers aware of the value of water, even a flat rate promotes water savings (for Tanzania, see van Koppen *et al.*, Chapter 6, this volume). But there is little evidence for this: on the contrary, farmers try 'to get as much as possible of the thing for which they have been taxed' (Moore, 1989; Bos and Wolters, 1990; Berbel and Gomez-Limón, 2000).

Pricing can thus conserve water only if supply is volumetric. Problems of volumetric measurement are well known (Moore, 1989; Sampath, 1992; Rosegrant and Cline, 2002). For historical, technical, financial and managerial reasons, measurement at farm level is rare and even then charges may not be based on measured volumes. In some cases (e.g. for paddy), measurement at the farm level is unworkable without major structural investment (Moore, 1989) and installing functional devices in flat gravity systems (e.g. in deltas) is impracticable. More generally, measurement at the farm level is prohibitively expensive in surface systems with thousands, if not hundreds of thousands, of small farms. Tampering is pervasive and the transaction costs of data collection, monitoring and enforcement are beyond the capacity of most agencies and control at farm level is an illusion: Cornish *et al.* (2004) conclude that 'in practice, volumetric methods of supply to individual farmers are probably not feasible in large parts of the developing world at present'.

Charging for bulk allocations – to a WUA, distributary organization or other scheme entity – is a way to circumvent the transaction costs of charging for individual supply (Carruthers *et al.*, 1985; Repetto, 1986; World Bank, 1986; Asad *et al.*, 1999) and is needed in any case for effective (volumetric) management. But, if bulk charges are to impact on water use, contractual or quasi-contractual agreements must be enforced (Fig. 2.3) which requires more than reforms based on little more than wishful thinking, as noted earlier. While enforcement and collection delegated down the system, closer to the farmer tends to promote participation and accountability, the critical point is to pass incentives on to farmers.

Is water demand elastic?

A second obstacle to effective conservation pricing is that the elasticity of demand for irrigation water at current charges is low or negligible (de Fraiture and Perry, Chapter 3, this volume). Bos and Wolters (1990) found that in all but one of the projects studied charges were less than 10% of net farm income and 'too low to have significant impact'. Latinopoulos (2005) found no relationship between charges and water use in a sample of 21 irrigation districts in Greece, and a study of nine Spanish schemes attributed differences in water use to other factors (soils, nature and abundance of the source, history, etc.), concluding that inelastic demand reflected the relatively low share of water in production costs and the lack of a substitute (Carles *et al.*, 1999). Some studies carried out in the USA indicate a similar lack of responsiveness to price (Hoyt, 1982; Moore *et al.*, 1994). Volumetric pricing is most often associated with pressurized systems and high-value crops, the very situations where efficiency is already high and water costs (hence elasticity) marginal (Albiac *et al.*, 2006).

That volumetric charges seldom impact significantly on farmer behaviour (Gibbons, 1986; Malla and Gopalakrishnan, 1995; Bosworth *et al.*, 2002; Rosegrant and Cai, 2002) is perhaps hardly surprising given that irrigation water is a subsidized intermediate input. There is probably always a range over which demand is elastic, with elasticity rising as charges approach full cost. However, such charge levels have been shown earlier to be unrealistic in uneconomic schemes where water is subsidized. At current levels, even large increases make little impact since other costs are relatively more important, and cross-elasticities determine water use. Water prices in Iran, for instance, would need to rise by a factor of 10 to be effective in curtailing demand (Perry, 2001). Given the political sensitivity of pricing issues governments cannot be expected to risk raising charges well above O&M costs, just for the sake of encountering elasticity.[17]

In contrast to inelastic demand at farmer level, autonomous irrigation entities should, in theory, behave like profit-maximizing industries and reduce use in response to all bulk charges. In developed countries, regulators require irrigation districts to cover costs but even then they often skimp on O&M and/or seek other income sources to avoid 'bankruptcy'. In developing countries, farmer resistance to enhanced charges is stronger, whether the system is managed by government agencies, canal organizations or WUAs. Evidence from China and elsewhere

[17]Although this is advocated by Brooks (1997): 'Most would argue that . . . water tariffs should be designed to encourage conservation, not just to recover costs (which implies that pricing should be high enough to move into the elastic portion of the demand curve).'

(see below) suggests that institutional reforms can strengthen main-system management and transfer costs to autonomous entities, but there are still few examples where bulk water charges as such have led to significant water savings.

Lastly, true elasticity of response is very hard to establish because there is so little information on the relationship between improving efficiency at the farm level and the costs of doing so for a given irrigation technology and a given pattern of supply (see de Fraiture and Perry, Chapter 3, this volume). All shifts involve costs, e.g. in increased drudgery, labour or capital, and depend, *inter alia*, on farmer strategies and on the opportunity cost of their labour[18] (Venot *et al.*, Chapter 10, this volume); but estimating such costs and the associated responses is complex. Modelling exercises almost invariably oversimplify and focus on induced changes in terms of crop mix or technology without recognizing all the costs involved. As a result, the estimates of elasticities tend to be crude and unconvincing (more on this later).

Water Pricing and Water Savings: Empirical Evidence

Dinar and Subramanian's (1997) cross-country review showed that water prices across countries are not related to relative water availability, suggesting either that the current objective for charging is not to manage scarcity, or that other factors come into play. That countries with higher scarcity are not 'more aggressive in reforming pricing schemes' also brings out that other mechanisms are preferred. This was confirmed by a 2000 review of the last 67 irrigation projects funded by the World Bank, which revealed that in none of the projects had water charging mechanisms been planned as incentive tools (Tiwari and Dinar, 2001). Since, in any case, relations between water use and prices can only be expected under conditions of volumetric management, we focus here on cases of bulk allocation and individual volumetric pricing.

Bulk allocation

Sri Lanka, Turkey, China and Mexico are amongst countries that have promoted bulk allocation and in some cases have also introduced charges for bulk supplies:

- Evidence from Mahaweli System H in Sri Lanka showed that allocation at block level can lead to lower diversions, but this is primarily due to stricter scheduling and improved main-system management, resulting in more predictable and uniform flows and reduced conflicts. Water charges are not differentiated at farm level, and though WUAs are charged in proportion to water allocations, charges are not based on volumetric measurement and are too low to provide incentives for water savings (IWMI, 2004).
- Similarly, in Turkey, major irrigation has largely been transferred to irrigation districts that receive bulk water at no cost though they are expected to meet O&M costs in their own area. Reliability of supply has improved and fee recovery has increased substantially (Yercan, 2003; Özlü, 2004), the transfer of the financial burden of O&M to farmers being the main objective of the programme (Ünver and Gupta, 2003). But flat-rate charges have no impact on water conservation at farm level and tertiary distribution remains deficient (Yercan, 2003).
- The transfer programme in Mexico goes a step further (Kloezen, 2002). The

[18]Such interventions include avoiding breaches in bunds or continuous irrigation (for rice farmers), fine-tuning cut-off time to avoid losses at the end of furrows or not using sprinklers on windy days. Other adjustments relate to changing cropping techniques, like resorting to rice dry-seeding (e.g. in the Muda scheme, Malaysia: Guerra *et al.*, 1998), using mulch in vegetable plots or reducing the length of furrows. Other responses are more capital-intensive, such as laser land-levelling, which allow reduced and more homogeneous application of water by gravity, and frequent renewal of drippers in micro-irrigation.

National Water Commission in consultation with user representatives determines allocations to Irrigation Districts on an annual or seasonal basis. Bulk charges are met out of an O&M charge assessed and collected by WUAs and passed to the Commission via the District. Although O&M charges are levied in proportion to the amount contacted to the farmer by the WUA, they remain fairly low (2–7% of gross product in the scheme studied by Kloezen) and reflect O&M costs rather than conservation objectives. Seasonal quotas are tradable amongst WUAs within a district, with trades usually triggered when a WUA cannot meet the contractual demands of their members (Kloezen and Garcés-Restrepo, 1998). Maintenance is often suboptimal, with many WUAs unwilling to incur major costs and raising revenues only as immediate needs arise (Pérez Prado, 2003).

- Lessons from China are masked by the diversity of physical and institutional settings (Lohmar *et al.*, Chapter 12, this volume). Water is usually delivered in bulk by basin and system organizations to township or village entities, WUAs and even private operators. Bulk water charges in some cases have contributed to reduced diversions as entities at each level seek cost savings. Generally, however, even if bulk water supplies are priced volumetrically, current pricing policies rarely effectively encourage water saving at farm level (see Fontenelle *et al.*, Chapter 7, this volume), in part because farmers may be unaware of how water charges relate to other rural charges. Farm quotas necessarily decline when diversions decline but the reform process still appears strongly government-controlled (Mollinga *et al.*, 2005).

These examples confirm that bulk allocation is primarily a mechanism for: (i) improving the predictability and reliability of deliveries at basin and main canal levels; and (ii) allowing partial financial and managerial autonomy to WUAs, thus shifting part of the O&M costs to them. Bulk water pricing can generate revenue, but even if farmer charges are assessed in relation to delivered quantities, they are seldom charged on a volumetric basis; and even if charged volumetrically, they are seldom high enough to promote conservation (Asad *et al.*, 1999; Tiwari and Dinar, 2001). Internal trading (as in Mexico) can improve scheme-level efficiency but, of the examples quoted, only in China is there evidence that some scheme managers have a clear incentive to reduce bulk diversions (Lohmar *et al.*, Chapter 12, this volume).

Individual Quotas and Irrigation on Demand

Technical control may allow volumetric monitoring at farm level, but only if water is supplied on demand can the full potential of water pricing be realized. There is a continuum from individual quotas to irrigation fully on-demand, depending on how constraining quotas are and how responsive the system is to user requests:

- In Morocco, farmers pay a minimum fee equivalent to 3000 m^3/ha (Ait Kadi, 2002). In most cases, water is distributed by rotation and farmers must pay the full amount. In practice, quotas are low and any savings would depend in effect on the adoption of micro-irrigation. The water charge is based primarily on cost recovery rather than on conservation criteria, though in pump schemes the water bill can be up to 65–70% of gross income (e.g. Souss Massa groundwater: Ait Kadi, 2002) and in these cases it undoubtedly influences farmer behaviour.
- In Jordan, quotas in the valley are assessed at individual level and based on crop type, thus promoting water savings (Venot *et al.*, Chapter 10, this volume). Despite pressurized systems over most of the area, water variability and canal capacity preclude arranged demand irrigation and water is rotated at block level. Charges are set in relation

to O&M costs rather than to regulate use, though higher charges may prompt crop shifts and raise water productivity. The (coming) Wahda dam (Courcier *et al.*, 2005) and on-farm reservoirs help offset the rigidities of rotational delivery.

- European countries – Italy, France, Spain – also provide examples of modern pressurized irrigation systems that handle scarcity in the first instance by quotas (which may be very low, e.g. 2000 m^3/ha in Capitanata (South Italy), Genil Cabral (Spain) and the Neste system (France)).[19] There is usually flexibility at the margin with the above quota-use penalized at rates as high as 10 times the variable component in Charentes in France, and 25 times unit cost in Genil Cabral (Maestu, 2001; Montginoul and Rieu, 2001). Water distribution is usually by 'arranged demand' rather than under direct farmer control, and rotational delivery is often required at peak periods or during droughts.
- In Israel, the small unified distribution system is almost fully reticulated and pressurized, and backed by storage in the Sea of Galilee and managed aquifers. In contrast to systems of 'arranged demand', cooperatives and farmers retain discretion over when to irrigate under normal conditions. However, they are subject to cooperative and/or individual quotas that are charged at rising block rates. This has contributed to regulating water demand at the margin (Kislev, 2001) so that average use has sometimes been below the quota. Quotas in principle are adjusted annually but, in practice, they are regarded as water rights (Plaut, 2000; Kislev, 2001).
- A system that comes close to fully on-demand is that operated by the Canal de Provence in France, where the main canal is dynamically regulated to meet agricultural and municipal demands. No formal quotas are announced and farmers are free to irrigate as they wish (although they have to subscribe to a given delivery discharge). Prices are set to recover costs rather than to control demand, but the price structure is complex (Jean, 1999), distinguishing differing periods and between peak and normal demand, and it can be assumed that there are some incentives for water savings.
- Other cases include California, Canada, Peru and China. During the 1990–1994 drought in California, Broadview's water supply had to be decreased by more than 50%. Instead of raising prices in order to reduce demand accordingly, it was found preferable 'to begin allocating water among individual farmers' proportionally to the size of their farms, while providing cheap loans to encourage farmers to purchase sprinklers and gated pipe irrigation systems (Wichelns, 2003). In one system of northern Peru studied by Vos (2002), pricing was volumetric but was not used to manage scarcity: rather in times of shortages the rules employed promoted equity and defined quotas that limited use. In Shangdong, China, the use of integrated circuit (IC) machines ensures that farmers cannot obtain irrigation water without paying (Easter and Liu, 2005) and seems to provide reliable on-demand water.
- In some countries (e.g. in western states of the USA, Chile, etc.) quotas are defined as individual rights and a legal framework has been developed for trading these rights. Management continues to be determined by quotas and water distribution is still, usually, by 'arranged demand'. However, water trading redistributes quotas and contributes to higher economic returns. System constraints, third-party concerns and regulatory aspects may confine trades to neighbouring farmers, with little impact on irrigation water use, but in some places water is traded out of agriculture (e.g. the Colorado-Big-Thompson scheme).

[19] See Mastrorilli *et al.* (1997), Altieri (2001), Berbel *et al.* (2001), Hurand (2001) and Maestu (2001).

Public and communal groundwater suffers many of the same constraints as surface irrigation. A study of collective wells in Mexico – which modelled crop and irrigation options – showed, for instance, that a 30% reduction in groundwater use would require water charges to be (unrealistically) raised by a factor of 4 (Jourdain, 2004). In contrast, private groundwater approximates to irrigation on demand. So long as groundwater is abundant and input and output markets remain undistorted, extractions are determined by costs or prices and the results can approximate to an economic out-turn. But, in contrast to surface systems subject to supply constraints and quotas, in the absence of these preconditions groundwater regulation is seldom feasible since the transaction costs usually prove insurmountable, given the number and dispersal of numerous small wells. Even where regulation is, in principle, feasible, for legal and historical reasons much groundwater continues to be unregulated.

Quotas versus Prices

Three main conclusions can be drawn from the above review. First, and most obviously, incentive pricing requires volumetric management and is thus precluded in the vast majority of developing country situations, at least at farm level. Second, even if volumetric supply is assured at farm level, in practice, price incentives are predominantly used at the margin to control use in excess of defined quotas or rights. This gives users some flexibility, whether water is distributed by 'arranged demand' or is under the control of users. This provides incentives for water saving, but falls short of true irrigation on demand. Third, even for systems that approach on-demand irrigation and have the capacity to meet peak demands, rights are capped by a quota and suspended (e.g. in favour of rotational distribution) during droughts since irrigation invariably receives low priority.

In other words, even in the rare cases where conditions are met to regulate demand through pricing, supply is instead invariably managed through administered quotas or water rights. Reasons for the predominance of quotas include: (i) transparency; (ii) ability to ensure equity when supply is inadequate; (iii) administrative simplicity and relatively low transaction costs; (iv) capacity for bringing water use directly in line with continuously varying available resources; and (v) limited income losses incurred (as compared with price regulation). 'When water is scarce, the surest and most common way to make customers use less water is to limit supply' (Cornish *et al.*, 2004) and this has been easily the most favoured solution for restraining demand (Bate, 2002).[20]

But quotas also have their drawbacks (Bate, 2002; Chohin-Kuper *et al.*, 2002; Tsur, 2005). While price or market regulation tends to promote economic efficiency at the cost of equity (Okun, 1975), quotas (when non-transferable) foster equity at the cost of efficiency: they can lack flexibility in response to changing circumstances, as in the case of settlement quotas in Israel.[21] Equity is also weakened in the case of conjunctive use of

[20]The virtues of rationing (in the short term) and/or the allocation of quotas (for long-term allocation) are getting more attention from the World Bank (2006) who reckoned that 'quotas work better than prices when water users are not very responsive to water price changes'. Bosworth *et al.* (2004) also concluded that 'getting the prices right' is not the most appropriate solution to managing scarcity.

[21]The Israeli case is instructive of the difficulty to readjust quotas once they have been defined and, at the same time, of the growing mismatch which can materialize between one village quota and its real use or needs (Plaut, 2000). The trajectories of kibbutzim and cooperatives depend not only on many factors, including ethnic composition, level of education and political linkages, but also on the links to markets, the availability of non-agricultural opportunities and the possible development of additional local resources (Lees, 1998). With time, some settlements (and some farmers within each settlement) tend to intensify agriculture, while others shift to partial farming. Resulting imbalances between quotas and needs have led to some inefficiency; in the 1980s, some farmers would irrigate carelessly so as to fully use their quota for fear of seeing it reduced (Lees, 1998); and trading within as well as between communities has emerged (Kislev, 2005).

canal water and groundwater, where quotas are rarely adjusted to rebalance overall combined supply (like in Morocco). In practice, quotas also often integrate pre-existing local systems of rights (see the Jordan valley in Venot *et al.*, Chapter 10, this volume). In the absence of an 'omniscient allocator', reallocation can be done either through rules that embody desired priority principles or by making quotas tradable, or by a combination of both in order to address equity concerns while promoting efficient allocations (Seagraves and Easter, 1983; Bjornlund and McKay, 1999; Johansson *et al.*, 2002).

It is true that management of quotas cannot fully simulate the economic scarcity signals of a market price. But, given the socio-economic and practical constraints, and the political costs of promoting irrigation pricing for managing scarcity, the management of quotas (the 'visible hand of scarcity') appears a far more satisfactory and practical solution to water savings in almost all real-life circumstances. Even in Europe, where pricing is being strongly promoted, Garrido's (2002) review concluded that 'irrigation pricing reforms should not expect significant reductions in farmers' water consumption' and that 'efficient allocation can be made without prices'. It should be noted that this conclusion does not rule out on-demand irrigation when feasible and cost-effective. Also, it does not rule out the development of regulated markets in water rights (or quotas) where willing buyers and willing sellers cooperate to transfer water from low-value to high-value uses (see later section).

PRICING AS AN ECONOMIC INSTRUMENT: CROP AND TECHNOLOGICAL CHANGE

Shifts in Cropping Patterns

Governments often seek to promote agricultural diversification. This may be to save water but the primary objective is to generally promote agricultural growth and raise farm incomes. Some equate the two, arguing that, if the price of water is raised (ideally to its opportunity cost), low-value crops are less attractive and farmers shift to higher-value crops (Rosegrant *et al.*, 1995[22]; Bazza and Ahmad, 2002). In principle, of course, it is true that water-intensive crops become increasingly less profitable relative to less water-using crops if water charges are increased. But in practice, because water costs usually comprise only a small part of farm costs, very high increases in water costs and attendant income reduction are necessary to make these less water-intensive crops more attractive. This is illustrated in Fig. 2.8. Assuming that coefficients are fixed, crop shifts are costless and other costs and prices remain the same, the charge per cubic metre at which crop A (net income 100, water costs of 10 deducted) becomes less profitable than crop B80 (initial net income 80% of crop A, water needs 50% of crop A) is five times the initial charge, while income is slashed by 40%.[23]

Possible 'crops B' will be available to the farmer only where these have a net income comparable to crop A and where water costs are already relatively (very) high. This is rare in practice but occurs in private pressurized irrigation with high fixed costs (Charentes, France: Moynier, 2006), particularly in some groundwater areas (e.g. in Spain, Varela-Ortega, Chapter 14, this volume) where the alternative is rain-fed agriculture.

Of course, a more favourable outcome would be to see farmers adopting higher-value crops instead of lower-value crops. Although such a shift is frequently expected from

[22]'We argue that valuation of water at its opportunity cost will provide incentives for farmers to shift from water-intensive rice to higher-valued, less water-intensive crops after wet-season rice; and in other environments to shift from field crops to fruits and vegetables' (Rosegrant *et al.*, 1995).

[23]For crops B60 and B40 which have initial net income of 60% and 40% of crop A, the increases are even more massive (see Fig. 2.8). Even in the case where water costs represent 30% of the initial net income (a very high value) crop B80 becomes more profitable after multiplying water costs by 2.3, *but* with an unchanged income loss (40%).

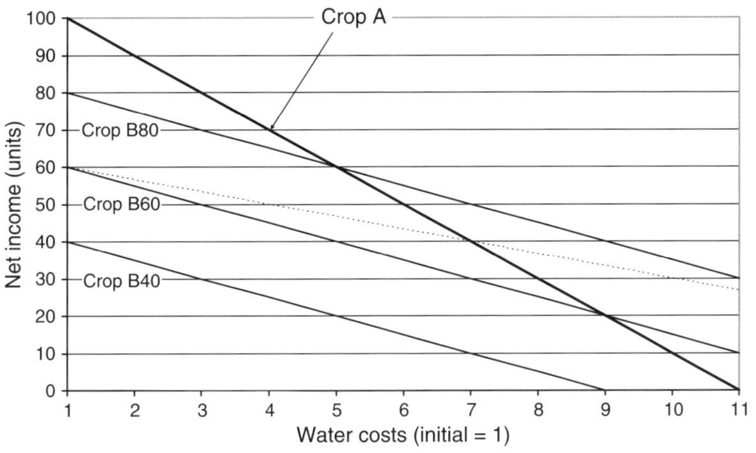

Fig. 2.8. Decrease of crop profitability with water costs.

increased prices, one may wonder in the first place why farmers would have neglected such an opportunity since it was already available to them, and why they would have to wait to see their benefits reduced by higher water costs before adopting it. This will enable us to get a closer scrutiny at farmer decision making regarding crop selection.

It must also be noted that high water use does not always imply low profitability and vice versa. 'Thirsty' crops with high returns include bananas (e.g. Jordan), rice (e.g. Egypt, Iran), sugarcane (parts of India) and qat (Yemen). Lucerne may consume a lot of water but does not have to be low-value, e.g. when in rotation with cereals. Above all, paddy is seldom grown *because* water is free or cheap (Falkenmark and Lundqvist, 1998) but in response to numerous environmental, social and other factors. Crops with lower requirements may not increase farmer incomes (and vice versa) and the impact on water productivity is far from self-evident. When high-value crops are also more water-intensive, higher prices may cause an increase in total demand for water, a phenomenon Dinar and Zilberman (1991) called 'the expansion effect'. In sum, the objectives of farmers (per hactare income), managers (reduce demand) or economists (water productivity) often do not coincide, although policies sometimes posit otherwise.

Economic growth, structural change and urbanization fuel demand for high-value products such as fruits, vegetables and meat (Rao *et al.*, 2004). Although the value of agricultural exports has risen dramatically, cereals continue to occupy more than 50% of the cultivated area worldwide, and fruits, vegetables and related high-value crops are confined to less than 7.5%. No doubt this share will rise but market constraints remain limiting, and cultivation must inevitably be confined to entrepreneurial farmers able to assume the costs and risks of high-return commercial agriculture. Access to groundwater greatly reduces water and related risks, but financial strength, entrepreneurial enterprise and credit access are still all required. Market volatility generates income instability (Hazell *et al.*, 1989; Quiroz and Valdés, 1995; Combes and Guillaumont, 2002) and most poor farmers cannot be expected to incur such risks, even if market volatility can sometimes be moderated by state interventions.

In addition to financial and marketing risk, crop choice is governed by a host of other well-identified factors.[24] These factors

[24]See, for example, Ellis (1998), Pingali and Rosegrant (1995) Quiroz and Valdés (1995), Pingali (2004), Arrojo (2001), Varela-Ortega *et al.* (1998), Dorjee *et al.* (2003); Barghouti *et al.* (2004), Gómez-Limón and Riesgo (2005), Binswanger and Rosenzweig (1986), World Bank (1988).

include: (i) labour constraints; (ii) lack of capital, credit or desire to get indebted; (iii) lack of information on market demand, quality requirements, agricultural techniques and agrochemicals, or adequate skills, etc.; (iv) land tenure uncertainty that hinders investments and adoption of perennial crops; (v) drudgery and health risk; (vi) soil, drainage or climatic constraints; (vii) high marketing costs due to poor transportation means (Delgado, 1995; World Bank, 2005a) and lack of infrastructure (cold storage trucking, refrigeration, etc.) (Barghouti *et al.*, 2004); (viii) the (un)reliability of irrigation supply and possible water quality constraints (Burt and Styles, 1999); and (ix) farmers' strategies, including food security considerations and many ageing farmers with exit strategies and no desire to take risk with new ventures, or to face increased drudgery.

This reminder serves here to dampen the enthusiasm that farm economic problems can be solved by a sweeping shift to high-value, capital-intensive and entrepreneurial agriculture. Another consequence is that farm models that seek to explain crop choice using fixed coefficients and oversimplified decision-making models fail to capture farmer responses, constraints and risks in full, with the implication that modelling approaches probably overstate the mobility of farming systems and their response to prices. Also, the responses are not confined just to farm practices. Farmers bring political pressure to bear when charges are raised and/or may refuse to meet obligations they consider punitive or unfair, break structures, tamper with metres or collude with field staff. Sanctions are difficult – even impossible – to enforce where control at the farm level is so often illusory.

In contrast to water charges, rationing and supply management can be very effective in influencing crop choice. The reasons are perhaps obvious. That water costs are seldom a critical issue does not mean that water is not a critical input. Farmers' indiscipline undermines supply management practices and, faced by shortages, deficit irrigation is a first response. But if schedules and quotas are strictly enforced, farmers perforce have to change their cropping patterns (or equipment) if basic water supplies are insufficient to meet minimum crop water requirements. Besides being a mechanism for managing scarcity and bringing supply and demand into immediate balance, supply management thus impacts on crop choice both in the short and (if sustained year to year) the long term.

Technological Change

By far the most important response to water scarcity has been the tube well revolution. Groundwater accounts for as much as 50% of agricultural value-added under irrigation, with much of it within the boundaries of surface irrigation schemes. Investment in water-saving technologies – buried pipes, sprinklers, micro-irrigation, land-levelling – represents a further response to water scarcity and to consequent high water costs. However, water is not the only factor involved. A profit-maximizing farmer, in principle, invests when (financial) capital and future O&M costs are justified in terms of anticipated increases in net income. Both farmers and conditions vary widely, and the decision to invest in costly equipment is seldom a straightforward response to water conditions but reflects a host of interconnected factors (Caswell and Zilberman, 1985; Green *et al.*, 1996; Schuck and Green, 2001; Moreno and Sunding, 2005). These may include[25]: (i) feasible crops; (ii) environmental conditions (soil quality, slope, plot size and shape, wind, water quality, etc.); (iii) the presence or absence of equipment suppliers and after-sales service; (iv) farmer education, skills, financial capacity and entrepreneurial spirit; (v) the amortization of existing material; and (vi) market opportunities, costs and risks.

[25]For discussion on the adoption of irrigation technology see also de Fraiture and Perry (Chapter 3, this volume), Green and Sunding (1997), Varela-Ortega *et al.* (1998), Dinar and Yaron (1990), Lichtenberg (1989), Sunding (2005), Green *et al.* (1996), Sumpsi Viñas (1998), Molle (2006), Green *et al.* (1996), Scheierling *et al.* (2006b), Dinar and Zilberman (1994), Schuck *et al.* (2005), Skaggs (2001), Shrestha and Gopalakrishnan (1993), Moreno and Sunding (2000).

Moreover, even discounting for risk and associated factors, profit maximization is not always the farmer's major preoccupation. Cropping in Jordan, for instance, can be explained in part by considerations of prestige and leisure (Venot *et al.*, Chapter 10, this volume).

Supply management and regulation of water use are sometimes used to dictate farm-level investments in water-saving technologies based on beneficial use or similar grounds. Some governments, supported in many cases by donor agencies, go further and subsidize such investments. Beyond initiating research and pilot schemes, however, such programmes are generally self-defeating, leading to overproduction, accentuated price volatility and discrimination against those who fail to obtain subsidies. Farmers are invariably the best judge of the investments justified in their own circumstances, and governments should limit their role to the regulation of water rights and water use so as to manage conflict, enable reallocation and promote environmental sustainability. Given extensive groundwater capacities, there is in particular little point in subsidizing modern water-saving technologies in massive surface systems which cannot compete with groundwater and which will inevitably remain largely for the production of cereals and other traditional crops.

Pricing, Crops and Technological Change: Empirical Evidence

Agricultural diversification and investments in water-savings technologies often go together, but are driven by market opportunities and total farming conditions rather than by water prices. Broad reviews at national level include that by Yang *et al.* (2003), who conclude that despite strong promotion of agricultural diversification 'the pace of this shift has not accelerated . . . [due to] constraints of marketing channels, processing and transport facilities, and market demand . . . particularly for perishable crops, such as vegetables and fruits'. With market saturation in many markets, they conclude that 'further raising irrigation charges are unlikely to lead to a substantial shift to cash crops'. Siriluck and Kammeier (2003) analysed a nationwide project aimed at fostering agricultural diversification in Thailand. They found that extension and credit packages may encourage some diversification but that 'blueprint' approaches insensitive to household diversity may push farmers into risky ventures and indebtedness. Artificially boosting output of specialty cash crops often sends market prices down, thus reducing the initial benefits of the shift and increasing the risk of bankruptcy.

Case studies provide similar conclusions. Both linear programming at farm and system level, and econometric models have attempted to capture the impact of pricing on cropping patterns and investments. Such models typically assume that farmers are profit-maximizing agents (Pinheiro and Saraiva, 2005), but differ greatly in their treatment of risk and other factors. Price elasticities and other outputs of such models heavily depend on the context, the assumptions made, the variables retained and the adjustments farmers are allowed to make (Ogg and Gollehon, 1989; Scheierling *et al.*, 2004). Most studies are from developed countries (western USA, Israel and southern Europe) and assume volumetric control and water on demand. In Spain, for instance, Varela-Ortega *et al.* (1998) show that to obtain a 10% reduction in water consumption 'irrigators of the Valencia region have to sacrifice up to 70% of their income, compared to 57% of their counterpart in the Castille region and a small 9% in Andalusia'. The low value in Andalusia is explained by the productive potential of this region, its large farms and the availability of alternative crops. Sumpsi Viñas (1998) obtained similar results for the Balbilafuente scheme, concluding that the elasticity of demand depends on farm size, initial water endowments, available crop alternatives and strategies of production (intensive or extensive), all of which differ regionally. Berbel and Gomez-Limón (2000) show for the Guadalquivir and Duero basins that farm incomes have to be decreased by 25% and 49%, respectively, before water demand decreases significantly. These and numerous other studies in Europe (Gómez-Limón and Riesgo, 2004a,b for Spain; Morris *et al.*, 2005 for the UK; Bazzani *et al.*, 2005 and Gallerani *et al.*, 2005 for Italy; Pinheiro and Saraiva, 2005 for Portugal), although undertaken in differing contexts with differing assumptions, hypotheses and coverage, tend to converge on a number of common conclusions:

- Response to price tends to be high for extensive and low for intensive high-value agriculture and depends on the number of crops that can be grown in any given region (which may be limited).
- Water savings due to crop or technological shifts only occur at price levels that severely dent farmers' incomes. If irrigation is extensive or has been developed as a social investment, large subsidies are needed to preserve farming after modernization.
- Water demand under micro-irrigation is inelastic. Once improvements in water-use efficiency have been achieved due to its adoption, further gains are increasingly unlikely.
- Water agency receipts often increase as water prices rise, though this is sometimes more than offset by reductions in water use.
- Because regions, and farmers within regions, are heterogeneous, nationwide policies will not be successful and have negative impacts on those who cannot adjust.

Many of these studies point to the adverse economic and political consequences of raising prices to levels that could impact on cropping and/or technology. Raising water prices sufficiently to impact on use and technology is not only a blunt instrument with widely differing regional impacts, but often results in irrigation becoming unprofitable. The decision on whether to provide subsidies forms part of a wider discussion on agricultural protection – the implication being that quotas are more effective in limiting water use if the concurrent aim is to preserve farm incomes and farming communities.

US studies have more mixed conclusions. While some are in agreement with these conclusions (e.g. Scheierling *et al.*, 2004 for South Platte; Scheierling *et al.*, 2006a,b; Hoyt, 1984; Caswell *et al.*, 1990), others suggest that technological change can occur in response to price (Caswell and Zilberman, 1985; Nieswiadomy, 1985; Negri and Brooks, 1990; Moore *et al.*, 1994). The reasons are unclear but some of the latter US studies appear to fail to establish a satisfactory level of causality between the water price and technological investment (Sunding, 2005), while others do not explore income losses and subsidies sufficiently to be comparable with the European studies. Be that as it may, there are many examples showing that water prices are seldom the primary driver in the adoption of water-saving technology since investment costs are almost invariably far greater than any savings in the water bill. Perry (2001a,b) shows, for central Iran, that the cost of reducing deliveries via such technologies is twice the actual cost of supply by the agency. In Gujarat, tube well farmers have complete flexibility and pay more than 30% of their net income for water, but there is little investment in improved technologies (Cornish *et al.*, 2004). De Fraiture and Perry (Chapter 3, this volume) conclude that 'empirical evidence shows that technology choice is hardly driven by water price' and Varela-Ortega *et al.* (1998) argue that 'the adoption of irrigation technology is not the most significant response to water pricing policies . . . technology adoption in highly productive regions can come about at zero water price rates'. In India (Shah *et al.*, Chapter 9, this volume) or in the Jordan valley (Venot *et al.*, Chapter 10, this volume), micro-irrigation developed when the price was very low, and Sunding (2005) concludes that 'water price is not the most important factor governing irrigation technology adoption' in San Joaquim valley; dissemination of centre pivots in California occurred when water costs were irrelevant (McKnight, 1983).

In practice, investment in water-saving technologies is linked to numerous other interacting factors (Dinar and Zilberman,

1994; Scheierling *et al.*, 2004). Diffusion of drip irrigation in Israel, for instance, was spurred by: (i) higher yields; (ii) subsidies; (iii) sandy soils; and (iv) the reuse of water savings to expand cultivation (Dinar and Zilberman, 1994). In other cases, produce quality (e.g. potatoes in the UK) and reduced labour costs are paramount. Calculations made by Sumpsi Viñas (1998) for vegetable and fruit production in several regions of Spain showed that impacts on yield, quality and labour use make drip and sprinklers more profitable than furrow irrigation. In Hawaii, drip irrigation was widespread in sugarcane because it increased yields, saved labour (and some water) and allowed expansion of cultivation on marginal and sandy soils (Shrestha and Gopalakrishnan, 1993). In Tunisia, although modernization targeted water saving, on-farm water use was not significantly altered, though higher yields and incomes were obtained (Al-Atiri *et al.*, 2004). García Mollá's (2000) study of Valencia in Spain and Carles *et al.*'s (1999) review of nine irrigation schemes also demonstrated that adoption of drip irrigation was motivated by reduced labour, enhanced quality, convenience and fertilizer saving.

Finally, contrary to common wisdom, the use of water-saving technology at the farm level does not necessarily mean that the fraction of applied water that is depleted (actually transpired or evaporated to the atmosphere) has been reduced. Soil evaporation is often reduced but crop evaporation is generally increased because of better and timelier application (Burt *et al.*, 2001; Perry, 2001a,b). Furthermore, evidence from arid and semi-arid regions, and more generally if land is not a limiting factor, suggests that water savings, to the extent they are obtained, are generally retained by the farmer or his neighbours to expand the cropped area. While benefits accrue to those expanding this area, the fraction of water depleted typically rises and return flows and aquifer recharge decline. García Mollá's (2000) study in Valencia revealed that districts adopting drip irrigation have attempted to maximize the area under cultivation. Similar situations have been described in countries such as Tunisia (Feuillette, 2001), India (Moench *et al.*, 2003), Spain (Carles *et al.*, 1999), Israel (Dinar and Zilberman, 1994), Morocco, the USA (Caswell, 1998; Huffaker *et al.*, 2000; Skaggs, 2001; Aillery and Gollehon, 2003; Huffaker and Whittlesey, 2003) and Hawaii (Shrestha and Gopalakrishnan, 1993). Public subsidies[26] aimed at improving efficiencies and releasing water for other uses are thus often counterproductive.

In sum, adoption of water-saving technology is seldom driven by water scarcity or water prices, but by an association of benefits that play out together: yield increases allowed by better and more homogeneous application of water, better quality and a more homogeneous product, bringing substantial increases in the market price, better application of fertilizers and chemicals, decreased labour costs, decrease in return flows contributing to reducing the leaching of fertilizer and pesticides and to controlling soil erosion are some of the associated benefits.[27] Further incentives are clearly linked to the possibility of using water savings to expand cultivation where land is not a constraint, and to that of capitalizing on existing pressurized supply when water is pumped from wells (Caswell and Zilberman, 1985; García Mollá, 2000; Becker and Lavee,

[26]Many countries subsidize micro-irrigation and farm-level improvement. In Morocco, for example, they are subsidized at a level of 30–40% and farmers are granted bonuses (Belghiti, 2005a) because technologies are too costly for farmers, but even then adoption is slow (Tizaoui, 2004). In Israel, micro-irrigation is generalized but the growth of 700% observed during 1975–1982 was spurred by heavy government subsidies that made the shift profitable (Shevah and Kohen, 1997). In the USA, the conservation of groundwater and surface water has been promoted by the Environmental Quality Incentives Program initiated in 1997, whereby cost-sharing may pay up to 75% of the costs of eligible conservation practices (Scheierling *et al.*, 2006a).

[27]For further discussion, refer to Caswell and Zilberman (1985, 1990), Dinar and Zilberman (1991, 1994), Caswell (1998), Morris *et al.* (2005), Wierenga and Hendrickx (1985), Carles *et al.* (1999), Skaggs (2001), Sumpsi Viñas (1998), McKnight (1983), Scheierling *et al.* (2006a), Becker and Lavee (2002) and García Mollá (2000).

2002). As a rule, these shifts generally result more from changes in market opportunities, output prices and subsidies (e.g. the Common Agricultural Policy (CAP) in Europe) than from changes in input costs.

PRICING AS AN ECONOMIC INSTRUMENT: ALLOCATION BETWEEN SECTORS[28]

Introduction

Urban growth and industrialization fuel rising water demands. According to the World Bank Strategy of 1993 'setting prices at the right level is not enough; prices need to be paid if they are to enhance the efficient allocation of resources' (World Bank, 1993); for Johansson (2000): 'The fundamental role of prices is to help allocate scarce resources among competing uses and users. One way to achieve an efficient allocation of water is to price its consumption correctly.' With higher prices that reflect opportunity cost, the reasoning goes, low-value activities are phased out, thus releasing water for high-value uses and raising social welfare.

As water shifts, *allocation stress*[29] moderates and economic gains are realized (Dinar, 1998; Rosegrant and Cline, 2002; Merrett, 2003; Hansen and Bhatia, 2004): 'supporting 100,000 high-tech California jobs requires some 250 million gallons of water a year; the same amount of water used in the agricultural sector sustains fewer than 10 jobs, a stunning difference' (Gleick, 2000). Elsewhere Gleick says: 'as much as half of all water diverted for agriculture never yields any food. Thus even modest improvements in agricultural efficiency could free up huge quantities of water.' But these and similar statements[30] need to be challenged. It is true that irrigation consumes much more water than urban uses, both absolutely and relative to diversions, but this is inherent to the activity (Abernethy, 2005) and it does not follow that increased 'agricultural efficiency' is a precondition for meeting other needs. To recapitulate:

- Irrigation may use uncontrolled and other marginal sources that may be unable to provide the security and quality needed by domestic or industrial users (Savenije and van der Zaag, 2002).
- There may be no hydraulic connectivity between irrigation and potential urban uses, and transfers and storage may be impracticable or prohibitively expensive (Smith *et al.*, 1997).
- Basin efficiencies are much higher than subsystem efficiencies (Frederiksen, 1996; Keller *et al.*, 1996; Perry, 1999; Molle *et al.*, 2004).
- Response to scarcity means that farmers use water more efficiently than is commonly assumed, adopting conservation measures and conjunctive use that offset the impact of reduced supply.

Moreover, if reallocation of water becomes necessary and is feasible, this almost invariably occurs, though not necessarily at lowest cost or in the most sustainable manner. Deficiencies in urban systems are thus primarily due to financial constraints and political priorities, and not to water being 'locked up' in 'inefficient' irrigation. The following subsections review these issues further under three headings: (i) allocation or financial stress?; (ii) transfer mechanisms; and (iii) implications. Issues associated with environmental externalities are discussed in the next section.

[28]This section is largely derived from Molle and Berkoff (2006), to which the reader is referred for further details.

[29]The allocation stress is typified by Bate (2002): 'The effect of under-priced water is that farmers use inefficient irrigation technologies to produce uneconomic goods at the expense of lucrative alternative economic activities.' The opportunity costs of this misallocation can be vast. See also Dinar and Subramanian (1997).

[30]See similar statements in Winpenny (1997), Simon (1998), IRN (2003), Postel (2001), Hansen and Bhatia (2004), ESCWA (1999) and Colby (1990), among others.

Allocation or Financial Stress?

Allocation stress

Allocation stress is said to occur when high-value sectors are deprived of water that is locked into lower-value activities. But the existence of a significant allocation gap is doubtful. In practice, farmers are 'losing out' (Winpenny, 1994), urban interests get the 'upper hand' (Lundqvist, 1993) and 'cities will continue to siphon water away from agriculture' (Postel, 1999). Transfers out of agriculture or ecological reserves (to the extent necessary and feasible) may be minor or major, gradual or outright, surreptitious or open, on the surface or underground, and with or without compensation, but by and large cities procure the water they need (Molle and Berkoff, 2006), in both the shorter and longer terms.

Priority in a drought is almost invariably given to urban uses, and to industry and services in particular. For example, shortages in industry and tourism in the 'Eastern Seaboard' near Bangkok have been quickly diffused by the implementation of six inter-basin transfers and drilling of 290 artesian wells for short-term relief (Samabuddhi, 2005).[31] Page (2001) cites a survey of the Hebei province that showed 'how local officials enforced restrictions on farmers but overlooked those on industry to lure projects from which they could profit'. Amman's supply was hardly impacted by the 2000/01 drought; the California State Water Project cut-off farmers in 1991, and the Bureau of Reclamation reduced supplies in the Central Valley by 75% (Anderson and Snyder, 1997); Jakarta's golf courses were supplied in the major 1994 drought; and in Cyprus farm supplies were cut by 50% in a 3-year drought but supplies to the 2 million tourists were maintained (Barlow and Clarke, 2003). Other examples where agriculture suffered first include Chennai, India (Ramakrishnan, 2002), the Guadaquiver basin in Spain (Fereres and Cena, 1997), the Alentejo region in Portugal (Caldas *et al.*, 1997) and Manila (McIntosh, 2003).

Whether longer-term investments in services and industry are constrained by water remains perhaps a matter of debate. Very high water-consuming industries, such as aluminium, are unlikely to settle in water-short areas, and suggestions have been made that water-intensive industries should be moved, e.g. inland from coastal China (Chan and Shimou, 1999). Many cities appear to be in the *wrong* place (Winpenny, 1994) and have to opt for more distant and costly transfers after exhausting nearby water supplies. But they can still continue to grow rapidly: Chennai, Mexico City, Las Vegas, Tianjin and Amman are widely differing cities that all illustrate this despite their very limited nearby resources. Ta'iz grew by 7.9% between 1986 and 1994, despite being one of the most water-stressed cities in the world. Even in water-abundant areas, cities outstrip proximate resources when located in upper catchments (e.g. São Paulo, Atlanta, Kuala Lumpur) or in small coastal catchments (e.g. Manila, New York, Boston). Although the costs of water vary greatly depending on local circumstances, there is little evidence that water constraints seriously impact on urban growth; and when this is the case it is rarely due to water being locked up in agriculture, except in situations where formal water rights may dictate so (e.g. western USA).

Financial and political stress

That cities, by and large, are able to obtain the water they need does not, of course, mean that water supply and sanitation (WSS) services have no deficiencies. Far from it. But these deficiencies reflect political priorities and financial constraints rather than water availability as such. In Europe for instance, in historic times, extension of WSS facilities beyond the affluent

[31] The Finance Minister is reported to have told senior bureaucrats that their 'heads are pledged as a guarantee, since this issue is a problem for the entire country . . . I don't want to hear again that industries along the Eastern Seaboard are facing water problems, whether it's this year or in any other year'.

can be attributed to a combination of the hygienist movement, a perceived 'threat from below' (Chaplin, 1999) and/or the need 'to preserve order, cleanliness and a healthy workforce' (Goubert, 1986). As early as the mid-18th century it was recognized that 'prevention of further environmental degradation was cheaper and more effective . . . than continuing with expenditure on poor relief' (Chaplin, 1999). Elites in Guayaquil (Swyngedouw, 2003) and Monterrey (Bennett, 1995) reacted in more recent times to social unrest. In contrast, Chaplin (1999) attributes the negative picture in India to a failure by the upper classes to pressure the government to invest. WSS investments differ in their political rewards and the key question is 'who will pay?' rather than 'where is the water?'.

Political considerations are compounded by financial and institutional constraints. Few cities in developing countries have been able to keep pace with inward migration (Lundqvist *et al.*, 2003) and the costs of collecting, conveying and disposing of water in line with city expansion have proven beyond their financial capacity. This has generally remained true throughout their history, when the population was far lower than now just as much as once the mega-cities of the present day had developed. Even in water-abundant regions, developing country cities have deficient WSS systems (e.g. Lagos, Dhaka, and Ho Chi Minh City). 'The root cause [of poor water supply to population] is our negligence and our resignation in the face of inequality' (Camdessus and Winpenny, 2003). Other documents addressing this issue similarly fail to refer to physical scarcity as a constraint (Anton, 1995; UNESCO, 2003). The question of 'who will pay' is key to understanding WSS conditions in cities. Capital cities are particularly well placed to access public funds (e.g. Mexico: Connolly, 1999) and how taxes are shared between local bodies, and state and federal governments, has an important bearing on the outcome. Some cities attract foreign subsidies (e.g. EU funds for Athens) or benefit from geopolitical considerations (e.g. Amman) or broad reconstruction factors (e.g. Phnom Penh). If society is receptive to privatization, the financial burden can be shifted to users, as in the UK, but elsewhere privatization and public–private partnerships have had mixed results in view of the risks, poor financial returns and political sensitivities (SIWI, 2004).

By and large, cities can secure necessary water resources. The mechanisms adopted to achieve the transfer, however, vary greatly. They depend, in particular, on the characteristics of the hydrological system, the nature and practice of government and on the strength of the regulatory and water rights systems. They are discussed below under three headings: expropriation (with and without compensation), opportunity cost pricing and markets.

Reallocation: Bureaucratic Expropriation, Administered Prices and Markets

Expropriation

An extensive literature review suggests that governments, urban utilities and industries commonly reallocate water by bureaucratic action (Molle and Berkoff, 2006). When successive urban projects take amounts that are small relative to river flows, reallocation can occur by stealth, with the impact on downstream farmers and ecosystems obscured by natural hydrologic variability. Even more prevalent than such reallocation of surface flows is the 'hidden' expropriation of groundwater resources as urban users deepen wells and increase pumping: approximately 1.5–2.0 billion people are said to rely on groundwater for domestic consumption, including 1 billion urban inhabitants in Asia (Foster, 1999), and industries often access groundwater directly because it is secure and needs no treatment. Where confiscation by stealth is impracticable, utilities may exercise *force majeure* – supported by politicians – and deprive farmers and other users outright. Since property rights are seldom clearly demarcated, confiscation may be legal in the sense that governments usually retain the final say on who receives water in the national interest. A further argument used to rationalize direct confiscation is that irrigation was a (heavily subsidized) gift of government in the first place. In cases where formal rights are effective, expropriation is precluded in the absence of financial compensation.

Expropriation is, in its nature, inequitable, depriving farmers of their traditional livelihood without recourse, accelerating the process of structural change and aggravating income inequities. Thus, although it is conceptually the simplest mechanism for effecting water transfers, direct expropriation can be problematic for any government, even an authoritarian one, especially in contexts where the local economy revolves around irrigated agriculture. This has led governments to consider compensation schemes on a case-by-case basis, even where formal property rights do not exist. This can take the form of *either* complementary action to ensure that the impact on irrigation is minimized *or* financial compensation for the losses incurred.

An example of complementary action was by El Paso which obtained water from the Rio Grande on condition that it reduced per capita consumption, recycled sewage water and eliminated leakage (Earl, 1996). Dongyang city obtained water from a dam managed by the Yiwu city, but had to finance an increase in the height of the dam and line irrigation canals (Liu, 2003). The 1998 agreement between the Imperial Valley Irrigation district and the Southern California Metropolitan Water Authority (MWA) included the lining of the All-American Canal by MWA with usufruct rights to the 100 Mm^3 thought to be 'conserved' passed to Southern California metropolitan area (Cortez-Lara and Garcia-Acevedo, 2000); similarly, the Upper Ganga canal was lined so that 'seepage losses' could be reallocated to Delhi. In both cases, however, these transfers were in practice at the expense of downstream groundwater users, who in the Californian case were Mexican farmers. Molle *et al.* (2004) use an example from Central Iran to show that in 'closed basins', where most or all resources are committed (often overcommitted), conservation measures do not save water, but merely reallocate it across the basin in a way that is not always perceptible.

Examples of compensation for water transfers include the buying out of agricultural wells around some cities (e.g. in Phoenix or Chennai); the diversion of water from neighbouring irrigation reservoirs to serve cities (e.g. Tsingtao in China where irrigation reservoirs were converted to urban use in preference to paying higher rates for Yellow River water); and the purchase of reservoir storage for hydro-generation from farmers during droughts in the Guadalquivir River basin, Spain. The merit of these and similar arrangements is that the transfer between irrigation and the utility can be adapted to specific local realities to the benefit of both sides. The government ultimately acts as mediator between the two and as the guarantor that the agreement will be honoured.

Opportunity cost pricing

Rather than expropriate water – with or without compensation – transfers can, in principle, be forced by full economic pricing of supply.[32] The World Bank's 1993 water policy and repetition by resource economists has disseminated the idea of the need for reallocation from low- to high-value uses, and this idea has been incorporated in national policy and legal documents. Zimbabwe's 1994 Irrigation Policy and Strategy, for example, states: 'Since water is scarce, its opportunity cost should be taken into consideration in determining price' (Nyoni, 1999). Despite these intentions and policies, however, charging economic prices

[32]While some see this as a desirable or compelling objective (although some phasing might be necessary to get there) (Khanna and Sheng, 2000; Rosegrant *et al.*, 1995; EU, 2000a; GWP-TAC, 2000; Plaut, 2000; Socratous, 2000; Saleth, 2001; Ünver and Gupta, 2003), others admit that it might be a far-fetched – or impractical – objective, especially when not even O&M costs are recovered) (Sampath, 1992; Smith *et al.*, 1997; Thobani, 1997; Asad *et al.*, 1999; Garrido, 2002; World Bank, 2003b).

has in practice remained elusive (Bosworth *et al.*, 2002; Kulshreshtha, 2002; ICID, 2004). Acknowledging the 'yawning gap between simple economic principles . . . and on-the-ground reality' that has prevailed for decades, the World Bank (2003) reconsidered the issue and singled out two main reasons for this gap: first, the impossibility 'to explain to the general public (let alone to angry farmers) why they should pay for something that doesn't cost anything to produce'; and second, the fact that 'those who have implicit or explicit rights to use of the resource consider (appropriately) such proposals to be the confiscation of property' (see Molle and Berkoff, Chapter 1, this volume).

A further reason why economic pricing is impractical (Asad *et al.*, 1999) and has seldom if ever been adopted (ICID, 2004) is that opportunity costs are location- and time-specific, and operate at the margin, falling off drastically once effective urban demand at any specific location has been satisfied (Savenije and van der Zaag, 2002). Moreover, the opportunity cost price does not equal the full opportunity value in urban uses but an intermediate value determined by the shape of the relevant demand curves given that a fixed amount of water must be allocated between competing uses when externality and other costs vary (Green, 2003). Even if this price could, in practice, be estimated, the implication is that high charges would be paid by those in irrigation schemes in direct competition with neighbouring urban areas, and that those further away and not in competition would pay much lower prices. As noted earlier, charging for opportunity costs would also be politically and socially self-defeating since the order of magnitude of these costs would bankrupt most of the irrigation activities affected (Bate, 2002; Tardieu and Préfol, 2002; The Economist, 2003[33]), especially when irrigation is inherently uneconomic (first section). Despite these impediments, two countervailing arguments are sometimes asserted:

- Stripped of normative content with regard to price fixing, the estimation of opportunity values in alternative uses sheds light on how much is recovered from users, paid by the state and left uncovered. This is a central argument of the EU's Water Framework Directive.
- Even if full opportunity cost pricing is impracticable, moving towards higher water charges might still instil a degree of market logic, promote structural shifts in the rural community, and favour those who can make the best use of available irrigation supplies.

Charging opportunity costs is nevertheless comparable to expropriation in that those who lose their water as a result of an inability to pay receive no compensation (Cummings and Nercessiantz, 1992) and this can be perceived as expropriation by those who have customary rights or who have bought land with the value of water incorporated in the price (Rosegrant and Binswanger, 1994; Garrido, 1999; World Bank, 2003a,b). Given also the potential for inefficiency and rent-seeking in the context of bureaucratic involvement, many point to water markets as a preferable solution to either expropriation or opportunity cost pricing to resolve allocation problems (Thobani, 1997; Bate, 2002).

Market reallocation

Small-scale water markets have long existed. The ancient markets of Alicante are well known (Maass and Anderson, 1978). More generally, community-based irrigation supplied by springs or qanats (Beaumont *et al.*, 1989) often has well-defined individual rights that lend themselves to temporary or permanent transactions. Most occur in 'spot markets': neighbours swap, lend, borrow, sell or buy water turns in order to fine-tune supply to individual demands. This also

[33] *The Economist* (2003) emphasizes that it is not 'politically plausible to suggest that farmers must always pay the full costs of their water. Water for irrigation is highly price-inelastic: since farmers have little alternative but to use the stuff, charging the full cost could simply drive them into bankruptcy'.

occurs in large-scale irrigation systems if supply is sufficiently defined in terms of time or discharge to permit quantitative estimation (a notable example being the warabandi systems of Pakistan and north-west India). Recently, groundwater markets have spread in South Asia and elsewhere although these are perhaps more akin to buying of a service than of the water itself (Shah, 1993). At these scales, transaction costs are minimized because users know each other (Reidinger, 1994), can readily communicate, and transfers are across short distances without costly infrastructure or significant losses. Permanent transfer of ownership is also socially controlled and local third-party impacts are easily identified.

Traditional markets reallocated water primarily within agriculture, although conversion of wells to water supply for tanker markets also occurs (e.g. in Jordan and India). Market reallocation has also sometimes performed well at a larger scale when the institutional conditions allow. Examples include trading of Rio Grande water in Texas (Chang and Griffin, 1992), the Westlands Water District in California (Brozovic *et al.*, 2002) and the Colorado-Big-Thompson scheme (Howe, 1986; Mariño and Kemper, 1999), where most transactions are spot transactions and rental (Carey and Sunding, 2001), but also include permanent transfers from agriculture to other sectors (Howe and Goemans, 2003). In South Africa's Orange River basin, trading has occurred between commercial farms (Backeberg, 2006). In Australia, transfers within and among distant irrigated areas have developed in the last 10 years (90% being temporary transfers) (Isaac, 2002; Turral *et al.*, 2004). Bauer's (2004) review of the Chilean experience describes active markets in the Limari basin (mostly short-term reallocation between irrigators supplied by the same reservoir), and in the Maipo and Mapocho basins close to Santiago (4% of all water rights were traded between 1990 and 1997, half being acquired by municipal utilities: Alicera *et al.*, 1999). In Mexico, trading occurs within large irrigation schemes, but interstate transfers are closely regulated (Simpson and Ringskog, 1997).

As the scale and number of users increase, however, water's well-known characteristics (see first section) make it prone to market failure (Livingston, 1995). Defining property rights can be very difficult; economies of scale invite natural monopolies (Easter and Feder, 1998); and the transaction costs associated with markets – information, regulation and enforcement – are typically large. Above all, third-party and externality effects are pervasive, and it is often very difficult to link particular flows with particular uses or users. Markets in the USA have, for instance, been constrained by the lengthy and costly litigation to which third-party impacts often give rise (Dellapenna, 2000; Kenney, 2003; Libecap, 2003). Market transactions within the Colorado-Big-Thompson system may work well, but this is partly because they are confined within one water district that holds the right to all return flows (Howe and Goemans, 2003; Libecap, 2003). China suspended an experiment in interprovincial trading once the return flow and environmental impacts became evident (Fu and Hu, 2002).

Moreover, water markets fail to account for scheme- and regional-level impacts of transfers. The transfer of some water rights to non-agricultural investors attached to acequias in New Mexico, for example, weakened management and maintenance of the system as a whole (Klein-Robbenhaar, 1996). Frederick (1998) reports that 'when farmers want to sell water to cities, irrigation districts resist, fearing the loss of agricultural jobs', while Wahl (1993) acknowledges that 'most agricultural water districts have viewed the potential for water transfers only very tentatively out of concern over the security of their water rights and potentially adverse effects on the districts and local communities'. The severity of impacts on the area of origin varies greatly (Gopalakrishnan, 1973; Charney and Woodward, 1990; Howe *et al.*, 1990). Sunk costs in social and non-irrigation economic infrastructure, for instance, may be a strong argument for preserving irrigation, but cannot be reflected in a market price.

Finally, markets may open the door for opportunistic and monopolistic behaviour. Bjornlund and McKay (1999) observed that in Australia, opportunistic buyers were able to exert undue pressure on sellers to obtain lower prices. Bauer (1997) and Hadjigeorgalis (1999) showed that in Chile, 'many small farmers are liquidity-constrained and often have sold rights to pay off large debts'; as 'land is of little value without water . . . it is not expected to observe farmers selling water rights unless they were exiting agriculture or facing liquidity constraint'. In Australia, on the other hand, 57% of water permanently traded was due to farmers having excess water or reducing their irrigation areas (Turral *et al.*, 2004). In California, presumably, transfers between large commercial farms reflect mere shifts in economic opportunities.

Although attractive in principle, the complexity of establishing markets for tradable water rights is formidable (CEPAL, 1995; Livingston, 1995; Siamwalla and Roche, 2001). Positive experience is confined to countries (e.g. the USA, Australia and Chile) having a sound knowledge of hydrology; a comprehensive and modern hydraulic infrastructure (notably of storage); strong legal, institutional and regulatory backgrounds; and relatively wealthy stakeholders. Proposals for the adoption of markets in tradable rights in countries where hydrologic data are scarce, physical infrastructure is lacking, water rights are ill-defined, farmers are numerous and small, and states have generally weak and ill-developed monitoring and enforcement capacity are unrealistic for the foreseeable future (see, e.g., Tanzania in van Koppen *et al.*, Chapter 6, this volume).

Implications

Differences between administrative and market allocation are not perhaps as large as sometimes stated (Mariño and Kemper, 1999). They both require considerable knowledge of the hydrology, control of the water regime, a command over who uses what water where and when and mechanisms for enforcement and dispute resolution. Differences in the effectiveness of regulatory structures may well reveal cultural or ideological values – even local idiosyncrasies (e.g. preference for licenses in Japan or France: Tardieu and Préfol, 2002 or market mechanisms in Chile), rather than degrees of efficacy.

Differences of opinion nevertheless persist between those who emphasize government failure and those who emphasize market failure. The former view state bureaucracies as at best inefficient and at worst subject to corruption and rent-seeking (Rosegrant and Binswanger, 1994; Holden and Thobani, 1996; Thobani, 1997; Easter *et al.*, 1999) and – in the USA – consider that public welfare and public trust doctrines destroy private property and hinder transfers towards higher value uses (Anderson and Snyder, 1997; Gardner, 2003). However, the majority of observers are doubtful that markets can constitute a major tool for the reallocation of water, no matter how theoretically desirable they may be, most especially in developing countries (Colby, 1990; CEPAL, 1995; Livingston, 1995; Morris, 1996; Gaffney, 1997; Frederick, 1998; McNeill, 1998; Dellapenna, 2000; Meinzen-Dick and Appasamy, 2002; Libecap, 2003; Kenney, 2006; Solanes and Jouravlev, 2006).

Markets can no doubt be facilitated at community and local level (Brown, 1997), but water allocation at higher levels requires a 'delicate interplay' between administrative and market control. This 'delicate interplay' would perhaps be best served by a more systematic adoption of compensation arrangements that recognize the economic benefits from reallocation – and the fact that urban interests will obtain their water needs – and also ensure transparency and that the interests of those deprived are taken into account. Ideally, the urban utility and the affected farmers would negotiate face to face, with both in effect faced by the opportunity cost of the water in dispute. The government regulator would, in principle, act as moderator and guarantor, and intervene more generally to safeguard farmers' interests and ensure that environmental externalities and third-party effects are taken into account. No doubt such a system would be open to abuse (government failure would not be abolished), but as regulation strengthens, negotiated compensation could increasingly approximate to regulated markets in which the particular circumstances of the water in dispute are taken into account.

PRICING AS AN ENVIRONMENTAL INSTRUMENT: WATER QUALITY AND SUSTAINABILITY

Introduction

So long as diversions are small relative to the water resource, consumptive and in-stream users are unconstrained in what they do and most water is left to the natural environment as the default *user of last resort* (see first section). But as diversions increase, especially for agriculture, and as in-stream users (e.g. hydroelectric dams) alter flow regimes, wetlands and deltas dry up, water tables and base flows decline, the natural ecology suffers and pollution is concentrated in the limited flow that remains. As a river basin closes, therefore, action must be taken to limit diversions if environmental flows and values are to be protected. What remains is typically diverted by irrigation, and agriculture rather than the environment becomes *the residual user*.

Both agriculture and urban uses contribute directly to pollution of streams and aquifers, sometimes making water unusable for domestic use. Direct agricultural pollution in the USA is said to be $9 billion per year (Bate, 2002). Despite 13 rivers flowing through the city, the degradation of their water due to agricultural and M&I uses has forced Jakarta to tap surface sources 78 km away (McIntosh, 2003); a similar situation is found in Seville because of pesticide and fertilizer residues in the Guadalquivir river; in Chinese cities (Bhatia and Falkenmark, 1993), including Chengdu, where water pollution and silt have forced the closure of two river intakes and the government is investing heavily in watershed rehabilitation (McIntosh, 2003). Irrigation is also responsible for waterlogging and soil salinization as water is diverted to poorly drained low-lying lands within, and at the tail of, irrigation schemes. Other externalities include the mobilization of silt due to catchment changes, which can have devastating impacts on river morphology (famously for the Yellow River), and the mobilization of toxic elements from the soil by leaching. Drainage of the Plain of Reeds in the Mekong delta, for example, releases acidity in waterways, while selenium in California has provoked high mortality of wild fowl in receiving wetlands (Wichelns, 2003).

With regard to groundwater, springs and wetlands fed by groundwater dry up in response to falling water tables (e.g. Azraq aquifer in Jordan) and base flows in rivers decline; falling water yields and water tables lead to higher pumping costs and to the expropriation of poorer farmers and others unable to afford ever-deeper wells (Kendy *et al.*, 2003 for China): falling water tables also aggravate salinity intrusion in coastal aquifers; especially in urban areas, land subsidence reduces aquifer storage and adversely impacts on infrastructure (Nair, 1991); and declining quality due to direct agricultural pollution compounds that from domestic use, industry and landfills (Sampat, 2000).

Environmentalists have vested high hopes in pricing mechanisms as a means of reducing excessive abstraction of water from ecosystems and of decreasing environmental degradation (de Moor and Calami, 1997; Avis *et al.*, 2000). Hodge and Adams (1997) argue that 'the price [of water] could be raised until the level of demand was consistent with the environmental constraints on supply'. Nevertheless, though there is an enormous amount of literature on valuing the environment, there has been limited work on how these values can be incorporated in irrigation pricing and few practical examples of where this has been attempted. As in the case of opportunity cost pricing (previous section), there appears to be little agreement as to how this should be done, and not much hope that farmers would have much understanding of why they should pay such costs. The discussion in this subsection is therefore relatively brief, reflecting as it does the limited evidence in the literature.

Environmental Pricing Mechanisms

The user-pays and polluter-pays principles embody the idea that quantity and quality externalities should be reflected in the price paid by water users as an incentive to reduce adverse environmental impacts and the emission of pollutants. These principles are much more forcefully applied in M&I (given the relative simplicity of volumetric charging and point-source pollution control) than in agriculture, given the problems of volumetric control in irrigation and the intractability of controlling and monitoring diffuse pollution from fertilizers and pesticides (UNEP, 2000).

The EU's Water Framework Directive goes some way in the direction of introducing environmental pricing in agriculture when it states that water charges should 'act as an incentive for the sustainable use of water resources and to recover the costs of water services by economic sector' (EU, 2000b) rather than be adopted for allocation purposes. Nevertheless, both full cost recovery and internalization of environmental externalities are widely seen as ambitious objectives and are, in many cases, impracticable. Modelling, for instance, suggests that much of Mediterranean irrigated agriculture would be jeopardized by strict application of the Directive (Berbel *et al.*, 2005). Mechanisms that have been suggested for irrigation pricing include both negative and positive incentives:

- *Resource charges*. Imposing a resource charge on irrigation equivalent to net externality costs has been suggested to limit diversions and protect the environment. Such charges, in principle, would be imposed on the scheme and passed down to the farmer as a component of the irrigation charge. In practice, however, charging even for recurrent O&M is difficult (as shown earlier) and resource charges have seldom been more than a small administrative fee aiming to recover the costs of resource management (in China, the UK, Spain, Peru, etc.). As far as is known, they have never been high enough to impact on irrigation diversions. Groundwater abstraction fees could, in theory, also be levied on a volumetric basis to limit abstractions to recharge or to some other defined sustainable level. In practice, however, they degenerate into a flat tax, and collection of volumetric charges remains an insurmountable issue, at least in developing countries (Albiac *et al.*, 2006).
- *Pollution charges*. Pollution charges are an incentive for reducing water use and pollutant discharge, though few countries have applied them in irrigation. Denmark is an exception where farmers are subject to the 1994 'Green Tax Reform' that imposes a water rate of €0.55/m^3 of raw water extracted. Further environmental fees are likely given concerns over pesticide contamination of groundwater. Green taxes also exist in Sweden, the UK, the Netherlands, Germany and Croatia (Berbel *et al.*, Chapter 13, this volume; Wright and Mallia, 2003). In France, farmers pay pollution fees for water used in cattle husbandry, but not in crop production. Income from such charges generally goes to the government budget rather than being used to resolve pollution issues, and are seldom high enough to alter behaviour significantly (Young, 1994).
- *Treatment or remediation charges*. Pollution charges may be more acceptable to farmers if used for remedial works within the scheme or in irrigation more widely – thus 'internalizing externalities' – for instance, to help resolve waterlogging, salinity and other problems that impact on scheme production. In South Australia, the government covers the costs of salinity management caused by irrigation projects constructed before 1988, but environmental externalities are charged for all subsequent projects in a two-part price structure. The environmental part of the charge is used to cover the cost of renovation or construction of infrastructure needed to reduce water quality-related externalities (Easter and Liu, 2005).

- *Taxes and rebates.* Rather than specific charges, pollution abatement programmes are more generally met through general taxes. These may, however, be limited to taxes on water users, introducing a degree of cross-subsidization, with the money collected used to treat the wastewater generated not only by the user but also by other dischargers, be they cities, cattle farmers or industries (as in the Basin Agencies in France). In Korea, in some upper catchments, pesticide and fertilizer use has been prohibited with 25% of the funds generated from domestic consumers along the river used as 'income compensation' for upstream farmers who suffer financial loss due to these environmental regulations (Min, 2004). Rather than being taxed, farmers may receive a tax rebate. In western Canada, for instance, rural municipalities have used the municipal tax system as a tool for encouraging specific behaviour by producers. They offered rebates to landowners who implement environmental practices on their land (e.g. grazing land) (Fairley, 1997).
- *Subsidies.* 'Delinking' farm subsidies from direct production payments under the EU reforms (Berbel *et al.*, Chapter 13, this volume) is a major attempt to build on existing programmes that have 'paid' farmers to adopt environmentally sustainable practices. Comparable payments are made directly to farmers in Switzerland who participate in three main ecological programmes: integrated production, organic farming and ecological compensation (extensive use of meadows). By 1996, 60% of agricultural area in Switzerland was farmed based on integrated production methods and 5% of the area met organic farming standards. The loss of income is said to be less than if the same effect had to be met through product price increases (Pfefferli and Zimmermann, 1997). In Germany, revenue from water taxes is often used to compensate farmers for restrictions on fertilizer use in vulnerable areas. This idea is also behind the wave of payments for 'environmental services', at the catchment level, for example.
- *Pollution permits.* Pollution permits for nitrogen or another pollutant are akin to quotas for water use. Restrictions on farm animal numbers are used in Europe as a proxy for pollution permits, e.g. in the Netherlands where the primary objective has been to limit groundwater contamination from pig and other intensive operations. As in the case of water quotas, 'permissions to pollute' are often more easily administered and have less implication in terms of welfare losses than a comparable tax on nitrogen utilization or on water use (Martínez and Albiac, 2004, 2006). Effluent permits can also, in principle, be made tradable although this is rare in agriculture. A programme in California with regard to selenium has been successful (Young and Karkoski, 2000) and, although comparable trading regimes have yet to be applied to irrigation or farming in Europe, they are being increasingly adopted in other sectors.

Water Pricing as an Environmental Instrument

Several conclusions can be drawn from this short review. Price incentives for the preservation and restoration of environmental sustainability and water quality have mostly been adopted in the non-agricultural sectors and generally in developed countries. While there have been major programmes that aim, for instance, to restore wetlands or tackle waterlogging and salinization in developing countries, these have almost invariably been funded by government and donors and pricing has seldom, if ever, been significant in controlling these ill-effects. With respect to nutrients and pesticide pollution, their diffuse nature makes them very difficult to measure and control, even in developed countries.

There are a variety of potential pricing schemes ranging from the straightforward application of the user-pays and polluter-pays principles, through partial or full cross-subsidizing by other water users, to full state subsidies. Implementation of the user-pays principle is constrained by all the issues related to irrigation charges discussed in earlier subsections, though any charge that limits water use should have some positive environmental impact. However, the feasibility of major additional environmental charges must be doubted. With regard to pollution, potential interventions are numerous although again problematic in developing countries. They vary from individual prevention incentives (stop the polluting activity) to individual remediation (do it better: use organic farming, extensive pastures, keep cattle sludge in farm reservoirs), to individual treatment (clean up your mess before releasing it), to collective treatment (state infrastructure funded by taxes on water users or the public).

Experience in developing countries suggests that negative incentives, though often feasible in the domestic and industrial sectors (where costs can be internalized within utilities and industrial firms), are often replaced by positive incentives in the agriculture sector whereby the polluter is subsidized to improve his environmental management: subsidies address either the cost of doing so, or the foregone benefits from abandoning polluting (but productive) practices. Payment for watershed services, again, is a good example of a positive incentive. Likewise, Varela-Ortega (Chapter 14, this volume) showed that among the various policies implemented to limit over-abstraction of groundwater in the Tablas de Daimiel, Spain, only the full compensation of farmers' foregone benefits proved to be successful (in contrast, compulsory quotas were not). Agriculture is in any case heavily subsidized and it makes sense to redirect subsidies away from incentives that tend to increase pollution (e.g. by rewarding higher yields) to those that promote good environmental management. Delinking of subsidy payments under the CAP is undoubtedly the most important and dramatic example of this trend, with the major underlying objective of promoting environmentally sustainable agriculture throughout the union (Berbel *et al.*, Chapter 13, this volume).

In conclusion, as in the case of opportunity cost pricing, there are severe practical difficulties of estimation, implementation and enforcement on the one hand, and of persuading farmers that they should pay for environmental externalities that – in their view – have only a tenuous connection with their activities on the other (World Bank, 2003a,b). Direct treatment measures can perhaps be 'internalized' but, with little agreement on how broader externalities can be valued, there is little prospect that farmers will be persuaded to pay for what they do not regard as their responsibility, and little prospect that politicians will impose such burdens under conditions of rising income inequalities and farmer unrest.

SYNTHESIS: CONTEXTUALIZING THE DEBATE AND SUGGESTING ANSWERS

An Emerging Storyline

This chapter has reviewed the different objectives of water pricing policies in agriculture. The overall picture that emerges is that of a gap between stated objectives and expected benefits on the one hand, and the actual and foreseeable impact of these policies on the other. Too often, stated objectives are based on analogy with the water supply and energy sectors. However, such an extrapolation can be very misleading given the particular characteristics of the irrigation sector.

An assumed correlation between low charges and low efficiency in surface irrigation has fuelled the chief narrative on water pricing. From this alleged causal link, it is inferred that raising prices would generate more careful practices and efficiency gains. Although generally valid for water supply and energy, this cannot be systematically assumed in irrigation. Reasons, in part, reflect the hydrological context and the characteristics of irrigation design and performance. In practice, most schemes and farmers are 'water takers', using whatever water is supplied to them, with the causes of uneven and unpredictable supply typically lying upstream of the scheme. Even when scheme supplies can be assured, it is deficiencies in scheme management that result in uncertainties and inequities at the farm gate rather than any price (dis)incentive. Farmers' responsiveness to price requires that charges are volumetric. Farmers have control over the quantity of water they take and the price is sufficiently high to correspond to the elastic portion of the demand curve. This combination of circumstances is, unfortunately, exceedingly rare.

Empirical evidence suggests that under conditions of scarcity: (i) farmers use water more efficiently, in particular, through conjunctive use; (ii) basin-level efficiency rises considerably; and (iii) surface water use is almost invariably regulated – in a more or less controlled manner – by rationing and quotas. The prevalence of quotas can be explained by their effectiveness in balancing supply and demand in response to variable supplies, while incurring far less loss in income than with price-based regulation; their relative transparency and equity; and the low infrastructural and transactions costs involved in their establishment. In a few modern systems, users have some latitude to use water above (or below) their quotas and in these cases water charges can be effective in influencing use at the margin. Markets at local level can also help balance supply and demand. Wider markets in quotas (water rights) can also promote high-value use, but have demanding technical and institutional preconditions and are seldom feasible in practice.

A more profound change than any of these has, however, been the spread of tube wells. By allowing farmer control, tube wells offset the risks, inadequacies and uncertainties not only of rainfall, but also of surface supply. Not only does this approximate to irrigation on demand – the holy grail of advocates of modernization and water pricing – but it also detracts from the need to deliver water on demand in surface systems since groundwater irrigation can (and in practice does) support a large part of the crop diversification and high-value farming that can be realistically envisaged. Ironically, and in contrast to surface supplies, it is the transaction costs of enforcing quotas that is prohibitive in the case of groundwater, and it is the long-term degradation of the resource that represents the major challenge in groundwater management.

What then is the role of irrigation water charges in surface irrigation? Figure 2.9 repeats the objectives suggested in Fig. 2.1, together with a summary of the constraints on achieving these objectives that have emerged in this chapter. They are briefly discussed below.

Economic theory suggests that, if the necessary preconditions are met, marginal cost pricing provides the signals to the farmer that optimizes his use of water. In contrast to the water supply and energy sectors, this chapter has suggested that marginal costs in irrigation should generally exclude initial capital costs. If so, direct marginal costs as a minimum comprise recurrent O&M, replacement and modernization costs. In principle, they should also reflect opportunity values in other uses and incorporate externality costs. The estimation and implementation of these measures is, however, fraught with difficulties. Moreover, marginal cost pricing is dependent on volumetric control, and in practice, pricing of water falls well short of full on-demand pricing.

Recovery of O&M costs is the most compelling reason for levying irrigation charges, notably if public funds are insufficient to operate and sustain the infrastructure. Cost recovery has understandably been the central objective of project design and national policies, and has become more pressing as irrigated areas have expanded and fiscal constraints have developed in many countries. Recovering just O&M costs has, however, proven much harder than expected and in the great majority of cases farmers are charged no more than a share of these costs. Moreover, defaulting is pervasive, especially in systems where supply is unpredictable and uneven and where staff has no incentives to enforce recovery. In a few cases, a share of capital cost is also recovered in addition to O&M, and/or farmers pay a management or a resource fee, or

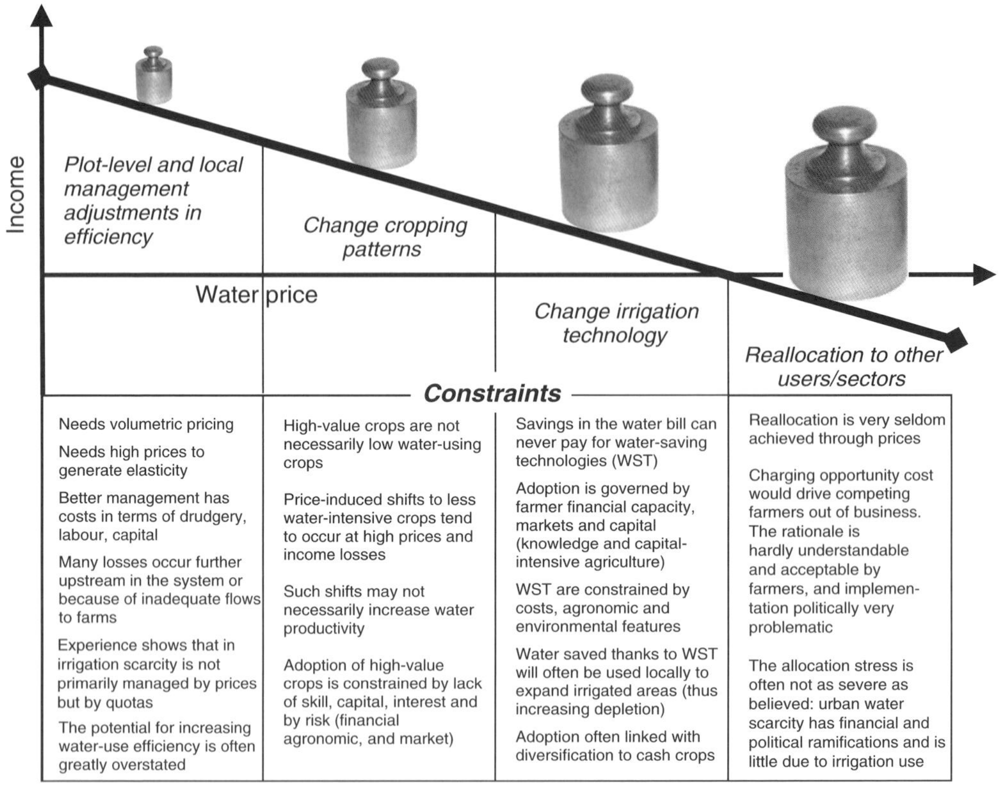

Fig. 2.9. Summary of constraints to using prices as an economic tool.

an environmental tax, but these seldom total more than about 10–25% of O&M costs. Charging for capital costs in new projects has the potential to ensure cost-effectiveness and users' interest and to crowd out politically motivated projects, but this is as yet seldom applied.

A wide array of benefits beyond sustaining the infrastructure is often anticipated for water charges, even when not warranted by the level or structure of the charge. This may reflect an improper understanding of charging mechanisms or be a means to justify the proposed policies. Chief among these are the view that raising prices will contribute to water conservation though, as discussed above, this is seldom valid. Charges may, however, have potential for eliciting longer-term shifts in crops and technology. Farm models often suggest that price-induced shifts and attendant water savings are possible but, as in the case of reducing water use, crop and technology choices are usually determined by other factors. Poor farmers irrigate low-value crops for many reasons (risk, capital, skill, markets, water supply, etc.) and, in particular, the risks to them of shifting to higher-value crops are considerable. Moreover, high-value cropping is inherently limited by market conditions and surface irrigators must compete with those having access to tube wells. If alternative crops or possible gains in efficiency are limited, farmers with extensive agriculture and low revenues will often revert to rain-fed farming, rent or sell out their farm, or just keep land fallowed, unless subsidies help them invest and intensify their practices. In practice, subsidies are often made available for such farmers.

High-value cropping often goes together with modern technologies, taking advantage of a host of positive factors beyond water savings, including higher yields, better product quality, fertigation, reduced labour, etc. Water costs are seldom the only or even the primary motivation for such shifts. In addition, water-saving technologies reduce return flows, but impact little on the fraction depleted by evaporation and transpiration; and in some cases, the water saved is used to expand the cultivated area, thus increasing depletion. In the latter case, promoting micro-irrigation can be counterproductive since the fraction consumed by crops increases at the expense of aquifer recharge, return flows and/or reallocation to other uses.

Low charges are also commonly taken to indicate a misallocation of resources that can be rectified by charging an opportunity cost. In practice, not only has opportunity cost pricing seldom, if ever, been attempted, but the very existence of an ‘allocation gap’ can be disputed. Priority is invariably given to M&I during a drought; over the longer term, most countries transfer water out of agriculture by stealth or administrative action; and there is little to indicate that urban and economic growth are eventually seriously constrained by water that is locked up in irrigation uses (except for some situations in the USA). Urban water and sanitation deficiencies are overwhelmingly due to political priorities and financial constraints rather than to lack of water. Moreover, opportunity cost is location-specific and, once effective demand in competing M&I uses is satisfied, opportunity cost falls off drastically. Opportunity cost pricing would drive those few farmers facing urban competition out of business, while most others would continue to obtain water at a much lower price. Markets are an attractive alternative, but the technical and institutional preconditions are daunting. Perhaps the most promising approach is negotiation on a case-by-case basis since, though government regulation is still required, compensation can be assured to those deprived in an open and transparent manner and in ways adapted to the particular conditions. Planning compensation mechanisms for temporary transfers *in anticipation* of drought will help avoid conflicts and turmoil when these occur.

Similar practical objections face the estimation and implementation of environmental pricing. Any charge that limits water use is likely to have some positive environmental impact but, given the constraints discussed above, imposing additional environmental charges on water use may not be feasible. It is therefore, perhaps, no surprise that while both the user-pays and the polluter-pays principles claim to internalize externalities by negative incentives at the source, in practice these externalities tend to be internalized at the system, basin or national level, through cross-subsidization from other users or the general taxpayers. Users get paid to control water losses or pollution, or even for the foregone revenue of not creating the externality, rather than being charged for the externality.

In conclusion, given the struggle to recover O&M and other recurrent costs in large-scale public irrigation, it is unlikely that water charges at levels much above O&M costs will ever become feasible. Participatory management, co-management, and autonomy can strengthen incentives for meeting the financial costs of supply, but irrigation charges are unlikely to have major impact on cropping patterns, technology or allocation between sectors; objections to opportunity and externality cost pricing will remain and, where farmers are given a say in the determination of charges, these are unlikely to be set much over O&M costs. In sum, whether management remains under state agencies or is shifted to farmer organizations, O&M will remain the reference ‘peg’. Pricing will be sometimes effective in groundwater use and as a mechanism to regulate use beyond the quota, wherever individual volumetric pricing is possible. Bulk allocation with innovative incentives may also, in the future, help achieve efficiency gains, as experimentation in China suggests. In other words, the consensus of the mid-1980s (see Molle and Berkoff, Chapter 1, this volume) still largely holds and much of the discussion on pricing instruments in public surface irrigation, and the hopes vested in them over the last two decades have been an unhelpful distraction. Physical sustainability and proper management remain compelling objectives and finding ways to strengthen financial autonomy and the reliability of supply remains paramount.

Cost Sharing with Power Sharing

Analysts in the 1980s appreciated that irrigation pricing policies had limited potential for promoting conservation and reallocation. Rather, they emphasized that farmer payments should be part of a wider realignment of roles and responsibilities in irrigation management. Irrigation charges could be the 'glue' of contractual arrangements between higher- and lower-level entities, down to the WUA. Autonomy at each level would create 'downward accountability', with payment made from the lower to the higher level in return for a negotiated service (defined as a certain pattern of supply). Each level would maintain and operate the infrastructure under its jurisdiction while contributing its share of system O&M costs. Under such conditions, user charges could help: (i) enhance availability of funds for O&M; (ii) strengthen accountability of managers to water users; (iii) increase involvement of water users in O&M; and (iv) improve the quality of investment decisions (Small, 1990).

This model has been constantly rediscovered and is deeply interwoven with strands of participatory management and turnover (Molle and Berkoff, Chapter 1, this volume). The nature and scale of what is transferred have varied widely. In some cases (Thailand, Sri Lanka, Pakistan and India) participation was based on tertiary canal user groups that were to federate. In practice, however, most were given too little power and fee collection has often failed (Merrey, 1996). Limitations in hydraulic infrastructure (Lankford and Gowing, 1997; Facon, 2002) have also been a constraint that often revealed the mistaken conception – perhaps inherited from domestic water supply – that it is possible to define a service in irrigation as 'simply' as in the domestic sector. In more successful cases (Mexico, Turkey and Argentina) O&M of the main system are retained by the public agency but WUAs are established at block and tertiary levels. In yet other cases, often smaller schemes with fewer richer farmers, the scheme has been entrusted wholly to farmers, with the state retaining a supervisory role (e.g. in Peru: Vos, 2002; Colombia: Vermillion and Garcés-Restrepo, 1998; Japan: Sarker and Itoh, 2001; and Catalonia: Fernandez-Urrutia, 1998).

The responsibilities transferred have also varied. WUAs are generally responsible for O&M within their area of jurisdiction, but some are only responsible for water management at higher levels. Their role in planning may be symbolic (allocations decided by the agency based on water availability), more proactive (with joint decisions on allocations to different areas) or even entail total responsibility. Financial contributions also differ (Spencer and Subramanian, 1997). Allotments to WUAs can be decided by the agency alone or jointly with WUAs; enforcement and monitoring of service can be more or less strict and with varied recourse by users; WUAs may trade allocations (as in Mexico); and in some cases charges levied also fund part of the agency's costs, while in others the agencies are subsidized by the state. Variations are inevitable and desirable and it is difficult to generalize. Nevertheless, empirical evidence collected over the last 20 years or so suggests a number of observations on the basic pattern.

The model is by and large valid but has exceptions

There is a strong relationship between the power devolved to farmers and their financial contribution. Where farmers are confined to tertiary-level activities, success has often been poor. When given management responsibilities besides O&M, they have often been able to take more substantive decisions, e.g. hiring field staff and deciding how to spend funds on maintenance (Mali: Aw and Diemer, 2005; northern Peru:

Vos, 2002; Argentina, etc). Where they are also contributing to the costs of running the public agency, their powers also tend to increase (Peru, Colombia), though this is not always the case (Vietnam: Fontenelle *et al.*, Chapter 7, this volume; Philippines). A farmer's financial contribution to O&M is no doubt necessary if farmers are to be given significant managerial powers, but is neither necessary nor sufficient for effective overall management and maintenance. In some cases (e.g. Morocco, Tunisia and Iran) farmers cover most or all of O&M costs and receive a reasonable service without strict accountability mechanisms. In contrast, the NIA in the Philippines illustrates the dangers of overestimating the capacity of supposedly autonomous agencies to ward off political interference. Moreover, NIA has responded to inadequate funds not by augmenting revenues, but rather by reducing costs and servicing only parts of the system (Kikuchi *et al.*, 2001; Oorthuizen, 2003). In the case of Taiwan (Moore, 1989; Lam, 1996) effective management by officials and farmers is achieved though user charges have long lost their significance, since the state re-established O&M funding in the early 1990s. Accountability is not supported by bureaucratic rules, but is embedded in social relationships and social control.

Narrow functionalism

Small and Carruthers (1991) recognized 'linkages existing between structural and managerial aspects on the one hand, with financial approaches on the other' (Small, 1990) but retained a functionalist view of agency–farmers arrangements: that charging linked to accountability could ensure transparent and effective cross-compliance and end the 'degradation vicious circle'. They have been criticized for overlooking the wider social and political dimensions that affect the level and utilization of charges independently of performance (Oorthuizen and Kloezen, 1995). Water charges are elements of negotiation in power struggles between farmers and their associations, and between WUAs and the agency or state. While these negotiations are bounded by hard-nosed realities, such as farmer financial capacity and the actual cost of supplying water (Lee, 2000), they also reflect competing interests, differing perceptions, the political clout and bargaining power of the different parties, and the various levels of accountability and dependency between them. They are permeated by the distribution of power within and across these groups (see case studies for the Philippines: Oorthuizen, 2003; Peru: Vos, 2002; Vietnam: Fontenelle *et al.*, Chapter 7, this volume; Taiwan, South-Korea, Japan: Sarker and Itoh, 2001; Tanaka and Sato, 2003). In other words, while 'money talks' and creates some dependency, accountability was shaped predominantly by inter-group and interpersonal relationships expressed in such factors as friendship, kinship, gifts, business partnerships, bribes, threats of violence, patronage, debts, asymmetries of power and information, and political allegiance. This warns us against simplified views of human organization and may help anticipate dysfunctions.

Second-generation problems

Encouraging financial and managerial autonomy of irrigation blocks or schemes coincides with the retreat of public agencies to higher levels of management. Autonomy has, in general, been successful in divesting the state of financial burdens but, according to many observers, has been largely neutral in terms of irrigation efficiency, water reliability and water productivity (Meinzen-Dick *et al.*, 1994; Vermillion, 1997). This in part reflects unrealistic expectations given that irrigation has always been more efficient than is commonly supposed and that farmers and managers have in any case adjusted to prevailing conditions. But it also reflects 'second-generation problems' that have gradually surfaced and have adversely affected performance including: the failure to adjust charges leading to deferred maintenance; the lack of data collection and analysis; imprecise rules governing asset ownership and management; and an unclear definition of water rights (Svendsen *et al.*, 1997; Vermillion and Garcés-Restrepo, 1998; Vermillion and Sagardoy, 1999). Among these, the most important problem has probably been the first: a short-term unwillingness to adjust fees upwards, to the detriment of long-term sustainability.

Opening up the model

The focus on financial autonomy has sometimes been superseded by more general participatory policies that emphasize reducing agency costs, or social engineering objectives. Nevertheless, there has also been renewed interest in the potential role of private operators and public–private partnerships (Frederiksen and Vissia, 1998) and in reviewing the whole spectrum of 'water service entities' from private to self-governing bodies (Lee, 2000; ICID, 2004; Frederiksen, 2005). Préfol *et al.* (2006) have pointed to the need for 'professional third parties' between farmers and government, irrespective of whether these are public or private. The crucial questions are accountability and incentive structures (Merrey, 1996). Promotion of volumetric management and bulk allocation is no doubt essential, but cannot ensure that incentives reach the individual farmer. Greater attention thus needs to be given to strengthening incentives at the tertiary and block levels. Interesting examples include the Philippines, where commissions are paid to WUAs that are successful in recovering charges (Ofrecio, 2005), and China where managers and subcontractors have both been given performance incentives (Lohmar *et al.*, Chapter 12, this volume; Li, 2006).

An alternative to the fiscal autonomy model patterned on utilities (O'Mara, 1990) takes up the idea of water delivery as 'co-production' (Lam, 1996; Ostrom, 1996). Under a 'co-production' approach, farmers and others participate in the production of public goods, in contrast to a 'service' approach under which they are merely passive 'clients'. It is argued that involving users at higher levels strengthens accountability and ensures that participants are aware of management constraints, existing inequities and actual available resources, the aim being to shift their role from that of 'selfish complainers' to co-managers of the whole system. According to this, the state must still inevitably retain supervisory powers, especially over financial management and maintenance standards, and in this regard it is lack of effective government capacity rather than lack of farmer and 'client' awareness that remains the major obstacle to creating self-sufficient entities (Frederiksen, 2005).

Perspectives for the Future

This review suggests that water charges can only achieve the objectives assigned to pricing as an economic tool (Fig. 2.1) in very special circumstances. But there is a continuum from projects with excess water and poor management at one extreme to those under volumetric management and – at the limit – irrigation on demand, at the other. Scarcity will continue to be dealt with by rationing in the large majority of cases, but price incentives can sometimes promote conservation and in a few cases regulate water use at the margin. The way forward is thus to expand the area served by volumetric management so as to facilitate extension of quota-cum-price regulation (Fig. 2.10), recognizing that this will be a slow process, given the structural and institutional changes needed, and that it may not always be appropriate or cost-effective to do so.

Such changes cannot be driven primarily by modernization investment or by social engineering that is inconsistent with the broader context. Effective financial mechanisms are predicated on the emergence of autonomous entities that vary with context but which entail genuine user empowerment. It should be recognized, however, that irrigation efficiency and water productivity are more about changes in irrigation management than changes in farmer behaviour;

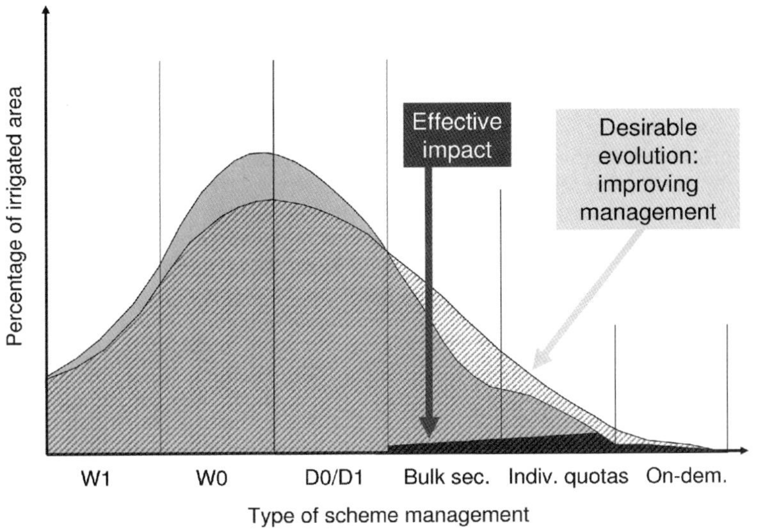

Fig. 2.10. Management types and desirable shifts.

more about designing cross-compliance arrangements and financial autonomy than simply establishing WUAs; (iii) and more about defining positive incentives to managers than introducing negative incentives to end-users.

Policies based on negative incentives alone are unlikely to have great success. The user-pays and polluter-pays principles thus need to be complemented by positive incentives. It may be more efficient (as well as more equitable) to buy out wells than to decree extraction quotas; to pay upstream farmers for not polluting water or deforesting watersheds than to tax these activities; and to negotiate compensation arrangements for water transfers than to expropriate them. The limited capacity of the state, and the political sensitivity of actions to modify behaviour that result in significant loss of income are major reasons why water and pollution charges have, in practice, been so difficult to introduce and enforce. Policy packages should ideally combine 'positive' and 'negative' instruments in ways that are adapted to circumstance (Bazza and Ahmad, 2002; Chohin-Kuper *et al.*, 2002; World Bank, 2005a). Since many factors other than water price so often determine water use, water policies must also be designed with due consideration to policies in other sectors.

Since individual metering is so problematic in surface irrigation, priority must be given to bulk allocation, all the more because it is consistent with strengthening co-management institutions and arrangements. Since financial incentives seldom impact directly on individual users, emphasis should normally be placed on management incentives (whether to private or community operators), while ensuring financial transparency. This is consistent with the fact that efficient management of supply is easier at block level than at individual farm level. There may be potential for trading in bulk allocations within the system, provided this is ultimately decided by stakeholders and can be effectively regulated, but intersector trading is likely to be feasible in only a few exceptional circumstances.

It must be recognized that much, if not most, surface irrigation, especially in countries with large irrigation sectors, will continue to be devoted to cereals and other relatively low-value crops. No doubt an increasing number of farmers will intensify and diversify output, often based on tube wells, but this is limited by market constraints and most farmers in surface irrigation are likely to remain relatively poor, at least as long as prices remain at current levels and until such time as economic development draws population off the land sufficiently to allow significant farm consolidation. This suggests caution in implementing expensive modernization and similar programmes that may not be justified by the production benefits. It also suggests the necessity of taking account of the deep social and political concerns raised by poor farmers. As stressed by Garrido (2002): '[N]o pricing policy will ever make progress if irrigators' benefits are severely compromised as a result of its full implementation. In the short and medium term, irrigation farms' economic survival is essential.' Economic policies pursuing efficiency will thus inevitably have to compromise with equity and social concerns and take into consideration the diversity of farming systems and regions.

Overemphasis on 'getting the prices right' (Svendsen and Rosegrant, 1994) has distracted attention from the nature of most of the irrigation in developing countries. Very few schemes can distribute water in a way approaching the on-demand supply model that typifies urban tap water. Farmers cannot be blamed for losses occurring upstream of their farm; nor can they be blamed for much of the waste arising out of a pattern of supply that is largely independent of their will. The importance of the old unglamorous issue of *managing supply* will thus continue to override that of *managing demand*. No doubt this will gradually change as irrigation moves along the continuum suggested in Fig. 2.10. But even then, developed countries' experience suggests that most efficiency gains are due to the numerous other factors involved in the shift from pragmatic to volumetric management; and that the task left to pricing even in the long term may well be far more modest than often assumed.

References

Abernethy, C.L. (2005) Financing river basin management. In: Svendsen, M., Merrey, D.J. and Shah, T. (eds) *Irrigation and River Basin Management: Options for Governance and Institutions*, CAB International, Wallingford, UK.

Abernethy, C.L., Sally, H., Lonsway, K. and Maman, C. (2000) *Farmer-Based Financing of Operations in the Niger Valley Irrigation Schemes*. Research Report 37. IWMI, Colombo, Sri Lanka.

Abu-Zeid, M. (2001) Water pricing in irrigated agriculture. *International Journal of Water Resources Development* 17(4), 527–538.

ADB (Asian Development Bank) (1986a) Review of bank operations in the irrigation sector 1966–1985. Manila.

ADB (1986b) Irrigation service fees. *Proceedings of the Regional Seminar on Irrigation Fees*, Manila, 21–25 July 1986. Manila.

ADB (1995) Sector synthesis of post-evaluation findings in the irrigation and rural development sector. Manila.

Ahmad, M. (2000) Water pricing and markets in the Near East: policy issues and options. *Water Policy* 2(3), 229–242.

Aillery, M. and Gollehon, N. (2003) Chapter 2.2: irrigation water management. In: Heimlich, R. (ed.) *Agricultural Resources and Environmental Indicators*, Agriculture Handbook No. (AH722). US Department of Agriculture (USDA), Economic Research Service, Washington, DC.

Ait Kadi, M. (2002) Irrigation water pricing policy in Morocco's large scale irrigation projects. *Hommes Terre & Eaux* 32(124), 25–33.

Al-Atiri, R., Faïza, R. and Bel Hassen, A. (2004) Les efforts de modernisation de l'agriculture irriguée en Tunisie. Project INCO-WADEMED, *Proceedings of the seminar 'Modernisation de l'Agriculture Irriguée'*, Rabat, Morocco, 19–23 April 2004.

Albiac, J., Martínez, Y. and Tapia, X. (2006) Water quantity and quality issues in Mediterranean agriculture. In: *Water and Agriculture: Sustainability, Markets and Policies*. OECD, Paris, Chapter 5, pp. 137–156.

Albinson, B. and Perry, C.J. (2002) *Fundamentals of Smallholder Irrigation The Structured System Concept*. Research Report 58. IWMI, Colombo, Sri Lanka.

Alicera, M.L., Brown, E.F. and Doña, J.E.N. (1999) *Bases Para el Análisis del Mercado de Derechos de Aprovechamiento de Aguas en la Cuenca de Rio Maipo*. VI Jornadas del CONAPHI-Chile. Santiago, Chile.
Altieri, S. (2001) Gestione tecnica ed administrative, in autogoverno, di un comprensorio irriguo pubblico. In: Leone, A. and Basile, A. (eds) *Proceedings of the Trans-National Workshop on 'Managing Water Demand in Agriculture through Pricing: Research Issues and Lessons Learned'*. CNR (National Research Council), Ercolano, Italy, pp. 213–219.
Anderson, T.L. and Snyder, P.S. (1997) Water Markets: Priming the Invisible Pump. Cato Institute, Washington, DC.
Anton, D.J. (1995) *Thirsty Cities: Urban Environments and Water Supply in Latin America*. International Development Research Centre, Ottawa. Available at: http://web.idrc.ca
Aquastat (2004) Available at: http://www.fao.org/AG/agl/aglw/aquastat/main/index.stm
Arriens, W.L., Bird, J., Berkoff, J. and Mosley, P. (1996) Towards effective water policy. In: *The Asian And Pacific Region. Vol. 1. Overview of Issues and Recommendations*. Asian Development Bank, Manila.
Arrojo, P. (2001) The impact of irrigation water pricing policy in Spain. In: *Pricing Water. Economics, Environment and Society. Conference Proceedings*. Sintra, European Commission, Brussels, pp. 177–184.
Asad, M., Azevedo, L.G., Kemper, K.E. and Simpson, L.D. (1999) *Management of Water Resources: Bulk Water Pricing in Brazil*. World Bank Technical Paper. World Bank, Washington, DC.
Avis, C., Tydeman, C. and Gelabert, E.R. (2000) *What Role for Water Pricing? Ten Actions for Internalising Sustainability*. WWF International, European Freshwater Programme, Brussels.
Aw, D. and Diemer, G. (2005) *Reforming the Office du Niger in Mali: Fostering Partnership between Farmers, Irrigation Agency and Government 1978–2000*. World Bank Technical Paper. World Bank, Washington, DC.
Backeberg, G.R. (2006) Reform of user charges, market pricing and management of water: problem or opportunity for irrigated agriculture. *Irrigation and Drainage* 55(1), 1–12.
Barakat, E. (2002) Cost recovery for irrigated agriculture: Egyptian experience. In: Hamdy, A., Lacirignola, C. and Lamaddalena, N. (eds) *Water Valuation and Cost Recovery Mechanisms in the Developing Countries of the Mediterranean Region*, Options Méditerranéennes: Series A. Séminaires Méditerranéens No. 49, CIHEAM-IAMB, Bari, pp. 73–90.
Barghouti, S., Kane, S., Sorby, K. and Ali, M. (2004) *Agricultural Diversification for the Poor. Guidelines for Practitioners*. Agriculture and Rural Development Discussion Paper 1. World Bank, Washington, DC.
Barker, R. and Molle, F. (2004) *Evolution of Irrigation in South and Southeast Asia*. IWMI Comprehensive Assessment Research Report 5. IWMI Comprehensive Assessment Secretariat, Colombo, Sri Lanka, vi.
Barlow, M. and Clarke, T. (2003) *Blue Gold*. LeftWord Books, New Delhi.
Bate, R. (2002) Water – can property rights and markets replace conflict? In: Morris, J. (ed.) *Sustainable Development: Promoting Progress or Perpetuating Poverty*? Profile Books, London.
Bates, R.H. (1981) *Markets and States in Tropical Africa*. University of California Press, Berkeley, California.
Bates, R.H. (1993) Urban bias: a fresh look. *Journal of Development Studies* 29(4), 219–228.
Bauer, C.J. (1997) Bringing water markets down to earth: the political economy of water rights in Chile, 1976–1995. *World Development* 25(5), 639–656.
Bauer, C.J. (2004) *Siren Song: Chilean Water Law as a Model for International Reform*. Resources for the Future, Washington, DC, 173.
Bazza, M. and Ahmad, M. (2002) A comparative assessment of links between irrigation water pricing and irrigation performance in the Near East. Paper presented at the International Conference 'Irrigation Water Policies: Micro and Macro Considerations,' Agadir, Morocco, June 2002.
Bazzani, G., Viaggi, D., Berbel, J., López, M. and Gutiérrez, C. (2005) Methodology for analysis of irrigated farming in Europe. In: Berbel, J. and Gutiérrez, C. (eds) *WADI – The Sustainability of European Irrigated Agriculture under Water Framework Directive and Agenda 2000*. European Commission, Joint Research Center, Institute of Prospective Technological Studies. Seville, Spain, pp. 26–48.
Beaumont, P., Borine, K. and McLachlan, K. (eds) (1989) *Qanat, Kariz, and Khattara*. School of Oriental and African Studies, London.
Becker, N. and Lavee, D. (2002) The effect and reform of water pricing: the Israeli experience. *International Journal of Water Resources Development* 18(2), 353–366.
Belghiti, M. (2005a) Gestion de l'eau d'irrigation au Maroc. Communication au Séminaire de promotion du SEMIDE et sur l'utilisation des systèmes d'information géographique pour la gestion et la protection des ressources en eau. Rabat, Morocco, 27–28 April 2005.
Belghiti, M. (2005b) Valorisation de l'eau et tarification dans les périmètres de 'grande hydraulique' au Maroc. Paper presented at the INCO-WATERMED seminar 'Modernisation de l'agriculture irriguée.' Rabat, Morocco, April 2004.

Bennett, V. (1995) *The Politics of Water: Urban Protest, Gender, and Power in Monterrey, Mexico*. University of Pittsburgh Press, Pittsburgh, Pennsylvania.

Berbel, J. and Gomez-Limón, J.A. (2000) The impact of water-pricing policy in Spain: an analysis of three irrigated areas. *Agricultural Water Management* 43, 219–238.

Berbel, J., Lopez, M.J. and Gomez Barbero, M. (2001) *Survey of Current Institutional Framework for Water Management in European Irrigated Systems: Spain*. Report for the WADI Project. University of Cordoba, Cordoba, Spain.

Berbel, J., López, M. and Gutiérrez, C. (2005) Spain (1): the case of the Guadalquivir river. In: Berbel, J. and Gutiérrez, C. (eds) *WADI – The Sustainability of European Irrigated Agriculture under Water Framework Directive and Agenda 2000*. European Commission, Joint Research Center, Institute of Prospective Technological Studies, Seville, Spain, pp. 49–70.

Berkoff, J. (2002) Economic valuation: why is it often unsatisfactory? And does it matter? With reference to the irrigation sector. Paper presented at the International Consulting Economists' Association meeting, 19 June 2002.

Berkoff, J. (2003a) A critical look at the irrigation sector. Presentation to World Bank Water Week, 4 March 2003.

Berkoff, J. (2003b) Prospects for irrigated agriculture: has the international consensus got it right? Paper presented to the Alternative Water Forum. Bradford University, Bradford, 1–2 May 2003.

Bhatia, R. (1991) Irrigation financing and cost recovery policy in India: case studies from Bihar and Haryana. In: Meinzen-Dick, R. and Svendsen, M. (eds) *Future Directions for Indian Irrigation*. Washington, DC, pp. 168–213.

Bhatia, R. and Falkenmark, M. (1993) *Water Resources Policies and the Urban Poor: Innovative Approaches and Policy Imperatives*. World Bank, Washington, DC, 47.

Bhattarai, M., Barker, R. and Narayanamoorthy, A. (2003) Who benefits from irrigation development in India? Implication of irrigation multipliers for cost recovery, and irrigation financing. In: *ICID Asian Regional Workshop, Sustainable Development of Water Resources and Management and Operation of Participatory Irrigation Organizations*, 10–12 November 2003. ICID, Taipei, Taiwan, pp. 285–296.

Binswanger, H.P. and Deininger, K. (1997) Explaining agricultural and agrarian policies in developing countries. *Journal of Economic Literature* 35(4), 958–2005.

Binswanger, H.P. and Rosenzweig, M.R. (1986) Behavioral and material determinants of production relations in agriculture. *Journal of Development Studies* 22(3), 503–539.

Bjornlund, H. and McKay, J. (1999) Do water markets promote a socially equitable reallocation of water? A case study of a rural water market in Victoria, Australia. Paper presentation, 6th Conference of the International Water and Resources Consortium, Hawaii.

Bos, M.G. and Wolters, W. (1990) Water charges and irrigation efficiencies. *Irrigation and Drainage Systems* 4(3), 267–278.

Bosworth, B., Cornish, G., Perry, C. and van Steenbergen, F. (2002) *Water Charging in Irrigated Agriculture. Lessons from the Literature*. Report OD 145. HR Wallingford, Wallingford, UK.

Bowen, R.L. and Young, R.A. (1986) Appraising alternatives for allocating and cost recovery for irrigation water in Egypt. *Agricultural Economics* 1, 35–52.

Brooks, D.B. (1997) Water demand management: conceptual framework and policy implementation. In: Brooks, D.B., Rached, E. and Saade, M. (eds) *Management of Water Demand in Africa and the Middle East: Current Practices and Future Needs*. International Development Research Centre, Ottawa.

Brown, F.L. (1997) Water markets and traditional values: merging commodity and community perspectives. *Water International* 22(1), 2–5.

Brozovic, N., Carey, J.M. and Sunding, D.L. (2002) Trading activity in an informal agricultural water market: An example from California. *Water Resources Update Issue No. 121: January 2002*. Universities Council on Water Resources, Southern Illinois University, Carbondale, Illinois.

Burger, R. (1998) *Water Legislation and Pricing in Kazakstan*. Environment Discussion Paper No. 40. NIS-EEP (Newly Independent States Environmental Economics and Policy) Project. Available at: http://www.epic.kz/hiid/finalwua.pdf

Burt, C.M. and Styles, S.W. (1999) *Practices in Irrigation Impact on Performance*. FAO Water Report No. 19. Rome.

Burt, C.M., Howes, D.J. and Mutziger, A. (2001) *Evaporation Estimates for Irrigated Agriculture in California*. ITRC paper P01-002. ITRC, Fresno, California.

Çakmak, B., Beyribey, M. and Kodal, S. (2004). Irrigation water pricing in water user associations, Turkey, *Water Resources Development* 20(1), 113–124.

Caldas, J.C., Sousa, P.L. and Pereira, L.S. (1997) Need for new approaches in valuing water: learning from a less successful case. In: Kay, M., Franks, T. and Smith, L. (eds) *Water: Economics, Management and Demand*. E & FN Spon, London. Cambridge University Press, Cambridge, pp. 137–144.

Camdessus, M. and Winpenny, J. (2003) Financing water for all. Report of the World Panel on Financing Water Infrastructure. Available at: www.wsscc.org/download/Camdessus%20document.doc

Carey, J.M. and Sunding, D.L. (2001) Emerging markets in water: a comparative institutional analysis of the Central Valley and Colorado-Big Thompson Projects. *Natural Resources Journal* 283(41), 288–289.

Carles, J., Abellá, L. and García-Mollá, M. (1999) Precios, costos y uso del agua en el regadío mediterráneo. In: *El agua a debate desde la Universidad*. The Congreso Ibérico sobre Planificación y Gestión de Aguas, Zaragoza.

Carruthers, I. and Clarck, C. (1981) *The Economics of Irrigation*. Liverpool University Press, Liverpool, UK.

Carruthers, I.D., Peabody, N.S. III, Bishop, A.A., LeBaron, A.D., Mehra, R., Ramchand, O., Peterson, D. and Wood, D.H. (1985) *Irrigation Pricing and Management*. Report to USAID. DEVRES, Washington, DC.

Caswell, M. (1998) Technologies as a water conservation tool. *Water International* 14(1989), 19–26.

Caswell, M. and Zilberman, D. (1985) The choices of irrigation technologies in California. *American Journal of Agricultural Economics* 67, 224–234.

Caswell, M.F. and Zilberman, D. (1990) The effects of well depth and land quality on the choice of irrigation technology. *American Journal of Agricultural Economics* 68, 798–811.

Caswell, M.F., Lichtenberg, M. and Zilberman, D. (1990) The effects of pricing policies on water conservation and drainage. *American Journal of Agricultural Economics* 72(4), 883–890.

CEPAL (Comisión Económica para América Latina y el Caribe) (1995) *Mercados de Derechos de Agua: Entorno Legal*. CEPAL, Santiago de Chile.

Chambers, R. (1988) *Managing Canal Irrigation: Practical Analysis from South Asia*. Cambridge University Press, Cambridge.

Chan, R.C.K. and Shimou, Y. (1999) Urbanization and sustainable metropolitan development in China: patterns, problems and prospects. *GeoJournal* 49, 269–277.

Chang, C. and Griffin, R.C. (1992) Water marketing as a reallocative institution in Texas. *Water Resources Research* 28(3), 879–890.

Chaplin, S.E. (1999) Cities, sewers and poverty: India's politics of sanitation. *Environment and Urbanization* 11(1), 145–158.

Charney, A.H. and Woodward, G.C. (1990) Socioeconomic impacts of water farming on rural areas of origin in Arizona. *American Journal of Agricultural Economics* 72(5), 1193–1199.

Chaudhry, M.G., Majid, S.A. and Chaudhry, G.M. (1993) The policy of irrigation water pricing in Pakistan: Aims, assessment and needed redirections. *The Pakistan Development Review* 32(4), 809–821.

Chohin-Kuper, A., Rieu, T. and Montginoul, M. (2002) Economic tools for water demand management in the Mediterranean. Paper presented to the forum on 'Progress in Water Demand Management in the Mediterranean' Fiuggi, 3–5 October 2002.

Colby, B.G. (1990) Transactions costs and efficiency in western water allocation. *American Journal of Agricultural Economics* 72(5), 1184–1192.

Combes, J.-L. and Guillaumont, P. (2002) Commodity price volatility, vulnerability and development. *Development Policy Review* 20(1), 25–39.

Connolly, P. (1999) Mexico city: our common future? *Environment and Urbanization* 11(1), 53–78. Contributing Paper to the World Commission on Dams. Available at: http://www.dams.org/

Cornish, G.A. and Perry, C.J. (2003) *Water Charging in Irrigated Agriculture. Lessons from the Field*. Report OD 150. HR Wallingford, Wallingford, UK.

Cornish, G., Bosworth, B., Perry, C. and Burke, J. (2004) *Water Charging in Irrigated Agriculture: An Analysis of International Experience*. FAO Waters Reports 28. FAO, Rome, Italy, p. xi.

Cortez-Lara, A. and Garcia-Acevedo, M.R. (2000) The lining of the All-American canal: the forgotten voices. *Natural Resources Journal* 40, 261–279.

Cosgrove, W.J. and Rijsberman, F.R. (2000) *World Water Vision: Making Water Everybody's Business*. Earthscan, London.

Courcier, R., Venot, J.-P. and Molle, F. (2005) *Historical Transformations of the Lower Jordan River Basin (in Jordan): Changes in Water Use and Projections (1950–2025)*. IWMI Comprehensive Assessment Research Report 9. IWMI, Colombo, Sri Lanka, pp. vi, 85.

Cummings, R.G. and Nercessiantz, V. (1992) The use of water markets as a means for enhancing water use efficiency in irrigation: case studies in Mexico and the United States. *Natural Resources Journal* 32, 731–755.

Davis, R. and Hirji, R. (Series eds) (2003) *Water Conservation: Irrigation*. Water Resources and Environment, Technical Note F2. World Bank, Washington, DC.

de Moor, A. and Calami, P. (1997) *Subsidizing Unsustainable Development: Undermining the Earth with Public Funds*. Earth Council, San Jose, Costa Rica.

Delgado, C.L. (1995) Agricultural diversification and export promotion in sub-Saharan Africa. *Food Policy* 20(3), 225–243.

Dellapenna, J.W. (2000) The importance of getting names right: the myth of markets for water (Fall, 2000). *William and Mary Environmental Law and Policy Review* 25, 317–377.

Dinar, A. (1998) Water policy reforms: informational needs and implementation obstacles. *Water Policy* 4, 367–382.

Dinar, A. and Saleth, R.M. (2005) Issues in water pricing reforms: from getting correct prices to setting appropriate institutions. In: Folmer, H. and Tietenberg, T. (eds) *The International Yearbook of Environmental and Resource Economics 2005/2006*. Edward Elgar, Cheltenham, UK.

Dinar, A. and Subramanian, A. (1997) *Water Pricing Experiences. An International Perspective*. World Bank Technical Paper No. 386. World Bank, Washington, DC.

Dinar, A. and Yaron, D. (1990) Influence of quality and scarcity of inputs on the adoption of modern agricultural Technologies. *Western Journal of Agricultural Economics* 15(2), 224–233.

Dinar, A. and Zilberman, D. (1991) The economics of resource-conservation, pollution-reduction technology selection: the case of irrigation water. *Resources and Energy* 13, 323–348.

Dinar, A. and Zilberman, D. (1994) Economía de las tecnologías modernas de riego: lecciones de la experiencia Israelí. *Revista de Estudios Agro-Sociales* 167(1), 155–183.

Dinar, A., Rosegrant, M.W. and Meinzen-Dick, R. (1997) *Water Allocation Mechanisms: Principles and Examples*. Policy Research Working Paper 1779. World Bank, IFPRI, Washington, DC, pp. ii, 41.

Dorjee, K., Broca, S. and Pingali, P. (2003) *Diversification in South Asian Agriculture: Trends and Constraints*. ESA Agricultural and Development Economics Division. Working Paper No. 03–15. The Food and Agriculture Organization of the United Nations, Rome.

Duane, P. (1986) Cost recovery in irrigation projects: perceptions from World Bank operation evaluation. In: FAO. Land and Water Development Division, USAID. *Water Management Synthesis II Project Technical Papers from the Expert Consultation on Irrigation Water Charges, Volume I, Rome, 22–26 September 1986*. FAO, Rome, Italy, pp. 122–157.

Earl, R.A. (1996) Sunbelt water war: The El Paso–New Mexico water conflict. *The Social Science Journal* 33(4), 359–380.

Easter, K.W. and Feder, G. (1998) Water institutions, incentives, and markets. In: Parker, D. and Tsur, Y. (eds) *Decentralization and Coordination of Water Resource Management*. Kluwer Academic, Boston, Massachusetts, pp. 261–282.

Easter, K.W. and Liu, Y. (2005) *Cost Recovery and Water Pricing for Irrigation and Drainage Projects*. Agriculture and Rural Development Discussion Paper No. 20. World Bank, Washington, DC.

Easter, K.W., Rosegrant, M.W. and Dinar, A. (1999) Formal and informal markets for water: institutions, performance and constraints. *The World Bank Research Observer* 14(1), 99–116.

Ellis, F. (1998) Household strategies and rural livelihood diversification. *The Journal of Development Studies* 35(1), 1–38.

Elston, J. (1999) *Water Pricing Study. Ministry of Water Resources and World Bank: Action Program Study for Water Resources (Australia-China Development Co-operation)*. Working Paper.

ESCWA (Economic and Social Commission for Western Asia) (1997) *Review of the Impact of Pricing Policy on Water Demand in the ESCWA Region, with a Case Study on Jordan*. (E/ESCWA/ENR/1997/6), Economic and Social Commission for Western Asia, Beirut, Lebanon.

ESCWA (1999) *Updating the Assessment of Water Resources in ESCWA Member Countries*. United Nations, New York.

EU (European Union) (2000a) *Pricing Policies for Enhancing the Sustainability of Water Resources*. Communication from the Commission to the Council, the European Parliament, and the Economic and Social Committee (COM(2000) 477 final). European Union, Brussels.

EU (2000b) *Pricing Policies for Enhancing the Sustainability of Water Resources*. Communication from the Commission to the Council, the European Parliament and the Economic and Social Committee (COM (2000) 477 final). European Union, Brussels.

Facon, T. (2002) Downstream of irrigation water pricing – the infrastructure design and operational management considerations. Paper presented at the conference 'Irrigation Water Policies: Micro and Macro Considerations,' Agadir, Morocco, 15–17 June 2002.

Fairley, B. (1997) Water Use and Water Quality in National Agricultural Policies: Canada's Experience. *OECD Workshop on the Sustainable Management of Water in Agriculture: Issues and Policies*. Directorate for Food, Agriculture and Fisheries and Environment Directorate, Athens, 3–6 November.

Falkenmark, M. and Lundqvist, J. (1998) Towards water security: Political determination and human adaptation crucial. *Water Resources Journal* 198, 12–28.

FAO and USAID (1986) *Water Management Synthesis II Project. Report on the Expert Consultation on Irrigation Water Charges, Rome, 22–26 September 1986*. FAO, Rome, Italy, pp. vi, 70.

Fereres, E. and Cena, F. (1997) Social benefits and environmental constraints of irrigation in an era of water scarcity. In: Kay, M., Franks, T. and Smith, L. (eds) *Water: Economics, Management and Demand*. E & FN Spon, London, pp. 128–136.

Fernandez-Urrutia, J.A.O. (1998) Spain: the institutional management of irrigation water in Spain. In: *Workshop on the Sustainable Management of Water in Agriculture*. The Athens Workshop, Case Studies. OECD, Paris, pp. 153–182.

Feuillette, S. (2001) Vers une gestion de la demande sur une nappe en accès libre: exploration des interactions ressources usages par les systèmes multi-agents; application à la nappe de Kairouan, Tunisie Centrale. Unpublished PhD thesis. Université de Montpellier II, Montpellier, France.

Foster, S. (1999) The interdependence of groundwater and urbanisation. In: *Proceedings of the 9th Stockholm Water Symposium*. 'Urban Stability through Integrated Water-Related Management,' 9–12 August, Stockholm, pp. 125–138 (cited in Lundqvist *et al*., 2001).

Frederick, K.D. (1998) Marketing water: the obstacles and the impetus. *Resources for the Future*, Issue 132. Available at: http://www.rff.org/resources_archive/pdf_files/132_water.pdf

Frederiksen, H.D. (1996) Water crisis in developing world: Misconceptions about solutions. *Journal of Water Resources Planning and Management* 122(2), 79–87.

Frederiksen, H.D. (2005) The future of rural water services in the developing countries and the governments' role. *Irrigation and Drainage* 54(5), 501–508.

Frederiksen, H.D. and Vissia, R.J. (1998) Considerations in formulating the transfer of services in the water sector. IWMI, Colombo, Sri Lanka, pp. ix, 80.

Freeman, D.M. and Lowdermilk, M.K. (1991) Middle-level farmer organisations as links between farms and central irrigation systems. In: Cernea, M.M. (ed.) *Putting People First: Sociological Variables In Rural Development*, 2nd edn. A World Bank Publication, pp. 113–143.

Fu, C. and Hu, Z.P. (2002) The practice on water rights allocation and trade-off in China. Paper presented to the River Symposium 2002, September 2000, Brisbane, Australia.

Gaffney, G. (1997) What price water marketing? California's new frontier – special issue: commemorating the 100th anniversary of the death of Henry George. *American Journal of Economics and Sociology* 56(4), 475–520.

Gaitskell, A. (1959) *Gezira – A Story of Development in the Sudan*. Faber & Faber, London.

Gallerani, V., Baani, G.M., Viaggi, D., Bartolini, F. and Raggi, M. (2005) The case of Italy. In: Berbel, J. and Gutiérrez, C. (eds) *WADI – The Sustainability of European Irrigated Agriculture under Water Framework Directive and Agenda 2000*. European Commission, Joint Research Center, Institute of Prospective Technological Studies, Seville, Spain, pp. 141–165.

Garcés-Restrepo, C. (2001) Irrigation management devolution in Colombia. IMT case study: Colombia. FAO, Rome. Available at: http://www.fao.org/ag/agl/aglw/waterinstitutions/docs/CScolombia.pdf

García Mollá, M. (2000) Análisis de la influencia de los costes en el consumo de agua en la agricultura valenciana. Caracterización de las entidades asociativas para riego. Unpublished PhD thesis. Universidad Politécnica de Valencia. Departamento de Economía y Ciencias Sociales, Valencia, Spain.

Gardner, B.D. (2003) Weakening water rights and efficient transfers. *Water Resource Development* 19(1), 7–19.

Garrido, A. (1999) *Agricultural Water Pricing in OECD Countries*. Organisation for Economic Co-operation and Development, Environment Directorate, Paris.

Garrido, A. (2002) *Transition to Full-Cost Pricing of Irrigation Water for Agriculture in OECD Countries*. Organisation for Economic Co-operation and Development, Environment Directorate, Paris.

Garrido, R. (2005) Price setting for water use charges in Brazil. *International Journal of Water Resources Development* 21(1), 99–117.

Gibbons, D. (1986) *The Economic Value of Water*. Resources for the Future, Washington, DC.

Gittinger, J.P. (1982) *Economic Analysis of Agricultural Projects*, 2nd edn. EDI Series in Economic Development. World Bank, Washington, DC.

Gleick, P.H. (2000) The changing water paradigm: a look at twenty-first century water resources development. *Water International* 25(1), 127–138.

Gleick, P.H. (2001) Making every drop count. *Scientific American*, February 2001. Available at: http://ic.ucsc.edu/~pchuang/eart006/sciam%20freshwater%20conservation.pdf

Gollehon, N., Quinby, W. and Aillery, M. (2003) Water use and pricing in agriculture. In: Heimlich, R. (ed.) *Agricultural Resources and Environmental Indicators*, Agriculture Handbook No. (AH722). US Department of Agriculture (USDA), Economic Research Service, Washington, DC.

Gómez, M., López, M.J., Gutiérrez, C. and Pujol, J. (2005) Water framework directive and irrigation. In: Berbel, J. and Gutiérrez, C. (eds) *WADI – The Sustainability of European Irrigated Agriculture under Water Framework Directive and Agenda 2000*. European Commission, Joint Research Center. Institute of Prospective Technological Studies, Seville, Spain, pp. 26–38.

Gómez-Limón, J.A. and Riesgo, L. (2004a) Irrigation water pricing: differential impacts on irrigated farms. *Agricultural Economics* 31(2004), 47–66.

Gómez-Limón, J.A. and Riesgo, L. (2004b) Water pricing: analysis of differential impacts on heterogeneous farmers. *Water Resources Research* 40 W07S05.

Gómez-Limón, J.A. and Riesgo, L. (2005) Spain (2): the case of the Duero river. In: Berbel, J. and Gutiérrez, C. (eds) *WADI – The Sustainability of European Irrigated Agriculture under Water Framework Directive and Agenda 2000*. European Commission, Joint Research Center. Institute of Prospective Technological Studies, Seville, Spain, pp. 71–94.

Gopalakrishnan, C. (1973) The economics of water transfer. *American Journal of Economics and Sociology* 32(4), 395–403.

Goubert, J.P. (1986) *La Conquête de L'eau*. Robert Lafont Collection, Pluriel, Paris.

Green, C. (2003) *Handbook of Water Economics: Principles and Practices*. Wiley, Chichester, UK.

Green, G., Sunding, D., Zilberman, D. and Parker, D. (1996) Explaining irrigation technology choices: a microparameter approach. *American Journal of Agricultural Economics* 78(4), 1064–1072.

Green, G.P. and Sunding, D.L. (1997) Land allocation, soil quality, and the demand for irrigation technology. *Journal of Agricultural and Resource Economics* 22(2), 367–375.

Grimble, R.J. (1999) Economic instruments for improving water use efficiency: theory and practice. *Agricultural Water Management* 40(1), 77–82.

Guerra, L.C., Bhuiyan, S.I., Tuong, T.P. and Barker, R. (1998) *Producing More Rice with Less Water from Irrigated Systems*. SWIM Paper 5. International Irrigation Management Institute, Colombo, Sri Lanka, p. v.

Gulati, A. and Narayanan, S. (2002) Subsidies and reforms in Indian irrigation. In: Water policy reform: Lessons from Asia and Australia. Proceedings of an International Workshop held in Bangkok, 8–9 June 2001. *ACIAR Proceedings* 106, 171–176.

GWP-TAC (2000) *Integrated Water Resources Management*. TAC Background paper no. 4. Global Water Partnership, Stockholm.

Hadjigeorgalis, E. (1999) Trading under risk and uncertainty in an agricultural water market in Chile. *American Agricultural Economics Association*, 12.

Hamdane, A. (2002a) *Irrigation Water Pricing Policy in Tunisia*. FAO Regional Office for the Near East, Cairo, Egypt.

Hamdane, A. (2002b) The National Strategy for Irrigation Water Conservation: The Tunisian Case, Mimeo.

Hanemann, W.H. (2006) The economic conception of water. In: Rogers, P.P., Llamas, M.R. and Martinez-Cortina, L. (eds) *Water Crisis: Myth or Reality*? Taylor & Francis, London.

Hansen, S. and Bhatia, R. (2004) Water and poverty in a macro-economic context. Report to the Norwegian Ministry of the Environment, United Nations, New York.

Harris, A.H. (1994) *Agricultural Taxation under Structural Adjustment*. Economic and Social Development Paper No. 128. Food and Agriculture Organization, Rome.

Harris, L.M. (2002) Water and conflict geographies of the southeastern Anatolia Project. *Society and Natural Resources* 15, 743–759.

Hazell, P., Jaramillo, M. and Williamson, A. (1989) *How Has Instability in World Markets Affected Agricultural Export Producers in Developing Countries?* Working Paper. Agriculture and Rural Development Department, World Bank, Washington, DC.

Hill, G., del Moral, L., Paneque, P., Pedregal, B., Spash, S. and Urama, K. (2003) Evaluation practices in water project decision-making processes. ADVISOR Project, Workpackage 2.

Hodge, I.D. and Adams, W.M. (1997) Allocating water for production and rural conservation: choices and institutions. In: Holden, P. and Thobani, M. (eds) 1996. *Tradable Water Rights: A Property Rights Approach to Resolving Water Shortages and Promoting Investment*. Policy Research Dissemination Center, Working Paper 1627. World Bank, Washington, DC.

Holden, P. and Thobani, M. (1996) *Tradable Water Rights: A Property Rights Approach to Resolving Water Shortages and Promoting Investment*. Policy Research Dissemination Center, Working Paper 1627. World Bank, Washington, DC.

Horst, L. (1998) *The Dilemmas of Water Division*. IWMI and Wageningen Agricultural University, Colombo, Sri Lanka.

Howe, C.W. (1986) Innovations in water management: lessons from the Colorado-Big Thompson Project and the Northern Colorado Water Conservancy District, Chapter 6. In: Frederick, K.D. (ed.) *Scarce Water and Institutional Change*. Resources for the Future, Washington, DC.

Howe, C.W. (2003) The functions, impacts and effectiveness of water pricing: evidence from the United States and Canada, Mexico. In: *Water Pricing and Public Private Partnership in the Americas*. Inter-American Development Bank, Washington, DC, pp. 70–184.

Howe, C.W. and Goemans, C. (2003) Water transfers and their impacts: lessons from three Colorado water markets. *Journal of the American Water Resources Association*, October 2003. Available at: http://www.zoeticzone.com/p/articles/mi_qa4038/is_200310/ai_n9331547

Howe, C.W., Lazo, J.K. and Weber, K.R. (1990) The Economic impacts of agriculture-to-urban water transfers on the area of origin: a case study of the Arkansas river valley in Colorado. *American Journal of Agricultural Economics* 72(5), 1200–1206.

Hoyt, P.G. (1982) *Crop-Water Production Functions and Economic Implications for the Texas High Plains Region*. Economic Research Service Staff Report. AGES-820405, US Department of Agriculture, Washington, DC.

Hoyt, P.G. (1984) *Crop-Water Production Functions: Economic Implications for Colorado*. Economic Research Service, Staff Rep. AGES-840427, US Department of Agriculture, Washington, DC (cited in Scheierling *et al*., 2006).

Huang, J.K. and Rozelle, S. (2002) *Trade Reform, WTO and China's Food Economy in the 21st Century*. Working Paper 01-E5, CCAP, Chinese Academy of Sciences. Available at: http://www.ccap.org.cn

Huffaker, R. and Whittlesey, N. (2000) The allocative efficiency and conservation potential of water laws encouraging investments in on-farm irrigation technologies. *Agricultural Economics* 24, 47–60.

Huffaker, R. and Whittlesey, N. (2003) A theoretical analysis of economic incentive policies encouraging agricultural water conservation. *Water Resource Development* 19(1), 37–53.

Huffaker, R., Whittlesen, N., Michelsen, A., Taylor, R. and McGuckin, T. (1998) Evaluating the effectiveness of conservation water-pricing programs. *Journal of Agricultural and Resource Economics* 23, 12–19.

Huffaker, R., Whittlesey, N. and Hamilton, J.R. (2000) The role of prior appropriation in allocating water resources into the 21st century. *Water Resource Development* 16(2), 265–273.

Hurand, P. (2001) La gestion opérationnelle d'un système hydrographique complexe: le Système Neste. Compagnie d'Aménagement des Coteaux de Gascogne, Tarbes, France. Anailable at: http://www.cacg.fr/pages/publi/pdf/Systeme%20Neste.pdf

Hussain, I. (2005) *Pro-Poor Interventions Strategies in Irrigated Agriculture in Asia. Poverty in Irrigated Agriculture: Realities, Issues, and Options with Guidelines*. Report for the Asian Development Bank. IWMI, Colombo, Sri Lanka.

Hussain, I. and Wijerathna, D. (2004) Implications of alternate irrigation water charging policies for the poor farmers in developing Asia: a comparative analysis. Draft paper, IWMI-ADB Project, 47.

Hydrosult (1999) *China: Strategic Options for the Water Sector*. Report prepared under TA No. 2817-PRC funded by the Asian Development Bank. Montreal.

ICID (International Commission on Irrigation and Drainage) (2004) *Irrigation and Drainage Services. Some Principles and Issues towards Sustainability: An ICID Position Paper*. International Commission on Irrigation and Drainage, New Delhi.

IRN (International Rivers Network) (2003) *A Crisis of Mismanagement: Real Solutions to the World's Water Problems*. IRN, Berkeley, California.

Isaac, M. (2002) The political economy of water reform feasibility in Australia. Paper presented at the World Water Congress, Melbourne, January. School of Political Science and International Studies, University of Queensland, St. Lucia, Australia.

Jean, M. (1999) Politique de tarification et application pratique: L'exemple du Canal de Provence. Paper presented at the conference 'Pricing Water' – Lisboa, Portugal, 6 et 7 Septembre 1999.

Johansson, R.C. (2000) *Pricing Irrigation Water: A Literature Survey*. The World Bank, Washington, DC, p. 80.

Johansson, R.C., Tsur, Y., Roe, T. L., Doukkali, R. and Dinar, A. (2002) Pricing irrigation water: a review of theory and practice. *Water Policy* 4(2), 173–199.

Jones, W.I. (1995) *The World Bank and Irrigation*. Operations Evaluation Department, a World Bank Operation Evaluation Study. World Bank, Washington, DC, 150.

Jourdain, D. (2004) Impact des politiques visant à réduire la consommation brute en eau des systèmes irrigués: Le cas des puits gérés par des collectifs de producteurs au Mexique. Unpublished PhD thesis. University of Montpellier I, Montpellier, France.

Keller, A., Keller, J. and Seckler, D. (1996) *Integrated Water Resources Systems: Theory and Policy Implications*. Research Report 3. IWMI, Colombo, Sri Lanka, p. 15.

Kendy, E., Molden, D., Steenhuis, T.S. and Liu, C. (2003) *Policies Drain the North China Plain: Agricultural Policy and Groundwater Depletion in Luancheng County, 1949–2000*. IWMI Research Report 71. IWMI, Colombo, Sri Lanka, p. v.

Kenney, D. (2003) Water allocation and management in the western United States: an overview. Paper presented to the International Working Conference on Water Rights: Institutional options for improving water allocation. Hanoi, Vietnam, 12–15 February 2003.

Kenney, D.S. (2006) Prior appropriation and water rights reform in the western United States. In: Bruns, B.R., Ringler, C. and Meinzen-Dick, R. (eds) *Water Rights Reform: Lessons for Institutional Design*. IFPRI, Washington, DC, pp. 165–182.

Keshavarz, A., Ashrafi, S., Heydari, N., Pouran, M. and Farzaneh, E.A. (2005) Water allocation and pricing in agriculture of Iran. In: *Water Conservation, Reuse, and Recycling: Proceedings of an Iranian-American Workshop*, pp. 153–174. Committee on US-Iranian Workshop on Water Conservation, Reuse, and Recycling, Office for Central Europe and Eurasia Development, Security, and Cooperation, National Research Council. The National Academies Press, Washington, DC.

Kikuchi, M., Fujita, M. and Hayami, Y. (2001) State, community and market in the deterioration of a national irrigation system in the Philippines. In: Aoki, M. and Hayami, Y. (eds) *Communities and Markets in Economic Development*. Oxford University Press, Oxford, pp. 265–317.

Kislev, Y. (2001) The water economy in Israel. Prepared for the conference on 'water in the Jordon Valley: Technical Solution & Regional Cooperation'. University of Oklahoma, International Programs Center, Center for Peace Studies, Norman, Oklahoma, 13–14 November 2001.

Kisler, Y. (2005) Personal communication by email, 19 May, 2005.

Klein-Robbenhaar, J. (1996) Balancing efficiency and equity: determining the public welfare in surface water transfers from Acequia Communities. *Natural Resources Journal* 36(1), 37–58.

Kloezen, W.H. (2002) Accounting for water: institutional viability and impacts of market-oriented irrigation interventions in Central Mexico. PhD thesis, Wageningen University, Wageningen, The Netherlands.

Kloezen, W.H. and Garcés-Restrepo, C. (1998) Equity and water distribution in the context of irrigation management transfer: the case of the Alto Río Lerma Irrigation District, Mexico. In Boelens, R. and Dávila, G. (eds) *Searching for Equity: Conceptions of Justice and Equity in Peasant Irrigation*. Van Gorcum & Comp, Assen, The Netherlands, pp. 176–188.

Krueger, A., Schiff, M. and Valdés, A. (1991) *Political Economy of Agricultural Pricing Policy*. Johns Hopkins University Press, Baltimore, Maryland.

Kulshreshtha, S. (2002) Water pricing under joint benefits: a case study of accounting for positive externalities. *Water International* 27(1), 105–118.

Lam, W.F. (1996) Institutional design of public agencies and coproduction: a study of irrigation associations in Taiwan. *World Development* 24(6), 1039–1054.

Lankford, B.A. and Gowing, J. (1997) Providing a water delivery service through design management interactions and system management: achieving control for water pricing. In: Kay, M., Franks, T. and Smith, L. (eds) *Water: Economics, Management and Demand*. E & FN Spon, London, pp. 238–246.

Lanna, A.E. (2003) Uso de instrumentos econômicos na gestão das águas no Brasil. *Bahia Análise & Dados Salvador* 13, 441–451.

Latinopoulos, P. (2005) Valuation and pricing of irrigation water: an analysis in Greek agricultural areas. *Global NEST Journal* 7(3), 323–335.

Lee, P.S. (2000) *ICID Survey on Funding of Operation, Maintenance and Management of Irrigation and Drainage Systems*. ICID, New Delhi.

Lees, S.H. (1998) *The Political Ecology of the Water Crisis in Israel*. University Press of America Inc. Lanham, Maryland.

Li, Y.H. (2006) Water saving irrigation in China. *Irrigation and Drainage* 55(3), 327–336.

Libecap, G. (2003) *Transaction Costs and the Reallocation of Water Rights from Agriculture to Urban and Environmental Uses*. Working Paper. Karl Eller Center, University of Arizona, Tucson, Arizona.

Lichtenberg, E. (1989) Land quality, irrigation development, and cropping patterns in the northern high plains. *American Journal of Agricultural Economics* 71(1), 187–194.

Lipchin, C. (2003) Water, agriculture and Zionism: Exploring the interface between policy and ideology. Paper presented at the 3rd conference of the International Water History Association, December 2003, Alexandria, Egypt.

Lipton, M. (1977) *Why Poor People Stay Poor: Urban Bias in World Development*. Temple Smith, London.

Liu, B. (2003) Water rights in China. Paper presented to the International Working Conference on Water Rights: institutional options for improving water allocation. Hanoi, Vietnam, 12–15 February 2003.

Livingston, M.L. (1995) Designing water institutions: market failures and institutional response. *Water Resources Management* 9(3), 203–220.

Loeve, R., Lin, H., Bin, D., Mao, G., Chen, C.D., Dawe, D. and Barker, R. (2003) Long term trends in agricultural water productivity and intersectoral water allocations in Zhanghe, Hubei, China and in Kaifeng, Henan, China. In: Shiang-Kueen Hsu (ed.) *Sustainable Development of Water Resources and Management and Operation of Participatory Irrigation Organizations*, ICID, Taipei, Taiwan, pp. 367–380.

Loeve, R., Dong, B., Molden, David: Li, Y.H., Chen, C.D. and Wang, J.Z. (2004) Issues of scale in water productivity in the Zhanghe irrigation system: implications for irrigation in the basin context. *Paddy and Water Environment* 2, 227–236.

Louw, D.B. and Kassier, W.E. (2002) *The Costs and Benefits of Water Demand Management*. Centre for International Agricultural Marketing and Development, Paarl, South Africa.

Lundqvist, J. (1993) Rural water for expanding cities: a case study on water re-allocation in Tamil Nadu, India. In: Tingsanchali, T. (ed.) *Proceedings of the International Conference on Environmentally Sound Water Resources Utilization*, II-41–49. Bangkok, Thailand, 8–11 November 1993. Vol. 1. AIT, Bangkok, Thailand.

Lundqvist, J., Appasamy, P. and Nelliyat, P. (2003) Dimensions and approaches for Third World city water security. The Royal Society, UK, 19 No 2003. *Philosophical Transactions of the Royal Society of London, Series B* 358, 1985–1996.

Maass, A. and Anderson R.L. (1978) *And the Desert Shall Rejoice. Conflict, Growth, and Justice in Arid Environments*. The MIT Press, Cambridge, Massachusetts.

Maestu, J. (2001) The political economy of the implementation of changes in pricing practices in Spain. What can we learn? In: European Commission (ed.) *Pricing Water. Economics, Environment and Society*. Conference Proceedings, Sintra. European Commission, Brussels, pp. 247–267.

Malla, P.B. and Gopalakrishnan, C. (1995) Conservation effects of irrigation water supply pricing: a case study from Oahu, Hawaii. *International Journal of Water Resources Development* 11(3), 233–242.

Mariño, M. and Kemper, K.E. (1999) *Institutional Frameworks in Successful Water Markets*. World Bank Technical Paper No. 427. World Bank, Washington, DC.

Martínez, Y. and Albiac, J. (2004) Agricultural pollution control under Spanish and European environmental policies. *Water Resources Research* 40(10).

Martínez, Y. and Albiac, J. (2006) Nitrate pollution control under soil heterogeneity. *Land Use Policy* 23(2006), 521–532.

Mastrorilli, M., Corona, P. and de Seneen, G. (1997) Italy: the *Capitanata* irrigation scheme – experiences in water sustainability. In: *Workshop on the Sustainable Management of Water in Agriculture, The Athens Workshop, Case studies*. OCDE, Paris, pp. 99–108.

McIntosh, A.C. (2003) *Asian Water Supplies: Reaching the Urban Poor*. Asian Development Bank and International Water Association, Manila.

McKnight, T.L. (1983) Center pivot irrigation in California. *Geographical Review* 73(1), 1–14.

McNeill, D. (1998) Water as an economic good. *Natural Resources Forum* 22(4), 253–261.

Meinzen-Dick, R. and Appasamy, P.P. (2002) Urbanization and intersectoral competition for water. In: *Finding the Source: The Linkages between Population and Water*. Available at: http://wwics.si.edu/

Meinzen-Dick, R., Mendoza, M., Sadoulet, L., Abiad-Shields, G. and Subramanian, A. (1994) Sustainable water user associations: lessons from a literature review. Paper presented at World Bank Water Resources Seminar, Lansdowne, Virginia, 13–15 December 1994, p. x.

Merrett, S. (1997) *Introduction to the Economics of Water Resources: An International Perspective*. UCL Press, London.

Merrett, S. (2003) The urban market for farmers' rights. *Irrigation and Drainage* 52, 319–326.

Merrey, D.J. (1996) *Institutional Design Principles for Accountability in Large Irrigation Systems*. Research Report 8. International Irrigation Management Institute, Colombo, Sri Lanka.

Min, B.S. (2004) A water surcharge policy for river basin management in Korea: a means of resolving environmental conflict? *Water Policy* 6(2004), 365–380.

Moench, M., Dixit, A., Janakarajan, M., Rathore, S. and Mudrakartha, S. (2003) *The Fluid Mosaic: Water Governance in the Context of Variability, Uncertainty, and Change*. Nepal Water Conservation Foundation, Katmandu.

Molle, B. (2006) Technical issues involved in modernizing farm irrigation system: some experiences in Mediterranean countries. Paper presented to the International Symposium on Irrigation Modernisation: Constraints and Solutions, Damascus, Syria, 28–31 March 2006.

Molle, F. (2004) Technical and institutional responses to basin closure in the Chao Phraya river basin, Thailand. *Water International* 29(1), 70–80.

Molle, F. (2007) Why enough is never enough: the societal determinants of river basin closure. *International Journal of Water Resource Development*.

Molle, F. and Berkoff, J. (2006) *Cities versus Agriculture: Revisiting Intersectoral Water Transfers, Potential Gains and Conflicts*. IWMI Comprehensive Assessment Research Report 10. IWMI Comprehensive Assessment Secretariat, Colombo, Sri Lanka, pp. vi, 70. Available at: http://www.iwmi.cgiar.org/Assessment/files_new/publications/CA%20Research%20Reports/CARR10.pdf

Molle, F. and Turral, H. (2004) Demand management in a basin perspective: is the potential for water saving overestimated? Paper prepared for the International Water Demand Management Conference, Dead Sea, Jordan, June 2004, p. 13.

Molle, F., Mamanpoush, A. and Miranzadeh, M. (2004) *Robbing Yadullah's Water to Irrigate Saeid's Garden: Hydrology and Water Rights in a Village of Central Iran*. IWMI Research Report 80. IWMI, Colombo, Sri Lanka, p. vi.

Molle, F., Wester, P. and Hirsch, P. (2007) River basin development and management. In: Molden, D. (ed.) Chapter 16: *Water for Food-Water for Life*. Comprehensive Assessment of Water Management in Agriculture. EarthScan, London.

Mollinga, P.P., Hong, G. and Bhatia, A.M. (2005) Leadership and turnover: the contradictions of irrigation management reform in the People's Republic of China. In: Shivakoti, G.P., Vermillion, D., Lam, W.F., Ostrom, E., Yoder, R. and Pradhan, U. (eds) *Asian Irrigation in Transition: Rising Challenges and Responses*. Sage Publications, India.

Montginoul, M. and Rieu, T. (2001) Irrigation water pricing reforms and implementing procedures: experience acquired in Charente and in Morocco. In: Pricing water. Economics, environment and society. *Conference Proceedings, Sintra*, European Commission, Brussels, pp. 256–267.

Moore, M. (1989). The fruits and fallacies of neoliberalism: the case of irrigation policy. *World Development* 17(11), 1733–1750.

Moore, M.R., Gollehon, N.R. and Carey, M.B. (1994) Multicrop production decisions in western irrigated agriculture: the role of water price. *American Journal of Agricultural Economics* 76, 859–874.

Moreno, G. and Sunding, D.L. (2000) Irrigation technology investment when the price of water is stochastic. *American Agricultural Economics Association* – 2000 Selected Paper, 2000 Annual Meeting, 30 July–2 August, Tampa, Florida, p. 13.

Moreno, G. and Sunding, D.L. (2005) Joint estimation of technology adoption and land allocation with implications for the design of conservation policy. *American Journal of Agricultural Economics* 87(4), 1009–1019.

Morris, J. (1996) Water policy: economic theory and political reality. In: Howsam, P. and Carter, R.C. (eds) *Water Policy: Allocation and Management in Practice*, pp. 228–234.

Morris, J., Weatherhead, E., Knox, J., Vasileiou, K., Vries, T., Freeman, D., Leiva, F. and Twite, C. (2005) The case of England and Wales. In: Berbel, J. and Gutiérrez, C. (eds) *WADI – The Sustainability of European Irrigated Agriculture under Water Framework Directive and Agenda 2000*. European Commission, Joint Research Center, Institute of Prospective Technological Studies, Seville, Spain, pp. 145–172.

Moynier, J.L. (2006) *Devenir des Exploitations Irriguées de Poitou-Charentes*. ARVALIS, Institut du vegetal, Saint Pierre d'Amilly, France.

Munasinghe, M. (1990) The pricing of water services in developing countries. *Natural Resources Forum* 14, 193–209.

Musgrave, W.F. (1997) Decentralized mechanisms and institutions for managing water resource: reflecting on experiences from Australia. In: Parker, D. and Tsur, Y. (eds) *Decentralization and Coordination of Water Resource Management*. Kluwer Academic, Boston, Massachusetts, pp. 428–439.

Nair, C. (1991) Bangkok's deteriorating groundwater. *Waterlines* 9(3), 21–24.

Negri, D.H. and Brooks, D.H. (1990) The determinants of irrigation technology choice. *Western Journal of Agricultural Economics* 15(2), 213–223.

Nickum, J.E. (1998) Is China living on water margin. *The China Quarterly* 880–910.

Nicol, J.-P. (2001) Irrigation water pricing at BRL. *Proceedings of the Conference: Economic Instruments and Water Policies in Central and Eastern Europe, Szentendre, September 2000*. The Regional Environmental Center for Central and Eastern Europe, Szentendre, Hungary.

Nieswiadomy, M. (1985) The demand for irrigation water in the High Plains of Texas, 1957–1980. *American Journal of Agricultural Economics* 67(3), 619–626.

Nyoni, M.J.M. (1999) Water pricing issues in a developing country: case of Zimbabwe. *International Journal of Social Economics* 26(1), 441–454.

OECD (Organisation for Economic Co-operation and Development) (2003) *OECD Economic Surveys 2002–2003*. OECD, Mexico, Paris.
Ofrecio, B.P. (2005) Participatory development and management: a cornerstone of Philippine irrigation program. Country paper presented at the 2005 Tsukuba Asian Seminar on Agricultural Education held on 8–14 November 2005. Agricultural and Forestry Research Center, University of Tsukuba, Ibaraki Prefecture, Japan.
Ogg, C.W. and Gollehon, N.R. (1989) Western irrigation response to pumping costs: a water demand analysis using climatic regions. *Water Resources Research* 25(5), 767–773.
Okun, A.M. (1975) *Equality and Efficiency: The Big Tradeoff*. The Brookings Institution, Washington, DC.
Olubode-Awosola, O.O., Idowu, E.O. and Van Schalkwyk, H.D. (2006) Assessing irrigation projects performance for sustainable irrigation policy reform. *Irrigation and Drainage Systems* 20, 303–315.
O'Mara, G.T. (1990) *Making Bank Irrigation Investments More Sustainable* (It is time to rationalize policy guidelines on Bank irrigation projects). Agriculture and Rural Development Department. Working Paper. World Bank, Washington, DC.
Onjala, J. (2001) *Water Pricing Options in Kenya: Cases of Mwea and West Kano Irrigation Schemes*. CDR Working Paper 01.9. Centre for Development Research, Copenhagen.
Oorthuizen, J. (2003) *Water, Works, and Wages: The Everyday Politics of Irrigation Management Reform in the Philippines*. Wageningen University Water Resources Series. Orient Longman, Hyderabad, India.
Oorthuizen, J. and Kloezen, W.H. (1995) The other side of the coin: a case study on the impact of financial autonomy on irrigation management performance in the Philippines. *Irrigation and Drainage Systems* 9, 15–37.
Ostrom, E. (1996) Crossing the great divide: coproduction, synergy, and development. *World Development* 24(6), 1073–1087.
Özlü, H. (2004) Participatory irrigation management (PIM) in Turkey. Paper presented at the 7th International PIM Seminar, 13–18 June 2004, Tirana, Albania.
Page, J. (2001) China's sinking city highlights water crisis. 19 January 2001. Reuters News Service. Available at: www.planetark.com
Parthasarathy, R. (1999) *Political Economy of Irrigation Water Pricing*. GIDR Working Paper No. 109. Gujarat Institute of Development Research, Ahmedabad, India, p. iii.
Pérez Prado, L.N. (2003) Mexico's Irrigation management transfer program: notes for a policy research agenda. *Journal of Environment and Development* 12(4), 373–388.
Perry, C.J. (1996) *Alternative to Cost Sharing for Water Service to Agriculture in Egypt*. IIMI Research No. 2. International Irrigation Management Institute, Colombo, Sri Lanka.
Perry, C.J. (1999) The IWMI water resources paradigm: definitions and implications. *Agricultural Water Management* 40(1), 45–50.
Perry, C.J. (2001a) Water at any price? Issues and options in charging for irrigation water. *Irrigation and Drainage* 50(1), 1–7.
Perry, C.J. (2001b) *Charging for Irrigation Water: The Issues and Options, with a Case Study from Iran*. Research Report No. 52. IWMI, Colombo, Sri Lanka.
Pfefferli, S. and Zimmermann, A. (1997) Switzerland: Swiss agri-environmental policy and water quality. *OECD Workshop on the Sustainable Management of Water in Agriculture: Issues and Policies*. Directorate for Food, Agriculture and Fisheries and Environment Directorate, Athens, 3–6 November.
Pigram, J.J. (1999) Economic instruments in the management of Australia's water resources: a critical review. *Water Resources Development* 4, 493–509.
Pingali, P.L. (2004) Agricultural diversification: opportunities and constraints. Paper presented to the conference 'Rice in Global Markets and Sustainable Production Systems,' 12–13 February 2004. FAO, Rome.
Pingali, P.L. and Rosegrant, M. (1995) Agricultural commercialization and diversification: processes and policies. *Food Policy* 20(3), 171–185.
Pinheiro, A.C. and Saraiva, J. (2005) The case of Portugal. In: Berbel, J. and Gutiérrez, C. (eds) *WADI – The Sustainability of European Irrigated Agriculture under Water Framework Directive and Agenda 2000*. European Commission, Joint Research Center, Institute of Prospective Technological Studies, Seville, Spain, pp. 173–200.
Pitman, G.K. (2002) *Bridging Troubled Water: Assessing the World Bank Water Resource Strategy*. World Bank Operations Evaluation Department. World Bank, Washington, DC.
Plantey, J., Tardieu, H., Mesny, M., Nicol, J.P., Rieu, T. and Verdier, J. (1996) Gestion de l'eau pour l'agriculture en France: Durabilité socio-économique et implication des usagers. Paper prepared for the Conférence Euro-Méditerrannéenne sur la Gestion Locale de l'Eau, 26 November 1996, Marseille, France.

Plaut, S. (2000) *Water Policy in Israel*. Institute for Advanced Strategic and Political Studies, Washington, DC.

Plusquellec, H. (2002) *How Design, Management and Policy Affect the Performance of Irrigation Projects*. FAO, Bangkok.

Postel, S. (1992) *Last Oasis: Facing Water Scarcity*. W.W. Norton, New York.

Postel, S. (1999) *Pillars of Sand*. W.W. Norton, New York.

Postel, S. (2001) Growing More Food with Less Water. *Scientific American*, February.

Prasad, K. and Rao, P.K. (1991) On irrigation water pricing in India. *Water Resources Development* 7(4), 274–280.

Préfol, B., Tardieu, H., Vidal, A., Fernandez, S., Plantey, J. and Darghouth, S. (2006) Public – private partnership in irrigation and drainage: need for a professional third party between farmers and government. *Irrigation and Drainage* 55(3), 253–264.

Quiroz, J.A. and Valdés, A. (1995) Agricultural diversification and policy reform. *Food Policy* 20(3), 245–255.

Ramakrishnan, N. (2002) Chennai water lorry operators on the back foot. *The Hindu Business Line*. 1 August 2002.

Rao, P.K. (1984) Comment on cost recovery and irrigation water pricing. *ODI Irrigation Management Network Paper*.

Rao, P.P., Birthal, P.S., Joshi, P.K. and Kar, D. (2004) *Agricultural Diversification in India and Role of Urbanization*. Markets, Trade and Institutions Division. International Food Policy Research Institute, Washington, DC.

Rap, E. (2004) *The Success of a Policy Model: Irrigation Management Transfer in Mexico*. PhD thesis, University of Wageningen, Wageningen, The Netherlands.

Reidinger, R. (1994) Observations on water markets for irrigation systems. In: Le Moigne, G., Easter, K.W., Ochs, W.J. and Giltner, S. (eds) *Water Policy and Water Markets*. Technical Paper No. 249. World Bank, Washington, DC, pp. 65–78.

Reisner, M. (1986) *Cadillac Desert: The American West and Its Disappearing Water*. Penguin, New York.

Repetto, R. (1986) *Skimming the Water: Rent Seeking and the Performance of Public Irrigation Systems*. Research Report 4. World Resource Institute, Washington, DC.

Richard, A. (2001) *La gestion des eaux du système Litani-Awali par l'Office national du Litani – Etat des lieux et préétude d'une modélisation pour une aide à la décision*. ENGREF, Montpellier, France.

Rieu, T. (2005) Water pricing for agriculture between cost recovery and water conservation: *where do we stand in France? OECD Workshop on Agriculture and Water: Sustainability, Markets and Policies, 14–18 November 2005*. OECD, Adelaide, South Australia.

Rinaudo, J.D. (2002) Corruption and allocation of water: the case of public irrigation in Pakistan. *Water Policy* 4(2002), 405–422.

Ringler, C., Rosegrant, M.W. and Paisner, M.S. (2000) *Irrigation and Water Resources in Latin America and the Caribbean: Challenges and Strategies*. EPTD Discussion Paper No. 64. Environment and Production Technology Division, International Food Policy Research Institute, Washington, DC.

Ringler, C., Vu, H.N., Cai, X. and Linh, H.G. (2002) Water allocation policies for the Dong Nai river basin in Viet Nam: an integrated perspective. Paper presented at the conference on irrigation water policies: micro and macro considerations, 15–17 June 2002, Agadir, Morocco.

Rodgers, C. and Hellegers, P. (2005) *Water Pricing and Valuation in Indonesia: Case Study of the Brantas River Basin*. EPT Discussion Paper 141. International Food Policy Research Institute, Washington, DC.

Rogers, P., Bhatia, R. and Huber, A. (1998) *Water as a Social and Economic Good: How to Put the Principle into Practice*. TAC Background Paper No. 2. Global Water Partnership, SIDA, Sweden.

Rosegrant, M.W. (1997) *Water Resources in the Twenty-First Century: Challenges and Implications for Action*. Food, Agriculture, and the Environment Discussion Paper 20. International Food Policy Research Institute, Washington, DC.

Rosegrant, M.W. and Binswanger, H.P. (1994) Markets in tradable water rights: potential for efficiency gains in developing-country water resource allocation. *World Development* 22(11), 1613–1625.

Rosegrant, M.W. and Cai, X. (2002) Water prices, environment, and food security. *International Conference on Irrigation Water Policies: Micro and Macro Considerations, Agadir, Morocco*. 15–17 June.

Rosegrant, M.W. and Cline, S. (2002) The politics and economics of water pricing in developing countries. *Water Resources Impact* 4(1), 6–8.

Rosegrant, M.W., Gazmuri Schleyer, R. and Yadav, S.N. (1995) Water policy for efficient agricultural diversification: Market-based approaches. *Food Policy* 20(3), 203–223.

Roth, E. (2001) *Water Pricing in the EU: A Review*. European Environmental Bureau, Brussels.
Saleth, R.M. (1997) India. In: Dinar, A. and Subramanian, A. (eds) *Water Pricing Experiences: An International Experience*. World Bank Technical Paper No. 386. World Bank, Washington, DC, pp. 54–60.
Saleth, R.M. (2001) *Water Pricing: Potential and Problems*. Policy Brief Focus 9. International Food Policy Research Institute, Washington, DC.
Samabuddhi, K. (2005) Mishandling of crisis leads to serious conflict. Bangkok Post, 19 July 2005.
Samal, C.K. and Kolanu, T. (2004) Water pricing and decentralized irrigation management in Andhra Pradesh – schism between objectives and realities. Paper presented at the *Deutscher Tropentag*, October 2004, Berlin.
Sampat, P. (2000) *Deep Trouble: The Hidden Threat of Groundwater Pollution*. Worldwatch Paper 154. Worldwatch Institute, Washington, DC.
Sampath, R.K. (1992) Issues in irrigation pricing in developing countries. *World Development* 20(7), 967–977.
Sarker, A. and Itoh, T. (2001) Design principles in long-enduring institutions of Japanese irrigation common-pool resources. *Agricultural Water Management* 48, 89–102.
Sarker, R., Meilke, K. and Hoy, M. (1993) The political economy of systematic government intervention in agriculture. *Canadian Journal of Agricultural Economics* 41, 289–309.
Savenije, H.H.G. (2001) *Why Water Is Not an Ordinary Economic Good*. Value of Water Research Report Series No. 9. IHE, Delft, The Netherlands.
Savenije, H.H.G. and van der Zaag, P. (2002) Water as an economic good and demand management: paradigms and pitfalls. *Water International* 27(1), 98–104.
Scheierling, S.M., Young, R.A. and Cardon, G.E. (2004) Determining the price-responsiveness of demands for irrigation water deliveries versus consumptive use. *Journal of Agricultural and Resource Economics* 29(2), 328.
Scheierling, S.M., Young, R.A. and Cardon, G.E. (2006a) Public subsidies for water-conserving irrigation investments: hydrologic, agronomic and economic assessment. *Water Resources Research* 42.
Scheierling, S.M., Loomis, J.B. and Young, R.A. (2006b) Irrigation water demand: a meta-analysis of price elasticities. *Water Resources Research* 42.
Schiff, M. and Valdés, A. (1992) The Plundering of Agriculture in Developing Countries. The World Bank, Washington, DC.
Schuck, E.C. and Green, G.P. (2001) Field attributes, water pricing, and irrigation technology adoption. *Journal of Soil and Water Conservation* 56(4), 293–298.
Schuck, E.C., Frasier, W.M., Webb, R.S., Ellingson, L.J. and Umberger, W.J. (2005) Adoption of more technically efficient irrigation systems as a drought response. *Water Resources Development* 21(4), 651–662.
Seagraves, J.A. and Easter, K.W. (1983) Pricing irrigation water in developing countries. *Water Resources Bulletin* 4, 663–671.
Sehring, J. (2005) *Water User Associations (WUAs) in Kyrgyzstan. A Case Study on Institutional Reform in Local Irrigation Management*. Working Paper No. 24. Zentrum für internationale Entwicklungs- und Umweltforschung (ZEU), Giessen, Germany.
Shah, T. (1993) *Groundwater Markets and Irrigation Development. Political Economy and Practical Policy*. Oxford University Press, New Delhi.
Shevah, Y. and Kohen, G. (1997) Economic considerations for water used in irrigation in Israel. In: Kay, M., Franks, T. and Smith, L. (eds) *Water: Economics, Management and Demand*. E & FN Spon, London, pp. 29–36.
Shiklomanov, I. (2000) World water resources and water use: assessment and outlook for 2025, World Water Scenarios, Analyses, World Water Council. Draft.
Shrestha, R.B. and Gopalakrishnan, C. (1993) Adoption and diffusion of drip irrigation technology: an econometric analysis. *Economic Development and Cultural Change* 41(2), 407–418.
Siamwalla, A. and Roche, F. (2001) Irrigation management under resource scarcity. In: Ammar, S. (ed.) *The Evolving Roles of the State, Private, and Local Actors in Rural Asia. Study of Rural Asia*. Oxford University Press, Hong Kong, pp. 183–212.
SIDA (Sindh Irrigation and Drainage Authority) (2003) Institutional reforms in the water sector. Available at: http://www.sida.org.pk/reforms.htm
Simon, P. (1998) *Tapped Out: The Coming World Crisis in Water and What We Can Do About It*. Welcome Rain, New York.
Simpson, L. and Ringskog, K. (1997) *Water Markets in the Americas. Directions in Development Report*. World Bank Group, Washington, DC.

Siriluck, S. and Kammeier, H.D. (2003) Government policy and farmers' decision making: the agricultural diversification programme for the Chao Phraya river basin, 1993–2000. In: Molle, F. and Srijantr, T. (eds) *Thailand's Rice Bowl: Perspectives on Social and Agricultural Change in the Chao Phraya Delta*. White Lotus, Bangkok, pp. 63–96.

SIWI (Stockholm International Water Institute) (2004) Report on the seminar on financing water infrastructure. Stockholm, 15 August 2004. Organized by the World Bank and the Stockholm International Water Institute. Available at: http://www.siwi.org/downloads/www-Symp/2004_www_SIWI_World_Bank_Seminar_Final_Report.pdf

Skaggs, R.K. (2001) Predicting drip irrigation use and adoption in a desert region. *Agricultural Water Management* 51(2), 125–142.

Small, L. (1987) *Irrigation Services Fees in Asia*. ODI paper 87/1c, London.

Small, L.E. (1990) Financial tools for improving irrigation performance. In: Sampath, R.K. and Young, R.A. (eds) *Social, Economic, and Institutional Issues in Third World Irrigation Management*. Westview Press, Boulder, Colorado, pp. 147–268.

Small, L.E. and Carruthers, I. (1991) *Farmer-Financed Irrigation: The Economics of Reform*. Published in association with the International Irrigation Management Institute. Cambridge University Press, Cambridge.

Small, L.E., Adriano, M.S. and Martin, E.D. (1986) Regional study on irrigation service fees: final report. Report submitted to the Asian Development Bank. International Irrigation Management Institute, Colombo, Sri Lanka.

Smith, L.E.D., Franks, T. and Melvyn, J. (1997) Water – an economic good? Theory and practice. *ICID Journal* 46(2), 1–13.

Socratous, G. (2000) *Water Pricing in Cyprus. Symposium Water Resources Management. Efficiency, Equity and Policy, University of Cyprus, Nicosia-Cyprus, 22–24 September, 2000*. Ministry of Agriculture, Natural Resources and Environment, European Commission DG, Brussels, XII.

Solanes, M. and Jouravlev, A. (2006) Water rights and water markets: lessons from technical advisoru assistance in Latina America. *Irrigation and Drainage* 55(3), 337–342.

Speck, S. and Strosser, P. (2000) Fifty points for a synthesis. In: The Regional Environmental Center for Central and Eastern Europe (ed.) *The Proceedings of the Conference 'Economic Instruments and Water Policies in Central and Eastern Europe'*. The Regional Environmental Center for Central and Eastern Europe, Szentendre, Hungary.

Spencer, G. and Subramanian, A.M. (1997) Capital cost recovery: water user organizations and irrigation operation and maintenance: financials aspects. In: Dinar, A. and Subramanian, A. (eds) *Water Pricing Experiences an International Experience*. World Bank Technical Paper no. 386. The World Bank, Washington, DC, pp. 154–159.

Squire, L. and van der Tak, H.G. (1976) *Economic Analysis of Projects*. Johns Hopkins University Press, Baltimore, Maryland.

Sumpsi Viñas, J.M. (1998) Efectos de las políticas tarifarías sobre la demanda de agua, renta agraria y recuperación de costes de la agricultura de regadío en España. Paper presented to the 'I Congreso Ibérico sobre Gestión y Planificación Aguas.' Zaragoza, Spain, 14–18 September 1998.

Sunding, D. (2005) The economics of agricultural water use and the role of prices. In: The National Academies (ed.) *Water Conservation, Reuse, and Recycling: Proceedings of an Iranian – American Workshop*. Committee on US–Iranian Workshop on Water Conservation, Reuse, and Recycling. Office for Central Europe and Eurasia Development, Security, and Cooperation, National Research Council. The National Academies Press, Washington, DC, pp. 120–152.

Sur, M. and Umali-Deininger, D. (2003) The equity consequences of public irrigation investments: the case of surface irrigation subsidies on India. *Proceedings of the 25th International Conference of Agricultural Economists, 16–22 August 2003*. Durban, South Africa.

Svendsen, M. and Rosegrant, M.W. (1994) Irrigation development in Southeast Asia beyond 2000: will the future be like in the past? *Water International* 19, 25–35.

Svendsen, M., Trava, J. and Johnson III, S.H. (1997) *Participatory Irrigation Management: Benefits and Second Generation Problems. Lessons from an International Workshop Held at Centro Internacional de Agricultura Tropical (CIAT) Cali, Colombia, 9–15 February, 1997*. Economic Development Institute of the World Bank and International Institute for the Management of Irrigation, Washington, DC.

Swyngedouw, E. (2003) Flows of power: the political ecology of water and urbanization. The case of Guayaquil, Ecuador (1890–1990). Paper prepared for the 3rd Conference of the International Water History Association, Alexandria, Egypt, 11–14 December 2003.

Tanaka, Y. and Sato, Y. (2003) An institutional case study of Japanese Water Users Association: towards successful participatory irrigation management. *Paddy Water Environment* 1, 85–90.

Tardieu, H. and Préfol, B. (2002) Full cost or 'sustainability cost' pricing in irrigated agriculture: charging for water can be effective, but is it sufficient? *Irrigation and Drainage* 51(2), 97–107.
Tarimo, A.K.P.R., Mdoe, N.S. and Lutatina, J.M. (1998) Irrigation water prices for farmer-managed irrigation systems in Tanzania: a case study of Lower Moshi irrigation scheme. *Agricultural Water Management* 38(1), 33–44.
TDRI (Thailand Development Research Institute) (1990) *Water Shortages: Managing Demand to Expand Supply*. Thailand Development Research Institute, Bangkok, 101.
Thakkar, H. (2000) *Assessment of Irrigation in India*. South Asia Network on Dams, Rivers and People, India.
The Economist (2003) Priceless. Available at: http://www.economist.com/displaystory.cfm?story_id=1906846
Thobani, M. (1997) Formal water markets: why, when and how to introduce tradable water rights. *The World Bank Research Observer* 12(2).
Tiwari, D.N. and Dinar, A. (2001) *Role and Use of Economic Incentives in Irrigated Agriculture*. Working Paper. World Bank, Washington, DC.
Tizaoui, C. (2004) Promotion de l'irrigation localisée dans le périmètre irrigué de la basse Moulouya au Maroc. Projet INCO-WADEMED, *Proceedings of the seminar 'Modernisation de l'Agriculture Irriguée'*, Rabat, Morocco, 19–23 April 2004.
Tsiourtis, N.X. (2002) Water charge, the Cyprus experience. In: Hamdy, A., Lacirignola, C. and Lamaddalena, N. (eds) *Water Valuation and Cost Recovery Mechanisms in the Developing Countries of the Mediterranean Region*. CIHEAM-IAMB, Bari, 2002, pp. 91–104. (Options Méditerranéennes: Série A. Séminaires Méditerranéens No. 49.)
Tsur, J. (2004) Introduction to special section on irrigation water pricing. *Water Resource Research* 40.
Tsur, Y. (2005) Economic aspects of irrigation water pricing. *Canadian Water Resources Journal* 30(1), 31–46.
Tsur, Y. and Dinar, A. (1995) *Efficiency and Equity Considerations in Pricing and Allocating Irrigation Water*. Policy Research Working Paper 1460. The World Bank, Washington, DC.
Tsur, Y. and Dinar, A. (1997) The relative efficiency and implementation costs of alternative methods for pricing irrigation water. *The World Bank Economic Review* 11(2), 243–262.
Turral, H.N., Etchells, T., Malano, H.M.M., Wijedasa, H.A., Taylor, P., McMahon, T.A.M. and Austin, N. (2004) Water trading at the margin: the evolution of water markets in the Murray Darling Basin. *Water Resources Research* 41(7).
UNEP (United Nations Environment Programme) (2000) Economic instruments for environmental management. A worldwide compendium of cases studies.
UNESCAP (United Nations Economic and Social Commission for Asia-Pacific) (1996) Guidelines for the establishment of irrigation. Environment and Sustainable Development Division (ESDD). ESCAP, Bangkok.
UNESCO (2003) *Water for People Water for Life*. World Water Assessment Programme. UNESCO, Paris. Berghahn Books, Barcelona.
Ünver, O. and Gupta, R.K. (2003) Water pricing: issues and options in Turkey. *International Journal of Water Resources Development* 19(2), 311–330.
USDA (United States Department of Agriculture) (2001) The road ahead: agricultural policy reform in the WTO. In: Burfisher, M.E. (ed.) *USDA Agricultural Economic Report No. 797*.
Vaidyanathan, A. (Chair) (1992) Report of the Committee on Pricing of Irrigation Water. Planning Commission, Government of India, New Delhi. Reproduced in the *Indian Journal of Agricultural Economics* 49(1), 107–133, Jan–March 1994.
Varela-Ortega, C., Sumpsi, J.M., Garrido, A., Blanco, M. and Iglesias, E. (1998) Water pricing policies, public decision making and farmers' response: implication for water policy. *Agricultural Economics* 19, 193–202.
Vermillion, D.L. (1997) *Impacts of Irrigation Management Transfer: A Review of the Evidence*. IIMI Research Report 11. International Irrigation Management Institute, Colombo, Sri Lanka, p. v.
Vermillion, D.L. and Garcés-Restrepo, C. (1998) *Impacts of Colombia's Current Irrigation Management Transfer Program*. Research Report 25. IWMI, Colombo, Sri Lanka.
Vermillion, D.L. and Sagardoy, J.A. (1999) *Transfer of Irrigation Management Services: Guidelines*. FAO Irrigation and Drainage Paper 58. FAO, Rome, Italy, p. x.
Vos, J.M.C. (2002) Metrics matters: the performance and organization of volumetric water control in large-scale irrigation in the north coast of Peru. PhD thesis. Wageningen University, Wageningen, The Netherlands.

Wade, R. (1982) The system of administrative and political corruption: canal irrigation in South India. *Journal of Development Studies* 18(3), 287–327.

Wahl, R.W. (1993) *Water Marketing in California: Past Experience, Future Prospects*. Reason Public Policy Institute, Los Angeles.

Wichelns, D. (2003) *Experience in Implementing Economic Incentives to Conserve Water and Improve Environmental Quality in the Broadview Water District, California*. World Bank, Washington, DC.

Wierenga, P.J. and Hendrickx, J.M.H. (1985) Yield and quality of trickle irrigated chili. *Agricultural Water Management* 9(4), 339–356.

Willardson, L.S., Allen, R.G. and Frederiksen, H.D. (1994) *Universal Fractions and the Elimination of Irrigation Efficiencies*. 13th Technical Conference. USCID, Denver, Colorado, 19–22 October 1994. Question 47, Irrigation planning and management measures in harmony with the environment.

Winpenny, J. (1994) *Managing Water as an Economic Resource*. Development Policies Studies. Routledge and Overseas Development Institute, London.

Winpenny, J.T. (1997) Demand management for efficient and equitable use. In: Kay, M., Franks, T. and Smith, L. (eds) *Water: Economics, Management and Demand*. E & FN Spon, London, pp. 296–303.

Wolfensohn, J.D. (2000) Remarks at the Second World Water Forum: 'From Vision to Action.' The Hague, 22 March 2000.

World Bank (1984) Financing operation and maintenance in Irrigation. Agriculture and Rural Development Department, World Bank, Washington, DC.

World Bank (1986) *World Bank Lending Conditionality: A Review of Cost Recovery in Irrigation Projects*. Report No 6283. Operations Evaluation Department, World Bank, Washington, DC.

World Bank (1988) *Rural Diversification in Asia. Agriculture and Rural Development*. Working Paper. World Bank, Washington, DC.

World Bank (1992) *Philippines: Irrigated Agriculture Sector Review*. Report No 9848-PH. World Bank, Washington, DC.

World Bank (1993) *Water Resources Management: A World Bank Policy Paper*. World Bank, Washington, DC.

World Bank (1997) *Nepal Irrigation Sector Project: Project Appraisal Document*. Report No. 17104-NEP NEPAL. World Bank, Washington, DC.

World Bank (2003a) International trade in agriculture: a developing country perspective. World Bank, Washington, DC. Available at: http://siteresources.worldbank.org/INTARD/214576-1112347900561/20424230/agtr.pdf

World Bank (2003b) *World Bank Water Resources Sector Strategy: Strategic Directions for World Bank Engagement*. World Bank, Washington, DC.

World Bank (2005a) *Shaping the Future of Water for Agriculture. A Sourcebook for Investment in Agricultural Water Management*. Agriculture and Rural Development, World Bank, Washington, DC.

World Bank (2005b) Priority issues for Indonesian agriculture. Policy briefs. January 2005. Available at: http://www.worldbank.or.id

World Bank (2005c) *Reengaging in Agricultural Water Management: Challenges, Opportunities, and Trade-offs*. Water for Food Team, Agriculture and Rural Development Department (ARD). World Bank, Washington, DC.

Wright, S. and Mallia, C. (2003) The potential of Eco-Taxes as instruments for sustainability. An analysis of the critical design elements. *The Journal of Transdisciplinary Environmental Studies* 2(2), 14.

WTO (World Trade Organization) (2000) Agriculture agreement, Geneva. Available at: http://www.wto.org/english/tratop_e.htm

WTO (2001) WTO agriculture negotiations: Proposal by Jordan. Committee on Agriculture Special Session. Available at: http://www.wto.org/English/tratop_e/agric_e/ngw140_e.doc

WWF (2002) Pricing as a Tool to Reduce Water Demand. WWF-Spain/ADENA's 'Alcobendas: water city for the 21st century' – a demonstration project. Available at: http://www.panda.org/downloads/europe/pricing2.pdf

Yang, H., Zhang, X. and Zehnder, A.J.B. (2003) Water scarcity, pricing mechanism and institutional reform in northern China irrigated agriculture. *Agricultural Water Management* 61(2003), 143–161.

Yercan, M. (2003) Management turning-over and participatory management of irrigation schemes: a case study of the Gediz river basin in Turkey. *Agricultural Water Management* 62(2003), 205–214.

Young, R.A. (1986) Why are there so few transactions among water users? *American Journal of Agricultural Economics* 68(5), 1143–1151.

Young, R.A. (1994) *Water resources: economics and policy*. Background paper for the Food and Agriculture Organization, Rome.

Young, R.A. (1996) Water Economics. In: Mays, L. (ed.) *Handbook of Water Resources*. McGraw-Hill, New York.

Young, T.F. and Karkoski, J. (2000) Green evolution: are economic incentives the next step in nonpoint source pollution control? *Water Policy* 2(2000), 151–173.

Zilberman, D., Dinar, A., MacDougall, K., Khanna, M., Brown, C. and Castillo, F. (1992) *Individual and Institutional Responses to the Drought: The Case of California Agriculture*. ERS Staff Paper. US Department of Agriculture, Washington, DC.

3 Why Is Agricultural Water Demand Unresponsive at Low Price Ranges?

C. de Fraiture and C.J. Perry

Introduction

With growing populations, increasing standards of living and growing concern for environmental issues, claims on water resources are intensifying. Competition between sectors is increasing and water allocation mechanisms currently in place, such as fixed allocations or rationing, may no longer be adequate. At the World Water Forum 2000, a large international conference, the majority of the international water community called for reforms in water allocation mechanisms (Cosgrove and Rijsberman, 2000). Proposed reforms relate especially urgently to agriculture. Worldwide, 70–80% of all developed water resources is used for agricultural production. In arid countries where rainfall is insufficient for rain-fed agriculture, this percentage may be as high as 90% (Gleick, 1998). Water use in agriculture is often heavily subsidized and trade in water is limited. Several studies report problems related to water scarcity and resources overexploitation in the USA, India, Pakistan, China, the Middle East and the Soviet Republics (Postel, 1999; Seckler *et al.*, 2000; Rosegrant *et al.*, 2002). They foresee that these problems will only intensify and spread to more regions in the near future, unless adequate action is undertaken to reform prevailing water management practices.

Economic incentives and mechanisms, such as water pricing and introduction of water markets, are often proposed as efficient and effective measures in demand management. According to Perry (2001), the three most common reasons for recommending water charges are:

- To recover the cost of providing water delivery service;
- To provide an incentive for efficient use of scarce water resources;
- As a benefit tax on those receiving water services, to provide potential resources for further investment to the benefit of others in society.

Cost recovery and tax purposes can be achieved through area- or crop-based pricing. These charging mechanisms are generally preferred to volumetric pricing because they are easier and cheaper to implement. To provide an incentive for more efficient use, charges must be a direct function of consumption.

Underpricing may lead to inefficient use of scarce water resources, and the introduction of volumetric water pricing may reduce water wastage and generate revenue to continue essential services in the future (Briscoe, 1996; Rosegrant, 1997; Huffaker *et al.*, 1998; Kumar and Singh, 2001). 'Getting the prices right', i.e. reflecting the economic and social value of the resource, is a desirable way to allocate water efficiently (Dinar and Subramanian, 1997; Johansson, 2000).

But it is debatable if volumetric pricing is an *effective* measure in water demand management. The development of the required institutional and physical infrastructure, lacking in many places, is a costly process. Externalities in water use, caused by recycling of drainage water, may render pricing less effective in reducing water use than foreseen by planners (Seckler, 1996). Perry (1997, 2001) shows that, in Egypt and Iran, costs of pricing to farmers and society outweigh projected benefits. Ray (2002) examines the implicit assumptions under which market forces can induce more efficient water use. She concludes that for India these assumptions are violated and that enforceable and transparent allocation rules may be more effective to curtail water demand. Molle (2001) reaches similar conclusions for Thailand. For the Middle East, Ahmad (2000) predicts that in the absence of well-defined water rights, economic measures may lead to higher water use rather than conservation of water.

Others argue that, especially in developing countries, there are millions of indirect beneficiaries such as the consumers who benefit as much as, or even more than, the direct beneficiaries of irrigation (i.e. farmers). It is therefore unjust to expect the farmers to bear the full burden. They argue that the cost of irrigation development should be legitimately shared by both consumers and producers (Sampath, 1983, 1992; Rhodes and Sampath, 1988).

Finally, several researchers claim that irrigation water demand is inelastic below a threshold price, and elastic beyond it (Varela-Ortega *et al.*, 1998; OECD, 1999). To induce a reduction in demand, considerable price increases are required (either in the general level of charging or through more complex multilevel charges). Political considerations may prevent such price increases (Perry, 2001; Ray, 2002).

For their analysis of policy impacts, economists rely on observed prices and market transactions to infer the value of a particular good. Commonly, the demand curve – as the basis of quantitative economic analyses – is determined through econometric curve fitting techniques using field data. This 'direct' approach is difficult in the analysis of water demand in agriculture. The price of water is only rarely determined in the market. Consequently, the value of water needs to be derived from modelling, starting from production functions and setting up the farmer's optimization problem. Examples of this analytical approach are found in Dinar and Letey, 1996; Rosegrant *et al.*, 2001.

Many analytical studies implicitly assume an ideal situation, free of price distortions and externalities. But the introduction of volumetric water charges as a demand management tool does not happen in a void. Water management practices already in place prior to the introduction of pricing have an important bearing on its effectiveness as a demand management tool. In this chapter two factors are explored: (i) the impact of technology; and (ii) the impact of prevailing rationing regimes.

The remainder of this chapter is organized as follows: the second section explores the impact of technology choice, application efficiency and scale; the third section examines the consequences of rationing; and the last section provides the conclusions and discussion.

Impact of Technology

Gardner (1983, cited in Ray, 2002) states that if water prices rise to reflect its opportunity cost, a rational farmer will have any or all of the four following responses: the farmer demands less water and leaves land fallow; applies less water to the crop accepting some yield loss; switches to less water-demanding crops; and/or invests in more efficient irrigation techniques. Literature provides evidence that farmers respond in all these ways. Examples are found in Ray and Williams (1999) for India; Bernardo and Whittlesey (1989) for Washington State; Hoyt (1984) for Texas; Berbel and Gomez-Limon (2000) for Spain; and Ogg and Gollegon (1989) and Weinberg *et al.* (1993) for the western USA.

The reduction in water use intended by more efficient irrigation depends to a large extent on the water application technology and its potential to substitute water for other inputs. Varela-Ortega *et al.* (1998) compare the price elasticity of water demand in three regions in Spain. They conclude that in the 'old' irrigation schemes where water application techniques are relatively inefficient, the response to increasing water charges is much higher than in the modern systems with drip systems. The authors conclude that the technical endowment in an agricultural district has a major effect on its response to water pricing.

Broadly speaking, three categories of application technology can be distinguished: surface, sprinkler and drip. The most capital-extensive but water- and labour-intensive technique is surface irrigation. Generally, sprinkler irrigation uses less water but requires more capital. Lastly, drip irrigation typically uses the least amount of water and labour but is the most capital-intensive technique.

Where water price is low, a rational farmer will substitute relatively expensive inputs – such as capital and labour – for cheap water.[1] For example, instead of manually weeding paddy fields, labour input is reduced by maintaining a water layer on the field to suppress weed growth, at the expense of additional water to cover evaporation and percolation losses. Conversely, where water charges are high, it may be cost-effective to invest in field canal lining to reduce seepage losses.[2] For each technology, the substitution potential, i.e. the scope of water savings through increased labour and capital input, differs. It is typically highest in surface irrigation. In drip irrigation systems, where water application efficiency is already high compared to surface systems, the scope of water savings is limited and comes at a relatively high incremental cost.

Theoretically, water pricing may impact both technology choice and the level of substitution. With increasing water charges, a farmer will operate the existing technique in a more water-efficient manner, until it becomes cost-effective to switch to a more advanced application technique using less water.

Technology choice

Empirical evidence, however, shows that technology choice is hardly driven by water price. It is mainly determined by structural factors, agronomic conditions and financial constraints (see Molle and Berkoff, Chapter 2, this volume). For example, on sloping fields the use of sprinklers may be more appropriate than flood irrigation which requires levelling. For reasons of erosion control and better fertilizer application, a farmer may opt for furrows or drip. Favourable subsidy schemes may induce a switch to drip because it gives higher yields per hectare, reduces labour input and is less prone to salinity problems. Lack of spare parts, knowledge and credit may prohibit the use of advanced technologies as sprinkler and drip. Crop choice may limit technology choice: tuber crops are best grown on furrows while cereals cannot be grown under sprinkler or drip. Caswell and Zilberman (1986) and Caswell *et al.* (1990) in their studies on California demonstrate that while the probability of drip irrigation adoption increases with higher prices, land quality and environmental considerations play a more prominent role. Green and Sunding (1997) find that technology choice primarily depends on land quality and crop choice. Varela-Ortega *et al.* (1998) arrive at similar conclusions for three irrigation systems in Spain. Hoyt (1984) notes that, in Texas, only dramatic price increases will induce capital investment in better technology.

Level of substitution

Within each application category, water can be substituted for capital and/or labour. For example, within the category of surface

[1] The potential of fertilizer and pesticides as substitute for water seems limited, at best. There is evidence that at some parts of the demand curve they behave as complements rather than substitutes.

[2] For the individual farmer this may lead to savings, but not necessarily at system level.

irrigation the most labour-extensive application is to simply flood the field, resulting in high water losses. Water application can be reduced dramatically at the expense of extra labour by field levelling, constructing bunds, using furrows or increasing the intensity of monitoring field conditions. Likewise, a labour-extensive way to operate a sprinkler system is to use a timing device so that the sprinklers are turned on at regular intervals. But this does not account for the rainfall that may occur during these intervals, and irrigation water may be lost. More water-efficient, but more capital-intensive, is to install moisture probes to determine the right time to sprinkle, based on actual water needs. This method does not account for rainfall that may occur in the days following irrigation. Even more efficient in terms of water use, but more capital-intensive, is a computerized system that uses actual water needs and weather forecast information.

There are clear limits to substitution. Below a certain point it is no longer possible or desirable to use more water to replace capital and labour. Too much water will damage crops, create erosion problems, cause waterlogging and flush away fertilizer. Consequently, there is a maximum amount of water a farmer will take, even if abundant water is available at zero cost. As a result, at low water prices water demand is not determined by price but by agronomic- and technique-related factors and water use is unresponsive to price. With the introduction of water pricing as a demand management tool, water use becomes elastic only beyond a certain threshold. The size of the threshold depends on initial water management practices and the substitutability of water for other inputs. The model developed in the following paragraphs explores the impact of these factors on water demand at low price ranges.

Demand curves

The water requirements of a crop depend on physical factors, such as climate, soils and crop characteristics. In general, the more the soil moisture is available to the crop, the higher the crop yield, up to a certain limit. At low water application rates an additional unit of water results in a substantial yield increase but the marginal product of water quickly declines at higher water levels. Beyond a certain level of water application crop yields suffer due to lack of aeration in the root zone. At that point, the marginal product of water becomes negative. A polynomial functional form, best captures the physical relationship between crop growth and soil moisture. Hargreaves (1977) proposes a cubic form. Following Dinar and Letey (1996) and Rosegrant *et al.* (2001), a quadratic functional form is adopted here:

$$Y_p = \beta_0 + \beta_1 W_c + \beta_2 W_c^2 \quad (3.1a)$$

$$Y_c = Y_r \cdot Y_p \quad (3.1b)$$

Where, Y_r stands for relative crop yield, Y_p is potential yield, Y_c is crop yield, βs are regression coefficients and W is the amount of crop evapotranspiration. The crop production function depends on crop characteristics, soil and climate and is unique for each crop and location. This is reflected by the intercept β_0. In the representation given by the equation (3.1a and 3.1b) inputs other than water (e.g. agrochemicals) are kept constant at an optimum level.

The variable W represents the amount of crop water evaporation. To get this amount to the plants it needs to be conveyed from source to fields and applied in the right quantities at the right time. The irrigation efficiency indicates the extent of water losses occurring in conveyance and application. Application efficiency at field level is defined as the amount of water beneficially used by crops (W) divided by the total amount diverted to the field (TotWat).

$$\text{Eff} = \frac{W}{\text{TotWat}} \quad (3.2)$$

Confronted with rising water charges, a farmer can reduce total water diversion by reducing the water layer on the field (W) through the adoption of deficiency irrigation or switching to a less water-demanding

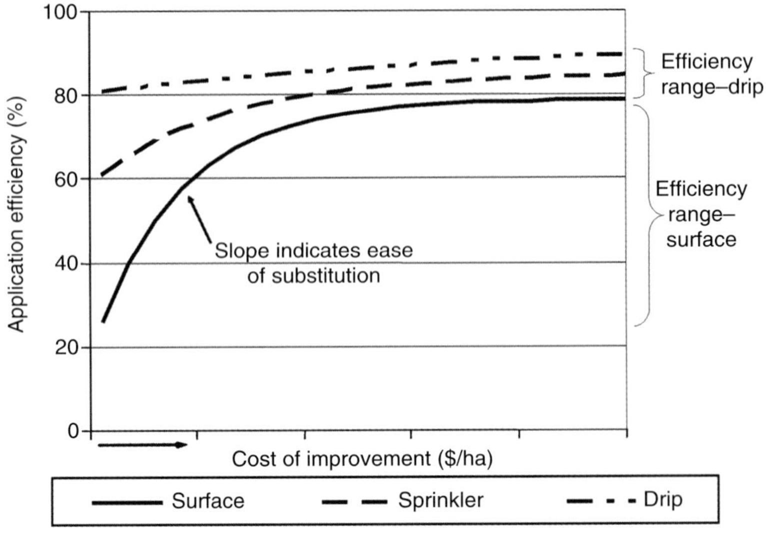

Fig. 3.1. Application efficiency cost curves.

crop.[3] Alternatively, a farmer can improve application efficiency (Eff) by substituting labour and/or capital for water, or, ultimately, leave land fallow. As explained above, for agronomic and technical reasons there is an upper limit to the amount of water a farmer takes, independent of price. Thus, water is applied with a minimum efficiency. An application efficiency of say 10% is undesirable because the large amount of water to meet crop water requirements (*W*) will cause problems as erosion, fertilizer loss, waterlogging and crop damage.

Figure 3.1 depicts the relation between application efficiency and cost of improvement for different technologies. The exact shape of these curves is site- and crop-specific and largely unknown. Three features are important for the discussion here. First, the curves do not intersect the *y*-axis at zero. In other words, an efficiency of zero does not exist and the minimum is well above zero. Second, additional labour/capital input exhibits a diminishing return. Third, the upper and lower bounds differ by technology. Efficiency in surface irrigation exhibits the widest range, while drip irrigation has the narrowest scope.

When these elements are incorporated in a simple farmer optimization model, the water demand curve reveals three zones (Fig. 3.2). At low ranges, price is not a determining factor in decisions related to technology choice and application efficiency and water demand is unresponsive to price. With increasing prices, the farmer may opt to slightly reduce the water layer on the field but because this will directly affect crop yields, demand is inelastic. Beyond a certain threshold, demand becomes elastic. At higher price ranges, demand becomes inelastic again, as water quantities approach the minimum amount needed for plant growth.

Price threshold

Several studies conclude that water demand becomes elastic only beyond a certain price threshold (Varela-Ortega *et al.*, 1998; OECD, 1999). Where prevailing prices are low relative to the threshold price, a considerable price increase is necessary to induce the desired reduction in demand. Political considerations may prevent such price increases

[3]Often crop choice is limited, due to climatic factors, the absence of marketing infrastructure, diet preferences and risks associated with other (cash) crops.

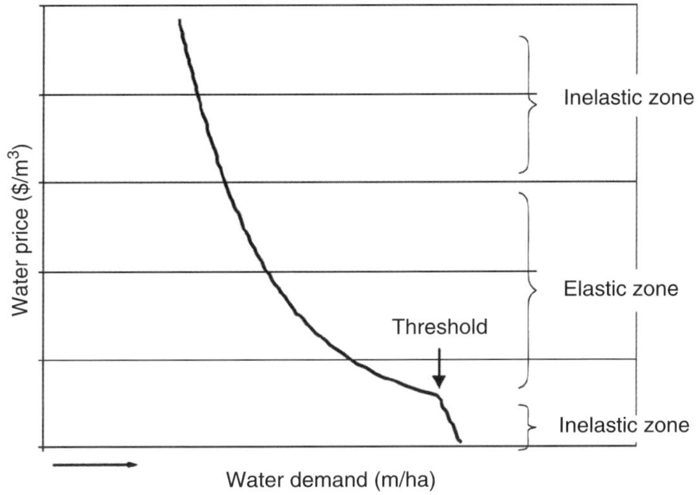

Fig. 3.2. Demand curve.

(Perry, 2001). To gauge the effectiveness of pricing as a demand management tool, it is thus essential to investigate the importance of the price threshold.

In the following paragraphs the sensitivity of technology on the threshold value is examined, using a numerical example using crop data from California. Crop production parameters are adapted from Dinar and Letey, 1996 and summarized in Table 3.1.

Little is known about prevailing application efficiencies and associated cost curves. This example, therefore, explores a wide range of values of substitutability, scope of improvement and initial efficiencies. Figure 3.3 presents a family of cost curves for an application technology of which the application efficiency ranges from 25% to 80%. That is, if farmers are free to take the amount of water they desire free of cost, they will choose to operate the system at 25% efficiency. The lowest curve represents a situation where efficiency improvements come at a high cost: $500/ha to increase efficiency from 25% to 50% (for comparison in this example, maximum crop revenue is $2500/ha). The 'high substitutability' curve indicates a low marginal cost of efficiency improvement: $150/ha to increase efficiency from 25% to 80%. Figure 3.4 depicts the resulting water demand curves. Water demand is elastic and thresholds are low and of minor importance, even in case of low substitutability of water.

Table 3.1. Crop data used in the numerical example. (From Dinar and Letey, 1996.)

Crop: cotton		Location: California	
Parameters			
β0	−0.13	Crop price ($/t)	1600
β1	2.30	Potential yield (t/ha)	1.7
β2	−1.20		

Note: In this table and throughout the book $ means US$.

The situation changes dramatically if the initial efficiency is set at 40% instead of 25% (Fig. 3.5). The dotted lines in Fig. 3.5 depict that part of the demand curve which is suppressed because of the high initial efficiency. The threshold level varies from negligible to considerable, depending on the ease of substitution. Figure 3.6 shows the family of demand curves for a technique whose scope of improvement is relatively limited (efficiency ranging from 60% to 80%). In this case, water demand is inelastic, unless the substitution of water comes at a very low cost.

This analysis makes clear that the threshold value depends on three interrelated

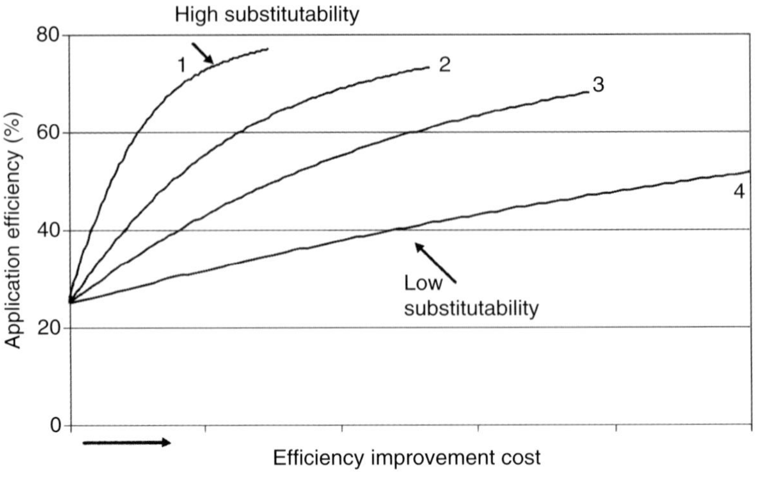

Fig. 3.3. Efficiency improvement cost curves (efficiency range 25–80%).

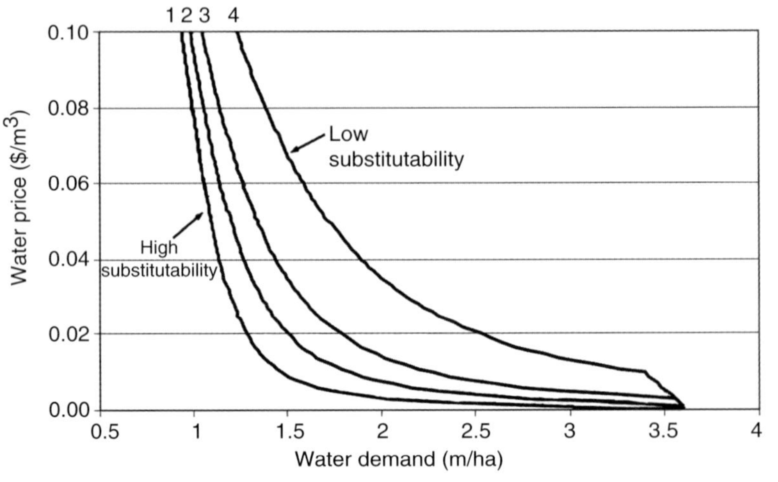

Fig. 3.4. Demand curves for efficiency range 25–80%.

factors, namely the prevailing application efficiency, the scope of efficiency improvement and the ease of substitution. These factors are, to a large extent, determined by technology choice and existing on-field water management practices, which are mostly unrelated to water price.

In this example, when the application efficiency is 25%, water demand is fairly elastic at low prices, even if efficiency improvements come at a relatively high cost. On the other hand, if the existing efficiency is 40% or 60%, reduction of water demand may require a substantial price increase depending on the ease of substitution.

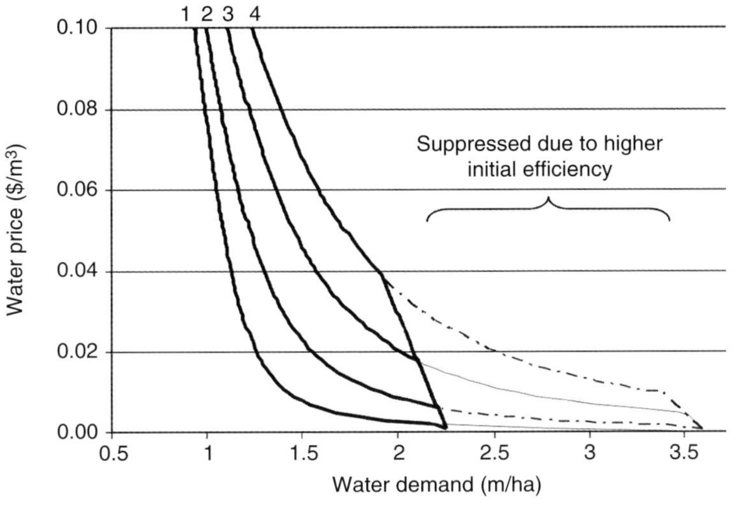

Fig. 3.5. Demand curves for efficiency range 40–80%.

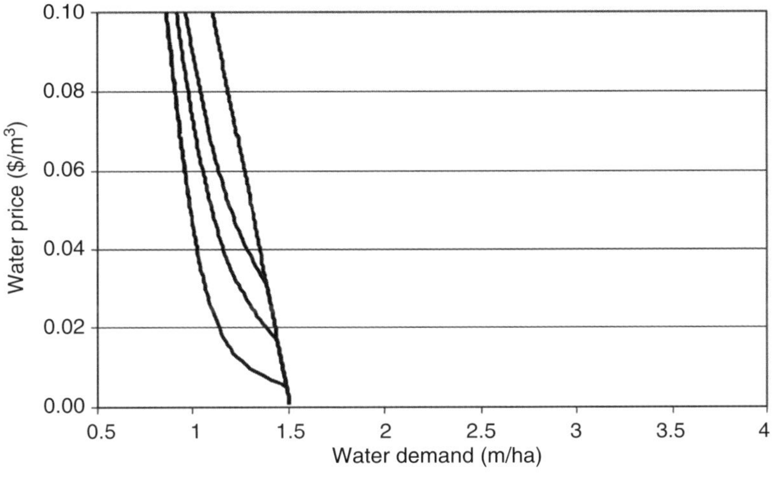

Fig. 3.6. Demand curves for efficiency range 60–80%.

Costs of water reduction

The existence of an inelastic section of the demand curve at low prices, or the lack thereof, has major implications for the cost of water reduction to farmers. Figure 3.7 shows the relation between water reduction and cost of water for the demand curves depicted in Fig. 3.5. Water reduction is expressed as a percentage of the maximum quantity demanded under price zero (i.e. 2.25 m/ha). Water costs, expressed as a percentage of total crop revenue, include water charges plus the costs of efficiency improvement.

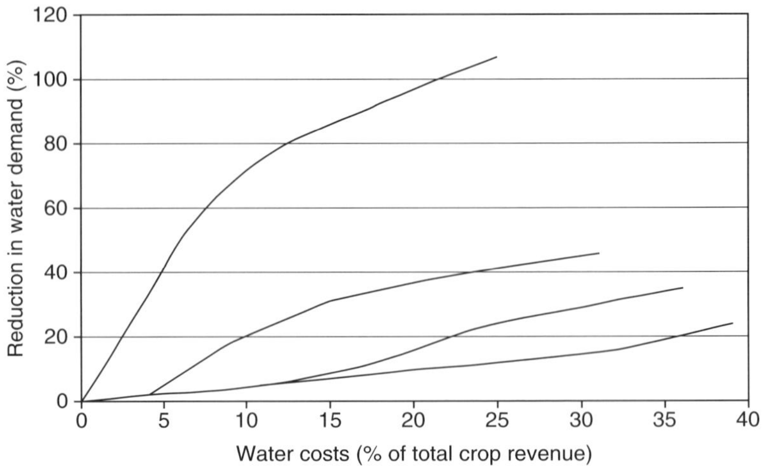

Fig. 3.7. Cost of water reduction.

Unless the ease of substitution is high, considerable impacts on farm income are implicit for using water pricing as a means to limit demand. Empirical evidence supports this finding. Perry (1997) estimates for Egypt that inducing a 15% reduction in water demand through volumetric pricing would decrease farm incomes by 25%. Berbel and Gomez-Limon (2000) estimate that farm income in Spain will decrease by 40% before water demand decreases significantly. Bernardo and Whittlesey (1989) and Hoyt (1984) conclude that in the Washington State and Texas farmers substitute water with labour, by switching to a more water-efficient mode of operation. But to induce these water savings by pricing (as opposed to restricting supply) results in a significant income loss to farmers and painful adjustments as some farmers may have to stop irrigating.

In countries where low-income farmers make up a large part of the voting population, pricing may not be a feasible demand management option from a social and political point of view.

Scaling up

Volumetric water pricing in agriculture is geared towards influencing water use behaviour of individual farmers. The aggregated impact of pricing at a scale larger than a farm may be governed by different processes and scaling up the impacts of pricing by aggregating individual responses may lead to erroneous conclusions.

Efficiency of water use is a scale-dependent concept. From a river basin perspective, drainage water from 'inefficient' farms is not necessarily lost, but can be reused by downstream users, water quality allowing (Seckler, 1996). Molden *et al.* (2000) show that, for Egypt, farm-level efficiency is as low as 40%, but overall basin efficiency is 90%. This implies that 90% of all diverted water is beneficially used for crop growth. Water 'wastage' is negligible and the scope for water savings, induced by pricing or other measures, is very small.

Although field efficiency is low, return flows from 'inefficient' users may be reused by downstream farmers, either by recapturing drainage flows or by pumping excess seepage. Pricing induces upstream farmers to use water more efficiently and thus create less return flows. Downstream farmers have to divert more water to compensate for this loss. Consequently, at the aggregate level of river basins, the reduction of water diversions as a result of pricing may be less than foreseen (Perry, 2001). A proper assessment of the impact of water pricing at basin scale requires a knowledge of hydrological interaction between users.

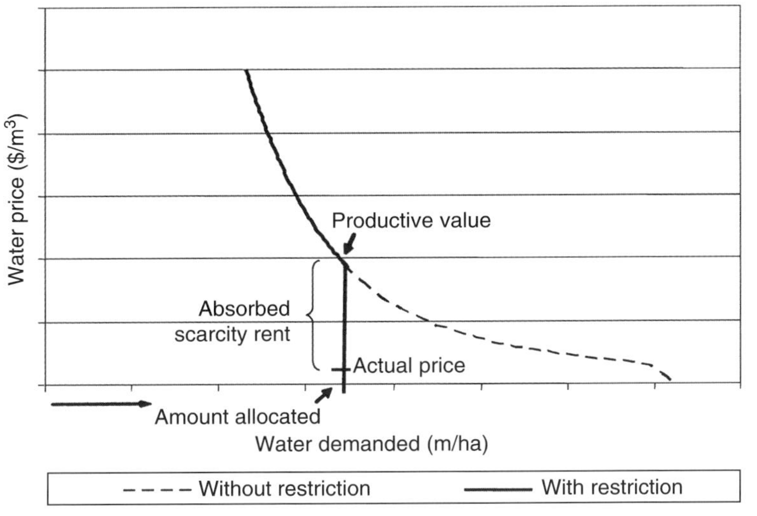

Fig. 3.8. Water demand and use under restricted supply.

Impact of Existing Rationing

In many parts of the world, farmers are not free to take the amount of water they prefer. Farmers' access to water is bounded by water rights or by fixed allocations. Also the size of canals, inlets or pipes may limit the amount of water a farmer can take (this could be called technological rationing as opposed to institutional rationing).

Where water is scarce and water prices low, the amount allocated is likely below the 'free market' amount (i.e. the amount of water that farmers would be willing to take at the prevailing price). A good example of an allocation mechanism in water-scarce areas is *warabandi*, which is practised on a large scale (over millions of hectares) in irrigation schemes in India and Pakistan. The system is designed to provide a rationed and equitable service (in proportion to landholdings) to all farmers under conditions of extreme water scarcity. Instead of planning for full irrigation of a small part of the area, the available water is spread over a large number of farms, thus giving farmers a choice between fully irrigating part of their land with water-intensive crops, or irrigating a larger area of less water-intensive crops, or deliberately underirrigating a still larger area. This approach encourages maximum output per unit of water, rather than maximum output per unit of land (Bandaragoda, 1998).

Figure 3.8 depicts the relation between water price, demand and actual use. The dotted line represents the demand curve. The solid line shows the actual use.

At low prices water use is constrained by rationing. Farmers optimize water use by choosing an appropriate crop, level of risk and efficiency according to its limited availability, independent of price. Consequently, water use is unresponsive to price. At a certain threshold, pricing becomes effective in reducing demand. This is the point where price equals the productive value of an additional unit of water (price equals marginal product).

If the price of water is set below the threshold and the maximum allocation is still in place, farmers start 'paying off the absorbed scarcity rent'. In other words, water diversions remain constant but farmer profit suffers substantially.[4] If the rationing system is fully replaced by water pricing allocation, and the price is set below the threshold, farmers will divert *more* water, until the gap between actual price and productive value is bridged.

These observations imply that where irrigation water is currently rationed, the

[4]Society as a whole may benefit depending on how water revenues are invested.

introduction of water pricing as a demand management tool is effective, *only if* the price is set above a certain threshold, i.e. the productive value of the last unit allocated under the rationing scheme. Depending on the initial water price and the size of the allocation, this threshold may be several times the original price. The lower the price actually paid and the more binding the existing allocation to farmers, the bigger is the gap between price and productive value. For Iran, Perry (2001) estimates that the productive value of water is $0.04, while the farmers at present pay $0.004. To induce water savings by pricing, a tenfold increase is required. Ray (2002) in her study on water pricing in India shows that in order to induce the water-conserving response under existing allocation practices, a sixfold price increase would be needed. She adds that under the prevailing political circumstances in India, this is very unlikely.

Conclusions and Discussion

The price of water is only rarely determined in the market. Consequently, models are needed to derive demand as a function of price. Many analytical studies implicitly assume an ideal situation, free of price distortions and externalities. But the introduction of volumetric water charges as a demand management tool does not happen in a void. Water management practices already in place before the introduction of pricing have an important bearing on its effectiveness as a demand management tool. This chapter explores the impact of technology choice, application efficiency and prevailing rationing practices on water demand elasticity.

At low water prices, farmers' decisions concerning technology choice and water use primarily depend on crop choice, land quality, agronomic considerations and structural factors (e.g. availability of capital and labour). Where water is restricted – either by institutional rationing or limits imposed by technology – farmers optimize water according to its limited availability. At prevailing (low) prices, the amount of water diverted is independent of price and water demand is unresponsive to price. It is only beyond a certain threshold that demand becomes responsive to price.

When prevailing prices are low relative to the threshold, considerable increases are necessary to induce the desired reduction in demand. Political considerations may prevent such increases. To gauge the effectiveness and feasibility of pricing as a demand management tool, it is crucial to investigate the importance of the threshold.

Where water is rationed, the threshold level mainly depends on the size of the allocation relative to the 'free market' amount (i.e. the amount of water farmers would be willing to take at prevailing prices). In water-scarce areas with low prevailing prices and very restrictive allocations, the required increase may be prohibitive.

The analysis presented in this chapter reveals that, where water is freely available, the threshold value depends on three interrelated factors: (i) the field application efficiency prior to the introduction of pricing as a demand management tool; (ii) the scope of efficiency improvement; and (iii) the ease of substitution (i.e. the marginal costs of efficiency improvement). These factors are, to a large extent, determined by technology choice and existing on-field water management practices, which are mostly unrelated to water price. When prevailing application efficiencies are low, say around 25%, demand is fairly elastic at low prices, even if efficiency improvements come at a relatively high cost. On the other hand, if the existing efficiency is 40% or 60%, reduction of water demand may require a substantial price increase, depending on the costs of substitution. This may lead to considerable income losses to farmers.

Although this conclusion may seem obvious, the implications are by no means trivial. Reliable information on field application efficiencies is not available, except for local case studies often implemented in an experimental set-up. Estimates are typically based on common perceptions and rules of thumb rather than on measurements. In this context, it is important to distinguish between field application and irrigation efficiency. The latter is substantially lower than the former because it includes conveyance and operational losses in the main irrigation system.[5] System losses are beyond the control of individual farmers, and thus unresponsive to water pricing charged to individual farmers. In large irrigation schemes, system losses may be more important than those occurring at the field level. Without reliable estimates on field application efficiencies prior to the introduction of pricing, its effectiveness as a demand management tool remains subject to personal judgements and opinions.

This issue is further complicated due to the scale dependency of irrigation efficiency. From a river basin perspective, drainage water from 'inefficient' farms is not necessarily lost, but can be reused by downstream users – water quality allowing. Pricing induces upstream farmers to use water more efficiently and thus create less return flows. Downstream farmers have to divert more water to compensate for this loss. Consequently, at the aggregate level of river basins, the reduction of water diversions as a result of pricing may be less than foreseen. A proper analysis of the impacts of water charges requires consideration beyond the individual farm level.

Results of this analysis depend on the model formulation, its underlying assumptions and parameter values. The model uses total seasonal demand curves without accounting for short-term rainfall variability. There may be short periods of zero responsiveness (after rain) or short periods of high elasticities (after unseasonal drought). The analysis here neglects these and provides an 'average' picture over the entire growing period. Further, the analysis is based on crop data for cotton in California. A sensitivity analysis revealed that as long as the crop production function is polynomial (with a clear maximum), the resulting form of the demand curves (with a threshold) does not change. The efficiency cost functions are assumed for want of data. The wide range of values tested most likely cover all plausible parameter values. The conclusions of this analysis are independent of the exact functional form of the efficiency function as long as efficiency has a clear upper and lower bound and the minimum efficiency is greater than zero.

The analysis in this chapter focuses on the impact of water management practices existing prior to the introduction of pricing. It does not include several potentially important factors influencing effectiveness of pricing, such as uncertainty in water supply (Perry and Narayamurthy, 1998), risk due to fluctuations in revenue (Bontemps *et al.*, 2001) and difficulties related to implementation (Tsur, 2000; Molle, 2001; Perry, 2001). The inclusions of these factors, which are considered outside the scope of this chapter, will improve the analysis but may not significantly affect its conclusions.

[5] $E_{ir} = E_{ap} \cdot E_{sys}$ where E_{ir} is irrigation efficiency, E_{ap} field application efficiency, and E_{sys} is system efficiency (conveyance and operational losses in main system). A rule of thumb for surface irrigation commonly used by irrigation engineers: E_{ap} = 50–70%, E_{sys} = 60–80%, resulting in an overall E_{ir} = 30–55%.

References

Ahmad, M. (2000) Water pricing and markets in the Near East: Policy issues and options. *Water Policy* 2, 229–242.

Bandaragoda, D.J. (1998) *Design and Practice of Water Allocation Rules: Lessons from Warabandi in Pakistan's Punjab*. Research Report 17. International Water Management Institute, Colombo.

Berbel, J. and Gomez-Limon, J.A. (2000) The impact of water pricing policy in Spain: an analysis of three irrigated areas. *Agricultural Water Management* 43, 219–238.

Bernardo, D. and Whittlesey, N. (1989) *Factor Demand in Irrigated Agriculture under Conditions of Restricted Water Supplies*. Economic Research Technical Bulletin no. 1765. US Department of Agriculture, USA.

Bontemps, C., Couture, S. and Favard, P. (2001) *Is the Water Demand Curve Really Convex?* FEEM Series 82,2001. Fondazione Eni Enrico Mattei, Italy. Available at: http://feem.it/web/activ/_activ.html

Briscoe, J. (1996) Water as an economic good: the idea and what it means in practice. Paper presented at the World Congress of the International Commission on Irrigation and Drainage. Cairo, Egypt.
Caswell, M.F. and Zilberman, D. (1986) The effects of well depth and land quality on the choice of irrigation technology. *American Journal of Agricultural Economics* 68, 798–811.
Caswell, M.F., Lichtenberg, M. and Zilberman, D. (1990) The effects of pricing policies on water conservation and drainage. *American Journal of Agricultural Economics* 72(4), 883–890.
Cosgrove, W. and Rijsberman, F. (2000) *World Water Vision: Making Water Everybody's Business*. Earthscan Publishers, London.
Dinar, A. and Letey, J. (1996) *Modeling Economic Management and Policy Issues of Water in Irrigated Agriculture*. Praeger Publishers, Westport, Connecticut.
Dinar, A. and Subramanian, A. (1997) *Water Pricing Experiences: An International Perspective*. World Bank Technical Paper No. 386. World Bank, Washington, DC.
Gleick, P.H. (1998) *The World's Water 1998–1999: The Biennial Report on Freshwater Resources*. Island Press, Washington, DC.
Green, G.P. and Sunding, L. (1997) Land allocation, soil quality and the demand for irrigation technology. *Journal of Agricultural and Resource Economics* 22, 367–375.
Hargreaves, G. (1977) *Water Requirements Manual for Irrigated Crops and Rainfed Agriculture*. Utah State University, Logon, Utah.
Hoyt, P.G. (1984) *Crop Water Production Functions and Economic Implication for Colorado*. US Dept. of Agriculture. Economic Research Service, Washington, DC.
Huffaker, R., Whittlesen, N., Michelsen, A., Taylor, R. and McGuckin, T. (1998) Evaluating the effectiveness of conservation water pricing programs. *Journal of Agricultural and Resource Economics* 23, 2–19.
Johansson, R.C. (2000) *Pricing Irrigation Water: A Literature Survey*. World Bank Report. World Bank, Washington, DC.
Kumar, M.D. and Singh, O.P. (2001) Market instruments for demand management in the face of scarcity and overuse of water in Gujarat, Western India. *Water Policy* 3, 387–403.
Molden, D., Sakthvadivel, R. and Habib, Z. (2000) *Basin-Level Use and Productivity of Water: Examples from South-Asia*. Research Report 49. International Water Management Institute, Colombo.
Molle, F. (2001) Economic tools for water demand management in Thailand: conventional wisdom and the real world. Paper presented to the ACIAR Water Policy Workshop. 7–9 June 2001, Bangkok, Thailand.
OECD (Organization of Economic Cooperation and Development) (1999) *Agricultural Water Pricing in OECD Countries*. OECD Working Papers 7, 33. OECD, Paris, France.
Ogg, C.W. and Gollehon, N.R. (1989) Western irrigation response to pumping costs: a water demand analysis using climatic regions. *Water Resources Research* 25, 767.
Perry, C.J. (1997) *Alternative Approaches to Cost Sharing of Water Services to Agriculture in Egypt*. Research Report 2. International Water Management Institute, Colombo.
Perry, C.J. (2001) *Charging for Irrigation Water: The Issues and Options, with a Case Study from Iran*. Research Report 52. IWMI, Colombo.
Perry, C.J. and Narayanamurthy, S.G. (1998) *Farmer Response to Rationed and Uncertain Irrigation Supplies*. Research Report 24. International Water Management Institute, Colombo.
Postel, S. (1999) *Pillars of Sand: Can the Irrigation Miracle Last?* W.W. Norton, New York.
Ray, I. (2002) Farm-level incentives for irrigation efficiency: some lessons from an Indian canal. *Water Resources Update* 121. January 2002. Universities Council on Water Resources.
Ray, I. and Williams, J. (1999) Evaluation of price policy in the presence of water theft. *American Journal of Agricultural Economics* 81, 928–941.
Rhodes, G.F. and Sampath, R.K. (1988) Efficiency, equity and cost recovery. Implication of water pricing and allocation schemes in developing countries. *Canadian Journal of Agricultural Economics* 36, 103–117.
Rosegrant, M. (1997) *Water Resources in the Twenty-First Century: Challenges and Implications for Action*. International Food Policy Research Institute, Washington, DC.
Rosegrant, M., Ringler, C., McKinney, D.C., Cai, X., Keller, A. and Donoso, G. (2001) *Integrated Economic–Hydrologic Water Modeling at the Basin Scale: The Maipo River Basin, Chile*. International Food Policy Research Institute, Washington, DC.
Rosegrant, M., Cai, X. and Cline, S. (2002) *World Water and Food to 2025. Dealing with Scarcity*. International Food Policy Research Institute, Washington, DC.
Sampath, R.K. (1983) Returns to public irrigation development. *American Journal of Agricultural Economics* 65, 337–339.

Sampath, R.K. (1992) Issues in irrigation pricing in development countries. *World Development* 20, 967–977.

Seckler, D. (1996) *The New Era of Water Resources Management: From 'Dry' to 'Wet' Water Savings*. Research Report 1. International Irrigation Management Institute, Colombo.

Seckler, D., Molden, D., Amarasinghe, U. and de Fraiture, C. (2000) *Water Issues for 2025: A Research Perspective*. IWMI's contribution to the 2nd World Water Forum. International Water Management Institute, Colombo.

Tsur, Y. (2000) Water regulation via pricing: the role of implementation costs and asymmetric information. In: Dinar, A. (ed.) *The Political Economy of Water Pricing Reforms*. Chapter 5. Oxford University Press for the World Bank, New York.

Varela-Ortega, C., Sumpi, J.M., Garrido, A., Blanco, M. and Iglesias, E. (1998) Water pricing policies, public decision making and farmers' response: implication for water policy. *Agricultural Economics* 19, 193–202.

Weinberg, M., Kling, C.L. and Wilen, J.E. (1993) Water markets and water quality. *American Journal of Agricultural Economics* 75, 278–291.

4 'Get the Prices Right': A Model of Water Prices and Irrigation Efficiency in Maharashtra, India

I. Ray*

Much of the world's freshwater is wasted. Governments are shying away from the answer: to price this valuable substance correctly.

(*The Economist*, 23 March 2000)

Introduction

In the name of food security for the nation and poverty alleviation for the rural population, every developing country provides its farmers with irrigation water at a fraction of its delivery cost (Repetto, 1986). However, growing populations, higher cultivation intensities, increasing urbanization and, of late, environmental concerns, have all combined to put pressure on global water resources. Irrigation is by far the largest consumer of freshwater, and the realization that this water is scarce and getting scarcer has forced a widespread rethinking of the 'cheap water' policy. Elementary economic theory says that farmers who pay next to nothing for water have no incentive to use it efficiently. They could use it to grow water-intensive but perhaps low-value field crops, irrigate carelessly using flood and furrow methods, neglect to maintain their field channels and overwater their standing crops. Water use inefficiency has therefore been cited, in developing and developed countries alike, as an argument for raising the price of irrigation water to reflect its scarcity value.

This policy is now under consideration (and partial implementation) in several countries, from Tunisia to Taiwan and from Botswana to Brazil, including India (Dinar, 2000, Table 1.1). Most country reports on water sector reforms mention – among other things – the need for higher water prices and the removal of flat, per unit-area irrigation charges. In short (irrigation) water is an economic good[1] and not a birthright, and wasteful water use can best be combated by 'getting the prices right'.

In this chapter, I examine the hypothesis that, in order to induce efficiency at the farm level, irrigation water prices should be raised. In the next section, I set out the rationale for opportunity-cost water pricing, citing modelling and empirical evidence in its

*An earlier version of this chapter appeared in *Economic and Political Weekly*, 13 August 2005. I thank François Molle, Jeremy Berkoff and A. Vaidyanathan for their valuable comments and suggestions on earlier drafts.

[1]'Water has an economic value in all its competing uses and should be recognized as an economic good' is one of the four *Dublin Principles* from 1992 (Solanes and Gonzalez-Villarreal, 1999).

favour. In the third section, I bring out the (often implicit) assumptions under which higher water prices at the farm level can, in fact, increase irrigation efficiency. The fourth section briefly describes the system of canal irrigation in Maharashtra, western India, and introduces the case-study canal. In the fifth section, I show that when these assumptions do not hold water prices have only limited impact on irrigation efficiency.[2] I illustrate the point with a programming model calibrated to the Mula canal in Maharashtra as a concrete example.[3] Finally, I analyse a different price policy – specifically, support prices[4] or procurement prices for particular crops – as an alternative means of conserving water.

Overall, this chapter does not claim that higher irrigation charges *cannot* induce water conservation, but that they will do so only under several preconditions. If these preconditions are far from ground reality, and I argue that in developing countries they often are, then water prices will not be the best way to save water or to increase its productivity. Transparent and enforced allocation rules may be more feasible, and output-price policy changes more effective, at least in the near term.

I should note that cost recovery rather than efficient irrigation is another important reason for charging higher water prices. Many developing country governments, India included, are considering higher water prices as a way to recover at least the operating costs of canal systems, and not as a way to reflect the opportunity cost of water. However, the rationale for cost recovery is financial, whereas the rationale for efficient pricing is economic. It is possible to raise water prices to the point where administrative costs are covered, and yet have them lower than the opportunity cost of water. In fact, an adequate per area-irrigated flat fee (that cannot induce efficiency) could cover the capital and operating costs of a canal system. Similarly, efficiency-inducing water prices can coexist with massive subsidies at the system level. The role of water prices for cost recovery purposes is not addressed in this chapter.

Opportunity-cost Pricing: The Evidence

If water prices rise to reflect its opportunity cost, a profit-maximizing farmer should have any or all of the following four responses (Gardner, 1983). He[5] can demand less water and leave some land fallow. He can cultivate all his land but stress the crop a little, thus maximizing his output per unit of water rather than output per unit of land. He can diversify out of thirsty but low water-productivity field and fodder crops into water-efficient crops such as vegetables. And finally, he can invest in efficient irrigation technologies, such as sprinkler and drip systems, which allow a larger fraction of diverted water to be used consumptively by the plant. Even a simple change such as shortening the length of the irrigation furrow could raise field-level irrigation efficiencies by up to 10%. The conclusions of both econometric analyses and

[2] The case study in my analysis is not meant to be 'representative' of canal irrigation all over India; rather, it illustrates the role of water prices in a context that shares many features with other canal-irrigated regions.

[3] I focus on canal water prices rather than on groundwater prices for three reasons: first, many analysts believe that canal water is used more inefficiently than groundwater (Dhawan, 1988); second, canal water prices are administratively set and so can be changed through public policy, while most irrigation wells are privately owned; and third, large canal irrigation schemes are the most significant users of freshwater. However, groundwater use in Indian agriculture is growing at a rapid rate.

[4] 'Support prices' are minimum prices that (usually) governments guarantee to farmers. These protect the farmer against low open-market prices. 'Procurement prices', on the other hand, are prices at which a farmer *must* sell a portion of his crop – usually to the government. These protect not the farmer, but the government and consumers, from potentially high open-market prices.

[5] I am using the term 'he' throughout the paper to refer to individual farmers, because most of the farmers I interviewed for this research were male. There are, of course, both male- and female-headed farm households throughout India.

mathematical programming models imply that farmers could respond to price-induced water scarcity in all these ways.[6]

Much of the literature on water prices is from the agriculturally rich, but water-short, western USA.[7] Using agronomically derived production functions for cotton, Ayer and Hoyt (1981) show that farmers in Arizona and New Mexico would reduce the water applied over the growing season as its price rises from $0.5 to $5 per acre-foot. Using Census of Agriculture data for several crops, Ogg and Gollehon (1989) derive downward-sloping, albeit rather price-inelastic, demand functions for irrigation water for the western USA. Caswell and Zilberman (1985), using an econometric analysis of several Californian water districts, find that the probability of adopting drip-irrigation technologies for perennial tree crops increases with increased water prices, among other factors such as land quality and crop type. Kanazawa (1988) asks: what range of price increases will induce conservation? For the Westlands Water District he finds that a three- to five-fold rise would take the price of water to its shadow value and beyond that, farmers would conserve.[8]

It should be noted that in most of these studies on water prices, the response of water use is rather low within *observed* price ranges. It is only when the price is projected to rise significantly, by a factor of 5, or sometimes 10, that the water demand is responsive. The consensus from the literature appears to be that the water-demand curve for agriculture is inelastic at low water prices. The elasticity is high when water prices are already high, and when it is cheap and feasible to substitute other inputs, such as labour, for water. For example, Levy (1982), a proponent of regulating water use through the price mechanism, agrees that the price elasticity of water is high when water is readily substitutable and when its share in total production costs is high. I shall revisit these points later in this chapter.

Programming models, which are not restricted to observed price ranges, can yield more elastic water-demand estimates. Many of these confirm the existence of low elasticities at low prices. In a modelling exercise, Weinberg *et al.* (1993) show that as water prices offered to the farmer rose from 0 to $50 an acre-foot, water-intensive crops were no longer optimal, and the amount of irrigation water applied fell. Hooker and Alexander (1998), in a programming model of San Joaquin valley, show that water demand is inelastic over a substantial price range, and steps towards conservation are taken only at certain threshold water prices. However, Howitt *et al.* (1980) have argued that including a demand function for the crop itself – not just one for water – should generate higher own-price elasticities.[9]

Implementing water trading – as opposed to implementing higher water prices – is another way in which market discipline can be brought to irrigation. Several agricultural regions of Australia are experimenting with intrabasin water trades, such as on the Murray–Darling basin. Spot markets are common in California, and interdistrict water trades, though less frequent than spot trades, do occur (Haddad, 2000; Chong and Sunding, 2006). In the developing country context, informal,

[6]There are no controlled experiments in natural settings that have tracked the response of farmers to progressively higher water prices while holding other key variables (more or less) constant. Therefore, the water pricing literature is largely made up of cross-sectional statistical analyses and modelling exercises. However, the latest OECD report on full-cost pricing of irrigation water in Europe questions the accuracy of demand elasticities derived from hypothetical price variations (OECD, 2002).

[7]For reports of European studies, see OECD (2002).

[8]These field studies measured water diverted, not water consumed. Therefore, the production functions used in such research could overstate or understate the yield response to water actually taken up by the crop. Molden (1997) points out that the marginal and average values of water should really be a function of water consumed. This distinction also has implications for how farm-level efficiency and system-level efficiency are measured, as I discuss later in the chapter.

[9]The logic works as follows: higher water prices raise the cost of production which is passed on as higher product prices to the consumer, thus lowering the demand for the product and finally bringing down the derived demand for water.

intrawatercourse trading exists in Indian and Pakistani canal systems (Bandaragoda, 1998). Short-term sales of groundwater are common although limited in their geographical scope. However, groundwater markets in Gujarat have functioned for many years (Shah, 1993; Dubash, 2002). Tradable water rights refer to longer-term commitments, for an entire growing season or longer. The most celebrated case of such rights comes from Chile, where agrarian reforms and the Water Code of 1981 formalized water rights, and allowed water sales to be separated from sales of land (Bauer, 1997).

In this chapter, I focus on water price policy rather than water trading as a tool for water conservation and irrigation efficiency. I note here, however, that many of the physical and managerial barriers to effective water price reform discussed in this chapter (and elsewhere in this volume) are equally barriers to effective water markets.

What Does It Mean to 'Get the Prices Right'?

The claim that increasing irrigation water prices is an effective means to irrigation efficiency is much more than a generic statement about downward-sloping demand curves. It contains many assumptions which are not always made explicit and thus need unpacking. These are:

1. Water costs are significant in the overall crop budget, and as a fraction of crop net revenues. (If they are not, the net effect of price increases may be so small that the water-demand function will barely respond.)[10]
2. There is a volumetric link between what a farmer pays and what he receives. (If water is charged by the hectare, as it usually is in developing countries, its marginal cost is zero and higher prices cannot induce efficiency.)[11]
3. Farm-level inefficiencies are significant in relation to overall system inefficiencies. (If this is not the case, the farm may not be the best place to look for water savings.)
4. Farmers irrigate using wasteful methods and/or grow low water-productivity field crops *because* water is so cheap. (If field crops are grown because local food or fodder markets are thin, or farmers overirrigate because their water deliveries are erratic, water price signals may not have the expected effect.)
5. The changes to the infrastructure that may be necessary to implement volumetric pricing, such as measuring devices, channels for conveyance, managerial and administrative changes, etc., are not prohibitively expensive. (If they are, any gains from more efficient water use will be neutralized by these implementation costs.)

The last item relates to the difficulties of implementing higher water prices on account of institutional or infrastructural barriers. It has borne the brunt of criticisms levelled at water price reform and water markets in the literature to date. Many reservations exist about the inadequate physical infrastructure of canal systems in developing countries, the administrative cost of introducing volumetric pricing (Perry, 1996), the difficulty of measuring water consumed rather than water diverted (Molden, 1997) and the possible third-party effects of water reallocation through pricing or trade (Rosegrant and Binswanger, 1994). The income losses that farmers could face on account of higher prices – especially small and marginal farmers on large canal systems – have also been critiqued on grounds of social equity (Chakravorty and Roumasset, 1991). In this chapter, I approach water prices as a means of water saving not from an infrastructural or administrative point of view, but from the point of view of the farmer – the actor who is supposed to do the saving.

I model an Indian canal system – the Mula canal system in Maharashtra – to ask: How effective are higher water prices as a means of curtailing a farmer's water demand, even if transaction and infrastructural costs are assumed not to be constraining? The model is a simplified representation of irrigation in the Mula canal system – simplified in order to isolate the effects of water prices on water use and productivity.

[10]If there are cheap and readily available substitutes for water, this condition need not hold (Levy, 1982).

[11]But they can aid in cost recovery, or cause farmers to withdraw from agriculture altogether.

Using a detailed, farming systems model of a median-sized farm, I analyse:

1. Whether higher water charges are the most feasible way to induce farm-level efficiency;
2. Whether farm-level efficiency is indeed as dismal as it is generally thought to be;
3. Whether water prices are the most relevant prices in a farmer's cropping decisions.

The price and input use data for the model, the pattern of water delivery over the agricultural year and the technical coefficients are all from my own 8-month-long fieldwork on the Mula canal system. The net irrigation requirements and the yield responses to water are from studies conducted at the Mahatma Phule Agricultural University at Rahuri, Maharashtra.

Irrigation in Maharashtra and in the Mula Canal System

The system of irrigation in Maharashtra is demand-based. Before the start of the irrigation season, the farmers who want water submit a demand statement which specifies the land they will irrigate and the crops they will grow. Depending on the water availability that year, the requests are fully or partially granted. The goal of canal systems in India was to ensure a reliable supply of food grains over a large area, even in drought-prone regions, to reduce the risk of famine and the dependence on food imports (Daines, 1985). Accordingly, canal command areas are extensive.[12] Annual grains and oilseeds are favoured for irrigation, while water-consuming cash crops such as sugarcane need a special 'sanction' (unless they are raised exclusively on groundwater). Dug wells are common in canal-irrigated tracts. Most of the Maharashtra plateau is underlain by basaltic rock; the basaltic layer keeps the water table high but the usable volume of groundwater low (Dhawan, 1986).

Canals in Maharashtra are fed by water stored in reservoirs, and are run on an 'on and off' basis (Gandhi, 1981). Only a subset of the watercourses is full of water at any given time. Each watering turn is called a 'rotation'. To compensate for the locational advantage of head-reach farmers, canals are generally (but not consistently) operated from tail to head. When a watercourse has its rotation due, the last field is watered first, and the irrigation turns move up the channel rather than down it. This system is known locally as *shejpali*.[13]

Traditionally, a farmer could irrigate until his field was 'adequately' wetted. Over time, and especially whenever irrigation demand exceeded the supply, this system came to be seen as too loose. From 1977 onward, the operational rules of major canal systems were gradually modified to a preset number of irrigation hours per hectare of land within each watercourse.[14] Only the lands and the crops for which the farmer has placed a demand are entitled to water, and this demand could differ from season to season and even from rotation to rotation within a season. The fixed irrigation entitlement, proportional to the area irrigated, is influenced by, but is not identical to, the *warabandi* system of North Indian canals.[15] It appears that this modification has introduced greater accountability and predictability in an otherwise overflexible system (Datye and Patil, 1987).

The Mula canal system in western Maharashtra has an irrigable command area of 80,000 ha[16]; the soils are medium–deep

[12]The command area is the area within gravity flow reach of the canal system. The irrigable command area (ICA) is the land that is actually expected to receive water within the command area. On average, major canal systems irrigate half of their official ICA.

[13]The word is derived from *shesh* (last) and *pali* (turn).

[14]As a reviewer points out, the more rigid system deprives farmers of the ability to adjust their water use to the actual soil moisture, which can vary by farmer and by season.

[15]The Northern canals are fed by perennial rivers and are run continuously all year. Every hectare in the canal command gets a few hours of water every week, on the same day and at the same time (Gustafson and Reidinger, 1971). This period is his fixed (*bundi*) turn (*wara*). Every farmer is entitled to water in every rotation; he need not submit an official 'demand'.

[16]But the area actually irrigated hovers around half that figure.

loams to dark clays; the average annual rainfall in the command is below 600 mm; the median landholding is between 1.6 and 2 ha, and even small farms produce crops for the market. The primary crops are sugarcane (a thirsty, lucrative cash crop), wheat and groundnut, followed by sorghum, chickpea and some cotton. Of late, sunflower has grown in popularity. Millet, a coarse grain that was once widely grown and eaten in the region, now occupies less than 10% of the gross cropped area. The arrival of year-round water has made other crops more profitable.

Water is allocated in the Mula canal system according to the modified fixed-turn system. As described above, it contains elements of Maharashtra's traditional shejpali system, and of the warabandi method of North Indian canals. As under warabandi, canal water is supposed to be delivered to farmers according to a preset rotation schedule – starting about the third week of July (unless it is still raining) and continuing through mid-June of the following year. As with shejpali, it is up to each individual farmer to place, or not to place, a water demand for each rotation of each season. The normal rotation interval – meaning the interval between two successive irrigations for any farm – is 21 days. Between March and June, when midday temperatures peak and the soils have no residual moisture, this interval is shortened to 14 days.[17] Each hectare is given a fixed duration of irrigation, e.g. 10 h/ha for a head-end farm and longer if the farm is at the tail end.[18] In practice, these durations are 'flexible' (sometimes intentionally and sometimes unintentionally).

The Mula is, in many ways, a typical South Asian canal. The water supply is more generous and more predictable at the head of the system than at the tail; upstream and downstream cropping patterns reflect both the soil variability and the uneven water delivery of the region; often, water does not reach the fields on time; much of the water released into the system is 'lost' in transit, or at least unaccounted for; there is a significant amount of unauthorized irrigation, especially in the upper half of the canal command; and the farmers pay a (small) per-hectare charge for the water they receive. This charge varies by the crop and the season, so there is some attempt, albeit a very loose one, to link water charges and volumes. The command area has several shallow wells, which are largely recharged by canal seepage, and which supplement canal water supplies. The water from these wells is also cheap because electricity for farm use is subsidized.[19] Irrigation professionals who are familiar with India will recognize many of these features even if they have never been to the Mula canal.

The Farming Systems Model

In this section, using a mathematical programming model written in GAMS, with numerical parameters calibrated to the upper–middle reaches of the Mula canal, I explore the role of canal water prices on the water use on a hypothetical medium-sized farm. Throughout

[17]The 14-day rule applies only to those parts of the canal system that are entitled to summer-season water. In the 1990s, the Right and Left Branch canals were allowed summer water, whereas a third branch, Pathardi, was restricted to an 8-month supply.

[18]The longer per-hour irrigation allowance at the tail end of the canal system is an attempt to compensate for the lower flow rate at the bottom third of long canal systems. The AI/DC ratio, which is the planned area irrigated per day cusec, is lower at the lower reaches and along the distributaries than at the higher reaches and along the main canal. (A day cusec is the volume of water flowing at 28.3 l/s for 24 h.)

[19]How to charge for groundwater is an ongoing debate in irrigation policy circles in India. Electricity is cheap and wells are often not metered. The Irrigation Department knows that the wells within a canal command are recharged by canal leakage and it frustrates them that farmers do not pay for groundwater. An obvious option is to raise electricity prices and meter the wells. Even if this were politically simple, which it is not, farmers could counter high electricity prices by switching to diesel-operated pumps. Diesel is subsidized too, but raising diesel prices would affect several other sectors (tractor power, transportation, residential electricity generation, etc.).

India, Maharashtra included, canal water is charged at a flat per-hectare rate. For modelling purposes, I have assumed that canal water is priced per ha-cm (water depth per hectare, expressed in centimetres) and have converted the relevant per-hectare charges to per ha-cm equivalents.[20] The 1.6 ha farm in the model is endowed with a male and a female adult, a specified allowance of canal water in each irrigation rotation, and a dug well. The farmer can irrigate from the canal, from the well or from both.[21] He is assumed to be profit-maximizing,[22] so the objective function maximizes the total on-farm profits over the agricultural year, subject to the constraints of land, family labour and the water available from the canal and the well.

The farm is modelled as a linear programme with eight crops (year-long sugarcane; monsoonal sorghum and millet; winter wheat, sorghum and chickpea; and summer season early- and late-sown groundnut) over one agricultural year. With year-round irrigation, the same piece of land can support two, or even three, crops a year. Data on the technical coefficients, output prices,[23] input prices for hired labour, fertilizers, draft power, etc., and the demand and supply constraints for family labour and water are from field observations and cost-of-cultivation surveys from 67 farm households. The resource constraints for labour and water are separately specified for each 14–21-day period to accommodate the water-delivery schedule from the canal and the seasonal nature of the agriculture.

As more water is made available per hectare, the yields of most crops increase, but at diminishing rates (Hillel, 1987). To keep the model linear and yet allow the production functions to exhibit diminishing returns, the concave water-response functions are broken up into between four and six linear segments. A crop with a lower water *availability* than its net irrigation requirement[24] is treated, in effect, as a separate crop with a lower water *requirement*, a lower yield and lower labour use. Crops have critical periods when water shortages cause a disproportionate fall in yields, which cannot be reversed by adequate irrigation at other times. For wheat, for example, the most water-sensitive stages are crown root initiation and pre-flowering. To reflect plant physiology as accurately as the data allow, the rotation-wise water requirements take into account any critical growth stage a crop might have. The final model has 36 crops from which the GAMS solver can choose.

The model entitles the farmer to a limited amount of canal water, proportional to his irrigated acreage, at very low crop-specific prices. This approximates the modified rotational water allocation rule in the Mula canal system. The farmer may use all, part

[20]Without this volumetric charge assumption, the marginal price of water would be zero, and the model solution would not respond to varying prices. Water prices in the model are lowest for grains and pulses, higher for summer-season crops such as groundnut, and highest for sugarcane – reflecting the official water charges. I have included taxes levied on irrigation water for sugarcane, for education and for the Employment Guarantee Scheme, as part of the 'price' of irrigation water.

[21]In keeping with the geo-hydrological conditions of the Maharashtra plateau, the model well is shallow. The water column varies with the season, and is lowest in summer when crop water needs are at their peak.

[22]A profit-maximizing farmer is, by definition, risk-neutral. The literature is divided on whether risk neutrality or risk aversion is a more realistic assumption when modelling the small farmer. My fieldwork on the Mula convinced me that risk neutrality was the more appropriate assumption for a median-sized (1.6–2 ha) farmer.

[23]All prices are quoted in the 1992 value of a rupee: $1 = Rs 30, approximately.

[24]The net irrigation requirement (NIR) is the crop-specific and location-specific water required for maximum yields, over and above effective rainfall and stored soil moisture (in a normal year). The seasonal NIRs for the crops are: sugarcane 190 cm, monsoonal millet 25 cm, monsoonal sorghum 30 cm, winter sorghum 38 cm, winter wheat 47 cm, gram 30 cm and groundnut 70–80 cm. These figures are from the Mahatma Phule Agricultural University and are averages calculated from three separate estimates. The crop-specific water-response functions in the model are derived from IARI (1977) and from unpublished studies at the Water and Land Management Institute, Aurangabad, Maharashtra.

or none of his canal water entitlement in each rotation. The model thus reflects the voluntary demand structure of the shejpali tradition as well as the per-hectare quota of warabandi. In order to analyse the effect of higher canal water prices, three further assumptions have been made. First, in addition to the cheap and limited canal water entitlement, the farmer can buy all the extra canal water he wants at a higher price. In effect, the farmer has access to a cheap baseline block of water and a second, higher-priced tier over and above the baseline entitlement. Second, the farmer can use canal water from either tier to irrigate his sugarcane crop, even if he does not have an official 'sanction' for this crop.[25] These assumptions are deviations from the actual irrigation rules, but the (hypothetical) effect of water prices on irrigation efficiency cannot be isolated if strict physical quotas and crop-zoning rules are binding constraints on the farmer's decisions.[26] Third, the model represents an 'average' year, without price and yield fluctuations. This assumption has been added to keep the model tractable as it is already rich in agronomic detail.

Water prices are significant in the overall crop budget

Canal water prices are heavily subsidized for the farmers on the Mula – so much so that water costs are insignificant in relation to the crops' per-hectare revenues. The surface flow rates in Maharashtra vary by crop so as to reflect (though loosely) the crop's water requirement as well as 'the ability of the crop to bear it' (Pawar, 1985). In 1985, water charges were supposedly fixed at 6% of the average gross income for food and non-cash crops and at 12% of the average gross income for cash crops. In practice, they have fallen far short of this goal. For example, water costs for sunflower are 0.77% of its (average) gross margins[27] per hectare; for winter wheat this proportion is 0.59%; for summer groundnut 1%; and for sugarcane 1.2%. Sugarcane, the most water-intensive of these crops, and the one to which critics of low water prices regularly refer, is in fact the least subsidized in terms of its relative water costs.

All the (previously cited) evidence on own-price elasticities suggests that water demand will not respond to price increases when the base price of water is so low. In addition, the existing system of per-hectare water prices means that the marginal cost of water is zero for each crop. It is true that higher water fees for water-consuming crops might induce a farmer to switch over to less water-intensive crops, or even to withdraw from farming altogether. However, prices would have to be raised by several hundred percent before water costs reach even 5% of a crop's gross margins.

An alternative proposal would be to physically ration the water given to agriculture, and to each irrigated hectare.[28] That is, no second tier of canal water could be bought. Recall that all the ways in which a farmer could respond to higher water prices – fallowing land, switching crops, etc. – require him to lower his total or his per-hectare water use. Rationing would directly force him into a lower, and potentially more efficient, water use pattern. By comparing the farmer's crop choices under low prices with rationing, and under successively higher water prices without rationing, we can ask:

1. At what price are the farm-level irrigation demands comparable with and without water rationing?

[25]Another interpretation of this assumption is that there is only very loose enforcement of the crop-zoning rules or the sugarcane sanctions. So once the canal water arrives, the farmer can use it as he wants. There is quite a lot of unsanctioned sugarcane in the Mula canal system, and many farmers do, in fact, supplement their well-irrigated sugarcane crops with canal water.

[26]This issue is often blurred in the literature on water prices. If price-based rationing and quantity-based rationing occur *together*, the physical limit rather than the price could well be the relevant constraint to water use.

[27]Gross margin means revenues minus variable costs, on a per-hectare basis.

[28]The difference between this proposal and current water-allocation laws is that current law calls for crop-pattern restrictions in addition to a water quota.

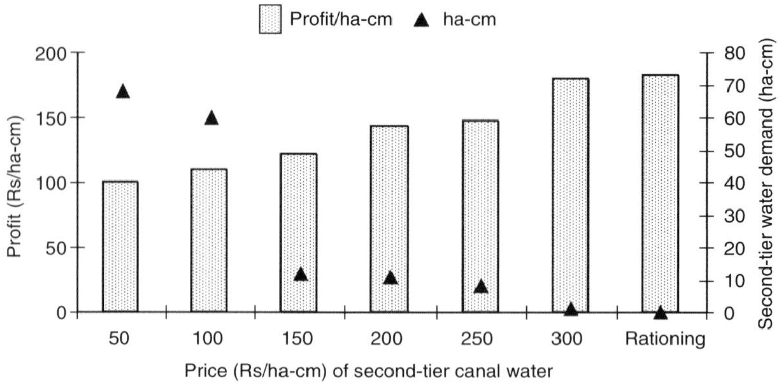

Fig. 4.1. Canal water prices and net revenues per unit of water.

2. Can we estimate the net revenues per unit of water applied[29] under various water price and crop choice scenarios?

Figure 4.1 plots the net revenues per unit of water, the price of canal water and the on-farm water demand from running the model at successively higher water prices. The *x*-axis shows the price per unit of canal water over and above the farmer's baseline entitlement. The secondary *y*-axis shows the model solution for the farmer's additional water use at the relevant price.[30] The primary *y*-axis plots the net revenues per unit of water applied, from the canal and the well, on the farm.[31] Sugarcane is the crop with the highest annual water requirement, and agronomic experiments show that sugarcane has low returns per unit of water used, but high returns per unit of land (Rath and Mitra, 1989). Hybrid grain varieties and oilseeds generally yield higher revenues per unit of water applied. Therefore, a cropping pattern that is water use-efficient should have less sugarcane and more seasonal crops such as wheat.

In each price scenario in Fig. 4.1, the farmer is allowed a cheap but limited volume of canal water (the first tier) which he can apply to any crop. In the rationing scenario, this is all he is allowed. The model solution shows that, when a farmer's water is rationed according to proportional allocation rules, a 1.6 ha plot would have 0.56 ha of sugarcane (which has a growing season of 12 to 14 months), and a winter–summer cycle of wheat followed by groundnuts on his remaining land. (This wheat–groundnut cycle is indeed common in the upper–middle reaches of the Mula.) If he can buy all the extra water he wants beyond the minimum entitlement, he grows 1.6 ha of sugarcane at a 'second tier' price of Rs 50/ha-cm and less

[29]'Net revenues per unit of water' means the annual total on-farm profits divided by the annual total quantity of irrigation water used.

[30]To explain in more detail, the *x*-axis shows the *additional price* of canal water for quantities above the baseline ration – i.e. it is the price of second-tier canal water. The *average price* of water actually paid by the farmer depends on the precise mix of baseline canal water, second-tier canal water and well water he uses. This average price will always be lower than the price of the above-the-baseline canal water shown on the *x*-axis. The primary *y*-axis shows the *average value* of water used on the farm – computed annually over all crops and using all three water sources. Ideally, we would like to compare the marginal price of water to its *marginal value*, but this rises and falls each month for each crop and could not be shown on a graph. We could also run this model for a farmer without a well, so that canal water prices would affect only canal water demand. But since most median-sized farmers of this region do have wells, and the use of well water is affected by availability of canal water, such a model would not have yielded a realistic cropping pattern.

[31]My assumption in the model is that the farmer's objective function is to maximize his total farm profits, not the output or economic returns per unit of water used. However, 'more crop per drop' or 'more value per drop' are the goals of water efficiency in agriculture, which is what we want to measure here.

and less sugarcane as water prices rise, and finally replicates the rationing crop pattern at a price of Rs 300/ha-cm. At Rs 150/ha-cm the water demand has dropped sharply, and at Rs 300/ha-cm the net returns to water are comparable to those under rationing. A rate of Rs 150/ha-cm represents a more than ten-fold increase over the average price of the baseline water block.

For the near future, such severe water-price hikes are unlikely to be suggested, let alone implemented. Farmers are numerous, and they vote. They object vociferously to price increases in water or electricity (The Economist, 1997), especially since such price hikes are usually unaccompanied by better or more reliable services. Price increases of this magnitude would have to be introduced in stages, and over time, at least in democratic regimes which are less able to implement swift policy changes (Dinar, 2000).[32] Nor would the urban population support rapid price increases, out of fear that their food costs would rise, or that national food security would be compromised. As Sampath (1992) points out, urban consumers of cheap food benefit at least as much from subsidized irrigation water as do the farmers. In short, in this region, significant price increases seem to be politically infeasible, and feasible price increases are economically insignificant.[33]

Finally, water fee collections on the Mula, as on most other Indian canals, are poor. Pawar (1985) estimates that major irrigation systems recover about 67% of their expected annual fees and minor systems recover just over 50%. The Irrigation Department's own (unpublished) records show that, from 1977 to 1990, collections on the Mula ranged from a low of 15% of the expected annual total to a high of 64%. Had the uncollected balances been rolled over from year to year in the accounts, these percentages would have been even smaller. If canals in India have been unable to recover their annual operation and maintenance costs, the state's inability to collect water fees is at least as much to blame as the low water charges themselves.[34]

Farm-level inefficiencies are a significant part of overall inefficiencies

If higher water prices are expected to improve irrigation efficiencies, it seems reasonable to ask how inefficient water use at the farm level really is, and what the relationship is between water prices, main system management and farm-level inefficiencies.

Farmers on the Mula canal – and in much of southern India – do flood-irrigate their sugarcane and grain crops, and they do allow water to spill beyond the borders of their fields. Rarely do they channel their water carefully through their furrows, or put a lot of labour into land preparation and levelling, as farmers trying to conserve water would do. The field channels are usually poorly maintained, allowing seepage and runoff losses, as even casual observation would reveal. These losses increase non-linearly down the system; seepage and evaporation reduce the flow rates to the tail end, and the slower flowing water then seeps out at an even higher rate.

It is now well understood that these local seepage and runoff losses are not necessarily lost to the basin. In a pioneering paper, Frederiksen (1992) distinguished farm- and project-level efficiency from system-level efficiency and argued that it was worth investing in irrigation efficiency in the lower reaches of a basin but not necessarily upstream. This is because seeped water re-enters the system as return flow

[32] I raised the issue of raising irrigation water prices (just enough to cover the annual operation and maintenance costs) at the Command Area Development Authority for the Mula canal. The response of the Chief Engineer was brief: 'You must be mad.'

[33] This situation is not unique to India. Empirical work on the Zayandeh Rud basin in Iran (Perry, 2001), for instance, had similar implications.

[34] During my fieldwork, new canal water rates were proposed for the state of Maharashtra. They were modestly higher than the existing rates, but some farmers in the Mula canal system were unhappy with the proposal. When I mentioned this to the Sub-Divisional Officer with whom I worked, he seemed genuinely surprised. 'Why are they angry? They don't pay us *anyway*.'

where it has instream uses or recharges the water table or can be diverted again. Thus, the water 'saved' in one part of the system, through price incentives or other means, may not be a net saving at all (Seckler, 1996). Bromley (2000) critiques the notion that irrigation water should be optimally used on the individual farm, and recommends that canal water be priced recognizing that it is a common property resource and that optimality is a system-wide concept.

Of course, some return flows become saline and unusable. On the other hand, water which recharges a well over which the farmer has complete control, and which can be used in the dry intervals between canal deliveries, has a very high marginal value.[35] The farming systems model shows that, in the parched month of May, one additional inch (2.5 cm) of well water had a marginal value equal to 1/12 of the profits from a hectare of groundnut.

But let us assume, for the sake of argument, that most of the seepage and runoff is irretrievably lost. What fraction of these losses occurs at the field level? Large canal systems in India consist of one or two main or major branches, then several distributaries that further divide up into minor branches, and finally a network of watercourses and field channels. Irrigation takes place at the level of the watercourses and field channels. Actual transmission losses are not measured (or at least, are not published) regularly in India, especially downstream of the distributary outlets. However, transmission losses in four canal systems of Maharashtra – just from the main canals to the distributary heads – have been estimated at between 10% and 59% (Rath and Mitra, 1989).

The Irrigation Department of Maharashtra measured the rates of flow down the length of the Mula canal to estimate its transmission losses – without taking into account any return flows – in the mid-1980s.[36] The cumulative measurements of conveyance, evaporation and other losses[37] along the canal were as follows: from the reservoir to the distributaries the flow had dropped by 35%; from these to the minor heads by 42%; and from the minors to the farms themselves by 65–70%. The farmer can be given price 'incentives' to be efficient with at most 30–35% of the irrigation water diverted from the reservoir. This is all the water that he has control over.

Farmers are inefficient in their water use *because* water is cheap

Locational asymmetry is a well-known phenomenon along major gravity-flow systems such as the Mula. Downstream farmers get less water than their upstream neighbours, and to make matters worse, their water deliveries are often delayed. For example, water from the Mula canal is supposed to arrive at 21-day intervals for the winter crop season, and 14 days apart in summer. In spite of the more frequent water supply in the hot season, this is a period of great stress. The clayey soils of the Maharashtra plateau are normally water-retentive but by April they are dry and cracking, and pan-evaporation rates can be as high as 15 mm/day (Lele and Patil, 1991). Despite these conditions, planned and actual water deliveries move further and further apart as they proceed down the canal. Table 4.1 shows the actual delivery intervals for one particular watercourse in 1991, which was not even a tail-end watercourse.

Many farmers openly admit that they take extra water and flood their fields generously when the water finally arrives. 'I just grab as much water as I can,' said a sugarcane farmer. 'The government says that's wasteful, that other people need water too. But what else can I do?' And in the words of a smaller farmer, lower in the system: 'The canal water

[35]The number of wells in the 360 ha study area increased from 22 to 183 within 15 years of the canal being extended to the region.

[36]The exact date is unclear. I obtained these data from unpublished reports at the offices of the Irrigation Department, Government of Maharashtra, in Ahmednagar.

[37]'Other' upstream losses include illegal water diversions, mostly for unauthorized sugarcane or for irrigation outside the official command area. Illegal irrigation is often not efficient, but, if it goes unchecked, it can hardly be made efficient through higher water prices (Ray and Williams, 2002).

Table 4.1. Example of irrigation delivery intervals on the Mula canal.

Winter irrigation No.	Interval (days) (planned = 21)	Summer irrigation No.	Interval (days) (planned = 14)
1	Inapplicable	1	Inapplicable
2	18	2	20
3	26	3	18
4	31	4	24
5	27	5	34
6	24	–	–

is like the rain. It may come, it may not come, it may come late. If it comes, we are happy. But my brothers and I, we can't rely on it.'

Farmers who do not know when to expect water, or have to plan for long dry intervals between irrigations, can be forced into stress-tolerant, possibly low-valued field crops. This is especially true of downstream farmers who typically have fewer opportunities for unauthorized irrigation, and of farmers without access to supplementary groundwater. The irrigation literature frequently implies that low water prices cause farmers to grow low-productivity crops such as lucerne and coarse grains, and that higher water prices would make them switch to, for example, vegetables and finer cereals. Water *is* cheap, and crops with low returns to water *are* grown, but such observations do not establish causation. An equally plausible hypothesis is that higher-productivity crops (such as groundnut or sunflower) need a steady supply of water at regular intervals, whereas crops such as millet or sorghum can make do with less water, less precisely timed.[38] To isolate the effect of delays in the water-delivery schedule, the original farming systems model was modified as follows:

1. The wells were taken out, so that the impact of canal water deliveries could be evaluated from the perspective of the most vulnerable farmers – those without supplementary groundwater. These farmers are entirely dependent on the canal, either because they are too poor to have a well or because the local hydrological conditions cannot support a well with reliable yields.
2. The arrival of water in a specific rotation was delayed, but compensated for in the next rotation. Therefore, the annual volumes of water delivered are unchanged from the original model.
3. Quantity restrictions were put in place; the farmers were not entitled to water over and above their first-tier allocations.
4. Water prices were kept low.

Three versions of the model were run, with delivery delays in March, April and May, respectively.[39] In each case only one rotation is delayed and the model treats the delay as anticipated. In reality, delays can be approximately known in advance (from past experience), or genuinely unexpected. In the second case, the effect on yields and revenues can range from a significant drop in yields to total crop failure. In the first case, which is modelled here, the farmer

[38]Other plausible hypotheses exist: field crops or coarse cereals are grown because of labour constraints, or a shortage of cash or credit to buy inputs for the more profitable crops, or are needed for home consumption if the local grain markets are thin. In this section, I analyse only the effect of irrigation delays. It can also be argued that poor farmers are risk-averse, that they choose crops with low returns to water and/or land rather than higher-productivity crops whose yields may fluctuate. The model solution shows that even risk-neutral farmers could choose to grow crops with low returns to water and/or land with untimely water supplies.

[39]I focus on these months because they are hot, and so the crops are most sensitive to delayed water deliveries. Delayed *kharif* season deliveries, by contrast, would usually be less damaging; and Irrigation Department records show that in the upper–middle reaches in the Mula canal system the irrigation demand for *kharif* season crops is, in any case, low.

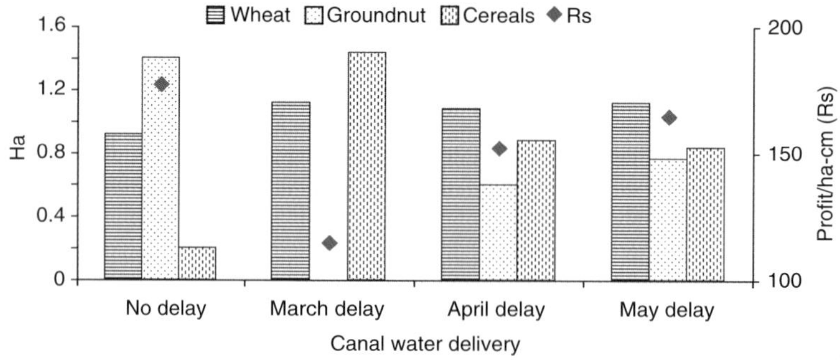

Fig. 4.2. Water-delivery regimes, crop patterns and profits.

can adjust his crop choices at the start of the season. The model solutions therefore represent the best-case delay scenarios. The solutions are explained in some detail to illuminate the connections between water deliveries and crop choices.

Figure 4.2 compares the cropping patterns and the net revenues per unit of water on a 1.6 ha farm under the planned water-delivery schedule, with those under late water arrivals in March, April and May. The water-delivery regime is shown on the *x*-axis. The optimal cropped areas under wheat, groundnut and coarse cereals (meaning, millet and sorghum) under each regime are plotted on the primary *y*-axis. These areas add to over 1.6 ha because of multiple cropping over three seasons. The returns per ha-cm of water are shown on the secondary *y*-axis. There is no sugarcane in these model solutions, not because of risk aversion or a desire for food security, but because of the high and year-round water needs of sugarcane. Sugarcane remained in the model as a crop choice, but with strict canal water rationing in place, it appears not to be a viable option without a well or a 'sanction'.

During informal conversations in the field, farmers without wells in the Mula command overwhelmingly preferred a winter–summer rotation of wheat (average gross margin Rs 7500 at 1992 prices)[40] and groundnut (average gross margin Rs 10,000). The coarser cereals (gross margins between Rs 2000 and Rs 4500) were mostly grown on rain-fed land or if the water supply was inadequate for a larger groundnut crop.[41] The model solution with no water delays reflects this ground reality, with its wheat- and groundnut-dominated cropping pattern. If the farmer expects a long dry spell in April or May, he opts for a smaller groundnut crop and a larger cereal crop – as well as a drop in his water productivity. But a delay in March is the most damaging of all. March is not a particularly water-demanding month, but that is when groundnut is planted, and when a pre-sowing wetting is really critical. Figure 4.2 shows that an irrigation delay in March cannot be made up by extra water in April, and that the farmer is forced into a monsoon–winter rotation of coarse staples followed by wheat – a low-value combination. Land records confirm that this monsoon–winter food grain pattern was common in the region before the arrival of canal irrigation.

If farmers overirrigate as a hedge against future shortfalls, or accept low returns to land or water because their canal water deliveries are untimely, they are not going

[40]The numbers themselves are location-specific, of course.

[41]Groundnuts are summer crops and coarse cereals are monsoonal crops. Nevertheless, they often compete for the same piece of land. If groundnuts are sown early, the land can be cleared in time for the monsoonal or *kharif* grain crop. If they are sown late, there is too short an interval between harvesting the summer crop and sowing winter (*rabi*) wheat to support a *kharif* crop. The model solutions accurately reflect the Mula farmers' preference for the wheat plus late-sown groundnut crop cycle.

to become efficient as a result of higher water prices. To what extent farm-level inefficiencies – which certainly exist – are significant in relation to, or are themselves a response to, main system inefficiencies is a very important question. Irrigation water prices can affect only that water over which the farmers have some control, and only those inefficiencies which are caused by low water prices.[42] In the current situation, higher water prices – if collected – are likely to lower farmers' net revenues, but have only a marginal impact on overall water-use efficiency.

Water Prices versus Crop Support Prices

Finally, if we must look to the price mechanism as a way to induce water efficiency, we should ask if water prices are the most relevant prices. In the Mula canal system, sugarcane is the cash crop of choice for both large and small landholders. The sugarcane-crushing mills, which are given a subsidy per tonne of sugarcane processed, guarantee a high support price to sugarcane producers. There is thus relatively little price risk with sugarcane compared to other cash crops such as sunflower or groundnut. In 1992, the average farm-gate price reported from this area was Rs 35/quintal.[43] The support price guaranteed by the state of Maharashtra was Rs 29/quintal. The average producer's cost, calculated from my own cost-of-cultivation surveys, was just above Rs 21.

Sugarcane is popular for its high and certain returns to land (the sugarcane-crushing factories pay farmers more than the government support price), for its resistance to pests, and for its low labour requirements compared to water-efficient crops such as vegetables, oilseeds or spices. The programming model of the representative farm was run again, this time keeping canal water prices at their low 'first tier' values, allowing the farmer to buy as much water as he desired at those low prices, letting him choose to irrigate from the canal, from his well or from both, and parametrically varying the price of sugarcane. The difference between this model and the version that varied canal water prices is that, in this version, first- and second-tier canal water has the same price. This model specification allows us to analyse the role of sugarcane prices in the absence of high water prices or water-quantity constraints.

The model solution shows that had the government not supported the price of sugarcane, or subsidized the sugarcane-crushing facilities, it would have been unprofitable for the farmers to grow sugarcane (Fig. 4.3). When sugarcane prices (shown on the *x*-axis) fall, the area under sugarcane (plotted on the primary *y*-axis) and the water used on the farm (on the secondary *y*-axis) both drop sharply. A 14% drop in the price of sugarcane triggers a 28% drop in the water demand[44] and the equivalent response would have required a nearly fourfold rise in the price of canal water charged at sugarcane rates. At sugarcane prices of Rs 25, even at low water prices, farmers would switch completely to a cycle of winter wheat followed by summer groundnut. That repre-

[42]The 1992 Planning Commission Report on irrigation pricing, chaired by Professor A. Vaidyanathan, in fact, concluded that irrigation charges should be raised, but that improving the physical condition of the main system, the timeliness of water deliveries and a higher rate of fee collection are preconditions for higher prices to be effective (GoI, 1992).

[43]A quintal is equal to 100 kg. At the time, this price represented an effective nominal protection coefficient (NPC) for raw sugarcane of almost 1.5. The NPC was computed through the procedure followed by the World Bank to estimate the unsupported price of sugarcane as a fraction of the international price for raw sugar.

[44]Although this hypothetical farm is endowed with a well, the model solution shows that the 28% drop in water demand is entirely from the canal. Well water is cheaper than canal water used for sugarcane, so the profit-maximizing farmer uses up his well water before buying canal water. Similarly, canal water is the first source of water he cuts if he reduces his overall demand. As the farmer opts out of sugarcane altogether, canal water for seasonal crops and well water can be used interchangeably since they cost about the same.

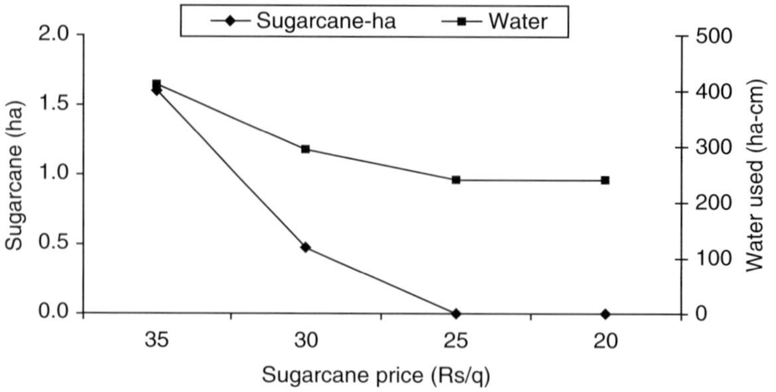

Fig. 4.3. Optimal sugarcane area and annual water use on a 1.6 ha farm; varying cane prices and low water prices.

sents a water-conserving choice *not* induced by higher water prices.

Maharashtra produces about 14% of India's sugarcane (by cane weight) and has approximately 12% of India's cropped area under sugarcane (Pant, 1999). If the government did attempt to remove the support price, it would find a powerful, well-organized and hostile opponent in the sugarcane-processing lobby (Attwood, 1985). Sugarcane-growing farmers, too, would be up in arms as the removal of price supports for raw sugarcane causes farmers' net incomes to fall. As I have earlier argued, drastic rises in water prices may not be feasible either – at least not over a short time period. A discussion on the comparative politics of higher water prices versus lower sugarcane prices is beyond the scope of this chapter. But the analysis presented here indicates that if we want to use price policy to reduce the demand for irrigation, or to induce efficient crop diversification, output rather than water prices may be an equally effective and a more direct route.

Conclusion

Economists are right when they point out that irrigation water prices are absurdly low compared with its scarcity value, and that at such low prices there is no incentive to conserve. However, it does not follow that raising water prices is the natural next step for developing countries such as India. From the perspective of the farmer who is supposed to save the water, I have suggested that there are two broad reasons for this conclusion. First, in the short to medium term, canal water prices cannot be raised to the point where they can significantly affect water demand. The negative impact on farm revenues would be too drastic and the policy would not find broad support. Second, low water prices are often not the main reason behind water-inefficient crop choices. Moreover, farm-level inefficiencies appear not to be the most significant inefficiencies on existing canals; nor are water prices the most significant prices driving irrigation demand.

A better first step would be to enforce simple allocation rules, such as a per-hectare ration that would make the scarcity value of water immediately obvious. This step, while hardly simple, could be more politically feasible than raising prices sharply, because quantity restrictions are *already* the basis of water allocation in most Indian canals. The rules are rather loosely followed at present (Wade, 1982; Ray and Williams, 2002), but a concerted attempt to implement them better would be perceived as fair, and would have the support of many middle- and tail-end farmers. There is also considerable field evidence that water users' associations (WUAs) could be helpful in implementing water allocation rules (Wade, 1988; Ostrom *et al.*,

1994), though WUAs are no guarantee against inefficiency (Vermillion, 1997). Physically rationed water shares that are transparent and enforced can also free up water to be transferred to urban areas, or to increase the number of farmers with access to canal water, or to meet environmental needs.

Proponents of water pricing certainly recognize that the price mechanism has to be embedded in a carefully designed institutional framework (GoI, 1992; Sampath, 1992; Saleth, 1997). From most of these analyses, however, it remains difficult to isolate the (efficiency) impacts of water prices from those of all the other recommended physical and institutional reforms. More research is also needed on whether enforcing simple allocation rules would be more, or less, costly to administer than a completely new tariff structure; it could be that the cost of restructuring water charges, under a range of conditions, is higher than the expected efficiency gains (Tsur and Dinar, 1997). Yet, over the last two decades, and especially since the Dublin Principles declared water to be an economic good, the mainstream literature on water sector reform has been significantly focused on the need for higher water prices and more water trades. In this chapter, I have argued that water may be cheap, and that water use in agriculture may be inefficient, and that these are indeed problems. But the case study of Maharashtra shows that low water prices are often not the most immediate causes for irrigation inefficiency, and so we cannot conclude that 'getting the prices right' is the most appropriate solution.

References

Attwood, D.W. (1985) Peasants versus capitalists in the Indian sugar industry: The impact of the irrigation frontier. *Journal of Asian Studies* 45, 59–80.

Ayer, H. and Hoyt, P.G. (1981) *Crop-Water Production Functions: Economic Implications for Arizona*. Technical Bulletin No. 242. University of Arizona, Tucson, Arizona.

Bandaragoda, D.J. (1998) *Design and Practice of Water Allocation Rules: Lessons from Warabandi in Pakistan's Punjab*. Research Report No.17. International Water Management Institute, Colombo.

Bauer, C.J. (1997) Bringing water markets down to earth: The political economy of water rights in Chile, 1976–1995. *World Development* 25, 639–656.

Bromley, D. (2000) Property regimes and pricing regimes in water resource management, Chapter 2. In: Dinar, A. (ed.) *The Political Economy of Water Pricing Reforms*. Oxford University Press, New York.

Caswell, M. and Zilberman, D. (1985) The choice of irrigation technologies in California. *American Journal of Agricultural Economics* 67, 224–234.

Chakravorty, U. and Roumasset, J. (1991) Efficient spatial allocation of irrigation water. *American Journal of Agricultural Economics* 73, 165–173.

Chong, H. and Sunding, D. (2006) Water markets and trading. *Annual Review of Environment and Resources* 31, 239–264.

Daines, S.R. (1985) *Irrigation Policy in India: Historical Trends and Public Policy Issues*. Irrigation Sector Strategy Review, Water Management Synthesis Report No.35. Utah State University, Logan, Utah.

Datye, K.R. and Patil, R.K. (1987) *Farmer Managed Irrigation Systems – Indian Experiences*. Centre for Applied Systems Analysis in Development, Pune, India.

Dhawan, B.D. (1986) *Economics of Groundwater Irrigation in Hard-Rock Regions*. Agricole Publishing Company, New Delhi.

Dhawan, B.D. (1988) *Irrigation in India's Agricultural Development: Productivity, Stability, and Equity*. Sage Publications, New Delhi.

Dinar, A. (2000) Political economy of water pricing reforms, Chapter 1. In: Dinar, A. (ed.) *The Political Economy of Water Pricing Reforms*. Oxford University Press, New York.

Dubash, N. (2002) *Tubewell Capitalism*. Oxford University Press, New York and New Delhi.

Frederiksen, H. (1992) *Drought Planning and Water Efficiency*. World Bank Technical Paper No.185. World Bank, Washington, DC.

Gandhi, P.R. (1981) *History and Practice of Management of Irrigation Waters in Maharashtra*. Water and Land Management Institute, Aurangabad, India.

Gardner, B.D. (1983) Water pricing and rent-seeking in California agriculture. In: Anderson, T.L. (ed.) *Water Rights, Scarce Resource Allocation, Bureaucracy and the Environment*. Ballinger Publishing Company, Cambridge, Massachusetts, pp. 83–113.

GoI (Government of India) (1992) *Report of the Committee on Pricing of Irrigation Water, Planning Commission*. Government of India, New Delhi.

Gustafson, W.E. and Reidinger, R.B. (1971) Delivery of canal water in North India and West Pakistan. *Economic and Political Weekly* 6, A157–A162.

Haddad, B. (2000) *Rivers of Gold: Designing Markets to Allocate Water in California*. Island Press, Washington, DC.

Hillel, D. (1987) *The Efficient Use of Water in Irrigation*. Technical Paper No. 64. The World Bank, Washington, DC.

Hooker, M.A. and Alexander, W. (1998) Estimating the demand for irrigation water in the Central Valley of California. *Journal of the American Water Resources Association* 34, 497–505.

Howitt, R.E., Watson, W.D. and Adams, R.M. (1980) A re-evaluation of price elasticities for irrigation water. *Water Resources Research* 16, 623–628.

IARI (Indian Agricultural Research Institute) (1977) *Water Requirements and Irrigation Management of Crops in India*. New Delhi.

Kanazawa, M. (1988) The economic efficiency of irrigation water supply by agricultural water districts: An econometric analysis. University of California at Berkeley, California. Draft.

Lele, S.N. and Patil, R.K. (1991) *Working for Farmer Participation in Irrigation Management in Major Irrigation Projects*. Centre for Applied Systems Analysis in Development, Pune, India.

Levy, Y. (1982) Pricing federal irrigation water: A California case study. In: *Energy: The Next Twenty Years*. (Report sponsored by the Ford Foundation, chaired by H.H. Landsberg.) Ballinger Publishing Company, Cambridge, Massachusetts.

Molden, D. (1997) *Accounting for Water Use and Productivity*. SWIM Paper 1. IIMInstitute, Colombo.

OECD (Organization for Economic Cooperation and Development) (2002) Transition to full-cost pricing of irrigation water for agriculture in OECD countries. Available at: COM/ENV/EPOC/AGR/CA(2001)62/FINAL. Downloaded 03/2006 from http://www.olis.oecd.org/olis/2001doc.nsf/

Ogg, C.W. and Gollehon, N.R. (1989) Western irrigation response to pumping costs: A water demand analysis using climatic regions. *Water Resources Research* 25, 767–773.

Ostrom, E., Gardner, R. and Walker, J. (1994) *Rules, Games and Common Pool Resources*. University of Michigan Press, Ann Arbor, Michigan.

Pant, N. (1999) Impact of irrigation management transfer in Maharashtra. *Economic and Political Weekly*, Review of Agriculture, 27 March–2 April.

Pawar, J.R. (1985) Recurrent cost study of operation and maintenance of irrigation systems in Maharashtra. In: Daines, S.R. and Pawar, J.R. (eds) *Irrigation Sector Strategy Review, Water Management Synthesis Report, Appendices Volume II*. Utah State University, Logan, Utah.

Perry, C.J. (1996) *Alternative Approaches to Cost Sharing for Water Service to Agriculture in Egypt*. Research Report No. 2. International Irrigation Management Institute, Colombo.

Perry, C.J. (2001) *Charging for Irrigation Water – the Issues and Options in Iran*. Research Report No. 52. International Water Management Institute, Colombo.

Rath, N. and Mitra, A. (1989) Economics of irrigation in water-scarce regions: A study of Maharashtra. *Artha Vijñana* 31.

Ray, I. and Williams, J. (2002) Locational asymmetry and the potential for cooperation on a canal. *Journal of Development Economics* 67, 129–155.

Repetto, R. (1986) *Skimming the Water: Rent-Seeking and the Performance of Public Irrigation Systems*. Research Report No. 4. World Resources Institute, Washington, DC.

Rosegrant, M.W. and Binswanger, H.P. (1994) Markets in tradable water rights; potential for efficiency gains in developing country water resource allocation. *World Development* 22, 1613–1625.

Saleth, R.M. (1997) India, Chapter 8. In: Dinar, A. and Subramanian, A. (eds) *Water Pricing Experience: An International Perspective*. Technical Paper No. 386. World Bank, Washington, DC.

Sampath, R.K. (1992) Issues in irrigation pricing in developing countries. *World Development* 20, 967–977.

Seckler, D. (1996) *The New Era of Water Resource Management: From 'Dry' to 'Wet' Water Savings*. Research Report No. 1. International Irrigation Management Institute, Colombo.

Shah, T. (1993) *Groundwater Market and Irrigation Development: Political Economy and Practical Policy*. Oxford University Press, Bombay, India.

Solanes, M. and Gonzalez-Villarreal, F. (1999) *The Dublin Principles for Water as Reflected in a Comparative Assessment of Institutional and Legal Arrangements for Integrated Water Resources Management*. TAC Background Paper No. 3. Global Water Partnership, Stockholm, Sweden.

The Economist (1997) Power struggle. 1 November, p. 44.

Tsur, Y. and Dinar, A. (1997) The relative efficiency and implementation costs of alternative methods for pricing irrigation water. *The World Bank Economic Review* 11, 243–262.

Vermillion, D.L. (1997) *Impacts of Irrigation Management Transfer: A Review of the Evidence*. Research Report No.11. International Irrigation Management Institute, Colombo.

Wade, R. (1982) The system of administrative and political corruption: Canal irrigation in South India. *Journal of Development Studies* 18, 287–327.

Wade, R. (1988) *Village Republics*. Cambridge University Press, New York.

Weinberg, M., Kling, C.L. and Wilen, J.E. (1993) Water markets and water quality. *American Journal of Agricultural Economics* 75, 278–291.

5 Thailand's 'Free Water': Rationale for a Water Charge and Policy Shifts*

F. Molle

Introduction

Despite the success claimed for the irrigation sector in contributing to falling food prices, food security and raising farm income, irrigation has, in the last two decades, elicited growing frustration in the community of aid agencies and development banks. A major reason for such sentiment is the low financial sustainability of the sector, which incurs recurrent rehabilitation expenditure and subsidies to operation and maintenance (O&M) that add to the large initial investment costs. A second reason is that agriculture accounts for 70% of the use of water and, despite growing shortages, is seen to be bedevilled by very low levels of efficiency (the water effectively used is only a small fraction of the water diverted) that seem unacceptable in a time of growing needs in other sectors. In addition, farmers often apply large quantities of water to irrigate crops that have both high water requirements and a low return (typically, rice in Asia).

These problems of perceived low efficiency, poor management and financial unsustainability have been addressed by a wide range of actions that include rehabilitation, modernization, improved technical management, participatory management, turnover and collection of water charges. The limited benefits obtained have spurred many proposals to tackle these problems with some economic tools and incentives, particularly in the aftermath of the Hague and Dublin meetings (Rogers *et al.*, 1997).

In Thailand, water is supplied to agriculture free of charge: water is best seen as a gift, traditionally linked to the good will or power of the absolute king, who mediates its supply from supernatural forces. *Chonlaprathan*, the Thai word for irrigation, embodies a notion of the royal gift. The *Loy Krathong* festival, in November, when offerings are put afloat on the waterways of the kingdom to thank the water spirits for the life that water brings, epitomizes the relationship between people and water. However, proposals for water pricing in the country can be found as early as 1903, in the *General Report on Irrigation and Drainage in the Lower Menam (Chao Phraya) Valley*, submitted to the Government of Siam by Van der Heide (1903), a Dutch engineer in charge of the Department of Canals:

> A water tax could be levied, in a manner similar to the paddy land tax, over the whole area at present cultivated and the future extension of this area, as far as the fields are benefited by the [irrigation] system . . . water rates could in general be assessed in some proportion to the quantity of water utilized, and would most probably be a suitable taxation for dry season crops and garden cultivation.

The logic for pricing water may have, at that time, been borrowed from practices in Java, India or other Asian countries under colonial rule. Likewise, in the post-World War II period when the International Bank for Reconstruction and Development funded the development of infrastructures in the Chao Phraya delta, the consultant in charge of the study saw no difference between irrigation supply and railways or electricity and stated that it would 'not be a misuse of language or an exaggeration to describe the position [of Thailand] as extraordinary. . . . The Irrigation Department is thus unique among the commercial departments of the Government in Thailand in deriving no revenue from its services and unique or nearly so in this respect, throughout the world' (IBRD, 1950).[1] Although, at the time, the Thai government had shown willingness to establish fees once the scheme would be completed and proper supply ensured to users (IBRD, 1950), the idea seems to have then vanished and only recently come to the fore. In the aftermath of the 1997 financial crisis, reform of the agriculture and water sectors was encouraged by both the World Bank and the Asian Development Bank (ADB), and the latter supported the definition of an ambitious plan aimed at introducing river basin management, service agreements between the Royal Irrigation Department (RID) and users, cost recovery dubbed as 'cost-sharing', and legal dispositions around a Water Law. This policy remained a dead letter for a set of reasons that cannot be easily untangled, but which includes resistance from line agencies, weak political support and the over-optimistic and often unrealistic nature of many of the proposals. Despite the setting of a policy matrix that defined commitment to successive milestones to be achieved, the process lost momentum before being eventually discontinued by the Thaksin administration.

In this chapter, I first examine the relevance of the arguments for establishing water charges in the particular context of Thailand, and most particularly that of the Chao Phraya delta, the rice bowl of the country (Molle and Srijantr, 2003). In the first section, I address successively the role of pricing as: (i) a means to signal to users the economic value of water and hence regulate its use and avoid wastage; (ii) an instrument to reallocate water to crops with higher water productivity or to non-agriculture sectors; and (iii) a cost recovery mechanism. In the second section, I briefly examine reforms that failed in the past, and attempt to draw conclusions on both the potential charging for water and the way a policy reform process should unfold. Although unsuccessful, these attempts at reforming the water sector provide useful lessons on the constraints commonly faced by water pricing policies, particularly when they fail to fully appreciate the context in which they are to operate.

Before turning to these points, it is useful to single out a few specific features of the Chao Phraya delta, on which the analysis will focus. Agriculture in the delta traditionally distinguishes between the wet season (where rain is abundant, sometimes in excess, and irrigation merely a complement) and the dry season (when irrigation is a prerequisite to agriculture). The hydrology of the delta is very complex, since it includes numerous side flows and return flows, canals serving for both supply and drainage, generalized use of pumps, predominance of paddy with common plot-to-plot systems of supply, vulnerability to flooding, use of waterways for navigation, domestic supply, dilution of pollution load, etc. This defines a context with numerous uses and users where it is difficult to clearly identify both the sources of supply and the uses, and which is therefore little amenable to quantitative regulative mechanisms. Many of these features apply to other Asian deltas, particularly those of the Cauvery, Ganges–Brahmaputra,

[1]The consultant also underlined the value of charging for water in order to limit wastage and to control society's demand for unsound projects: 'Mankind values the things it has to pay for and thinks little of and uses wastefully the things it gets free. Moreover if water is supplied free, farmers who get no water will be unable to see why their neighbours should and the Government will be embarrassed by pressure to carry out schemes regardless of whether they are sound or not.'

Irrawaddi and Mekong rivers. On the other hand, the delta includes Bangkok and enjoys good transportation networks and rather efficient linkages to urban and export markets.

Water Pricing and Its Potential Roles in Thai Irrigated Agriculture

Dealing with unacceptable water wastage?

The statement that water is wasted when it is free or underpriced probably appears in one form or another in all papers and reports that address the issue of water pricing (see Molle and Berkoff, Chapter 2, this volume). This simple axiom has been disseminated widely by analysts like Sandra Postel (1992), who observes that 'water is consistently undervalued, and as a result is chronically overused', by development banks and agencies (e.g. World Bank, 1993; ADB, 2000), as well as by many academics. In Thailand, an endless number of observers[2] have taken it for granted, notably TDRI (1990) and Christensen and Boon-Long (1994), who posit that 'since water is not appropriately priced, it is used inefficiently, and consumers have no incentive to economize'. Several reasons, related to both theoretical assumptions and constraints to implementation, showing that such statement may be misleading are reviewed here.

That rising water fees may be conducive to water saving is shown by numerous experiences in the domestic and industrial water sectors (Gibbons, 1986; Dinar and Subramanian, 1997; Dinar, 2000). Since individual meters can be easily installed on pressurized pipe networks, volumetric charging is practical and users' behaviour is generally affected by rising charges although, beyond a certain point, the elasticity of water demand falls drastically. The facts that *volumetric* charging is a prerequisite and that it is not feasible in the short run in most large-scale irrigation schemes of Asia are well recognized in the literature. Yet, in Thailand, where most of the hydraulic structures are rather crude, this evidence is generally glossed over and the potential benefits of volumetric charging are often assumed implicitly for pricing *in general*, as illustrated by the various statements collected in footnote 3.

Since volumetric pricing at the individual farm level is unrealistic, 'water wholesaling' in which water is attributed to groups of users, for example, to the farmers who are served by the same lateral canal, appears to be an attractive option. This alternative has the advantage of encouraging farmers to act collectively to achieve reduced demand within the command area of their canal, and shifts on them the burden of solving conflicts and collecting a water charge. However, the effectiveness of such an arrangement rests on the possibility of: (i) defining and registering who the beneficiaries are; (ii) designing a transparent allocation mechanism at basin, project and farm levels; (iii) ensuring water supply to groups in accordance with an agreed service; and (iv) having Water User Groups that are in a position to perform all the tasks entrusted to them. Therefore, the wholesaling of water appears more like an option that would be made possible by a series of critical reforms spanning technical, legal, managerial and political domains, than a measure that can be put forward in a 'non-mature' context. In the case of Thailand, few, if any, of these prerequisites are met.

The policy framework supported by the ADB in the 1999–2001 period (see later section) laid some foundations for establishing 'cost-sharing' and defining 'service agreements' between the RID and users that could amount to a kind of bulk allocation. Attractive in its design, the policy probably much

[2]How popular wisdom emerges can be sensed from the following declarations. An official of the Ministry of Agriculture said: 'Water should be priced in order to increase the efficiency of its use in the farm sector' (*The Nation*, 2000, 21 April); 'Agricultural experts agree that water-pricing measures would help improve efficiency in water use among farmers' (*The Nation*, 1999, 17 February); the Director of the National Water Resources Committee observed: 'In reality water is scarce, and the only mechanism to save water and encourage efficient use is to give it a price' (*The Nation*, 2000, 23 April); etc.

underestimated[3] both the technical difficulties to define and ensure service agreements and the institutional/political transformations required (Molle *et al.*, 2001). Even where bulk allocation was implemented as part of a programme of management transfer (as in Mexico and Turkey), was credited with some success and contributed to a better fee collection and financial situation, there is little evidence that significant water saving in land or water productivity or gains have resulted from these reforms (Murray-Rust and Svendsen, 2001; Samad, 2001).

Even if some kind of volumetric pricing were possible, prices would have to be set at a level high enough to have a bearing on farmers' behaviour. There is, indeed, overwhelming evidence from the literature that tariffs which reflect O&M costs and are economically feasible are in too low a range to have any significant impact on behaviour (Gibbons, 1986; de Fraiture and Perry, Chapter 3, this volume; Ray, Chapter 4, this volume). An average water fee of B(baht)120/rai (one rai = 0.16 ha) as proposed by the ADB policy (H&P and A&E, 2001) would amount to 5–7% of the farmer's net income per rai. While not negligible, such a value would be unlikely to affect behaviour at the margin, assuming – for the sake of demonstration – *volumetric* and individual pricing, saving, say, 30% of water would increase the revenue per rai by only 2%, a value much under the opportunity cost of the additional labour necessary to achieve such water savings at the plot level. It can therefore be safely concluded that the proposed fee, based on area and set at half the estimated O&M costs, would have no impact on water use whatsoever, despite repeated claims to the contrary.

The second issue considered here is whether water is indeed wasted, and whether significant savings could be achieved, through pricing or other means. Recently, the Director-General of the Royal Irrigation Department on a Thai national television channel declared somewhat contritely that water efficiency was very low in Thailand and that this had to be remedied in the face of the water shortages experienced by the country. International agencies (and sometimes, in their footsteps, local officials) commonly report that Thai farmers are *guzzling* water or are showing *water greed* (*The Nation*, n.d.), furthering the general idea that efficiency in large state-run irrigated schemes is often as low as 30% (TDRI, 1990), and sticking to this overall vision without questioning it any further. Yet, research conducted in recent years has shown that water basins tend to 'close' when demand builds up: most of the regulated water in the basin is depleted and little water is eventually 'lost' out of the system *when it has value* (downstream requirements and environmental services taken into account). There has been widespread recognition that focusing on relatively low irrigation efficiency at the on-farm or secondary levels could be totally misleading (Keller *et al.*, 1996; Perry, 1999; Molle, 2004). When analysed at the basin level, closing systems are eventually found to *operate with a high overall efficiency* during the dry season.

In-depth investigations in the Chao Phraya river basin (Molle *et al.*, 2001; Molle, 2004), most particularly in the delta, have shown that users and managers have not been passive when confronted with water scarcity but, on the contrary, have responded to it in many ways. Farmers have developed conjunctive use, dug farm ponds, drilled wells, closed small drains and invested in an impressive pumping capacity to access these sources. Dam managers have come under pressure to avoid dam releases that are in excess of downstream requirements and have improved management. Reuse of water along the basin and within the delta has developed to the point that, in the dry season, only an estimated 12% of the water released by the dams is lost to non-beneficial evaporation or outflow – effectively recycling the 'losses' from excessive water diversions

[3]One of the consultants involved considered that the policy was not optimistic but 'simply stated what, ideally, *ought to be done*, without claiming that it *would* be done'. This, however, implies that proposals are made on a prescriptive and idealized mode without taking into consideration the institutional and political context in which they are supposed to be inserted.

in exactly the way that research elsewhere has found and predicted. Because of the tendency to focus on state-designed policies, all the endogenous adjustments to water scarcity that accompany the closure of a river basin are generally overlooked (Molle, 2004).

Irrespective of whether they pay for water or not, farmers are aware that water is valuable and scarce because they are directly confronted with the consequences of its scarcity, and have made significant investments in pumps, wells and ponds to tackle it. To squander water, farmers should first be in a position to access *more* water than they need, which is contradictory to the situation in the dry season, where cropping intensity is around 60% and where water shortages push farmers to actively look for alternative sources of water.

In the wet season, patterns of water use often differ. In many instances water management is geared towards getting rid or controlling the potential damage, of excess water, rather than saving water. Water use at the farm level may be wasteful, but this only reflects the fact that supply is continuous and abundant (with a zero opportunity cost) and that the water 'wasted' was destined to flow back to the river anyway. Indeed, abundant water can ease management both to farmers and operators so that 'wasting' water may be the economic optimum given its zero opportunity cost.

Finally, stating that water is 'free' misses the point that the majority of farmers have to resort to pumping to access water in the dry season (when saving water is an issue), to offset both the lack of water and the uncertainty in delivery. Because of the costs incurred by these water-lifting operations, there is little likelihood that farmers (80% of farmers in the lower Chao Phraya basin have at least one pump set) will squander water (Bos and Wolters, 1990).

Shortages and crises are not due to a hypothetical low efficiency but to the insufficient control over interannual regulation, water allocation and distribution. The lack of strong technical criteria in managing dams and in allocating water to irrigation, the uncontrolled planting[4] by farmers and the irresistible political pressures to which competition for water gives rise, lead to escalating risk and sporadic shortages. This does not dismiss the fact that efficiency gains are desirable but draws our attention to the inconsistency of the commonly stated relationship between farmers' efficiency and water shortage.

Overall, it emerges that both the empirical and theoretical justifications advanced to support the use of water pricing as a regulatory tool for saving water do not hold in the present case. On the one hand, water is not squandered as commonly assumed (adjustments to de facto scarcity occur), the overall efficiency of water use is high (reuse of return flows), and most farmers incur costs to access water that is, therefore, neither free nor wasted. On the other hand, theoretically, savings could be expected if pricing was volumetric and high enough to affect farmers' behaviour, but this has not been verified.

Pricing as a reallocation tool

Improving irrigation efficiency is only one aspect of better using scarce water resources. Another potential benefit from water pricing could be to encourage a shift towards crops that are less water-intensive, and/or that display a better water productivity ($/m^3), or towards non-agricultural uses. Volumetric pricing would directly penalize crops with high consumption of water, but it could also be possible to establish water charge differentials based on crop type, that would

[4]The hopelessness of officials is apparent in public declarations: The Deputy Agriculture Minister reported in early 1998 that 'plantations in Nakhon Sawan, Tak and Kamphaeng Phet had increased to more than 670,000 rai from a target of 190,000' (*Bangkok Post*, 1999, 13 January), while the RID Director admitted that 'things are out of control', with 330,000 rai under cultivation, against a limit set at 90,000 rai (*The Nation*, 1999, 8 January). 'Our major concern is that we have no effective measures to control the use of water by rice growers. The only thing we can do is ask for their cooperation to cut down rice cultivation.'

encourage farmers to grow crops with lower water requirements. This runs into the same difficulties exposed in the preceding section regarding the elasticity of water use, the impact on farm income, and the constraints to metering volumes (crop-type-based fees escape this last constraint but face costs in monitoring effective land use). This rationale on crop selection often implicitly assumes that farmers do not diversify into field crops, vegetable or fruit crops *because* water is cheap or free, leading them to favour water-intensive crops (e.g. rice or sugarcane). This assumption also needs to be put in context.

In Thailand, the possibility of achieving water conservation by inducing a shift away from rice to field crops, which consume (ET) only 50–80% of the amount of water needed for rice, has long been underlined by policy makers and has formed the cornerstone of state projects aimed at fostering agricultural diversification (Siriluck and Kammeier, 2003). This was already a recommendation of the FAO as early as the 1960s, as well as the alternative that 'received the most attention' from Small (1972), in his study of the delta. Such a concern has been constantly expressed for at least four decades. Even nowadays, it is not rare to hear officials complaining off record, that 'farmers are stubborn', that 'they lack knowledge and only know how to grow rice' and that 'they oppose any change', despite being shown the benefits they might expect from it. Crop selection, however, is a more complex issue than merely choosing the crop with higher return to land or water.

First, the rationale for induced shifts in land use is generally – implicitly or explicitly – based on average farmers' income, overlooking the aspect of risk, which is crucial in shaping farmers' decision making. Even for irrigated agriculture, where yields are deemed to be more secured, risks in production are not negligible and include both agronomic hazards (diseases, pests, etc.) and a higher risk in marketing, further compounded by the higher requirements of cash input demanded by commercial crops. As a general rule, the potential return of capital investments is strongly correlated to the level of risk attached to the activity undertaken (Molle *et al.*, 2002). This is clearly exemplified by Szuster *et al.* (2003) in their comparative study of rice and shrimp farming in the delta. In other words, while cash crops may generate higher average returns, they are also subject to more uncertainty, either in terms of yields or farm-gate prices. Thus, only those farmers with enough capital reserve to weather the losses experienced in some years can afford to benefit from the average higher returns; others become indebted or go bankrupt. Shrimp farming in the delta, again, provides a good example of such a situation.

It could be argued, however, that the price of rice in Thailand is also unpredictable and that rice production suffers from uncertainty as much as other crops do. If the rice price does fluctuate, its crucial importance for the rural economy brings it under more scrutiny. Despite recurring complaints, echoed in newspapers, that rice farmers lose money when producing rice, the political ramifications of possible low prices and the outcry they instantaneously generate, largely shield them in reality from dropping under the break-even threshold.[5] Ad hoc public interventions are always implemented when such a risk arises (even though their impact generally falls short of expectations, and benefits tend to be captured by millers and other actors in the rice industry). This does not hold, however, for secondary or marginal crops (that invariably include the desirable 'cash crops'), and complaints of scattered producers have little chance of being heard in case of depressed prices. A typical example of such a cash crop is chilli, a rather capital- and labour-intensive crop, which can fetch B25/kg in one year (providing a high return) and B2 or B3/kg in the following year (with a net loss for farmers).[6]

[5]In addition, rice can also be readily stored and used for own consumption, or provided to relatives and friends.

[6]This situation differs significantly from that of western agriculture, where floor prices or 'intervention schemes' are generally established to compensate for economic losses when these occur. In addition, western farmers generally benefit from insurance (against exceptional yield losses) that comes with stronger cooperative and professional structures.

Second, several other constraints to diversification related to production factors are faced by farmers: labour may be lacking; for example, the harvest of mung bean, a typical supplementary crop with no additional water requirements, is often a problem because of labour shortage; capital is often required to transform the land (e.g. conversion to shrimp farms or orchards) or to invest in microirrigation; specific skills are necessary and not easily acquired by an ageing farming population; markets may be limited or the farmers not linked to them. Third, the delta agroecology, including heavy soils with little drainage and flood risk, is overall not favourable to growing field crops especially if neighbours are all growing rice. Fourth, the overextension of irrigation facilities, fostered by considerations of regional equity and by political patronage, makes it impossible to confine them to high-return agriculture only.

The last point is noteworthy. Farmers are expected to behave as rational profit-maximizers and they are not directly concerned with water productivity ($/m^3) but, rather, by the net income per unit of land ($/ha) as well as by the risk attached to a given crop or activity (Wichelns, 1999). There are several alternative crops to rice. A first group – vegetables, fruits and flowers – fares better in terms of income, water productivity and absolute water consumption. A second group – field crops, such as groundnut, mung bean and maize – uses less water, and may have better water productivity, but is generally less profitable and/or riskier with regard to selling prices. A third group – fruits in raised beds, aquaculture – includes crops with better income and water productivity but higher consumption of water. Considering these various options it is clear that water productivity may or may not be increased by a profit-maximizing cropping pattern.[7]

Siriluck and Kammeier's (2003) study of a large-scale public programme aimed at encouraging crop diversification in Thailand showed that such interventions are met with mixed success and are not flexible enough to adapt to different physical and socio-economic environments. In many instances, the attempt by extension workers to meet the 'targets' ascribed by the project has led to inadequate investments and choices, sometimes resulting in debts or bankruptcy. It is doubtful that 'pushing' for more diversification is eventually beneficial. Decisions should be made by farmers, based on their own appreciation of their environment and left to market mechanisms, in order to avoid exposing non-entrepreneurial farmers to bankruptcy. Evidence of the dynamics of diversification in the delta (Kasetsart University and IRD, 1996; Cheyroux, 2003; Molle and Srijantr, 2003) points to the fact that farmers display great responsiveness to market changes and opportunities (a point definitely confirmed by the recent spectacular development of inland shrimp farming: Szuster *et al.*, 2003). Good transportation and communication networks allow marketing channels to perform rather efficiently. Farmers will shift to other productions if uncertainty on water and sale prices is lowered. Time and again, Thai farmers have shown dramatic responsiveness to constraints on other production factors, such as land and labour for example (Molle and Srijantr, 1999), and have already sufficiently experienced the scarcity of water to adapt their cropping patterns, should conditions be favourable. Inducing crop shifts by raising differential fees to the level where they might be effective would substantially impact on farm income and critically raise economic risk, which is precisely the main factor that hinders diversification. While some potential may remain unrealized it is very unlikely that water would be a main constraint, or that pricing it would result in any significant shift.

The reallocation of water towards more beneficial uses can also occur across sectors. The issue is somewhat simpler, as few object to the fact that domestic and industrial uses are to receive priority over irrigation. Here

[7]An example of this contradiction can be found in Iran, or in Egypt, where rice appears as a productive and profitable crop, while being water-intensive, presenting a 'headache issue' (El-Kady *et al.*, 2002) to managers.

again, differential prices could theoretically help reallocate water, although water markets[8] are generally seen as being more efficient in theory. While the literature seems to underscore that there are significant potential economic gains to be expected from such transfers, it is apparent that in Thailand, this reallocation *does* occur and that non-agricultural activities are very little constrained, if at all, by lack of water. While the impact of the transfer of water out of agriculture is an important question (Howe *et al.*, 1990; Rosegrant and Ringler, 1998), leaving open the question of compensation, reallocation is taken care of by the state in several ways, as shown by the case of Bangkok Metropolitan Area (BMA): the growth of BMA generated a rise in demand from 0.46 million m^3/day in 1978 to approximately 7.5 million m^3/day in 2000, a 16-fold increase in 22 years (Molle *et al.*, 2001). This has been made possible not only by increasing the share of the Chao Phraya flow allocated to the city (up to 45–50 m^3/s) but also by using groundwater, with an average extraction around 3 Million m^3/day (TDRI, 1990). Future demand will be met by a recently completed canal which transfers water from the adjacent 'water-rich' Mae Klong basin (with a planned capacity of 45 m^3/s to be reached in 2017).

This shows, first, that the priority given to Bangkok has readily translated into an increased diversion of surface water (to the detriment of irrigation to the extent that it reduces the amount available in the dry season), and, second, that the impact of the shift has been mitigated by allowing industries to mine deep aquifers (at the cost of land subsidence and sustainability). Water from the Mae Klong basin will allow Bangkok to face future growth in demand, although possibly at a higher capital cost in economic terms than might have been possible if more water had been diverted out of agriculture in the delta area. This illustrates that Bangkok's needs are attended to in priority and that – despite its larger share in total water use – agriculture largely gets the leftover water in the system. Commentators, however, keep on asserting that the state has proved inefficient in centrally allocating water to the most beneficial use.[9] It is interesting to note the ubiquity of this argument even in settings where this problem has been handled relatively successfully.

Pricing and Cost Recovery

Justifications for cost recovery are diverse. One argument is that irrigators form a segment of society that has benefited from a specific capital investment by the state and, as such, are expected to channel back to the nation a part of the profit generated. If this logic of 'reimbursement' is often justified by notions of equity (redistribute part of the profits of those benefited), ideology (state involvement should be limited) or financial clarity (activities must be turned autonomous), shifts in public policy are generally motivated by more mundane reasons of 'financial drought'. We will examine here the rationale for cost recovery, as applied to the case of Thailand.

[8]A market is unrealistic in the present situation given the lack of control over volumes and of connectivity between users. The assertion that 'if the price of rice is low, [Thai] farmers would be happy to cede their right to industrialists' (Wongbandit, 1997) not only runs counter to the evidence that industrialists or cities are anyway served first, but also that physical constraints make such a reallocation impossible. How would the 'rights' of a group of farmers in, say, Kamphaeng Phet (middle basin) be transferred to a given golf course or factory in the suburbs of Bangkok?

[9]A typical example is provided by Christensen and Boon-Long (1994): '[A] concern which could raise problems in the area of basin management involves the authority of the basin authorities to impose allocation priorities. . . . The burden of proof for such an initiative is to show that command and control could result in better allocations and less market failure.' Israngkura (2000), for his part, considers that 'the returns on the irrigation dam investment have been low due to the lack of effective water demand management that could prevent less productive water utilisation'. This suggests that the assumed low return of irrigation has deprived other potentially more productive use, whereas irrigation is, in fact, largely allocated the leftover in the system.

Equity, redistribution and the overall arithmetic of rice production

A first line of debate is about whether, indeed, irrigated agriculture can be said to have benefited from a preferential treatment within the national economy and, thus, whether – out of a concern for equity – water pricing as an additional government tax is justifiable as means to: (i) return part of its value-added to government coffers; or (ii) allow, in particular, further investments in the non-irrigated agriculture sector (FAO, 1986).

It is necessary, therefore, to examine whether irrigated agriculture, and in particular rice cultivation, is – overall – subsidized or taxed. Thailand has long chosen to tax its agricultural exports (Schiff and Valdés, 1992) and to recover her investments in irrigation through indirect mechanisms (Small *et al.*, 1989). The revenues siphoned by the state off rice cultivation through the mechanism of the rice premium, between 1952 and 1986, have been estimated at 25% of all rural income (Ingram, 1971; Phongpaichit and Baker, 1997) and it is clear that rice farmers have indirectly paid back more than any realistic water fee. It was estimated that in 1980 these indirect revenues amounted to three times the O&M costs (Small *et al.*, 1989) while capital cost recovery reached uncommon levels. Indirect taxation may be inequitable but is quite efficient since it avoids the costs of collection and the possible corruption that may come with it (Hirschman, 1967). Because declining food prices in the last two decades (driven, in large measure, by the increase in reliable production from irrigation investments) have depleted the surplus that could be extracted from agriculture, these indirect revenues have now dwindled down, being captured as consumer surplus.

This questions the rationale used by consultants to support cost recovery: 'Thai taxpayers are paying B35 billion a year to run RID. If this is worthwhile to the farmers then why should the taxpayers have to pay for RID?' (H&P and A&E, 2000). This question stems from a narrow definition of what 'taxpayers' pay for and ignores the more global arithmetic of sectoral taxes, subsidies and cross-subsidies, not to mention the distribution of benefits to consumers and multiplier effects in the economy. Indeed, rice farmers have probably contributed more to the rest of society than they have received from it, both through taxation and impact on rice market prices.

One might argue, however, that this holds for the past but that the situation has changed. Leaving aside the argument that the water subsidy could be seen as a (small) compensation for the past pattern of indirect, yet heavy taxation, a water fee could be now construed as a charge reflecting the costs of providing irrigation water. This argument differs, depending on whether one considers that: (i) the disappearing of the premium reflects an increasing rice supply in the international market and a decline in real price (squeezing farmers' income and rendering the extraction of surplus unsustainable); or (ii) it stemmed from the growing political clout of a rent-seeking rice sector. Since the evidence unambiguously points to the first interpretation (Isvilanonda, 2001), this can be taken as an indication that rice incomes are now squeezed and that further taxation would have substantial socio-economic and political implications.

Another major argument regarding equity is that of discrimination against rain-fed agriculture, resulting from both the subsidies in capital costs and the supply of free water, since the irrigated sector can produce more per unit of land than rain-fed agriculture and better absorb the impact of declining rice prices driven by overproduction (and, initially, by taxation). Such concern for equity is often mentioned by officials and ADB consultants ('60% of the budget of the Ministry of Agriculture went to 20% of farmers' provided with irrigation). This militates for closing the gap between the two sub-sectors, for example, by having irrigators bearing the cost of water delivery. This argument is valid when applied to the initial phase of irrigation development, when rain-fed farmers disproportionately bore the costs of the rice premium and low prices, although this was smoothened by the fact that rain-fed production was mostly for home consumption and little for the market. In addition, initial differences have now been evened out by the evolution of farming systems: in the mid-term, average farm size and the degree of farm fragmentation at inheritance appear to be in line with the average income derived from a unit of land. Molle *et al.* (2002) have studied three sub-areas of the Chao Phraya delta where cropping intensities and return to land per year markedly differ. The study showed that differences in annual land productivity were largely compensated over time (albeit not fully) by growing differences in farm size, family size (linked to the rate of migration) and pluri-activity which partly rebalance final farm incomes.

Rice as a global commodity

Another relevant issue is the international dimension of subsidies, as many of these commodities, notably rice, are traded in international markets. The insistence on having farmers pay the 'real' cost of water can first be questioned when European and American agriculture is admittedly heavily subsidized (Sarker *et al.*, 1993; Baffes and Meerman, 1997; CRS, 2002). This applies especially for crops that compete in international markets – here the price is substantially set by the lowest (net)-cost producers – and it is not clear why developing countries should adopt policies which are not part of the agenda of their western or East Asian competitors. The US Congress, for example, provided $24 billion between October 1998 and 2001 to shield growers against low prices and crop disasters (The Nation, 2001). In May 2002, another 10-year $190 billion farm bill was signed by President Bush. This concerns, in particular, rice production whose revenue includes a share of 50% of subsidies (USDA, 2001, web site). Complying with orthodoxy (full operational cost recovery and 'real' factor prices), on the one hand, and disregarding it entirely, on the other, through intervention when benefits get squeezed by declining prices, illustrate that a real-cost regulated market is not yet in place for reasons that are far broader than water pricing.

An additional difficulty for Thai rice farmers comes from their wide linkage with international markets. Whereas in many markets a change in input prices is readily passed on to the consumers, albeit partly depending on the structure of the market, this does not easily occur for commodities where producers mostly operate as 'price takers', for example, because of links to international markets. In the case of rice, the Thai farm-price elasticity relative to the world-market price is 0.8 (Sombat Saehae, by e-mail, January 2000, personal communication). It follows that farm-gate prices are predominantly driven by the world market and that internal balancing mechanisms to reflect changes in factor prices are critically constrained, to the detriment of producers.

O&M expenditures, financial drought and payment for service

The need for 'cost-sharing', however, may become more pressing when the government is faced with financial squeeze and seeks to reduce expenditure, while the deterioration of irrigation facilities impinges on productivity and farm income, and gives way to costly recurrent rehabilitation programmes. Such deterioration appears relatively limited in the present case (RID's maintenance, especially in the Central Region, can be considered quite good if compared with many other countries), and there is no evidence that financial squeezes, even after the 1997 economic crisis, have drastically altered RID budgets or its capacity to carry out maintenance work. In Thailand, O&M costs are said to correspond to a 'huge drain on the national budget' (H&P and A&E, 2001) but these costs must be put in context[10]: the potential gains from the cost-sharing policies proposed represent 0.37% of the value of Thai agricultural exports, 0.27% of Thai government expenditures or 15% of the

[10]The proposal by ADB's consultants was to set up a tentative fee of B120/rai in pilot projects. This value was intended as a compromise derived from the total estimated O&M costs: B522/rai, out of which B210 were true direct costs (H&P and A&E, 2000).

RID budget itself. Savings of 0.27%, not considering the transaction costs corresponding to the collection of fees, may be not negligible but certainly not considerable when compared with the political risk attached to it. Thus, it seems that the financial squeeze that was one of the major drivers of the Philippine NIA and of the Mexican reforms is not (yet) a crucial incentive to change in the Thai case.

An important distinction must be made between cost recovery that goes to the government coffers, and irrigation financing, that is the provision of funds actually used for irrigation costs (Small, 1996). Surprisingly, the Royal Irrigation Act of 1942 recognized this fact early. It made it legally possible to charge users for water (despite fixing unrealistically low limits), but stipulated that collected money could not be considered as state revenue and should constitute a special fund to be put back into the development of irrigation. If this is the case, and if users are granted partial or total control of the allocation of these funds, then incentives to pay and limit degradation are created and a sense of 'property' may emerge. More generally, it is the potential role of pricing at the interface between line agencies and users, which deserves emphasis (see next section).

Raising fees that only contribute to the government income is a measure that is not conducive to internal improvement and is, therefore, a decision pertaining to the design of the tax system as a whole: making users bear a part of O&M costs is helpful in internalizing costs from the point of view of the government, but shifting this financial burden has to be reasoned, based on wider public objectives of poverty alleviation and wealth redistribution, sectoral policies, possible treasury difficulties and political risks, which are all dependent upon the context of each particular political economy. Schiff and Valdés (1992) showed how governments are caught up in a web of contradictory goals, including protecting farmers, protecting consumers from high food prices, raising revenues through taxation and ensuring the competitiveness of economic sectors in the world market. This makes decision making more complex than just embracing the principle of cost recovery. The question raised here is how governments can change their policy, for example, from providing public goods for free to charging for it, without providing compensation.

To conclude this section it is interesting to draw a parallel between charging for irrigation water and charging for groundwater use. Charging for groundwater use is backed by strong economic justifications because of the critical costs of overdraft in terms of land subsidence and increased flood risk and damage. Yet the constraints faced in establishing such charges illustrate what is at stake. Groundwater use mostly concerns industries in BMA and has remained admittedly underpriced, largely because of the political clout of both the Federation of Thai Industries.[11] All in all, charging for irrigation water use may be a more difficult business – both socially and technically – than charging for groundwater, which lends itself much more easily to control and volumetric charging.

Recent Attempts to Reform the Water Sector and Future Prospects

Further to the 1997 financial crisis, Thailand obtained a $600 million loan from both the ADB and the Japanese Bank for International Cooperation under the name of Agriculture Sector Program Loan (ASPL), conditional upon acceptance of some principles and a reform of the water sector (RWS). A policy matrix was defined, showing commitment and successive milestones to be achieved. The RWS was designed by consultants to the ADB and issued in March 2001. It included several components (H&P and A&E, 2001):

- Strengthening of the Office of the National Water Resources Committee (ONWRC) and transforming it into an apex body;

[11]The federation opposed a gradual rise of the groundwater price (from B3.5 to 8.5/m^3, in an attempt to catch up with tap water at B12.5/m^3), stating that a price of B5 would 'lead to hardship'. Recently, the Thaksin administration seems to have adopted a more energetic stance and given deadlines for the phasing out of wells in areas where pipe water is available.

- Decentralization of water management to river basins;
- Watershed protection strategy;
- Setting of performance indicators and service standards;
- Participatory irrigation management (PIM) and definition of farmers as clients rather than beneficiaries;
- Cost-sharing of O&M;
- Reorganization, decentralization and privatization of RID.

In parallel, the National Water Resource Committee was drafting a Water Law which was supposed to encapsulate many of the crucial aspects of this ambitious reform, notably the establishment of River Basin Committees (RBCs), and the separation of the water policy, management and O&M functions. It is beyond the scope of this chapter to discuss the merits of the proposed reform but the aspects of cost-sharing, service agreements and participatory management are relevant to our current discussion.

The RWS aimed at establishing a contractual relationship between RID as *provider* and farmers as *clients*. It was expected that such agreements duly defined through established standards and monitored through performance indicators would significantly increase the quality of delivery, thus justifying the principle of cost-sharing put forth (as opposed to cost recovery). This would set in motion a virtuous circle whereby farmers would get financial autonomy and better service, while participating fully in the definition of operational targets and maintenance priorities. This virtuous circle is well identified in the literature (Small *et al.*, 1989; Small and Carruthers, 1991; see Molle and Berkoff, Chapter 1, this volume) but it has several prerequisites that were overlooked in the RWS.

The first crucial weak point of the reform was that there was no provision to ensure that RID will deliver water, following standards of service agreed upon. By failing to link RID's financial income to such service, no drastic pressure would be put on RID to reform its management and it is highly doubtful that raising their awareness of the necessity of change by seminars or capacity building would be sufficient to ensure this. When fees contribute significantly to the salary of the officials of the agencies, or are used to pay field staff who are selected by the users themselves, there is a real change in the governance pattern of irrigation. This, of course, was the most contentious part of a reform and the one that was likely to be compromised.

Service agreements were supposed to be established between users and RID but little was said about whether the existing human and physical capacity needed to achieve this, exists or not. After the early overemphasis on structural aspects, it has now become all too common to disregard the physical dimensions of management and to overlook their impact on reforms (Briscoe, 1997; Facon, 2002). Water management in the Chao Phraya basin is constrained by various aspects, including the lack of control over abstraction along the waterways, the occurrence of side flows, the crude technical design of most hydraulic regulation structures and the development of conjunctive use by farmers (Molle *et al.*, 2001; Molle, 2004). This makes the definition of service agreements at lower levels extremely problematic. The RWS made no provision to ensure that hydraulic regulation was up to the task envisaged. It just assumed that 'farmers will receive improved irrigation service delivery. Farmers need to feel confident that service is being improved' (H&P and A&E, 2001).

Initial service agreements were to be developed at the project level between RID and Water User Groups (WUG): '[A]s soon as WUG get ready . . . as federation of water users moves up the system, to IWUGs and WUAs, service agreements will move with them.' This was the second weak point of the reform. As is the case in many failed reforms of PIM, farmer organizations are first built at the tertiary level. This is easily accepted by irrigation agencies because they usually have no interest in what is occurring beyond the tertiary turnout and blame for deficiencies can then be placed if required on the farmers themselves. Since certainty in supply at the tertiary level generally depends on allocation and distribution at higher levels in the system and cannot be fully ensured, farmers soon discover that there is nothing to be managed and that they are wasting their time. Present reforms still consider water management at the tertiary level and maintenance as crucial issues but these may actually have lost importance in the eyes of farmers. As a result of the ongoing decentralization process, local administrations have seen their budget increasing and are now using the resources under their control to fund maintenance (notably mechanical ditch dredging). Likewise, the organizational needs of water management have been radically changed further to the introduction of direct seeding in lieu of transplanting, the development of secondary water sources and the spread of pumps. This has weakened the exigency of collective action and fostered individual strategies.

In contrast, the issue that has gained prominence in a context of water scarcity is the allocation of water in the dry season (Molle, 2004). The process towards involving users in management should be initiated by allowing a transparent allocation process in which users would have representatives at each level (main canal level, scheme level, plus the delta and basin levels for farmers in the Chao Phraya delta). The definition of (seasonal) entitlements in which users have a say (as a first step to defining water rights) is the preliminary step to the definition of service agreements. Such agreements must be accompanied by a technical capacity to operationalize them, to monitor distribution and to assess whether the actual and the agreed supply match. This, again, has technical, managerial, legal and political implications that need combined support from the government, the political class and the society, which does not seem to be forthcoming. A part of RID officers' foot-dragging in considering the issue might be linked to the fact that establishing service agreements and a water charge may eventually backfire, in that farmers would be given 'the legal standing to bargain forcefully with the water conveyance bureaucracy for timely and efficient service' (Rosegrant and Binswanger, 1994).

The reform process initiated under the ASPL has been phased out during 2002 and 2003. Pilot projects have been implemented partly, and without supervision, leading to no real change. Cost-sharing policies and service agreements have disappeared from the front scene. The draft Water Law has been shelved. The restructuring of RID has been limited to measures such as the non-replacement of retiring staff. Only the setting of RBCs has proceeded, under the guidance of the ONWRC. At present, however, RBCs still lack the formal recognition that would give them more importance than a mere consultative forum. The failure of the reform can be partly attributed to some of its internal weaknesses (over-optimism, structural constraints to the definition of service agreements, misplaced emphasis on building from the tertiary level, etc.) but was chiefly undermined by the lack of support from the Thai side, from both bureaucratic and political quarters. Its final dismissal came with the decision of the Thaksin administration to discontinue loans from the ADB. This failure exemplifies disregard of what Briscoe (1997) considers the first requirement for reform: that there be a demand for it. However sound and well intentioned they may be, reforms decided and imposed by external institutions have little chance of succeeding.

In addition to the lack of strong political commitment and support, and of structural rehabilitation, the reform failed to ensure the crucial point of financial autonomy. Financial autonomy makes the water charge a 'glue factor' in a wider process of transfer of responsibility to users, who can decide on the hiring of staff and the priorities in maintenance which are ensured by their own funds. This factor, crucial in the Mexican reform, was absent from the ASPL and raises the question of whether a partial reform can achieve partial benefits or whether it is doomed to failure because of the absence of crucial linkages in the virtuous circle to be created.

Conclusions

Pricing mechanisms are often held as a potential tool to help 'rationalize' the use of water in ways that increase the economic efficiency of both water use and allocation. Application of such measures has been met with some success in the domestic and industrial water sectors but has so far failed to produce convincing examples in the large-scale public-irrigation sector of developing countries. In the particular case of Thailand, both the rationale and the applicability of such measures were found to be problematic.

The idea that water waste would be a consequence of the non-pricing of water was little supported by evidence. The closure of river basins, most notably the Chao Phraya basin, is accompanied by reductions in losses, both at the farm and the basin level, with only 12% of dam releases in the dry season lost to non-beneficial use: a reality which contrasts sharply with the image of outright waste that is routinely conjured up to justify pricing as a way to induce water savings. The technical impossibility of establishing volumetric water deliveries, as well as the wholesaling of water in the present context, removed the possibility of influencing users' behaviour through pricing. Even if this is possible, there are indications that the elasticity of water use is very low at the range of prices proposed to meet appropriate cost recovery objectives, in addition to the political difficulties in implementing them.

The possibility of inducing land-use shifts towards crops with higher water productivity runs into similar difficulties. It was shown that farmers' decision making gives much emphasis to risk, and that water savings or water productivity objectives do not necessarily coincide with income maximization. To assume that there are substantial gains to be expected from shifts in cropping patterns if water is priced is to misunderstand the dynamics of, and constraints to, diversification. If much higher profits could readily be made through diversification, farmers would not wait for this. To penalize rice because of its higher water needs would only raise the vulnerability of the main crop, without making alternatives more secure or removing the other constraints to diversification, particularly the need of stable markets. Likewise, few economic gains can be expected from intersectoral reallocation of water, as non-agriculture sectors are already given de facto priority.

The principle of cost recovery is generally propped up by an image of irrigators who have unduly benefited from government largesse and are expected to pay back the 'taxpayers'. This was confronted with the net transfer of wealth from agriculture to other sectors, symbolized in Thailand by 30 years of rice premium, and with the multifaceted benefits of irrigation accruing to the society. It was also recognized that political considerations and national challenges, such as food security, rather than mere aspects of return to capital, dictated earlier priorities in state investments and that shifts in policy are not easily justified and implemented.

A water charge would be akin to a flat tax that would decrease farm income, without effectively sending a signal of water scarcity, and decrease international competitiveness (especially with regard to western countries that continue their policy of subsidy), while it would not be easily passed on to the consumer because of the strong linkages between domestic and world rice markets. While reductions in price subsidies in developed countries are compensated for by adequate income policies, the latter are generally omitted in developing countries (partly due to the difficulty in implementing such income-support schemes). Shifting, even partly, the O&M costs to the users is helpful in internalizing costs from the point of view of the government and signalling to all concerned the real cost of system O&M. It may help ensuring financial sustainability if public budgets happen to be lacking, but has socio-economic and political implications that need to be addressed.

Beyond 'the obsessive traditional concern on the part of resource economics with correct pricing levels for irrigation water' (Svendsen and Rosegrant, 1994), water pricing is made more attractive when it is construed as a binding element of a wider mechanism that redefines relations between users and the agency (Small and Carruthers, 1991; Bromley, 2000). It gains sense if a full reform is implemented that includes a degree of turnover and financial autonomy whereby water delivery service is paid for by users and linked to the quality of service. Service agreements should include definition of the allocation of resources and of the timing of the distribution of allotments. In both processes, the users should have a say, given their importance in a context of scarcity. Modifying the status of public agencies and civil servants in order to link their salary to performance and to the payment of users requires a much more ambitious reform in the direction of which the government has so far taken no unequivocal steps.

The failure of the ASPL reform illustrates several lessons that failed to be learnt, in particular, the importance of infrastructure in the design of service agreements or bulk allocation, as well as the necessity to muster internal and political support for the reform. Emphasis thus, should be placed on paving the way for a thorough reform, ensuring in particular, the technical and managerial capacity to define and operationalize services, as well as the legal framework and the political/public support for changes in line agencies. Failing to alter the pattern of governance jeopardizes reforms which remain generally restricted to isolated components, backed by arguments that are turned invalid. It is not clear, therefore, whether 'half-measures' provide 'half-benefits', and must be seen as 'second-best' options, as economic parlance suggests, or if they are likely, because of the absence of linkages and invalid supporting assumptions, to fail and lead to an overall negative result, rather than to the theoretical gains envisioned. All in all, it appears unwise to propel water pricing to the fore of the reform, as a symbol of restored economic orthodoxy, when it is expected to play a more crucial and later role in a wider and longer reform process.

References

ADB (Asian Development Bank) (2000) Water for all: the water policy of the Asian Development Bank. Available at: http://www.adb.org/documents/policies/water/water.pdf

Baffes, J. and Meerman, J. (1997) *From Prices to Incomes: Agricultural Subsidization without Protection?* World Bank, Washington, DC.

Bangkok Post (1999) No charge for using river water, 13 January.

Bos, M.G. and Walters, W. (1990) Water charges and irrigation efficiencies. *Irrigation and Drainage Systems* 4, 267–268.

Briscoe, J. (1997) Managing water as an economic good: rule for reformers. In: Kay, M., Franks, T. and Smith, L. (eds) *Water: Economics, Management and Demand.* E & FN Spon, London.

Bromley, D.W. (2000) Property regimes and pricing regimes in water resource management. In: Dinar, A. (ed.) *The Political Economy of Water Pricing Reforms.* Oxford University Press, New York, pp. 141–166.

Cheyroux, B. (2003) Fruits and vegetables in Thailand's rice bowl: the agricultural development of poldered raised bed systems in the Damnoen Saduak area. In: Molle, F. and Srijantr, T. (eds) *Thailand's Rice Bowl: Perspectives on Social and Agricultural Change in the Chao Phraya Delta.* White Lotus, Bangkok, pp. 157–176.

Christensen, S.R. and Boon-Long, A. (1994) *Institutional Problems in Thai Water Management.* Thailand Development Research Institute, Bangkok, p. 54.

CRS (Congressional Research Service) (2002) Agriculture policy and farm bill briefing book. Available at: http://www.cnie.org/nle/crsreports/briefingbooks/agbill/ebagr8.cfm

Dinar, A. and Subramanian, A. (1997) *Water Pricing Experience: An International Perspective.* World Bank Technical Paper No. 386. World Bank, Washington, DC.

Dinar, A. (ed.) (2000) *The Political Economy of Water Pricing Reforms.* OUP, for the World Bank, New York, X, pp. 405.

El-Kady, M., Moustafa, M. and Zhu, Z. (2002) Water demand management: adopted polices in Egypt. Draft.

Facon, T. (2002) Downstream of irrigation water pricing – the infrastructure design and operational management considerations. Paper presented at the conference 'Irrigation water policies: micro and macro considerations', Agadir, Morocco, 15–17 June 2002.

FAO (Food and Agriculture Organization of the United Nations) (1986) *Report on the Expert Consultation on Irrigation Water Charges.* FAO, Rome.

Gibbons, D.C. (1986) *The Economic Value of Water*. Resources for the Future, Washington, DC.

H&P (Halcrow and Partners) and A&E (ARCADIS/Euroconsult) (2000) Sharing the cost of irrigation. Draft final report Vol. 9, Capacity building in the water resources sector project ADB-TA 3260-THA.

H&P and A&E (2001) Component C: reorienting and reorganising service delivery operations in irrigation, Final report Volume 3/3, Capacity building in the water resources sector project ADB-TA 3260-THA.

Howe, C.W., Lazo, J.K. and Weber, K.R. (1990) The economic impacts of agriculture-to-urban water transfers on the area of origin: a case study of the Arkansas River in Colorado. *American Journal of Agricultural Economics* 72(5), 1200–1204.

Hirschman, A. (1967) *Development Projects Observed*. The Brookings Institution, Washington, DC.

IBRD (International Bank for Reconstruction and Development) (1950) *Technical Report on Chao Phraya Irrigation, Drainage and Communications Project in Thailand*. Project Vol. 1. IBRD, Washington, DC.

Ingram, J.C. (1971) *Economic Change in Thailand 1850–1970*. Oxford University Press, Kuala Lumpur, London, Singapore.

Israngkura, A. (2000) Why can't Thailand afford more irrigation dams? *TDRI Quarterly Review* 15(3), 3–7.

Isvilanonda, S. (2001) Rice supply and demand in Thailand: the recent trend and future outlook. A paper presented at the workshop on 'Medium and long-term prospects of rice supply and demand in the 21st Century,' 3–5 December, 2001, Los Baños, The Philippines.

Kasetsart University and IRD (1996) *Identification of Agricultural and Irrigation Patterns in the Central Plain of Thailand: Prospects for Agricultural Research and Development*. Bangkok.

Keller, A., Keller, J. and Seckler, D. (1996) *Integrated Water Resources Systems: Theory and Policy Implications*. Research Report 3. International Irrigation Management Institute, Colombo.

Molle, F. (2004) Technical and institutional responses to basin closure in the Chao Phraya river basin, Thailand. *Water International* 29(1), 70–80.

Molle, F. and Srijantr, T. (1999) *Agrarian Change and the Land System in the Chao Phraya Delta*. DORAS Project. Research Report No. 6, Kasetsart University, Bangkok, p. 191.

Molle, F. and Srijantr T. (eds) (2003) *Perspectives on Social and Agricultural Change in the Chao Phraya Delta*. White Lotus, Bangkok.

Molle, F., Chompadist, C., Srijantr, T. and Keawkulaya, J. (2001) *Dry-Season Water Allocation and Management in the Chao Phraya Delta*. Research Report No. 8. DORAS Project. Kasetsart University, Bangkok.

Molle, F., Srijantr, T. and Latham, L. (2002) Balance and imbalances in village economy: a three village study in the Chao Phraya Delta. Paper presented to the 8th Thai Studies International Conference, Ubon Ratchasima, Thailand.

Murray-Rust, H. and Svendsen, M. (2001) Performance of locally managed irrigation in Turkey: Gediz case study. *Irrigation and Drainage Systems* 15, 373–388.

Perry, C.J. (1999) The IWMI water resources paradigm: definitions and implications. *Agricultural Water Management* 40(1), 45–50.

Phongpaichit, P. and Baker, C. (1997) *Thailand: Economy and Politics*. Oxford University Press/Asia Book, Singapore.

Postel, S. (1992) *Last Oasis: Facing Water Scarcity*. W.W. Norton & Co., New York.

Rogers, P., Bhatia, R. and Huber, A. (1997) Water as a social and economic good: how to put the principle into practice. Paper prepared for the meeting of the TAC of the Global Water Partnership.

Rosegrant, M.W. and Binswanger, H.P. (1994) Markets in tradable water rights: potential for efficiency gains in developing-country water resource allocation. *World Development* 22(11), 1613–1625.

Rosegrant, M.W. and Ringler, C. (1998) Impact on food security and rural development of transferring water out of agriculture. *Water Policy* 1(6), 567–586.

Samad, M. (2001) *Impact of Irrigation Management Transfer on the Performance of Irrigation Systems: A Review of Selected Asian Experiences*. ACIAR Water Policy Workshop, Bangkok.

Sarker, R., Meilke, K. and Hoy, M. (1993) The political economy of systematic government intervention in agriculture. *Canadian Journal of Agricultural Economics* 41, 289–309.

Schiff, M. and Valdés, A. (1992) *The Plundering of Agriculture in Developing Countries*. The World Bank, Washington, DC.

Siriluck, S. and Kammeier, H.D. (2003) Government policy and farmers' decision making: the agricultural diversification programme for the Chao Phraya river basin, 1993–2000. In: Molle, F. and Srijantr, T. (eds) *Thailand's Rice Bowl: Perspectives on Social and Agricultural Change in the Chao Phraya Delta*. White Lotus, Bangkok, pp. 63–96.

Small, E.L. (1972) An economic evaluation of water control in the northern region of the Greater Chao Phraya Project of Thailand. PhD thesis, Cornell University, Ithaca, New York.

Small, L. and Carruthers, I. (1991) *Farmer-Financed Irrigation: The Economics of Reform*. Cambridge University Press, Cambridge.

Small, L.E., Adriano, M.S., Martin, E.D., Bhatia, R., Shim, Y.K. and Pradhan, P. (1989) *Financing Irrigation Services: A Literature Review and Selected Case Studies from Asia*. International Irrigation Management Institute, Colombo, p. ix.

Small, L.E. (1996) Financial tools for improving irrigation performance. In: Sampath, R.K. and Young, R.A. (eds) *Social, Economic, and Institutional Issues in Third World Irrigation Management*. Westview Press, Boulder, Colorado, pp. 147–268.

Svendsen, M. and Rosegrant, M. (1994) Irrigation development in South-east Asia beyond 2000: will the future be like the past? *Water International* 19, 25–35.

Szuster, B.W., Molle, F., Flaherty, M.S. and Srijantr, T. (2003) Socio-economic and environmental implications of inland shrimp farming in the Chao Phraya delta. In: Molle, F. and Srijantr, T. (eds) *Thailand's Rice Bowl: Perspectives on Social and Agricultural Change in the Chao Phraya Delta*. White Lotus, Bangkok, pp. 177–194.

TDRI (Thailand Development Research Institute) (1990) *Water Shortages: Managing Demand to Expand Supply*. Thailand Development Research Institute, Bangkok, p. 101.

The Nation (1998) Water greed threatens Asian farms.

The Nation (1999), 8 January.

The Nation (1999) Government to consider ADB terms, 17 February.

The Nation (2000) Groups against farmers paying to use water, 21 April.

The Nation (2000) Water pricing test project to start soon, 23 April.

The Nation (2001) USA struggles to help farmers within rules, 10 January.

USDA (US Department of Agriculture) (2001) Electronic outlook report from the economic: Research service, July 2001, p. 10. Available at: www.ers.usda.gov and complementary information from Nathan Childs.

Van der Heide, H. (1903) *General Report on Irrigation and Drainage in the Lower Menam Valley*. Ministry of Agriculture, Bangkok, p. 149.

Wichelns, D. (1999) Economic efficiency in irrigation water policy with an example from Egypt. *Water Resources Development* 15(4), 542–560.

Wongbandit, A. (1997) *Legal Aspects, Annexe C of the Report 'Chao Phraya Basin Water Resources Management Strategy'*. Binnie & Partners, Bangkok, p. 74.

World Bank (1993) *Water Resources Management: A World Bank Policy Paper*. Washington, DC.

6 Water Rights and Water Fees in Rural Tanzania

B. van Koppen, C.S. Sokile, B.A. Lankford,
N. Hatibu, H. Mahoo and P.Z. Yanda

Introduction and Background

Aim of the chapter

Tanzania is an agrarian country, which ranks 151th out of 173 on the Human Development Index (UNDP, 2002) and 80% of its 34 million inhabitants live in rural areas, where agriculture constitutes their primary economic mainstay. Agriculture contributes 48% to the gross national product (GNP). Physical water resources are relatively abundant in the coastal and highland areas, which receive well over 1000mm of rainfall/year, but most parts of the drier interior receive less than 600mm. An estimated 50% of all annual surface runoff flows into the Indian ocean and the large lakes (URT, 2002). However, temporal and spatial variability in rainfall and surface flows is high. Yet, Tanzania's level of infrastructural development to harness water and to mitigate nature's variability is still very low, primarily because of the lack of financial, technical and institutional resources to bridge the infrastructural gap. It is estimated that the naturally available land and water resources are sufficient for 2.3 million, 4.8 million and 22.3 million ha of high-, medium- and low-irrigation potential areas in the country respectively. However, currently, the total area under irrigation is only 191,900ha, out of which 122,200ha (64%) falls under traditional irrigation schemes (JICA/MAFS, 2003). The remaining 36% are medium-sized centrally managed irrigation schemes, owned by public and private institutions, primarily for sugarcane, rice and tea. More than 60% of energy produced in the country is from hydropower plants located in the Rufiji and Pangani basins, downstream of smallholder irrigators. Other economic sectors that utilize the underdeveloped water resources include livestock, forestry, mining, tourism, industry and fisheries (URT, 2002).

The priority in Tanzania's National Water Policy of 1991 was to further develop water resources for domestic and productive uses nationwide to boost socio-economic development. However, this changed drastically in the mid-1990s, when the Tanzanian government amended the national water rights system and, anticipating a redrafting of the entire water law, started implementing pilot experiments of this system in the Rufiji and Pangani basins. The amendment increased fees charged for the mere use of water, in addition to the fees users paid for service delivery through public infrastructure construction, operation and maintenance. The twofold aim of this new fee was cost recovery for basin-level water management services and fostering the wise use of what was seen as a scarce 'economic' resource (World Bank, 1996). The new fees system concerns anyone who diverts and abstracts even the smallest quantities of surface and groundwater for productive uses and also includes all water users who invest privately in water infrastructure. In state-supported irrigation schemes, the fee is additional to the partial or full cost recovery of infrastructural construction, operation and maintenance – the latter type of fee is not further addressed in this chapter. Related to this fee payment is that all water users or groups are obliged to register with the Ministry of Water and Livestock Development to obtain a 'water right'. This is a certificate indicating the purpose and an annual volume of water resources to which the right holder is entitled. Water users have to pay an application fee at the moment of registration of the water right equivalent to $40,[1] plus an annual 'economic water user fee', proportionate to the volume allocated and depending upon the purpose of the water use. The minimum flat rate for uses up to 3.7 l/s for the annual economic water fee is $35.

In this new policy and law, the government also started advocating stronger user participation in the river basin Water Boards, which were fully governmental up to the mid-1990s. It further strengthened the establishment of Water User Associations (WUAs) at the lowest tiers, which were expected to manage water for multiple uses at village and ward level and were to be represented at higher levels, up to the basin level (World Bank, 1996).

With all ingredients present for what was then, at abstract level, seen as the best practice of integrated water resources management, but nowhere in sub-Saharan Africa really implemented as yet, the first results of implementation in the early 2000s appeared disappointing, at least among small-scale informal users, who constitute the large majority of water users in Tanzania. This was apparent in the Upper Ruaha catchment in South West Tanzania, the focus of this chapter. The Upper Ruaha catchment is a sub-basin of the Rufiji basin where, together with the Pangani basin, the amendment was implemented through the River Basin Management (RBM) project, funded by a loan from the World Bank. In the Upper Ruaha catchment, neither of the two goals of cost recovery for water management services by the government nor wiser water use to solve the water-scarcity problem has been achieved, at least among the majority of small-scale users. In contrast, fee payment by the few large users did contribute to achieving the goal of cost recovery for basin management.

This chapter analyses the implementation and impacts of the new water rights and fees system in the Upper Ruaha, which encompasses farmer-managed irrigation through river abstractions, the typical mode of irrigation in 64% of Tanzania's irrigated area. The second section analyses how the Tanzanian government, advised by the World Bank, suddenly abandoned its agenda of water development in the early 1990s. Justified by basin-specific, localized conflicts over water in the dry season, a water regulation agenda was introduced that put water scarcity and conservation nationwide at the centre stage. It describes how the new water administration that was put in place to effectuate that regulation agenda was grafted upon the formal legal framework that was inherited from the colonial powers since 1923. These colonial roots explain why water management has ever since been implemented by highly centralized water authorities. However, up till 1994, the administrative system of water rights remained rather dormant, and reached only few formal, large-scale users. The revival of that system, expansion of its implementation nationwide to also include the informal rural majority, and the drastic increase of the fees to obtain water rights were to generate revenue and self-finance government and the expanding basin-management institutions and activities. Payment and valuing water as an economic good were put forward as effective ways to stimulate water conservation and saving.

The three subsequent sections evaluate the implementation processes and impacts on the ground in the Upper Ruaha basin, distinguishing the three components of the water rights system: registration, cost recovery and water allocation. The third section discusses the strengths and weaknesses of registration. As elaborated in the fourth section, the weaknesses of the registration render the system a shaky foundation for volume-based cost recovery among many small users. The fifth section highlights how the new water rights and fees system completely failed as a water allocation tool and aggravated upstream–downstream conflicts in the dry period. The sixth section concludes the chapter by identifying the adjustments required in the current water law in order to reach logistically realistic registration, cost recovery that generates net benefits for government, and government intervention in the water allocation issue that effectively support conflict mitigation during the dry season.

Background of the Upper Ruaha catchment

The Upper Ruaha catchment covers an area of 21,500 km^2 and forms the headwaters of the Great Ruaha river – itself forming a major sub-basin of the Rufiji river (Fig. 6.1). The catchment can be broadly divided into a surrounding high escarpment, the lower slopes and a central plain, named the Usangu plains. The plain receives 600–800 mm of average annual rainfall with a peak of 1500 mm observed on the high escarpment. There are five perennial rivers and a large number of seasonal streams draining from the escarpment. Most of the rain falls in one season from mid-November to May. The dry season is from June to November.

The population in the Upper Ruaha catchment which stood at 1.3 million in 1996 in this area has grown extremely rapidly, mainly because of a continuous influx of migrants. By 1990, 55% of the population consisted of migrants from at least 20 different ethnic groups – especially cultivators from the southern highlands. In-migrating livestock herders from central and northern Tanzania constituted 18% of the population,

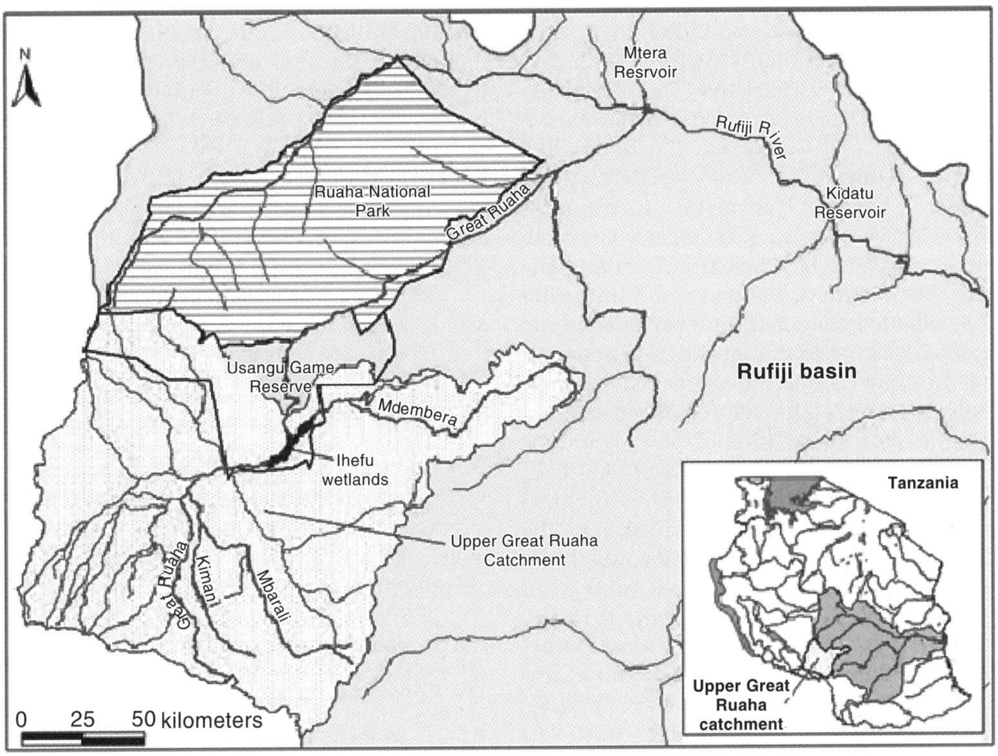

Fig. 6.1. The Upper Ruaha catchment.

and today they own the majority of herds in the area. They are concentrated in the downstream plains (SMUWC, 2000a,b, 2001). Since the government's gazetting (a notification of its legal status as a game reserve), and closure of the wetland area situated at the lowest point in the plains in 2002, pressure on land and water resources in the other parts of lower plains further increased. While the clans of settler-cultivators located upstream have kept their social structures somewhat intact in spite of Ujamaa villagization and the growing influence of local governments, the social cohesion among dispersed communities in the downstream plains is weaker.

Since the early 20th century, the original settler societies and the in-migrating cultivators started taking up irrigated agriculture in both the wet (paddy) and dry seasons (paddy and other crops, albeit in small areas) by abstracting water from the many streams. By blocking these streams with *dindilos* (seasonal weirs of wood and grasses), water is diverted into earthen diversion canals (Lankford, 2004). In the last two decades, external support was provided to replace some of these seasonal structures with permanent concrete structures. This saved the communities the recurrent efforts of rebuilding the seasonal weirs after the floods had washed them away. Unfortunately, these structures have not been made with a view of providing an easy and transparent way to apportion water between the canal and the river. Sluices are rudimentary and if it is not clear whether the maximum capacity of the intake is related to real needs, it is apparent that they have not been designed based on an analysis of the catchment overall supply and demand (nor on an idea of how to reduce diversions in times of shortage). In total, there are an estimated 120 offtake structures in the catchment, 70 of which are in the Mkoji sub-catchment. More than two-thirds of the intakes were constructed after 1970 (SMUWC, 2000a,b; Sokile and van Koppen, 2004). In the 1970s and 1980s, three state-owned rice schemes were initiated for smallholder cultivation at the lower slopes: Kapunga (3000 ha), Mbarali (3200 ha) and Madibira (3000 ha). In addition, 'valley bottoms' of small streams are cultivated in the high and medium catchment towards the south and west of the area. Recently, favourable markets for irrigated crops further increased demand for irrigated land and water. While prices for the original non-irrigated crops such as coffee and pyrethrum fell, prices and markets for irrigated vegetables and maize improved. Currently, the total wet-season irrigated area ranges from 20,000 to 40,000 ha depending on the annual rainfall. Most irrigated land is farmer-managed.

Farmers' own irrigation development has been accompanied by effective customary water management arrangements within and between schemes of a common stream. Community-based user groups govern the construction and maintenance of dindilos and diversion canals, and water distribution within the local schemes. Customary water management principles that contribute to this efficacy include water rights based on labour contributions, rotational water allocation within a scheme and, at times, some forms of rotation among upstream and downstream schemes, consensus building and conflict resolution before escalation, consideration for the weakest community members, and peer control with low transaction costs (Maganga, 1998; Sokile and van Koppen, 2004; Sokile, 2005). In the dry season, rotation between the respective schemes covers villages along long stretches of the common stream (Sokile, 2005).[1]

However, customary water-sharing arrangements between upstream and down-

[1]An example of customary interscheme water rotation (locally known as *zamu*, or 'turn') is the Mlowo tributary to the Mkoji river. At the beginning of the critical dry period, local leaders and canal committee members from four villages, other formal and informal water right holders, two private farmers, the government-owned Langwira seed farm, the NARCO ranch and representatives of pastoralists from four villages further downstream – Mahongole, Mhwela, Mwatenga and Kilambo – come together to agree on a weekly rotation. On Sundays, water is left to flow for domestic uses, including brick-making along the banks, and for the users further downstream that are not part of the *zamu*. The distance between the upstream and downstream participants is 24 km.

stream users came under considerable stress in the last two to three decades and they could not prevent the rapid growth in water abstraction in the upper catchment, and also by the three public schemes, which increasingly deprived the further downstream areas of the dry-season flows they used to have in the past. Some downstream dindilos and schemes have even been abandoned for this reason, while former perennial flows now dry up for some weeks in the dry season. Initially, village elders from the downstream areas organized official delegations to upstream communities and the public schemes, but without much effect (Video, 'Talking about Usangu', 2001). Some downstream farmers sought individual solutions and started to rentirrigable land in upstream farmer-managed irrigated areas.

Further downstream are the Ruaha National Park which requires a minimum flow of water for wildlife and tourists, and the Mtera and Kidatu hydropower plants. The Great Ruaha river fills these two dams with the floods during the rainy season; the contribution of the small dry-season flow is very limited. Hence, the remaining sections exclude the hydropower plants as stakeholders in the upstream–downstream conflicts in the Upper Ruaha catchment.

The Crafting of the New Water Rights and Fees System

Water legislation in Tanzania up till the 1990s

The system of water rights and fees designed in the 1990s and implemented in the pilot World Bank-funded RBM project builds on three key aspects of the formal water law and institutions introduced to Tanzania by German and British colonial settlers in the early and mid-1900s. First, the ownership claims to water by the state, rooted in the colonial origins of water appropriation, legitimize an even more far-reaching claim stipulated in the new water rights and fees system, which is that the government as owner of the nation's water resources is, therefore, also fully entitled to charge its citizens for the use of the resource. Initially, the settlers developed water rights systems in areas where they intensified their own agricultural water use, for example around Kilimanjaro. This enabled the regulation of their own local water use but, implicitly, this also entailed the exclusion of others without such water rights from formal entitlements. In 1948, the then colonial state enshrined this appropriation of water within the then prevailing colonial boundaries into formal law. The Water Ordinance of 1948, Chapter 257, stipulates in section 4 that 'the entire property in water within the Territory is hereby vested in the Governor, in trust for His Majesty as Administering Authority for Tanganyika'. After the independence in 1961, the new government under Julius Nyerere continued this principle, declaring that 'all water in Tanganyika is vested in the United Republic' under the Water Utilization (Control and Regulation) Act 1974, section 8.

A second aspect of the new water rights and fees system that has its roots in the colonial design of water management is the highly centralized, top-down nature of government institutions for water management. This absolute central state authority is delegated and expanded to lower tiers of regional- and basin-level water management institutions and Water Officers who are only accountable upwards. Since the Water Ordinance of 1959, the Minister has been appointing national Water Officers, vested with the almost absolute authority to make decisions regarding the allocation and changes of water rights. The Water Ordinance of 1959 and the Water Utilization (Control and Regulation) Act of 1974 prescribe regional officers below the national Principal Water Officer, all to be appointed by the Minister. From 1981 onwards, basin boundaries have been introduced to gradually replace the regional boundaries (URT, 1981). In the Pangani basin a Water Office was opened in 1991, supported by NORAD of the Norwegian government. In the Rufiji basin, the Water Office started in 1993 with government funds. These two basins were selected because of their importance for the nation's hydropower generation. Over the years, the central Principal Water Officer and his delegates at regional or basin level had almost absolute powers in carrying out their key tasks of assessing whether new entrants applying for a right could be approved or not, and of issuing these water rights with or without attached conditions. Up till 1997, a Water Officer had only to 'consider', but was 'not bound to follow the advice'[2] of regional- and later basin-level government-appointed (Advisory) Water Boards. From 1997 onwards, the duties of the Water Officer became more specified and uniform (Water Utilization (General) Regulations of 1997).[3] Also since 1997, members of the Central and Basin Water Boards were to be drawn from public, private, NGO and women's organizations, instead of exclusively from governmental bodies. The National Water Policy of 2002 expresses the intention to further devolve authority for water rights allocation to Basin Water Sub-Offices at the 'catchment' level or even to local WUAs (URT, 2002), but this has not been implemented as yet.

Third, the core of the administrative water rights system through which government seeks to manage water has hardly changed either since the early 20th century. Registration to obtain a paper water license, permit or right from the recognized water authority of the area was already practised under German law, and then stipulated in the Water Ordinance of 1923 and every revision thereafter. With each legal revision, registered rights under any former Water Ordinance were continued in one form or another. Besides white farmers since the early colonization, other water users seeking registration included large-scale governmental and often foreign private-irrigated farms and forestry estates, and the Tanzanian Electricity Supply Company (TANESCO). Urban water supply was 'protected' under other specific legislation. Thus, 'water rights' strengthened the claims of large-scale rural and urban governmental and private enterprises of a predominantly colonial rural and later urbanizing formal economy, at least on paper.

The obligation to register played an important implicit role in the legal recognition, or not, of small-scale rural water uses under customary arrangements by the inhabitants of Tanzania. In the colonial era, the law gave some legal status to these existing uses, albeit a secondary status with specific conditions. Sections 3 and 5 of the Water Ordinance of 1948, Chapter 257, recognize earlier rights including those 'under the 1923 Water Ordinance, lawful mining operations, some claims under the Indian Limitation Act, and native law and custom'. For the latter, however, only the 'duly authorized representative' of natives is recognized (section 13 (9)). Moreover, under some conditions, natives are only recognized in addition to the District Commissioner (section 33 (9)).

This secondary status shifted into 'illegal use' once registration for water rights was made compulsory for all those who 'divert, dam, store, abstract and use' water. In the next Water Ordinance of 1959 (sections 11, 12 and 14) the option of registration was also extended to native water users, leaving the legal status of those who did not register their water use somewhat undetermined. However, the Water Ordinance (Control and Regulation) Act No. 42 of 1974 (section 14) rendered registration obligatory. It stipulated that registration for a right was the only way for any Tanzanian to ensure that his or her water use was considered formally legitimate (Maganga *et al.*, 2003). Hence, any de facto unregistered customary small-scale water use became de

[2]Water Ordinances 1959 5–(4); Water Utilization (Control and Regulation) Act 1974 6–(2).

[3]From 1997 onwards, through the Water Utilization (General) Regulations of 1997, the obligation to check comments about new entrants among those affected was further formalized, e.g. by stipulating that the Water Officer has to announce new applications through the Gazette, by notifying those who may be affected and those who are nominated in the Water Boards, and through announcements at the District Commissioner's office. This law also harmonized criteria and registration by promulgating uniform water rights application forms, which specify the purpose of water use and also the volumes allocated (not the volumes used) and annual or, if further detailed, half-yearly averages.

jure illegal and susceptible to legal prosecution (Kabudi, 2005). On paper, the formal water law declared the large majority of mostly illiterate rural 'traditional' small-scale water users, who were completely ignorant of the law, as offenders.

In practice, though, up till the 1990s, this water rights system remained largely dormant and served primarily as an incomplete register. Even though registration was formally obligatory, there was a silent consensus among water professionals that it was meaningless to impose the bureaucracy of registration designed for a few large-scale users on many small-scale users whose water use was fully legitimized under customary arrangements. For them, registration would hardly serve as a water allocation tool and certainly not as a fee collection tool. There was hardly any link between the registered use of water and actual state intervention for water allocation and conflict management, even for the larger users who did register. The certificate usually only mentioned the purpose of water use and conditions, if any, attached to the right such as water quality or obligatory return flows. The assessment of any volume of water allocated, if stipulated at all, was typically the Water Officer's best subjective guess of an average annual volume as measuring devices were virtually non-existent. Water Officers could regulate new entrants and stipulate conditions to be attached to certain water rights. They also had the formal power to curtail excessive water abstractions by title-holders and manage water-scarcity situations. Evidently, average annual volumes were of no use to regulate the low flows during the dry season, when scarcity problems are most acute. Underdeveloped infrastructure in most areas implied that there were hardly any devices to control water according to any agreement. Hence, Water Officers' intervention in water allocation and conflict resolution itself was subjective and largely based on top-down state authority. There was no formal accountability either, as the water law mentioned that 'nothing in any such water right shall be deemed to imply and guarantee that the quantity of water therein referred to is or will be available' (Water Ordinance, 1959, pp. 16-4; Water Utilization (Control and Regulation) Act 1974, pp. 15-4).

Similarly, up till the 1990s the colonial and post-colonial governments had never used the authority ascribed to itself to 'prescribe the fees payable in respect of any application or other proceeding under this Ordinance' since the promulgation of the Water Ordinances Chapter 257 of 1948 (35(d)). This authority was reproduced in the Water Ordinance of 1959 (38-2b), and the Water Utilization (Control and Regulation) Act of 1974 (38-2) but actual fees for registration were absent or nominal and they were only charged at the moment of registration in order to cover some of the administrative costs. No other fees were applied.

In sum, for decades this water rights system had remained a rather dormant administrative measure. The few large-scale rural and urban water users who registered could declare their own existing and expanding water uses as more legitimate than that of all (potential) water users who failed to register: typically the small-scale water users and the original rural inhabitants of Tanzania.

Legal reform in the 1990s

Blanket revival of the system with new fees

In 1994, a Subsidiary Legislation (Government Notice No. 347 of 1994 under section 38 (2) of the Water Utilization (Control and Regulation) Act No. 42 of 1974) was promulgated. This new piece of law not only revived the dormant registration system but also used the formal authority to charge fees and introduced, at once, a fixed once-off payment for registration of $40, plus the 'economic water users fees'. The annual economic water user fee was proportional to annual volumes of water allocated (in absolute volumes [m^3] or flows [l/s]) and depended upon its use. Three years later, in the Water Utilization (General) Regulations of 1997, a Schedule of Fees for much higher amounts was promulgated. The tariffs were slightly revised in the Water Utilization (General) (Amendment) Regulations, 2002 (see Tables 6.1 and 6.2 in the Annex). The main difference with the list of tariffs of 1997 was that for small uses below 3.7 l/s, charges were not volume-based anymore. Instead, a flat rate of $35/year was applied, irrespective of the actual flow or volume used. The motive for this decision was that one uniform legal system for all was pursued, while the majority of water users in Tanzania fell under this category and 'one cannot exempt a majority from cost recovery' (senior water manager, Iringa, 2004, personal communication). Charging a flat rate for those who would otherwise be exempted from any payment would increase tax collected while avoiding the hassle for the Water Officers of setting rates for lower amounts than the minimum flat rate – but at the expense of the small users who now had to pay disproportionate amounts. Besides including these drastic new fees in the national administrative water rights system on paper, implementation of the full-fledged system was effectively taken up in the Pangani and Rufiji basins.

Drivers of the reform: from water development to water regulation

This change of a dormant administrative system for a few large-scale water users into a blanket cost recovery system for water management fell in a period in which Tanzania also introduced cost recovery for many other government services, such as domestic water supply, health services and education – radically breaking with the socialist past, in favour of structural adjustment and privatization programmes. Similarly, a much larger portion of operation and maintenance costs of irrigation schemes was transferred to the irrigators, although investments in capital costs are still seen as at least a partial government responsibility. The new water fees were one of the several new financial burdens for Tanzania's citizens. The simultaneous decrease in government's own financial resources increased the attractiveness for the government to explore options for raising money out of 'its' water resources.

A driving force behind the transformation in the water sector was the Rapid Water Resources Assessment in 1994/1995 supported by the World Bank and DANIDA (URT, 1995). Findings of this mission were used as inputs into the Staff Appraisal Report (World Bank, 1996) for the formulation of the River Basin Management and Smallholder Irrigation Improvement Project (RBMSIIP) that started in 1996 with a loan from the World Bank. The design and pilot-testing of the legal reform under the RBM component is implemented by the Ministry of Water and Livestock Development.[4] The drafting of the new National Water Policy of 2002 is part of the same project and reflects the same assumptions.

The diagnosis in these various documents is that there would be an urgent need, nationwide, to shift away from the water

[4]The River Basin Management component 'would fund interventions designed to improve water management both at a national level and in the two target basins. Activities to be funded include: (i) strengthening national water resources management by reforming the regulatory framework to improve stakeholder participation in basin management, strengthen the water rights concepts and management, increasing penalties and raising fees for water use and improving information gathering and analytical capabilities at the national level; (ii) improving both the regulatory capabilities and the information and resource monitoring capabilities at the basin offices in Rufiji and Pangani; and (iii) rehabilitating the hydrometric network in the Rufiji and Pangani basins' (World Bank, 1996, Section 2.8). The loan disbursed for the River Basin Management and Smallholder Irrigation Improvement Component to the Tanzanian Treasury under the Development Credit Agreement amounts to $18.2 million Special Drawing Rates (equivalent to $26.3 million) to be reimbursed within 30 years after the first payment, due in October 2006. Till 2016, the interest rate is 1% and after that, 2%. The RBM component comprised slightly more than $10 million. The latter was to support the national Water Resource Department and office of the principal Water Officer through specialized equipment, vehicles, training and technical assistance. The two basin offices were to be provided with equipment, training and technical assistance (Section 2.18). After project closure in 2004, part of these costs is to be borne by the basin inhabitants.

resources development agenda of the National Water Policy of 1991, which the Government of Tanzania had just adopted. Instead, a regulatory agenda would be needed, because in two of the nine basins, the Pangani and Rufiji basins, 'there are serious user conflicts, deterioration of resources due to misuse and lack of comprehensive planning and management mechanisms' (URT/MOW, 1995). In the Rufiji, upstream water use was believed to have reduced electricity delivery by the Mtera–Kidatu power plants, which caused electricity rationing in Dar es Salaam in 1993.[5] Hence:

> [A] framework is needed for preventing and resolving conflicts among competing users and for regulating demand. The conflicts surrounding the inflow and use of water in the Mtera reservoir crystallize the issue. With . . . an emphasis on drainage of wetlands, so land can be used productively and other water development and flood control structures, the 1991 National Water Policy may result in actions which further degrade environmental quality in Tanzania. The Bank and the Government would collaborate on the refinement of the National Water Policy under the project.
> (World Bank, 1996, section 1.27)

'The conflict in the demand for water can only be resolved through more transparent, structured allocation and control mechanisms for basin water resources' (World Bank, 1996, section 2.1). Even though only two of the nine basins are mentioned as having water-scarcity problems, the shift from the water development agenda to the water regulation agenda was seen as a matter of new national policy and a new uniform nationwide framework, without any explanation.[5]

[5]Recent studies by Machibya *et al.* (2003) and Yawson *et al.* (2003) show that the reduced electricity production in 1993 had no relationship with upstream water flows, but had been caused by deviating from the originally designed management arrangements of dam storage within the stretch between Mtera and Kidatu.

Fee payment to recover costs and deter water use

The solution to this growing competition over water proposed in the RBM project was to further increase the 'economic water users fees' that had been introduced in 1994 and 'which is recommended to be redefined as a tax on water use assessed to cover the costs of operation and maintenance of basin monitoring and regulatory facilities' (World Bank, 1996). According to the Staff Appraisal Report, the key weakness of the existing law had been that neither the economic water users' fees for all productive water uses nor the service charges only for those using public infrastructure cover the true cost of managing the resource. According to the report, this had caused two problems:

> In both the water supply sector as well as in irrigation, insufficient revenues are generated to cover operation and maintenance costs. The quality of the service and of the water received is undermined. A second problem is that the low tariffs encourage inefficient use of water and waste by industry, consumers and irrigators.
> (World Bank, 1996, section 1.28)

The introduction of economic fees was expected to solve these two problems at the same time. First, such fees would enable self-financing of basin and catchment Water Offices and Water Boards. In other words:

> With regard to the 'economic water users fees' to be collected by basin Water Officers, it is proposed under this project that these rates be raised to a level which would provide sufficient funds to support the administration of basin Water Offices, including the collection of information on water quality and availability, the enforcement of pollution standards, and the administration and monitoring of water rights.

Functions of Basin Water Boards encompass:

> the issuing of water rights and registration, regulation and enforcement, but also water resources exploration, assessment,

> pollution control, monitoring and evaluation, environmental protection, basin planning and development and other cross-sectoral activities.
> (URT, 2002, p. 50)

Second, payment of fees was expected to contribute to managing water as an economic good. Volume-based fee payment would enable 'the allocation of water as a public good and as an economic good with a value in all its competing uses and the use of a water user fee as a means of encouraging efficient use of the resource and for meeting the cost of regulatory functions'. The National Water Policy expresses the same expectations of fee payment for cost recovery. 'Economic instruments include water pricing, charges, penalties and incentives to be used to stimulate marketing mechanisms, and serve as an incentive to conserve water, and reduce pollution of water sources' (URT, 2002, p. 7). Further, 'decision making in the public sector, private sector and in civil society on the use of water should reflect the scarcity value of water, water pricing, cost-sharing, and other incentives for promoting the rational use of water' (URT, 2002, p. 21). 'Economically, trading of water rights, application of economic incentives and pricing for water use, shall be gradually built into the management system as a means or strategy for demand management and water conservation' (URT, 2002, p. 30).

The practical implementation of the proposed 'enhancement of water fees and pollution charges as incentive for water conservation and pollution control, and as a source of funds for water regulation activities, catchment conservation, and water resources monitoring' (World Bank, 1996, annex A) would be via the Water Officers.

> The basin Water Offices will be mandated to collect revenue such as fees and charges and to be used to meet the cost of regulatory functions and financing of water resources assessment services. The Minister of Finance has already authorized the basin Water Officers to collect user fees and operate a bank account for the use of such funds. The basin Water Offices and basin Water Boards will be required to account for the use of these funds, which will also be audited annually by Government auditors as is occurring with other public funds.
> (World Bank, 1996, annex A)

Thus, by 1996, referring to the economic water fees, 'plans were in effect to progressively increase water tariffs throughout Tanzania and to be continued under the present project' (World Bank, 1996, section 1.29), and 'it was agreed that Government will, by December 1996, revise existing regulations so as to increase the water user fee to a level sufficient to cover operating costs of the river basin offices' (World Bank, 1996, section 2.17). These plans led to the above-mentioned schedules in the Water Utilization (General) Regulations of 1997 and its amendment of 2002, and were also reflected in the National Water Policy which seeks to 'ensure financial sustainability and autonomy of Basin Water Boards' (URT, 2002, p. 26), especially 'by charging water use for productive purposes' (URT, 2002, p. 50).

Water use registration system as the basis for fee payment and water allocation

The existing administrative water rights system was welcomed as a good and readily available basis for fee payment and also actual allocation and regulation. The system was expected to perform well; it just needed to be implemented.

> The conceptual framework for integrated river basin management is already laid out in the 1974 Act, as amended in 1981. However, the legislation has never been effectively implemented. The Government has submitted a letter of Water Resources Management Policy outlining measures to be taken to update the legislation and improve management of this resource.
> (World Bank, 1996, section 2.13)

The expectations regarding the effectiveness of the existing administrative water rights system as a water allocation and regulation tool were also high:

> The administrative system, centralizing information for the river basin, should: (i) be in a position to control withdrawals of surface and groundwater by issuing and revoking water rights; and (ii) know at all the times the quantity of water available in the basin, and its use, by monitoring both the sum of water rights granted, and physical availability.
> (World Bank, 1996, section 2.24)

Similar optimism about the existing system as an effective tool to curtail water use was expressed in the National Water Policy of 2002. The key legal instruments to be adopted would 'include restrictions and all prohibitions imposed by the regulatory body and the Government. These are individual licenses for abstractions and their revisions' (URT, 2002, p. 7). Yet, some problems in implementing the new legal framework were anticipated. It was recognized that 'water rights applications required a fairly lengthy procedure' (World Bank, 1996, section 1.24) and that 'data on precipitation, hydrometric data and actual abstractions for irrigation is inaccurate and sketchy' (World Bank, 1996, section 1.25). Six years later, the problem is still serious. 'Currently the data collection networks are in a state of near total collapse due to lack of adequate resources and tools' (URT, 2002, p. 35).

Problems in registering, charging fees and managing water allocation among many scattered small-scale water users under customary water management arrangements were also foreseen. Three possible solutions were mentioned. First, long-term government measures would include 'encouraging smallholders to form groups, especially smallholder farmers, which will make it easier to collect the fee from the groups, rather than from individual users' (World Bank, 1996, annex A). Second, a review of the institutional framework was foreseen that would address:

> the strengthening of the water right concept by: (i) clarifying how the vesting of all water in the State, with the Government sanctioning all uses, affects customary water rights, exercised by riparians or livestock owners or other traditional users, who have not sought, nor been given water rights under the law; (ii) clarifying the cases in which the State is entitled to modify or withdraw this water right (now very broadly defined, and permitted whenever water is required for a public purpose).
> (World Bank, 1996, section 2.15)

A third solution was to introduce fee payment in a phased manner.

> The actual user fee will be levied first on economic activities such as hydropower production, and large farms, followed by levies on smallholder farmers. The related fees will gradually be built into the management system that touches all users with the ultimate objective of promoting conservation and minimizing abuses.
> (World Bank, 1996, annex A)

The Water Utilization (General) (Amendment) Regulations of 2002 already include all water users as proposed for this last stage.

'At the start, we thought it would be easy', commented a senior Tanzanian staff member of the RBM project in 2003. The findings of the factual implementation of the new water rights and fees system in the Upper Ruaha catchment demonstrate that none of the above-mentioned assumptions are valid with regard to small-scale water users in that area, and most probably elsewhere in Tanzania. However, among the few large-scale water users, the new system appeared to work for fee payment for cost recovery.

Registration Tool: Limited Information

Available data: names and uses

The Rufiji Basin Water Office in Iringa has started to compile a considerable list of names of water users and the purposes of their water use. By mid-2003 the database contained 990 water rights issued in the entire Rufiji basin, with 40% of the titles held by governmental agencies, 12% by Brooke Bond Tea Company and 8% by various Catholic dioceses. The remaining 40% of registered users include private irrigation schemes, such as those belonging to Baluchistani and other Asian immigrants who were brought by the British colonialists (Sokile, 2005).

Of all water rights 14% were issued between 1955 and 1960. The number steadily increased over the years. From 1995 onwards, registration intensified with more than 29% of the rights administered under the new Rufiji Basin Office, though these are largely still in the stage of application or with a provisional status. The right holders utilize water mostly for domestic purposes, followed by irrigation, but often also in combination. Livestock is sometimes explicitly mentioned, and sometimes considered under domestic purposes. Water rights for hydropower constitute 3% of rights, while industrial use constitutes only 2%. The cadastre of the Rufiji basin also stipulates the status of the water right, which includes those who abandoned the use of their water right. As many as 47% of the registered rights are 'not operated' any more. The proportion is highest for the oldest rights, and may be related to the outflow of Germans, Baluchis, Greeks after independence in 1961 and the Arusha Declaration in 1967, which announced further nationalization. However, even in the most recent applications, abandonment of the water right occurs (Sokile, 2005). Probably, other cases of abandonment of water rights, e.g. by people who have died or moved out of the region, have not been notified as yet.

In the Upper Ruaha catchment, requests for water rights are first processed in the catchment sub-office, before being brought to Iringa, 300 km away, for final approval by the Basin Water Officer and incorporation in the register. In this catchment, more than 100 water rights have been issued, including water rights for individual farmers and farmers organized in Water User Associations. Slightly more than half (56) of the water rights are in the Mkoji sub-catchment, and are mainly issued for irrigation purposes. Most rights in this sub-catchment, especially those among smallholders, were issued in the late 1990s or recently under the RBM project, especially since the opening of the Rufiji Basin Water sub-office for the Upper Ruaha catchment in Rujewa, Mbarali district, in 2001 (Sokile, 2003).

An inventory in the whole Rufiji basin of the 990 names of the individual or collective water users, their main uses and the operational status of the right are an obvious first step for any cadastre. However, many actual water users have not been registered as yet. Recently, an inventory of unregistered water users in the Rufiji basin was conducted, which estimated that the number of unregistered users is 573, so more than half of the registered users (Msuya, 2003).

Estimates: Sites

Any information other than names and purposes of water use becomes much more problematic. Information about the sites where water is used is only documented in the register by mentioning names of the larger streams and the nearby villages and wards. There are no detailed maps, coordinates or map references to provide more precise information attached to the cadastre. While water rights would still allow estimating water availability to some degree at aggregate levels, this lack of clarity of the sites of water rights renders formal water rights a meaningless, if not a counterproductive tool, if it is used in localized water disputes. Indeed, in one dispute, the issue at stake was the location of the water right, which, according to the disputant, differed from the site mentioned on the certificate (Maganga *et al.*, 2003).

Lack of data: volumes

An even weaker part in the registration system concerns the figures for annual volumes of water use. Only 28% of the rights registered have any specified volume at all. However, even for this portion, the variation in annual volumes allocated shows that mistakes have been made, for example, in registering and entering the place of the commas and the number of decimals. As yet, there is hardly any registration of half-yearly average volumes, differentiating the rainy and dry seasons (Sokile, 2005). This lack of reliable and accurate data on volumes of water allocated, let alone volumes of water used, is inevitable. The high seasonal and annual variability of runoff, streams, and water abstracted and the general lack of any measuring devices render any estimate a subjective guess. Even if the few permanently constructed intakes that divert water from the streams were fully operated according to their technical design, which is never the case, fluctuations of abstractions during flooding and dry spells cannot be captured in half-yearly and yearly average abstractions. Moreover, water abstractions vary with the quantities of direct rainfall on farmers' land, evaporation rates, cropping patterns, changes from grazing land to cropland, etc. Return flows are equally variable. In fact, even the most sensitive hydrological models, based on information from ample flow monitoring devices, can only generate rough estimates for aggregated annual uses in major streams, and certainly not for each individual along such streams, especially in the dry weeks. Therefore, there are no grounds at all for the assumption that the administrative system – or even hydrological models – would 'know at all the times the quantity of water available in the basin, and its use, by monitoring both the sum of water rights granted, and physical availability' (World Bank, 1996). It is only if water resources are fully developed into highly (large-scale) controlled systems that volumes can be sufficiently known and manipulated – a rare situation even in developed countries.

Costs of maintaining cadastres

While the current computerized spreadsheets of the water register only include names of some of the water users and approximate streams or communities where they are located, the costs of maintaining even this simple system in rural Tanzania are much higher than in most other places in the world. This is due to the generally low levels of literacy among small-scale users, the distance to many scattered hamlets, bad roads especially in the rainy seasons, expensive vehicles and fuel, the lack of affordable telecommunications, no way of writing to water users, and minimal computer and software facilities. The costs of compiling and maintaining an administrative cadastre may be justified when it only concerns a few large users. However, among all water users in a basin, costs of just noting the names of users and updating changes are extremely high. The question is whether the costs of blanket registration are justified in light of the limited benefits of the registration system as a basis for water resources planning, charging fees, and allocating water and water conflicts (see elaborated next).

Cost Recovery Tool: Subjectivity by Design and Costing Public Funds

Subjectivity by design

Before the 1990s none of the water lawyers drafting the administrative water rights system had ever thought of using the system for charging volume-based fees. Indeed, insurmountable problems arose as soon as this administrative system became the foundation for volume-based blanket tariff setting and fee collection to finance the government's water management services. First, the lack of objective and transparent procedures incorporates 'subjectivity by design' into the new system of water rights and fees in at least four ways: in rate setting; enforcement of fee payment; handling of public funds; and in discouraging genuine organization of water users. Second, among small users, the system appeared to drain public funds, instead of generating funds. Third, it met with fierce protest on the ground.

Arbitrary rate setting

Volume-based rate setting may seem objective and fair. However, in the absence of any objective basis to assess the volumes allocated and, thus, to set volume-based rates, Water Officers can only rely on their subjective judgement. Even setting tariffs relatively by ranking structures according to their sizes appeared difficult. In the Mkoji sub-catchment, for example, the volumes and related fees for the larger structure of Inyala A were initially set at lower rates than for a nearby smaller structure of Inyala B. The water users complained. In this case, the Water Officer accepted the complaints and changed the fees the other way around. However, generally there is enormous confusion among small- and medium-scale users in the Upper Ruaha about the amounts to be paid (Sokile, 2003).

The recently introduced threshold below which a flat rate has to be paid may mitigate the problem of rate setting along some range of volumes, but it hits the smallest, often poorest, users hardest. Punishing small water users by charging disproportionately high rates because of administrative problems is difficult to justify on social grounds and, once they have paid, would certainly fully justify that they start using as much water as possible. Significantly, among private larger water users, rates were not set on the basis of water volumes used, but rather negotiated with the Water Officers. Payment followed promptly (Sokile, 2003). So willingness and ability to pay seem a sounder basis for rate setting than highly contestable hypothetical water volumes.

Arbitrary and weak enforcement

Significantly, 92% of private companies/ estates, such as Brooke Bond Tea Company Ltd or Tanzania Wattle Company Ltd, appeared to fulfil their duties (Sokile, 2003). In fact, enforcement of payment appeared most difficult vis-à-vis other government agencies. Only 38% of the government agencies holding water rights (e.g. local government for domestic supply and state farms) regularly pay fees. In the Mbarali and Kapunga State Farms, in particular, the arrears in payment are among the highest and the cash instalments paid during each trip are typically small. In these schemes, where the Water Officers have control over scheme operational devices to cut water use, enforcement still remains extremely difficult. These and other government agencies use the argument 'why should the government pay the government?' to justify their refusal to pay the water fees, but this jeopardizes the goal of cost recovery for the functioning of the basin offices.

The degree of payment varied among smallholders, livestock keepers and other water users. The main threat that the limited staff on the ground can use is intimidation that defaulters will be brought to court, which mainly works in the case of the least powerful. However, in case of reluctance to pay, time and transport costs of repetitive reminders are high, let alone the costs of initiating a court case. The threat to cut access to water in case of non-payment can hardly be implemented because there are hardly any sluices, gates or other water control structures that the Water Officer can operate. And even if he locked any of the few improved intake structures, farmers would break them as soon as he left the village.[6] Obviously, subjectivity by design, combined with strong delegated state power, invites corruption and abuse of power.

Arbitrariness in water user associations as tax collectors

As already proposed in the RBM project Staff Appraisal Report, the remedy to high costs for individual registration and fee collection was to promote the formation of new WUAs by smallholders who were irrigators. As water rights can be either individual or collective, any number of water users sharing a common water source could apply collectively for one water right, for example, as an existing farmer association or by forming a WUA. The water users would save on individual application fees, while the government would win the most by shifting most transaction costs for fee collection to these local bodies.

[6] Collecting and transfer of public money is a new task for Water Officers. Water Officers are accountable by writing receipts for taxes received. Further, when submitting the collected funds from the sub-catchment office to the basin office in Iringa, the accountant notes the amounts in the books. A public auditor is supposed to check the various amounts, but, for the moment, the public auditor's key interest is in the publicly allocated funding from the government, and not parallel funds for basin offices. This administrative system for fee payments is separate from the computerized spreadsheet of registered water users. An alternative is to include water cost recovery in the mandate and implementation channels of the Tanzania Revenue Authority, which has much more experience in these matters.

More than 24 new WUAs have been formed in the Upper Ruaha catchment (Sokile, 2003). Although the WUAs are still too young for impacts to be assessed, the risks are real that the rapid 'organization' into some form of committee revives the same type of rent seeking that existed under government-imposed villagization and cooperative building, as also prevailed in the Upper Ruaha sub-basin. Committee leaders have more power than government officials to effectively cut water of those who do not pay their share of the government taxes. If seen as powerful, they can more easily interfere in the customary irrigation arrangements or threaten to do so. Thus, the commonly shared water resources risk becoming a source of income for the few more powerful – again hitting the most powerless the hardest. Moreover, the incentive for organization is low indeed if it mainly implies that one has to pay fees.

Draining public funds

Contrary to expectations, charging fees for cost recovery among small users appears to be a drain of scarce government human and financial resources. Government officials from the lowest to the highest level with whom this issue was discussed admitted that the transaction costs of charging scattered smallholders in farmer-managed irrigation schemes without telephone, e-mail, post office or bank account facilities are considerably higher than any net revenue gained from this category. A simple calculation illustrates this point. For an immediately paying small-scale water user at only 15 km distance from the sub-basin office, the income of $35–40 breaks even with the estimated fuel costs, according to government tariffs, which are 2 × 2 × 10 km × $0.75/km = $45. However, the Water Officer typically needs to make two or three trips to smallholder areas, one for announcement, one for the collection of fees and, often, one trip as a reminder. Moreover, the distances in the Rufiji basin from the Water Office to the water users, even if one can reach various water users within the same trip, are much longer. The average distance from either the Iringa or the Rujewa basin offices is estimated at 87 km (Sokile, 2004). So, the fuel costs for collecting taxes from small-scale water users typically requiring three trips/year amount, on average, to $392, divided by the number of water users that can be reached during one trip. Evidently, there are many more costs than fuel alone, such as the costs of the four-wheel drive vehicle purchase and maintenance, the salaries and per diems of the Water Officer, driver and assistants, plus all other administrative costs.

This stands in sharp contrast with the very minimal transaction costs of taxing large users. For example, TANESCO pays an annual Royalty Fee directly to the Ministry by bank transfer. After billing, large users such as the Brooke Bond Tea Company, Kilombero Sugar Company, Kilombero Valley Teak Company, District Governments and the Dioceses normally pay by cheque or bank transfer. For the rare payments in cash, one trip to such large-scale users is usually sufficient. The Rufiji Water Office estimates the negotiated average fee paid by large-scale users at $100, which is three times the minimum flat rate (Sokile, 2004). This amount is negotiated independently from any water volume allocated or used in reality as those volumes are not given in the registers.

Currently, the annual fees for basin management collected in the Rufiji basin amount to $50,000, as estimated by the Basin Office (Sokile, 2004). TANESCO's royalty payment of $165,500 for the hydropower works in both the Rufiji and Pangani basins is not included in this because it remains at national level. Overall expenditures of the Rufiji basin office are estimated at nearly $225,000 (see Table 6.3 in the Annex; Sokile, 2004).

In sum, taxing scattered small-scale water users has not contributed to achieving the goal of self-financing of the Rufiji basin office. The huge implementation costs of taxing this majority of water users were insufficiently anticipated during the design of the new water rights and fees system. Promoting WUAs and Water Officers merely as tax collectors is no solution either. However, collecting a net income appeared feasible among large-scale water users. This is also justifiable on the ground that small users are primarily subsistence farmers with limited land, while large-scale companies are undertakings with large water abstraction and considerable benefits.

Lack of legitimacy

The government's new water fees for basin management have met with fierce local opposition among smallholders and livestock keepers in the Upper Ruaha catchment. The well-intended explanations of the Water Officer that money is needed for the vehicles, fuel, construction and office costs of the Rufiji Basin Water Office did not impress the protesters. Their main complaint was that there has been no improvement in services delivered in return for what they perceive as taxation and rent-seeking. Rural water users contest the government's claims of ownership that would also entitle them to charge for water use. According to their customary notion of property claims, water is given by God, and use rights are only established on the basis of their own efforts to build infrastructure. Given this widespread opposition, one could have expected a categorical rejection of the new system. Ironically, the reason for its partial acceptance can be found in the new conflicts and divisions that emerged between upstream and downstream users, where the former use the new system to strengthen their own claims to water at the expense of the latter, as described below.

The legitimacy of the new taxation system has also been questioned at national level. In the budget speech of June 2003, the government abandoned the proliferation of rural cost recovery, realizing that the costs for collecting small, rural taxes are often higher than the amount collected; that they tend to discourage economic activity; and that they often meet with widespread resistance, among others by opposition politicians (O-H. Fjeldstad, e-mail, 2004, personal communication). The trend of abolishing existing taxation is diametrically opposite to the efforts of the Ministry of Water and Livestock Development to introduce new rural taxes. Last but not least, charging up to $35 or $40 from individuals or groups of organized poor people earning a dollar or two a day merely aggravates poverty.

Conclusion

Imposing a blanket fee payment system on small-scale water users failed to achieve the expected goal of self-financing governmental basin management. Instead, it cost the government its scarce resources. The new system lacks legitimacy at local and national level because there is no improvement in government service delivery and because fee payment for basin management is at odds with both national poverty eradication and rural taxation policy. Government credibility is further weakened by the arbitrariness of the new system. At the same time, the ability and willingness to pay fees for basin management services of large-scale private users who derive considerable benefits from water use appeared effective.

The straightforward implication is to continue taxing the large users who make the highest profits from water and can easily be reached logistically. However, for informal, small-scale users the lose–lose scenarios for both water users and government is to be avoided. Taxation of these users should, in any case, be phased according to logistical capabilities – as also proposed by the designers of the RBM project. However, the real challenge for the government is to deliver tangible services in return to the taxes, in order to achieve willingness to pay and reduce transaction costs for fee collection in a sustainable way. As discussed below, the oversimplistic connection between claims to water and payment is certainly to be thoroughly revisited. This is even likely to *save* water.

Water Allocation Tool: Increasing Water Use and Inequities

The expectations of the RBM project and the National Water Policy of 2002 that an administrative water rights and fees system would, by itself, serve as a tool to allocate water and mitigate conflicts and 'be in a position to control withdrawals of surface and groundwater by issuing and revoking water rights' (World Bank, 1996) were high. While the registration and taxation component of the new water rights system worked at best partly, issuing water rights and making people pay for water failed completely as a water allocation tool, and even aggravated downstream water scarcity.

The above-mentioned lack of water measuring and control devices that prevented Water Officers from effectively controlling access to water and the lack of implementation capacity to enforce state authority undermined the obligatory registration and fee payment. Moreover, water certificates with, at best, an average annual volume specified appeared to have no meaning at all for the key water problem in the Upper Ruaha, which is the dry season in which fractions to be used are much smaller than any average, certainly for downstream users. These implementation weaknesses are the Achilles heel for any water rights system that solely depends on the government's authoritative and practical ability to curtail water use.

Ironically, the newly introduced payment of water 'as an economic good' even exacerbated water scarcity downstream during the dry season. The Water Officer had started issuing water rights to the upstream irrigators. They were somewhat wealthier and already quite well organized. In that area, irrigation expanded rapidly, for example up to 40% as in the Inyala village, where land values doubled as well. This rapid expansion was triggered not only by market and other opportunities but also by the newly constructed intake structure under the RBMSIIP project, which increased water security in the dry season. Reluctantly, these irrigators registered and paid fees. The Water Officer hardly contacted and informed the more distant and largely unorganized livestock keepers and the fragmented in-migrating communities in the plains downstream. Not a single WUA has been established in that area. In the initial days of implementing the water rights system, the promise of the Water Officer that those who registered and paid the new fees would be better supported in water conflicts than those who had not paid as yet helped to convince them and others. It certainly facilitated the Water Officer's job of achieving quick registration and fee payments.

As a result, the irrigators in the Inyala village argued that 'since 2000 they had bought water for $100' – in their perception of water as an economic good – to strengthen their claims to exploit this precious resource to the maximum. So contrary to the assumption of the RBM project and the National Water Policy of 2002 that paying for water leads to reduced water use, it increased the water use of upstream users. This was with immediate detriment to the downstream users as registration and tax payment did not generate any extra drop of water in the zero-sum game of dividing a limited pie during the dry season in the Upper Ruaha catchment.

Significantly, in 2003 the Water Officer of the Upper Ruaha realized the likely repercussions of 'selling unrealistic expectations', and started emphasizing how the water law itself stipulates that the government does not provide any guarantee that issued water rights, for which taxes are paid, are actually delivered (Msuya, 2003), as mentioned earlier.[7] The Water Officer protected himself by emphasizing the disconnection between fee payments and water allocation. Recently, the Water Officers stopped issuing water rights altogether. They now first finalize the identification and registration of all significant users that should have taken place at the onset. Crude and unmonitored water rights are inadequate tools to regulate upstream–downstream water conflicts in such a context.

In order to address water scarcity during the dry season in the Upper Ruaha catchment, the government does not rely anymore

[7]As mentioned in Section 2, the Water Ordinance 1959, Part IV 16 (4) and its literal repetition in the Water Utilization (Control and Regulation) Act 1974 Part IV 15 (4) stipulate: 'Nothing in any such water right shall be deemed to imply any guarantee that the quantity of water therein referred to is or will be available.'

on paper water rights but catalyses the formation of negotiation fora. The newly established Rujewa Sub-catchment River Basin Water Office, supported by the Sustainable Management of the Usangu Wetlands and Its Catchment (SMUWC) project, brought the managers of the three smallholder irrigation schemes, TANESCO and the Ruaha National Park together into what is now called the Planning Group. In 2003, the River Basin Water Office supported the introduction of a 'River Basin Game', developed by RIPARWIN (Lankford *et al.*, 2004a,b) to foster dialogue between upstream and downstream users, to raise awareness about downstream deprivation during the dry season, and to elicit remedial options, such as the further exploitation of groundwater or construction of small dams to hold storm water and floods during the rainy season for use during the dry weeks. For example, a small dam is proposed in the Ndembera river in the Upper Ruaha catchment, which would provide the minimum flow required for wildlife in the Ruaha National Park during the dry season. Significantly, FAO already proposed this in the 1960s, but the plan was shelved ever since because the discourse shifted away from water development to water regulation for the reasons mentioned in this chapter. Rotations along streams and building upon customary practices are also elaborated. Also, an encompassing legal infrastructural framework for catchment apportionment is proposed. This allows rebuilding the concrete intakes in the upstream part in such a way that, during the dry season, less water is diverted upstream in order to leave more water in the flows for downstream use. Water fees for the respective irrigation schemes would be based on abstractions during the wet season as concretized in the technical design (Lankford and Mwaruvanda, 2006).

Conclusions and Recommendations

The foregoing analysis illustrates, above all, how well-intentioned reforms that are governed by ideological principles in vogue (centralized formal water rights and cost recovery for water as an economic good) and an unsound scientific analysis of the complexity of the real world, combined with a lack of meaningful prior consultation with stakeholders, can get it wrong.

Grafted upon a dormant colonial system of water rights, Tanzania supported by the World Bank, introduced increased water fees with two objectives: managing water resources and cost recovery for water resources management functions. Relative failure to achieve the first goal of water resources management was primarily due to the heroic assumptions on the regulation capacity of the state. However, in the Upper Ruaha, as in many rural areas in sub-Saharan Africa, the state manages only a few of the structures, reservoirs and large public schemes. It has only direct control over the water regime through the canal regulation programme on two out of more than 150 intakes. This lack of 'reach' was compounded by the hydrological complexity of many catchments, high resource variability and unpredictability, the lack of hydrological knowledge, the multitude of small unorganized users, and the inaccessibility of the dindilos. Moreover, the concrete structures to replace the indigenous dindilos were built without full acceptance of the hydrology and uses, and without a view on how dindilos and dindilo-type structures could practically and technically apportion water. These new structures now hinder water-sharing arrangements even more. The cart was put before the horse again by distributing 'rights' *before* knowing about use, users and resources. So, managing water appeared illusory. However, even if the state had been able to sufficiently control and manage the streams and registers would have been well maintained, water rights based on registered average annual volumes are of little help in sharing and prioritizing water resources during dry-season scarcity. Not only was the goal of improved water management not achieved at all, but new upstream–downstream conflicts were created. These experiences suggest that it is more reasonable and effective to entrust management of water to sub-catchment decision-making networks, building on already existing customary arrangements. Their tasks would be, first, to regulate allocation in times of low flows, with constraints to ensuring downstream flow determined by the RBO and, second, to find arrangements for the increasing demands by new users. For example, a 'catchment water master' could be appointed and paid by the catchment users. This could also start a mechanism of fee collection with a clear objective and benefit, which can be extended to a wide range of basin and water services once benefits are received from that level. For managing water in a case like the Mkoji sub-catchment and the many similar sub-catchments, formal collective rights rather than individual rights would be most appropriate.

The second objective, raising net revenue for the River Basin Water Office, was not achieved in the Mkoji sub-catchment because of the disproportionate costs of registration and cost recovery from many small users compared to the amounts gained. Moreover, users had little incentive to pay from the perspective of water assurance or service in exchange though they had more incentive from the perspective of the ill-defined threats of not being legally entitled to the water. In contrast, taxation of the few large users in the Rufiji basin did generate net revenue for the Basin Water Office. From these experiences, it can be concluded that cost recovery should be limited to the users who derive large benefits from high water diversions and allow the government to recover costs. The national government would considerably support the process if government schemes were also forced to pay and if the TANESCO contribution stayed at basin level instead of going to the central government. Given the different state interventions at stake, especially in informal settings with limited physical water control, water allocation and water taxation, a clearer separation of the goals and means to reach the goals of both measures, would contribute to the rationality, transparency and effective implementation of both. Above all, water allocation would recognize and build upon the many strengths of existing customary practices.

Last, Tanzania is a country still with a very low per-capita storage capacity. In many instances, the option of year-round storage development for improving the water supplies to all is still open. Instead of suggesting that localized and temporal absolute scarcity issues are the nation's key concern, more resources should be allocated to solve the primary issue: economic water scarcity.

Acknowledgements

This chapter is based on research carried out by the IWMI Water Resources Institutions and Policies Theme and by the Raising Irrigation Productivity and Releasing Water for Intersectoral Needs (RIPARWIN) project – a collaborative project implemented by the Soil Water Management Research Group (SWMRG) of the Sokoine University of Agriculture, Morogoro, Tanzania; the University of Dar es Salaam; the International Water Management Institute (IWMI) and the Overseas Development Group of the University of East Anglia, UK, with funding from DFID and IWMI. The authors gratefully acknowledge the support and insights provided by the officials at all tiers of the Ministry of Water and Livestock Development, Tanzania. The authors especially thank Francois Molle and Jeremy Berkoff for their valuable comments on earlier drafts of this chapter.

Annex

Table 6.1. Fees according to Water Utilization (General) Amendment Regulations (2002).

Item of water use	Application fees ($)	User fees ($/year)	
		Flat rate	Increment rate
Domestic/livestock	40	35	0.035/100 m^3 above 3.7 l/s
Small-scale Irrigation	40	35	0.035/1000 m^3 above 3.7 l/s
Fish farming	40	35	0.035/100 m^3 above 3.7 l/s
Large-scale irrigation	150	70	0.070/100 m^3 above 3.7 l/s
Industrial	150	35	0.035/100 m^3 above 1.11 l/s
Commercial	150	35	0.15/100 m^3 above 0.94 l/s
Mining	150		0.17/100 m^3

Table 6.2. Non-consumptive water use fees in Tanzania.

Use	Charge ($/year)
TANESCO – Power royalty	165,500
Power royalty fees per 1 MW installed capacity	300
Transport in inland water bodies (less than 5 t)	10
Transport (above) for every additional tonne	2.2

Note: Exchange rate (2004): $1.00 = TSh 1000.

Table 6.3. Estimated costs of the Rufiji Basin Office. (From Sokile, 2004.)

Cost element[a]	Estimated amount ($)
Remuneration – Basin Officer	8,640.00
Remuneration – Resource Management staff (2)	7,200.00
Remuneration – Quality Management staff (2)	6,000.00
Remuneration – Operations staff (5)	4,800.00
Remuneration – Corporate services	5,000.00
Remuneration – Casual labour	13,860.00
Institutional support (including resolving conflicts)	11,900.00
GIS data capture	12,100.00
Water quality analysis/hydrology sampling and analysis	9,200.00
Fixed overheads	4,500.00
Travel and subsistence	37,000.00
Printing and photocopies	8,700.00
Communication	11,000.00
Bills (electricity, water)	3,900.00
Consultants	–
Sundry and contingency	6,700.00
Interest and finance costs	5,000.00
Total	155,500.00
Other expenditures (occasional)	
Improvement of intakes	37,300.00

Continued

Table 6.3. *Continued*

Formation of WUAs	7,800.00
Water resources analysis	5,400.00
Board meetings	6,240.00
Water resources management strategy	11,200.00
Total	67,940.00
Grand total	223,440.00

[a]Costs exclude assets such as buildings, furniture, computers, photocopiers, motor vehicles/bikes and laboratory equipment.

References

Legislation and project documents published by the United Republic of Tanzania (URT) listed chronologically.

Water Ordinance 1923.

Water Ordinance of 1948, Chapter 257. Government Printers, Dar es Salaam, Tanzania.

The Water Ordinance of 1959. The Governor of Tanganyika, Dar es Salaam, Tanzania.

Water Utilization (Control and Regulation) Act of 1974. The United Republic of Tanganyika, Dar es Salaam, Tanzania.

Act No. 10 of 1981. The United Republic of Tanzania, Dar es Salaam, Tanzania.

Subsidiary Legislation. Government Notice No. 347 of 1994 under section 38 (2) of the Water Utilization (Control and Regulation) Act No. 42 of 1974. The United Republic of Tanzania, Dar es Salaam, Tanzania.

Ministry of Water (1995) *Rapid Water Resources Assessment. Vols. 1 and 2*. DANIDA/World Bank.

World Bank (1996) Staff appraisal report, Tanzania. In: *River Basin Management and Smallholder Irrigation Improvement Project. Report No. 15122-TA*. Agriculture and Environment Operations, Eastern Africa Department, Washington, DC.

Water Utilization (General) Regulations of 1997. The United Republic of Tanzania, Dar es Salaam, Tanzania.

Ministry of Water and Livestock Development (2002) *National Water Policy*. The United Republic of Tanzania, Dar es Salaam, Tanzania.

Water Utilization (General) (Amendment) Regulations (2002) The United Republic of Tanzania, Dar es Salaam, Tanzania.

Ministry of Agriculture and Food Security – Japan International Cooperation Agency (2003) *National Irrigation Master Plan*. Ministry of Agriculture and Food Security, Government of the United Republic of Tanzania, Dar es Salaam, Tanzania.

Other References

Kabudi, J.P. (2005) Challenges of legislating for water utilization in rural Tanzania: drafting new laws. In: Van Koppen, B., Butterworth, J.A. and Juma, I.J. (eds) *African Water Laws: Plural Legislative Frameworks for Rural Water Management in Africa. Proceedings of a Workshop Held in Johannesburg, South Africa, 26–28 August 2005*. IWMI, Pretoria. Available at: www.nri.org/waterlaw/workshop/papers.htm

Lankford, B.A. (2004) Irrigation improvement projects in Tanzania: scale impacts and policy implications. *Water Policy* 6(2), 89–102.

Lankford, B.A., van Koppen, B., Franks, T. and Mahoo, H. (2004a) Entrenched views or insufficient science? Contested causes and solutions of water allocation; insights from the Great Ruaha River Basin, Tanzania. *Agricultural Water Management* 69(2), 135–153.

Lankford, B., Sokile, C., Yawson, D. and Lévite, L. (2004b) *The River Basin Game: A Water Dialogue Tool*. Working Paper 75. International Water Management Institute, Colombo.

Lankford, B.A. and Mwaruvanda, W. (2006) Legal infrastructure framework for catchment apportionment. In: Van Koppen, B., Butterworth, J. and Juma, I.J. (eds) *Community-Based Water Law and Water Resource Management Reform in Developing Countries*. CAB International, Wallingford, UK.

Machibya, M., Lankford, B. and Mahoo, H. (2003) Real or imagined water competition? The case of rice irrigation in the Usangu basin and Mtera/Kidatu hydropower. Paper presented at the Tanzania Hydro-Africa Conference, 17–19 November, at AICC, Arusha, Tanzania.

Maganga, F.P. (1998) Indigenous knowledge and irrigation development in Tanzania: experiences from Msanzi, Nyeregete and Musa Mwijanga farmer-managed irrigation schemes. Paper presented at the Scandinavian Seminar College on African Perspectives on Policies and Practices Supporting Sustainable Development, October 1998, Harare, Zimbabwe.

Maganga, F.P., Hilda, L., Kiwasila, I., Juma, H. and Butterworth, J.A. (2003) Implications of customary norms and laws for implementing IWRM: findings from Pangani and Rufiji basins, Tanzania. Paper presented at the WARFSA/WaterNet international symposium, October 2003, Gaborone, Botswana.

Msuya, I. (2003) The role of water rights. Unpublished paper presented at the Seminar Ruaha + 10, 11–12 December, Morogoro, Tanzania.

SMUWC (Sustainable Management of the Usangu Wetlands and Its Catchment) (2000a) *Interim Report Supporting Volume A Water, Supporting Report 2, Water Management*. For the Directorate of Water Resources, Ministry of Water, Government of Tanzania, The SMUWC Project, Mbeya Region, Tanzania.

SMUWC (2000b) *Interim Report Supporting Volume B Water, Supporting Report 2, Water Management*. For the Directorate of Water Resources, Ministry of Water, Government of Tanzania, The SMUWC Project, Mbeya Region, Tanzania.

SMUWC (2001) *Final Project Reports*. Directorate of Water Resources, Ministry of Water, Government of Tanzania, Dar es Salaam, Tanzania. The SMUWC Project, Mbeya Region, Tanzania.

Sokile, C. (2003) Fact sheet on water rights. Unpublished paper. In: *Raising Irrigation Productivity and Releasing Water for Intersectoral Needs (RIPARWIN) Project*. Sokoine University of Agriculture, University of Dar es Salaam, IWMI, University of East Anglia, Morogoro, Tanzania.

Sokile, C. (2004) Estimated costs. Rufiji Basin Office. Field Notes, PhD research.

Sokile, C. (2005) Analysis of institutional frameworks for local water governance in the Upper Ruaha Catchment PhD thesis. Institute of Development Studies, Dar es Salaam, Tanzania.

Sokile, C.S. and van Koppen, B. (2004) Local water rights and local water user entities: the unsung heroines to water resource management in Tanzania. In: Nhapi, I., Mashauri, D.A., Sakupwanya, J., Mudege, N. and Swatuk, L.A. (eds) Water, Science and Technology and Policy Convergence and Action by All. *Physics and Chemistry of the Earth* 29(15–18). Elsevier Publishers, Devon, UK, pp. 1349–1356.

The Ministry of Water and Livestock (2001) Talking about Usangu. Produced and directed by The Ministry of Water and Livestock, Dar es Salaam, Tanzania. Video cassette.

United Nations Development Report (2002) *Deepening Democracy in a Fragmented World*. United Nations Development Program, New York.

Yawson, D.K., Kashaigili, J.J., Mtalo, F.W. and Kachroo, R.K. (2003) Modelling the Mtera–Kidatu reservoir system to improve integrated water resources management. Report presented at the Hydro-Africa Conference, 17–19 November, Arusha, Tanzania.

7 Who Will Pay for Water? The Vietnamese State's Dilemma of Decentralization of Water Management in the Red River Delta

J.-P. Fontenelle, F. Molle and H. Turral

Introduction

Many state-run large-scale irrigation schemes worldwide have long been financially supported by public funds. Because of stretched public finances and a general trend to hand over the management of irrigation schemes to farmers, an emphasis is often placed on both cost recovery and the financial autonomy of these schemes. Water fees in most countries generally cover only a part of operation and maintenance (O&M) costs and amount to a small percentage of the agricultural gross product, typically less than 10%. In some other countries, water supply is free and is considered to be a state obligation. However, in situations where irrigation and drainage operations demand the use of pumping devices, operational costs are generally significantly higher, as they include the costs of energy and the maintenance of equipment, and consequently water fees also tend to be higher. This is the case in the Red River delta, where thousands of pumps of all capacities are used in water management.

The Red River delta is also well known for having one of the highest rural population densities of the world. Consequently, agricultural production is extremely intensive, cropping intensity is high and the proper management of water is paramount in achieving social welfare and food security. The relationship between the state and the farming population has seen dramatic changes, from colonial times to the recent liberalization, throughout the collectivist period. The question of financing irrigation must therefore be addressed as a particular aspect of a changing political economy, where the taxation system and the roles and responsibilities of the different actors are being redefined. With all these changes, the pumping costs of irrigation and drainage have yet to be covered. This warrants an investigation into how water pricing is conducted in the Red River delta and who eventually pays for what.

The first section of this chapter describes the political changes which induced the technical and institutional evolution of this delta's water control systems, as the organization of the operation and even the technological nature of these systems were influenced by national political choices. The second section describes the management framework and the financial organization of the delta's water control systems. In-depth studies conducted at local level provide a better understanding of the present situation.[1] Water management in the Red River delta appears to be strongly organized by the state into successive nested levels, from the central level of the Ministry of Agriculture to the local level of the cooperatives. This structure has been challenged by the emergence of local pumping stations and water management practices, which have superimposed themselves upon this bureaucratic structure. It is shown that the mismatch between administrative and hydraulic units adds to the complexity of the definition of both the financing and the management of hydraulic operations. The third and last section of this chapter examines the financing of the different operators, the amount and use of the water fees paid by farmers, and questions the process of water management decentralization and 'privatization'[2] in the delta. While there is scope for improving downward accountability to farmers, the present system of bulk pricing and nested levels of subsidiarity allows a relatively high rate of cost recovery and a relative financial self-sufficiency.

The Evolution of the Red River Delta Water Control Systems

With a population of more than 75 million and a total area of 331,700 km^2, of which only one-third is covered by plains, Vietnam shows much concern for its food security (Cuc *et al.*, 1993; Fforde and Sénèque, 1995). Fertile and crowded plains, notably the Mekong and Red River deltas, play a key role as the country's rice bowls. The Red River delta is the smaller and more densely populated of the two deltas (Fig. 7.1). It has a gross area of 1.5 million ha (or 4.5% of the total area of Vietnam) and a total population of 20 million (27% of the total population of Vietnam) (Le Ba Thao, 1997). This represents one of the world's highest rural population densities, with more than 1300 inhabitants/km^2 in some areas. This explains why agricultural intensification, anchored in a strong security against climatic vagaries provided by irrigation, drainage and flood-protection infrastructures, is such a vital issue for the Government of Vietnam.

Water control before collectivization

High population density is not a new feature of the Red River delta. Population density was already above 400/km^2 at the beginning of the 20th century (Dumont, 1935; Gourou, 1936). This delta is an area of ancient human settlement where reclamation by paddy growers has been proved to date back to more than 2000 years (Sakurai, 1989). Early and dense settlements are quite conspicuous, judging from the unfavourable natural conditions faced by the population living in this delta: dangerous river floods and occasional typhoons, as well as droughts, are common during summer monsoons. During dry winter and spring seasons, the main concern is accessing water for irrigated agriculture. To minimize the impact of these constraints, large-scale water control works, such as dykes and canals, were initiated by the imperial state more than eight centuries ago and developed during the 19th century before the arrival of the French (Chassigneux, 1912). Dykes protected the Vietnamese population from floods during the monsoon[3]; during the dry season, canals could receive water

[1]This description is based on the results of the 'DELTAS' INCO-DC research project, funded by the European Union (DG XII). In Vietnam, the project was implemented by GRET (Paris) and the Vietnam Agricultural Sciences Institute (Hanoi) between 1998 and 2000. Additional information is sourced from ACIAR (Australia)-funded project 9404 'Integrated management of pumped irrigation systems in the Red River Delta – 1995–1998', carried out with the Vietnam Institute of Water Resources Research.

[2]The term 'privatization' may be ambiguous and is understood here as the emancipation from the state of groups of users who are able to manage their pumping stations and irrigation schemes independently. However, these undertakings are communal, theoretically non-profit-oriented, and have often been made possible thanks to public funds.

[3]The monsoon in northern Vietnam is characterized by high precipitation and frequent typhoons (Taillard, 1995; Le Ba Thao, 1997).

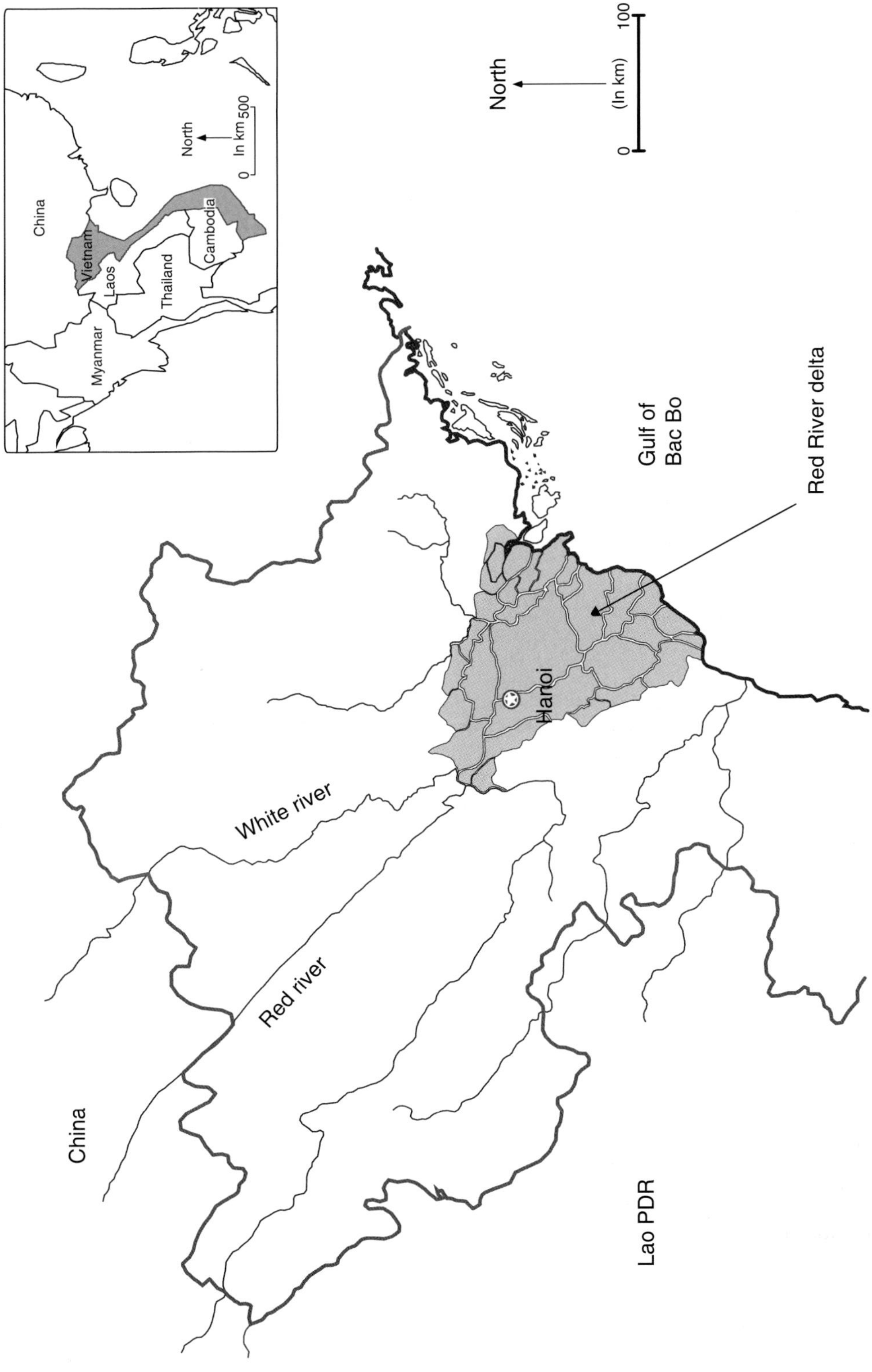

Fig. 7.1. The Red River delta in Vietnam.

from the river (through sluices in the dykes) and channel it to the lowest paddy fields, gravity allowing. To secure and intensify paddy agriculture, individual irrigation equipments such as water-lifting baskets and tripod scoops were introduced through Chinese influence, which lasted in Vietnam for 1000 years.

The imperial state took responsibility for the construction of dykes, water gates and main canals along river banks by mobilizing local (forced) labour. The responsibility of irrigation was left to the villages (*lang xa*) (Fontenelle, 1998). During the French colonization, state investment in hydraulic works increased dramatically, with the improvement and completion of the Red River delta system of dykes, gates and the network of main canals. Although the combined action of the central state and farming communities had already gone a long way in developing intensive agriculture in this delta, the farmers' situation remained uncertain due to the occurrence of droughts and floods, as well as the imposition of taxes and the burden of forced labour (Hémery and Brocheux, 1995). Poor drainage within the polders[4] resulted in continuously saturated conditions and a predisposition to rapid flooding, as the water levels in the river were (and still are) higher than in the surrounding paddy fields during the rainy season. As regards irrigation, low levels in the river made manual water lifting necessary and hindered rice development during the dry season.

The centralized modernization of water control

The modernization of water control in the Red River delta began in the 1960s under the policy of agricultural collectivization and with the establishment of cooperatives. The modernization of water control was considered a strategic mission, as a necessity towards the collectivization of agriculture. The combined effects of collective mobilization for hydraulic works and the improvement of agricultural conditions were supposed to encourage popular participation in the new cooperative system (Yvon-Tran, 1994).

The state placed great emphasis on mechanized drainage and irrigation. In 1962, 9.8 million man-days of labour were recorded against 2.3 million in 1959. In the Hung Yen province alone, 4000 km of canals were dug at the end of 1963. More than 80% of the direct investments in agriculture by the state were dedicated to the improvement of water control. Large drainage and irrigation schemes were created with a comprehensive network of canals, from the primary to the tertiary level, channels connecting polders to rivers, and large-scale irrigation and drainage pumping stations. Between 1961 and 1965, more than 2500 pumping stations were reportedly[5] set up in the Red River delta (Vo Nhân Tri, 1967 quoted in Yvon-Tran, 1994). By 1966, 73% of the cultivated area of the Red River delta was equipped with electrically powered irrigation and drainage pumping stations. Thus water could be extracted and supplied without human labour (Lê Thanh Khoi, 1978). These works, combined with the introduction of improved paddy varieties and chemical fertilizer, led to the further intensification of agriculture and to the double cropping of rice throughout the delta. Beyond the mere modernization of infrastructure, the way in which the Government of Vietnam intended to manage water supply also changed. From a situation where local management at the village level prevailed, water management was transferred to the state, provincial and district water services. Water distribution was organized according to strict irrigation turns among all cooperatives belonging to a single irrigation scheme, and farmers were effectively excluded from the water distribution process (Fontenelle, 1999).

[4]The Red River delta is divided into 30 independent hydraulic units, which are fully dyked and surrounded by arms of the river. These are called polders or *casiers*, in this chapter.

[5]These official statistics are subject to caution. However, there is no doubt that the 1960s witnessed a massive development of large-scale pumping stations.

However, the improvement of food security in the Red River delta did not last long. Between 1960 and 1975, the population resisted and resented the move towards collectivist economy and cooperatives. This, combined with the dysfunctional centralized management, appeared to have compounded an emerging economic crisis (Kerkvliet, 1999). The situation worsened at the end of the 1970s, when the government tried to sustain the collectivist economy through further heavy investments in water control equipment and stronger centralization of production management. Drainage capacities were upgraded through investment in new pumping stations with higher discharge capacity. Most village cooperatives were aggregated into commune cooperatives. Districts became responsible for all production aspects, including the establishment of the crop calendar, choice of rice variety and the management of hydraulic structures. This policy failed dramatically and the very poor living conditions of farmers sometimes degenerated into starvation (Nguyên Duc Truyên, 1993). The food crisis faced in this delta at the end of the 1970s was not the result of a lack of production capacity or funds, since water control infrastructures were well developed by then. This crisis appeared to be due to excessive state intervention, which undermined the capacity of farmers to innovate in, and take control of, production. The crisis was political rather than technical (Tessier and Fontenelle, 2000).

Liberalization reforms and decentralization of water control

This situation lasted until the beginning of the 1980s, when Vietnamese authorities recognized the failure of the 'great socialist agriculture' and proposed, through the *Khoan 100* (Directive 100), a new contract for production with farming households. This contract, in which paddy land was leased to households for a fixed contribution and the surplus of production left to farmers, arose in a context of an economic crisis compounded by farmers' rejection of collectivism (Beresford, 1988; Kerkvliet, 1995). The directive resulted in a boom in agricultural production and encouraged farmers to claim fuller responsibility for agricultural production, including the supply of water. The aspirations of farmers could not be satisfied through the strict rotation of irrigation turns which prevailed in centrally managed schemes. First, individual land management created the need for a specific access to water for each small field leased to farmers, in contrast to the former organization of water supply on large collective plots[6] (Mai Van Hai, 1999). Second, a strict organization with the establishment of a collectively fixed crop calendar did not allow for the diversification of crops and paddy varieties (Fontenelle and Tessier, 1997). The negative impact of this constraint was reinforced in the case of droughts or power cuts. In order to improve local irrigation conditions, farmers and cooperatives had to free themselves from their dependency on centralized irrigation systems. Farmers deepened existing tertiary canals to store water for a few days after pumping and to gain some flexibility in irrigation at the farm level (Dang The Phong and Fontenelle, 1995).[7] Cooperatives set up local pumping stations to get direct and autonomous access to water supply (Fontenelle and Tessier, 1997). These pumps were financed by revenues from cooperatives and subsidies from the state. Local pumping stations abstracted water from *arroyos*[8] and from the canal networks built by the state in the 1960s. Local irrigation

[6]Reality, however, is diverse. In other areas, like Dan Hoai and La Khe polders, the water management groups still do the bulk of in-field water management. Because of the disaggregation of holdings due to land reform, farmers find it too difficult to manage water individually.

[7]This added to significant secondary storage and in-drain storage within command areas, which tapped direct supply or return flow from the 'centralized' irrigation systems, which had long been developed.

[8]The term 'arroyo' is used to define the network of natural drainage channels and the lands lying in between that, in most cases, are now bounded by dykes.

schemes thus emerged as fragments of the old centralized irrigation schemes.[9]

The construction of local pumping stations increased during the 1980s taking advantage of further political reforms initiated by the government. In 1984, through Directive 112/HDBT, the central government decreased its involvement in water management, not only partly devolving management of water control services but also strengthening mechanisms aimed at balancing revenues and costs under strong provincial control. A new actor, the Irrigation and Drainage Management Company (IDMC), was created in each polder. The IDMCs are public companies owned by the state, which were supposed to balance their accounts through the collection of a water fee paid by the cooperatives. They are essentially bulk water suppliers. Furthermore, the *Doi Moi* reform in 1986, which resulted in the abolition of subsidies and in the liberalization of production activities, the *Khoan 10* (Directive 10) in 1988 and the Land Law in 1993, which governs the redistribution of land to farming households, created new conditions for water management and agriculture. Finally, in 1996, the state issued a law on cooperatives aimed at improving their management in a way reminiscent of the 1984 reform of the IDMCs. Cooperatives were no longer considered responsible for production and were supposed to provide service to farmers, for which they could charge a fee. They were still responsible for the collection of water fees paid by the farmers.

Agriculture became more diversified and intensive, as farmers gained the freedom to manage their production individually. Farmers diversified the number of paddy varieties they used, adopted direct seeding techniques, and increased commercial crop production, especially during the winter season (Lê Duc Thinh and Fontenelle, 1998; Bach Trung Hung *et al.*, 1999). These changes had an impact on water demand, both in terms of overall requirements and frequency of supply (Mai Van Hai, 1999). To meet these requirements the cooperatives increased the number of local pumping stations in order to get more autonomy and flexibility in water supply. These stations now serve approximately half the irrigated area of the Red River delta. High population densities do not seem to have jeopardized food security in the Red River delta as it did in the past, as agriculture now provides more than 300 kg of paddy per head per year (Dao Thê Tuan, 1998). Agriculture is very intensive and the paddy production of this delta accounts for up to 22% of all Vietnamese rice production. The people of this delta seem to successfully combine a high population density with intensive agriculture and strong water control measures.

Institutional and Financial Framework of Water Management

This section focuses on the example of the Bac Hung Hai (BHH) polder. It is the largest polder and the first in which hydraulic modernization was implemented at the end of the 1950s. With an extension of 210,000 ha, 185,000 ha of which are protected by the dyke system, 126,000 ha cultivated and 100,000 ha irrigated, the BHH polder makes up 13% of the total area of the delta. It includes 15 districts from four provinces: Hanoi Metropolitan area (1), Bac Ninh province (2), Hung Yen province (6) and Hai Duong province (6) (Fig. 7.2). In 1996, the number of pumping stations in BHH totalled 1022, including 698 local stations.

National and provincial administrative levels

In 1995, the former Ministry of Water Resources, the Ministry of Agriculture and Food Industries, and the Ministry of Forestry were combined into a new Ministry of Agriculture and Rural Development (MARD). The Department of Water Resources within

[9]Built along arroyos from which they abstract water, local pumping stations also benefit from the presence of former centralized irrigation canals built on the side of arroyo banks. Therefore, centralized irrigation canals are cut into several reaches, which become primary irrigation canals of the new local irrigation systems.

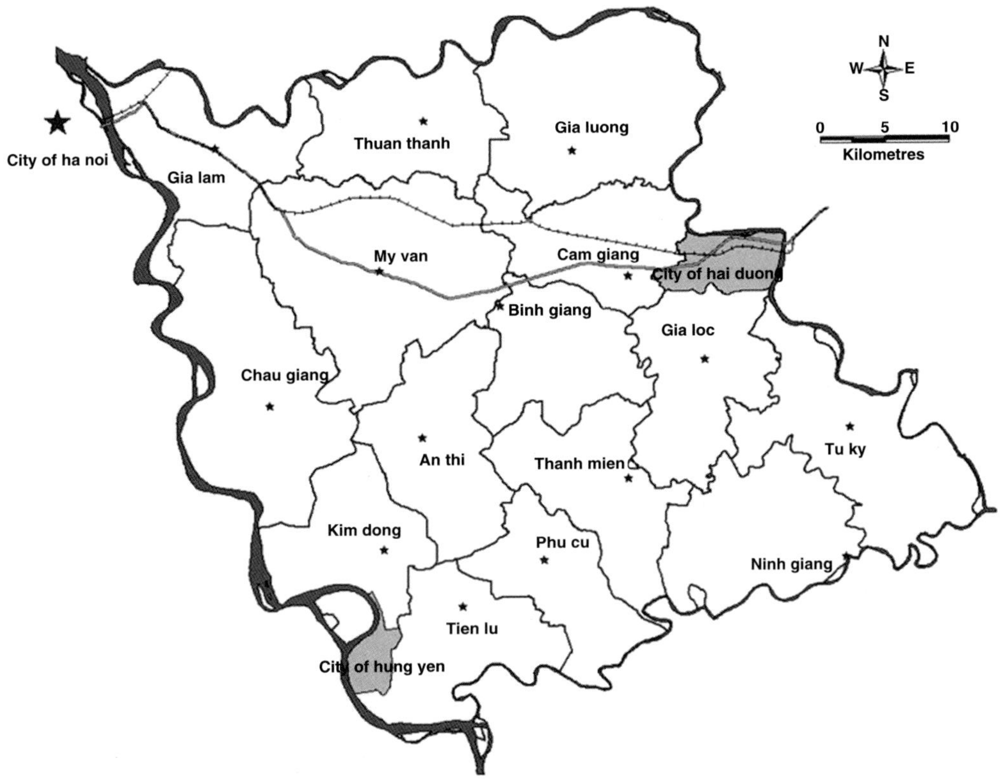

Fig. 7.2. The Bac Hunh Hai polder and administrative boundaries.

this ministry is responsible for the planning, design, construction and funding of major irrigation projects larger than 150ha. It fixes the national guidelines for the calculation of the water fee according to the type of irrigation (gravity, one or two pumping operations) and drainage (gravity and/or pumping).

The responsibility for managing existing public irrigation and drainage systems, and planning and executing smaller projects is delegated to the province under the leadership of the Provincial People's Committees (PPCs). The PPCs provide policy advice and funds and oversee the work of technical services, set provincial water rates based on national guidelines, allocate subsidies for local water resources projects, and make investments in local infrastructure. The provinces have established Water Resource Services (WRS) to handle these water-related responsibilities. There are ten WRS involved in the water management of the Red River delta, since the delta overlaps ten provinces. WRS are line agencies of the provincial governments. Their duties are similar to those of the central Department of Water Resources in terms of planning, design and construction, but are focused on smaller projects below 150ha.

Additionally, they shoulder the responsibility of calculating water fees paid by farmers, in consultation with PPCs and the party bureaucracy, and oversee the District Enterprises (DEs), which operate irrigation systems within polders. Water fees and their calculation were originally based on a national decree that the government cabinet promulgated in August 1984 (112 HDBT, 1984). Following national policy, the total water fee cannot exceed 8% of each province's average paddy yield for the last five consecutive seasons, for spring and summer seasons. The fee calculation is based on three subsidiary fees which correspond to rice nursery irrigation, paddy field irrigation and paddy field drainage operating costs. The maximum value of the fee for these different services depends on whether water is supplied by gravity, or through one or two pumping operations: the irrigation fee, for example, includes a 'diversion' fee which is paid to the company in all cases (operation of the main system), a pumping fee if such an operation is necessary and a field application fee. The diversity of situations leads to a great complexity in the calculation of the fees. Even though farmers now generally pay in cash, the fee is expressed in kilograms of paddy, and the PPCs determine every year an official rate for 1kg of paddy in order to insulate the calculation of the fee from the price fluctuations in the paddy market.

IDMC at the polder level

IDMCs are provincial state companies established under the WRS to identify and design water resource projects, to construct and repair civil works and to manage irrigation water. Most often, an IDMC has responsibility for all existing public irrigation in a primary hydraulic unit (or polder). Several IDMCs can respond to the same WRS when the province encompasses more than one polder. Unlike the Department of Water Resources and WRS, the IDMC level is not based on an administrative division but on the polder division. There are 30 IDMCs in the delta, managed by 10 provincial WRS (Fig. 7.3). In larger polders, which extend over more than one district, the IDMC is assisted by several sub-companies[10] (otherwise known as District Enterprises or DEs), one per district concerned. In 1995, 14 DEs were recorded in BHH, the largest polder in the delta.[11] Each IDMC or DE is structured based on irrigation stations, called *cum*, each of these being responsible for approximately 1000ha. Hydraulic cum work with an average of 3–5 cooperatives to manage water, maintain facilities and collect the water fee. Hydraulic cum are responsible for the O&M of schemes, from the pumping station to the secondary canal.[12] Overall, the mismatch between hydraulic units (polders, irrigation units) and administrative ones (province, districts, communes) generates a complex set of nested structures. Management practices, financing and accountability will have to be defined at all levels and made compatible.

With the 1997 national Directive 56/CP, IDMCs (and DEs) were transformed into public utilities. They were expected to cover the costs of water diversion, O&M of irrigation and drainage and depreciation, through the collection of the water fees paid by the farmers. However, IDMCs do not have control over their income and are, in particular, not allowed to raise service fees or keep surplus funds, except for minimal maintenance. In case of climatic hazards, such as typhoons and droughts, state subsidies are supposed to be granted in order to compensate for extra drainage and field application costs, while water fees are reduced in case of paddy losses from flooding. Implementing Directive 56/CP is the responsibility of each PPC, which adapts the directive to its own situation and issues provincial circulars on this issue.

The DEs are normally responsible for the main pumping infrastructure located at the head of the main canals and for operating the main drainage stations within an irrigation system (usually a large sub-polder). The DEs are nominally district-level organizations, but in practice they may often cover multiple districts within one province. The IDMC operates the main hydraulic infrastructure on the river system and the DEs, which are owned by the individual provinces, pay a bulk water fee to the IDMC. At Bac Hung Hai, the DEs tap water from, and discharge it into, the natural channel and

[10]In the case of BHH, this distinction is important since the DEs are managed by the individual provinces, but the IDMC is managed by a consortium of provinces and the MARD centre: the Director of BHH does not report to the provinces but to MARD, and is in fact usually in conflict with the provinces over the payment of bulk service charges by the DEs.

[11]The two districts of Bac Ninh province have a joint DE. This is why there are only 14 DEs for 15 districts in the BHH polder, which overlap with four different provinces (Fig. 7.2).

[12]Their formal responsibility ends at the tertiary turnout, which is where the responsibility of the cooperatives' water management groups begins. However, in practice, the cooperatives often control the secondary channels and even sometimes control the operation of the secondary head gates, but this usually brings them into conflict with the company and the *cum*.

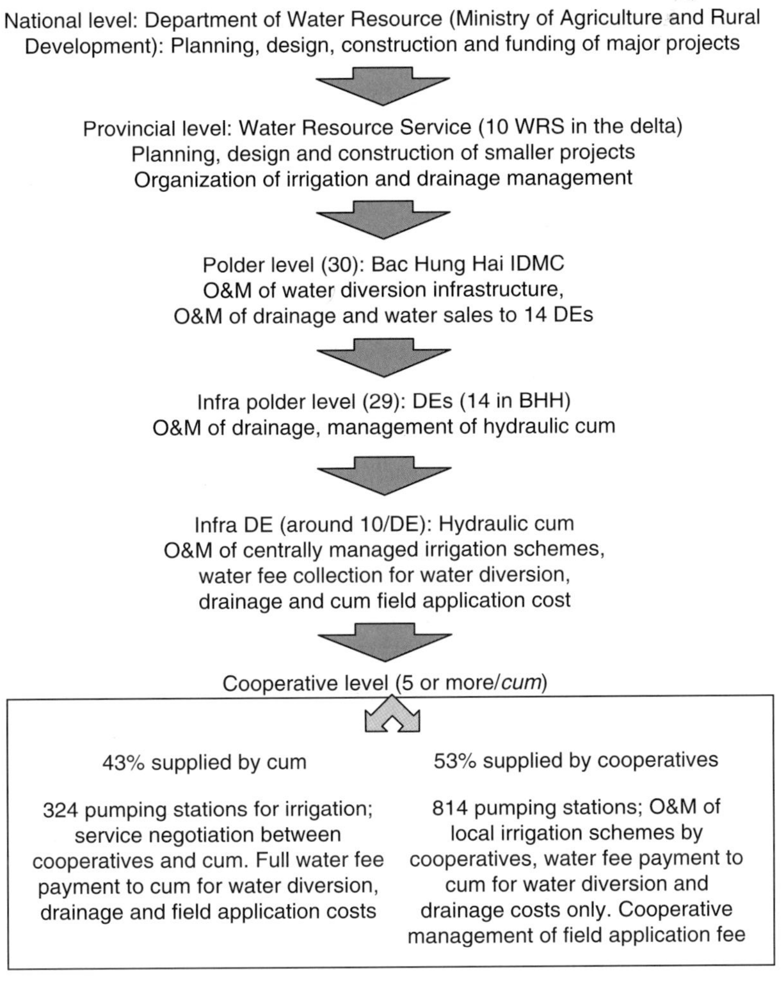

Fig. 7.3. Water management organizational framework in the Bac Hung Hai polder.

main canal network, which is operated by the BHH IDMC. Where the IDMC operates only within a province, payment of bulk water charges is enforced by the Provincial WRS (and the Economic Court) and it has not been a major problem. However, in BHH, the IDMC was jointly owned by four provinces and was then taken under the Ministry's jurisdiction because of financial losses amounting to around $1.00 million per annum over the period 1994–1998. Underpayment of bulk water charges by DEs has been a significant contributing factor to this situation, and it is still unresolved.

Because of the size of the BHH polder, BHH IDMC constitutes a special case: before 1998, it was supervised by the Hai Hung provincial WRS.[13] Nowadays, BHH IDMC is supervised by a System Management Council, constituting representatives from the four provincial WRS concerned, and chaired by the Director of the Department of Water Resources. BHH IDMC is responsible for water diversion and transportation from the river through the dual-purpose central canal network on the whole BHH polder, and for

[13]The Hai Duong and Hung Yen provinces formerly formed the Hai Hung province. The 1997 reform led to the division of several provinces and districts in Vietnam and resulted in the BHH overlapping with four provinces.

the operation of most tail-end drainage facilities (pumping stations and gravity gates) which discharge outside the dykes of the BHH polder.

Within the BHH polder, the situation of each district depends on the province it belongs to: the DEs from Hung Yen and Hai Duong provinces (which make up 85% of the BHH-supplied area: Fig. 7.1) pay, based on actual supplied area (36 kg/ha for the spring season and 24 kg/ha for the summer season), while DEs of Hanoi and Bac Ninh provinces pay a percentage of BHH IDMC annual expenditures equivalent to the share of area covered by each DE (3% for Hanoi DE and 12% for Bac Ninh DE). Table 7.1 indicates the breakdown of revenues and expenditures of the IDMC as dictated by the national regulation and its evolution in the Hai Duong province after decentralization measures started to be enacted.

Table 7.1. Annual revenues and expenditures of IDMCs and DEs in the Hai Duong province.

		National regulation	Hai Duong provincial regulation
		Circular 90/1997/TTLT/TC-NN	Decision 1854/1998/QD-UB
Incomes	Water fee average level	From 3% to 8% of the yield Directive 112/HDBT (1984) Circular 90/1997/TTLT/TC-NN (1997)	From 1.6% to 5.9% of the yield Decision 1132/QD-UB (1993) Decision 283/QD-UB (1998) - when yield decreases >30%
	Public subsidies	- when yield decreases >30% - when income < expenditures (from national budget) - when drainage cost > average ratio kWh/ha	- when income < expenditures (from national/provincial budget) - when drainage cost > average ratio kWh/ha - permanent subsidy to decrease water diversion cost to farmers
	Commercial activities	Directive 112/HDBT (1984)	Directive 112/HDBT (1984)
	Gasoline and electricity	<50%	Circular 16/DM-XN (1989)
	Salaries	<8% of total expenditures Circular 06/NNPTNT (1998)	<8% of total expenditures Circular 06/NNPTNT (1998)
Expenditures	Social and health insurance	19% of salaries Directive 112/HDBT (1984)	19% of salaries Directive 112/HDBT (1984)
	Diversion cost paid to BHH IDMC	Directive 112/HDBT (1984)	Decision 1132/QD-UB (1993)
	Depreciation of equipment	Decision 1062TC/QD/CSCT (1996)	Decision 1062TC/QD/CSCT (1996)
	Exceptional repairs	18–20% Decision 506TC/DTXD	16–19% Circular 06/TL (1990)
	Ordinary repairs	20–30% Decision 211/BNN (1988)	14–16% Circular 06/TL (1990)
	Water fee collection	2–3%	<3%
	Management overheads	5–6%	<5%

Regarding public subsidies, the Hai Duong provincial decision No. 283/QD-UB stipulates that altogether 136,000 kWh are annually needed to cover the electricity costs of drainage stations. When drainage needs are higher than this rate, subsidies are granted by the provincial WRS (no longer by the finance ministry) to the IDMC to compensate for the losses. Moreover, a permanent (but small) subsidy is given to the IDMC to decrease the cost of water to farmers. Finally, commercial activities also contribute to the company's income. They include transport fees for boats using the primary canal network, and the maintenance fees for the main works directly carried out by the company. An analysis of the period 1995–1999 showed that, on average, diversion fees paid by DEs amounted to 87% of the BHH IDMC annual revenue, while subsidies and commercial fees represented only 2% and 11%, respectively (Nguyen Thi Hong Loan, 2000). Table 7.1 also specifies expenditures in terms of percentage of the revenue. The larger share goes to maintenance work, while salaries plus health-care costs have to remain approximately below 10%.

Cooperatives and farmers

Cooperatives[14] are the lowest formal administrative level involved in irrigation and they are collective bodies supposed to represent all the farmers who depend on their agricultural services now mainly concentrated on water and electricity. They are managed by commune officials only, and access to membership (with corresponding rights) is restricted to volunteer farmers (members of the Party or of the Farmers Association). The relationship between cooperatives and DEs, via a hydraulic cum, depends on the existence and the location of local pumping stations. Every year, each cooperative signs a service contract with a cum, which acts on behalf of the district. These contracts are established on a seasonal or annual basis by mutual agreement and signed between each cooperative director and the staff in charge of the cum, or by the DE's director directly. The contract specifies the seasonal or annual water fee to be paid by the cooperative. For the spring season, the area cultivated by the cooperative is indicated and the supplier specified: water can be either provided by the cum or by a local pump of the cooperative itself. For the area to be supplied by the cum, more details are given: these include the kind of crop (rice, rice nursery, food crops or industrial crops), and the kind of irrigation, which is provided (direct gravity irrigation, single or double pumping, 'hand lifted' irrigation). For each type of crop and irrigation, a water fee rate is given in kilograms of paddy per hectare, based on provincial regulations. These rates are multiplied by the area of each type of crop and irrigation, and then aggregated. The sum gives the amount of irrigation fee, including the water diversion costs, to be paid by the cooperative to the cum. For the summer season, an additional fee for drainage is calculated on the basis of the whole area cultivated by the cooperative. The date, place and nature of payment are specified too. Contracts vary according to the water-supply situation of each cooperative, as explained below:

- When there is no local pumping station, cooperatives are responsible for distribution of water and maintenance of irrigation canals, from secondary canals to quaternary canals. They collect a water fee from farmers, which is equivalent to water diversion, drainage and field application costs. Of the fee, 98% is paid to the hydraulic cum, which supplies them with water, and 2% is kept by the cooperative for field-level water management.
- When there are local pumping stations built along one of the channels of the polder, cooperatives have to operate and maintain their system from the pumping station down to the quaternary canals. They collect a fee from farmers as explained above but they do

[14]Cooperatives are established at the commune or village level. In the latter case, the village cooperatives are subsidiaries of the Economic Development Committee of the commune. In any case, cooperatives are closely linked to commune authorities.

not pass the total on to the cum. They only pay for water diversion and drainage costs and keep the irrigation fee (adjusted so as to incorporate the cooperatives costs) for themselves.

- When there are local pumping stations that withdraw water from primary (raised) irrigation canals supplied by a pumping station of the cum, cooperatives have to operate and maintain their local systems from the local pumping station to the quaternary canals. The field application fee is increased, since some of it is kept by the cooperative to cover the cost of its own irrigation pumping operations, while the standard fees for diversion, drainage and field application are paid to the cum.

Some cooperatives are fully independent while others still rely on centrally managed pumping stations for a percentage of their irrigated area, ranging from a few hectares to the whole cooperative-irrigated area. Combinations of two of these three cases can also be found within the same cooperative, as sub-areas may have different statuses: in such cases, the costs of supplying water to farmers differ but they are averaged in order to come up with a uniform fee per hectare. In the BHH polder, there are only a few cases of double pumping which are not recorded in DE's statistical data. The official figures indicate that 53% of the BHH irrigated area is supplied by cooperative stations and 43% by DEs (Table 7.2).

Finally, farmers have to pay part of their annual individual water fee to the cooperative twice a year, after spring and monsoonal rice harvests. The amount they pay reflects the situation of the cooperative regarding irrigation and drainage facilities. They all pay the same amount per unit of area, irrespective of the location of their plots. The water fee is paid together with other levies such as the land tax and several local taxes established by the commune (maintenance of local roads, field surveillance, taxes on houses, gardens and ponds, solidarity tax, etc.). As a result, only a few farmers know the exact amount paid for the irrigation and drainage service (Fontenelle and Tessier, 1997).

Table 7.2. District area supplied by DEs and cooperatives in the spring, 1996.

District	DE (ha)	Cooperative (ha)	Total (ha)	DE (%)	Cooperative (%)
Gia Lam[a]	1,665	132	1,892	88	7
Thuan Thanh	4,312	1,761	6,073	71	29
Gia Loc	3,796	3,367	7,163	53	47
Chau Giang	4,934	2,129	9,675	51	22
An Thi	3,238	3,651	6,889	47	53
My Van	5,391	6,094	11,719	46	52
Tien Lu	2,061	2,733	4,794	43	57
Thanh Mien	2,642	4,499	7,141	37	63
Kim Dong	1,547	2,749	4,296	36	64
Cam Giang	1,877	3,338	5,215	36	64
Gia Luong	3,282	6,094	9,376	35	65
Phu Cu	1,671	3,244	4,915	34	66
Binh Giang	1,492	4,245	5,737	26	74
Tu Ky	1,949	5,196	8,119	24	64
Ninh Giang	1,633	5,255	7,101	23	74
Total BHH	41,490	54,487	100,105	43[b]	53

[a]Cooperatives outside BHH polder are not taken into account.

[b]The total does not amount to 100%: a 4% difference is due to missing data.

The Intricacies of Water Pricing

Overlapping rationalities

The emergence of local pumping-irrigation stations in the Red River delta led to the creation of a dual system where two kinds of irrigation stations, with different technical characteristics, supply fragments of the same original network. In the BHH, there were 814 local stations in 1996 supplying 54,487 ha, and 324 centralized stations supplying 41,490 ha. Figure 7.4 provides the example of Van Giang DE, which includes four cum (the average size of local schemes, 67 ha, is, therefore, half that of the present (reduced) size of centralized schemes, 128 ha). Local pumping stations had a higher per-hectare pumping capacity than centralized stations when they were constructed (Table 7.3). Their investment cost per unit area is higher but, on the other hand, they provide several benefits to farmers (Fontenelle and Tessier, 1997; Mai Van Hai, 1999) as listed below:

- *Satisfaction of water requirements.* Technical surveys conducted on irrigation efficiency at scheme, plot and field levels in the An Binh cooperative, in the Nam Thanh district, showed that crop water requirements were met. This contrasts with the former situation of centrally managed stations where downstream cooperatives could not access water in time (Bousquet *et al.*, 1994; Dang The Phong and Fontenelle, 1995).
- *Flexibility/autonomy.* Field surveys conducted in 13 communes of the Nam Thanh district have shown that farmers did not want an irrigation interval longer than 7 days (Dang The Phong and Fontenelle, 1997). On local irrigation schemes, there is no delay between the decision to pump and the arrival of water. During the rice season, the full supply by local irrigation units is achieved within a day. Farmers can now complete their land preparation within 2 days, instead of 11, as earlier, which allows them more flexibility in terms of

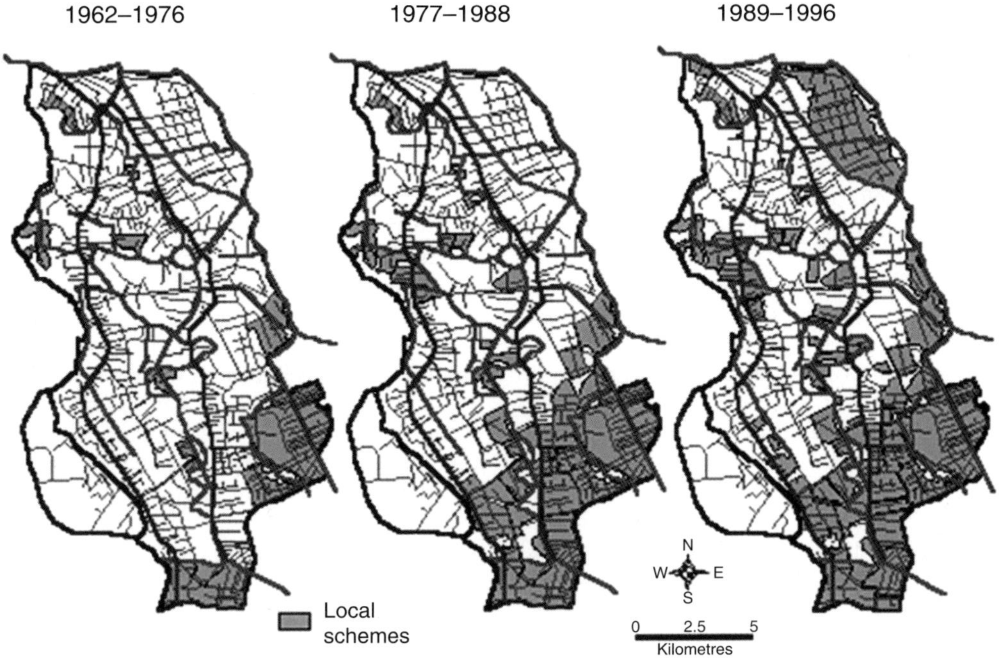

Fig. 7.4. Growth of local irrigation schemes in the Van Giang centralized scheme (Chau Giang district).

Table 7.3. Comparison of irrigation duration for local and centralized stations.

	24h–average	Land preparation	Rice-season irrigation
	Continuous flow	Supply of 100mm (night and day, 20h)	Supply of 30mm (12h maximum per day)
Local station	7.0l/s/ha	40h: 2 days	12h: 1 day
Centralized station	1.2l/s/ha	231h: 11.5 days	69h: 6 days

Table 7.4. Average volumes pumped per hectare during spring season 1996.[a]

	Land preparation (m^3/ha)	Rice-season irrigation (m^3/ha)	Seasonal consumption (m^3/ha)
Local station	1600	2400	4000
Centralized station	3900	5900	9800

[a]Monitoring of ten pumping stations in the Nam Thanh district, Hai Duong province. Rainfall during the spring season is 405mm on average. For more information on field water balance in the Red River delta see Dang The Phong and Fontenelle, 1995.

cropping patterns and choice of rice variety. The cooperatives' decisions to pump are triggered by the actual water status in paddy fields and not based on a fixed pumping calendar. Managers and users of local schemes are from the same village, or even from the same hamlet. They define their water supplies and rules among themselves, without DE intervention. Localities commonly share irrigation benefits and constraints within their boundaries, as it was the case before the agricultural collectivization of the 1960s (Fontenelle, 1998, 1999).

- *Efficiency*. The design command area of local schemes is smaller than in centrally managed schemes, below 100ha instead of 1000ha or more.[15] Canals are shorter and less water is wasted compared with centrally managed schemes, which suffer from water losses and illegal water diversions (Bousquet *et al.*, 1994; Fontenelle, 1999). As a result, local stations pump less water per unit of irrigated area than central ones, as can be seen from Table 7.4.[16] Differences in water use are due, in part, to the fact that local management is more efficient, but higher per-hectare consumption rates of companies are also due to some illicit arrangements between cooperatives and staff of cum pumping stations. In some instances, staff of pumping stations 'sell' water to cooperatives (which under-report their irrigated areas) in order to increase their income. This increases the total volume delivered per hectare, which puts further pressure on the DE to balance its books, since it cannot revise the charges per unit area.

However, reasons for investing in local pumping stations are not technical only. First, before the end of the 1980s, it was a way to justify and obtain the electrification of a village (*thon*) or a commune (*xa*) (Do Hai Dang, 1999). Second, the establishment of a local pumping station in

[15]The original area of the centrally managed Van Giang scheme was 14,000ha.

[16]These values are based on two combined approaches. One consisted of the monitoring of date and duration of each pumping. The other consisted of power readings. In both cases they represent actual volumes pumped and do not represent billed amounts.

each village of a commune is sometimes the sign of political competition between influential persons (*notables*), who all want to have a local station serving their village. Effective continuous flows of 5 l/s/ha may be technically acceptable but they sometimes reach 10 l/s/ha, which are clearly unnecessary as far as paddy cultivation is concerned. Beyond the mere technical question of crop water supply, local water management and investments embody local competition for prestige and power political struggles among commune and village leaders.

Costs to farmers

To assess the cost of water to farmers, six cooperatives were surveyed in two districts of the BHH. Two were fully responsible for their irrigation and two others partly responsible, while the last two were supplied by the central pumping stations of the company for all their irrigated area (Table 7.5). Results show that when pumping stations are managed by the cooperatives themselves, the calculation of the water fee can be based either on actual costs paid by the cooperatives or on fixed rates chosen by each cooperative. When the water supply to the cooperative depends on central stations the water fee calculation is based on provincial regulations only.

Table 7.6 specifies the amount of water fees paid by farmers and shows significant differences between cooperatives. These can be due to the natural or hydraulic conditions of each cooperative, such as the necessity of double pumping in the Tan Lang commune. But differences should not appear within each type of water supply, since rates are based on the same provincial directives and national decrees. For instance, single pumping fees range from 395 to 473 kg/ha/year in the same province of Hai Duong (cf. Table 7.6), which is 'officially' impossible. The highest levy was paid by farmers from the Tan Lang cooperatives, where all irrigated areas are supplied through two consecutive pumping operations. It amounted to 639 kg of paddy per hectare per year (paddy/ha/year). The lowest fee was paid by farmers from the Hung Thai cooperative, in which water supply of all types was cheaper than in other surveyed cooperatives. For example, a

Table 7.5. Irrigation type and water fee calculation system for six surveyed cooperatives.

Name of the cooperative	Scale	Number of pumping stations	% of the area supplied by local stations	District	Province	Basis of fee calculation
Tan Vinh	Village (since 1982)	2	100	Ninh Giang	Hai Duong	Cooperative effective expenditures (CEE)
Tan Lang	Commune	9	100	Gia Luong	Bac Ninh	Cooperative fixed rates (CFR)
Hung Thai	Commune	3	61	Ninh Giang	Hai Duong	Provincial fixed rates (PFR)
Dong Tam	Commune	3 (+ 4 + 12)[a]	60[b]	Ninh Giang	Hai Duong	CEE for pumping and PFR for water diversion
Ngo Phan	Village (since 1992)	0	0	Gia Luong	Bac Ninh	Provincial fixed rates
Kim Thao	Village (since 1987)	0	0	Gia Luong	Bac Ninh	Provincial fixed rates

[a]There are four local collective diesel pumping stations and 12 individual petrol pumping stations for 28% of the total cooperative area.
[b]Including the 28% supplied by diesel and petrol pumps.

Table 7.6. Water fee paid per type of water supply (in kg of paddy/ha/year).

Name of cooperative	Type of access to water existing within the cooperative			Remarks
	Single pumping	Double pumping (DE + Local)	Diversion by gravity	
	Total (company + cooperative)			
Tan Vinh	619 (473 + 146)	–	324 (178 + 146)	Including 146 kg of extra fee for maintenance and new construction
Tan Lang	–	639	–	
Hung Thai	464 (408 + 56)	345 (289 + 56)[a]	253 (197 + 56)	Including 56 kg of extra fee for maintenance and new construction
Dong Tam	478 (395 + 83)	–	280 (197 + 83)	Including 83 kg of extra fee for maintenance and new construction
Ngo Phan	475	475	475	Fee averaged for all, for simplicity and equity concerns
Kim Thao	586 (475 + 111)	–	–	Including 111 kg of extra fee for maintenance and new construction

[a]In fact, in this cooperative there is only one pumping from the DE. The second lift is done manually by farmers.

single pumping operation by a local station costs farmers 464 kg of paddy/ha/year in the Hung Thai cooperative, which is 28% cheaper than in the Tan Lang cooperative.

But beyond these differences between cooperatives, another difference is introduced by local extra water fees defined, collected and used by cooperatives to improve the quality of their service (extra costs for local maintenance) and to develop their capacity (capitalizing for new investments in local stations). Extra fees, also referred to as 'exceptional levies', range from 56 to 146 kg/ha, i.e. between 12% and 45% of the total fee.

These differences are a manifestation of their autonomy but they also create a degree of inequity among farmers, who do not benefit from the same production conditions depending on the cooperative they belong to. However, compared with the annual production of paddy (an average of 8 t of paddy in two seasons, plus an additional crop in one-third of the area), water fees appear to be quite small. Even in the Tan Lang cooperative, they do not exceed 8% of the annual paddy production (not considering the benefit of the winter crop). In most areas of BHH water can be supplied by a single pumping operation. Therefore, water fees paid by farmers in these cooperatives (including extra fees established by cooperatives) range from 5.8% to 7.7% of their annual paddy production, which is reasonably expensive.

The point is that most farmers do not know the details of the calculation of the water fee. This information is withheld by the village chief who is in charge of tax collection on behalf of the Commune People's Committee, and downward accountability linkages are weak (Small, 1996). Ambiguity also results from the complexity of the breakdown of the water fee, depending on local conditions and is further strengthened by the fact that land taxes are usually assessed and collected at the same time. Farmers only know how many kilograms of equivalent paddy they have to pay at the end of each rice season, and even if they know the amount of the water fee they are not in a position to ascertain whether the extra fees collected are justified or not and what their exact utilization is (Fontenelle and Tessier, 1997). Sometimes, there is an ambiguity between irrigation services and the provision of electricity to households, which also allows some illicit gains to the cooperatives.[17] All this lack of clarity is embedded in kinship and patronage relationships and tends to engender mistrust in the villages (Do Hai Dang, 1999). Altogether, the annual taxes paid by farmers amount to 20–25% of the value of the annual paddy production (Bousquet *et al.*, 1994). They include not only the water fee but also the land tax and several other taxes (house, field watching, cooperative fund, construction of local roads, health, labour insurance, construction, crop damages, 'solidarity tax' for pioneer settlements or solidarity with Cuba, and the police). More than an issue of only taxation, farmers' difficulties are due to the low economic return of paddy production. Production costs (not considering labour and water fees) amount to 25% of the annual gross value of paddy production. Added to the water fees and other taxes, almost 50% of farmers' annual gross paddy production value evaporates.

The cooperatives: balanced but non-transparent accounts

The financial situation of the cooperatives surveyed was analysed using data communicated by the cooperatives themselves, except for the Kim Thao village cooperative, where information was not made available (Table 7.7). On the basis of the available information, it appears that the breakdown of expenditures varies from one cooperative to another.

The number of staff is obviously larger in commune cooperatives than in village cooperatives but it seems that there is no economy of scale as the share of management costs is higher in the former than in the latter. With the available information, it is difficult to interpret correlation between this share and the degree of dependence on the company. The amount paid to the DE is directly correlated to the percentage of area supplied by the central pumping stations, ranging from 30% (100% locally irrigated) to 75% (100% centrally irrigated) of total costs. On average, repairs amount to 15% of total expenditures, and investment in new construction or savings for depreciation of the equipment are not frequent. With the exception of the Tan Lang cooperative, and on the basis of the available values, cooperatives seem to balance their accounts, which are not the cases of IDMCs and DEs, as will be shown later. The main point about these values is that no justification is given for them. Cooperative managers do not present their accounts with more detail than the data provided in this table. Moreover, in three of the surveyed cooperatives no information was provided on the amount of fees collected. Financial transparency is not the rule.

IDMC's finances

The analysis of annual fee recovery of BHH IDMC and four DEs showed a cumulative financial deficit. Table 7.8 first illustrates the situation encountered in four DEs, one from each of the provinces overlapping the BHH polder for four consecutive years. These included Ninh Giang DE from the Hai Duong province, Chau Giang DE from the Hung Yen province, Gia Lam DE from the Hanoi province and Gia Thuan DE from the Bac Ninh province.

The water fee collected annually by each of these four DEs never reached the expected income, but fee recovery from the cooperatives nevertheless exceeded 92%,

[17]The electricity for pumping (irrigation) is billed 30% cheaper than domestic electricity by the company in charge of this service. The cooperatives sometimes apply only the higher tariff to all types of consumption.

Table 7.7. Annual water management average expenditures and balance (years 1998 and 1999).

Name of cooperative	No. of staff	% Management costs	% Paid to DE	% Electricity	% Repairs	% Invested	% Depreciation	% (Income – expenditures)/ incomes
Tan Vinh	9	9.7	30.1	11.5	18.7	30.0	0.0	+7.3
Tan Lang[a]	27	9.3	29.1	44.4	17.2	0.0	0.0	–81.4[b]
Hung Thai	17	23.0	43.6	18.8	11.9	0.0	2.7	+4.0
Dong Tam	32	29.9	50.0	12.5	7.6	0.0	0.0	+2.5
Ngo Phan[a]	9	5.9	74.8	0.0	19.3	0.0	0.0	–0.8

[a]Average on 1997, 1998 and 1999.
[b]Electricity costs are much higher for this cooperative because of double pumping.

Table 7.8. Water fee, incomes and expenditures (DE).

DE	Year	Water fee (in billion dong[a])		Incomes	Expenditures		Cost/ income
		Due	Collected	%	(in billion dong)	(in billion dong)	%
Ninh Giang	1996	3.0	2.9	97	3.0	3.5	117
(Hai Duong	1997	2.5	2.4	96	2.9	3.6	124
province)	1998	2.7	2.4	89	3.0	3.3	110
	1999	2.3	2.2	96	2.9	?	?
Chau Giang	1995	2.4	2.4	99	2.6	2.7	104
(Hung Yen	1996	2.6	2.4	92	2.6	3.1	119
province)	1997	2.1	2.1	98	2.2	2.6	118
	1998	2.8	2.7	96	2.8	3.3	118
	1999	2.5	2.5	97	2.7	3.2	119
Gia Lam	1995	2.5	2.2	88	2.5	3.0	120
(Hanoi	1996	3.2	3.0	94	3.0	3.6	120
province)	1997	3.2	2.8	88	2.8	4.3	154
	1998	3.1	2.6	84	3.3	4.3	130
	1999	3.9	3.2	82	3.7	4.1	111
Gia Thuan	1995	6.7	6.3	94	7.9	6.7	85
(Bac Ninh	1996	8.4	7.4	88	11.0	9.5	86
province)	1997	7.1	6.5	92	8.1	11.5	142
	1998	8.0	7.4	93	9.0	11.5	128
	1999	8.1	7.7	95	9.1	11.0	121
Average	5 years			92			118

[a]Note: $1.00 = Dong 15,980.

which is quite remarkable.[18] However, this is partly achieved though the manipulation of areas and income to be able to report such a recovery rate. The analysis of the annual effective expenditures compared to annual effective incomes (fee + subsidies + commercial activities) shows that the situation of the DEs is really unbalanced, with expenditures exceeding incomes by 18% on average (Table 7.8).

[18]The reasons for defaulting are not clear. It is possible that cooperatives which receive poor services decide to withhold part of the fee. Since these figures come from the DEs, it is also possible that these have interest in showing a shortfall. Interestingly, there is a recent move towards establishing contacts between the cum and the cooperatives which are not based on area but on real pumping hours and days. The gains, however, may not reach farmers as they are unaware of the nature of the contracts.

Table 7.9. Cumulative debt of four DEs (up to 1999, included).

	Ninh Giang (in million dong)	Chau Giang (in million dong)	Gia Lam[a] (in million dong)	Gia Thuan (in million dong)
Electricity cost	116 (17%)	1349 (58%)	1126 (61%)	1754 (41%)
BHH water diversion fee	444	968	697	1975
Other (repairs, maintenance, etc.)	141	0	25	517
Total	701	2317	1848	4246
In percentage of annual income	24	90	60	47

[a]These figures account for the totality of the Gia Lam district (communes inside and outside BHH included).

This situation is due to the incapacity or unwillingness of the provinces to provide subsidies to compensate for the loss, as dictated by the regulation. The shortfall thus corresponds to debts incurred with BHH IDMC and electricity companies, as specified in Table 7.9. On average, the cumulative debt of these companies exceeds 55% of their annual income, with important differences from one company to another. The status of each company is strongly correlated to the importance of the cumulated electricity debt rather than to the BHH water diversion fee, which amount is known by each DE and does not vary much from one year to the next. This does not apply to electricity costs, which depend on annual rainfall and farmers' practices. These differences between incomes and expenditures show that the present regulation does not allow the financial equilibrium of the activities of companies without the provision of subsidies by the national or provincial levels, and the granting of loans by the banks.

A similar analysis was done for BHH IDMC. Table 7.10 shows that for the 5 years studied the company could not collect the full water diversion fee owed by the 14 DEs. The fact is that DEs do not pay their diversion fee to the BBH IDMC as they should (80% at the most).

This financial imbalance has a direct impact on BHH IDMC activities. Every year, the company has to submit its activity plan to the authorities. Priority is given to operational activities to the detriment of maintenance and repairs. Financial resources cover priority costs, such as salaries for IDMC staff, electricity and petrol for station operations, costs of water fee recovery and interest on loans. Maintenance and repair activities depend on the annual collected income, on cash flows and loans made with public organizations (banks and public companies). Figures for major repairs show that differences between planned and achieved activities are very large every year (see Table 7.11). It was only in 2 years, 1996 and 1999, that the company could mobilize enough funds to cover the cost of the planned repairs. This is because, in 1996, BHH IDMC got a loan of 3.7 billion dong from

Table 7.10. Comparison between due and collected water diversion fee (IDMC).

Year	Water diversion fee due (in billion dong)	Water diversion fee collected (in billion dong)	Percentage
1995	7.6	4.4	58
1996	8.4	6.7	80
1997	7.4	4.3	58
1998	11.1	6.1	55
1999	9.2	6.6	72

Table 7.11. Comparison between planned and achieved main repairs (BHH IDMC).

Year	Main repairs planned (in billion dong)	Main repairs achieved (in billion dong)	Percentage
1995	3.7	1.6	43
1996	3.4	3.7	109
1997	3.2	1.5	47
1998	3.8	0.9	24
1999	2.9	3.2	110

the dredging (public) company and in 1999 it got a subsidy from the Ministry of Agriculture and Rural Development. This shortfall in income weakens the capacity of the IDMC to meet its annual O&M costs.

Institutional contradictions and difficulties

From a functional point of view, the relevant unit of an irrigation scheme is the hydraulic unit. But decisions on water management (and, in a large part, on financial issues) are based on administrative decisions and on administrative units. This is a classical problem with irrigation schemes which also applies to the IDMCs and DEs, which are under control of the Water Resource Services of the province. Moreover, some IDMCs, as was the case in the BHH polder, are under the control of more than one province. When water is provided to hydraulic units that span different provinces, the level of fees and subsidies can be different for the same service. Currently, there are four different directives governing the level of the water fee paid by farmers living in the BHH unit, and policies on subsidies vary from one province to another. This situation leads to inequity in water fees paid by farmers, depending on the province they belong to.

The level of fees is determined by People Committees, under an overall framework fixed by the state. They are based on a percentage of the yield, depending on the kind of water service that is provided. At national and even more at provincial levels, the determination of fees is based more on political considerations than on the economic analysis of water service costs. For example, the level of fees did not follow the huge increase in electricity costs which took place between 1986 and the early 1990s.

The companies have limited control over their income, which depends on the area actually irrigated and drained, and on the level of the fees. Even if they collected 100% of the fees, Table 7.8 suggests that only half of their deficit would be covered. Officially, provincial subsidies are supposed to cover the differences between income and expenditure. Moreover, the reference for the fees is supposed to be the average yield for the past 5 years. But often, this reference has not been revised since 1984, even if real yields have dramatically increased. In addition, provincial WRS did not add a third irrigation fee for the winter-season crop, even when some irrigation supply was required. Instead, they decided that the cost of the third crop would be covered by subsidies as a political measure to promote intensification of agriculture. This makes DEs reluctant to supply water in winter, which encourages farmers to develop their own pumping schemes. Considering the actual agricultural production (paddy yields and a third winter crop) of farmers, the effective water fee they pay to the companies is lower than the maximum nominal official percentage (5.9% of yield for Hai Duong).

The main operating costs of companies are electricity and maintenance, along with salaries. The electricity bill depends on the year (and especially on the amount of drainage pumping done) but companies have to meet it even if they jeopardize their annual financial balances, for fear of occasional power cuts. They cannot stop drainage or irrigation when the electricity expenses are above the provisional budget. Most company charges are defined and fixed by the administration. Decrees on water management specify how many people have to be employed for each kind of work. Depending on its power, a pumping unit must have a head, a worker and, maybe, a third person. Therefore, the number of persons working for the company is broadly defined by the structure of the scheme. Some officials at the central level say that these norms are too high and that it is not necessary to have so many persons. Salaries, social security contributions, etc., are also fixed by the administration. Even if they wanted to, companies could not significantly reduce the cost of labour. This cost, in all instances, if we trust official statistics, remains under 10%. It is not as high as expected but might reflect employees' very low salaries. Thus, it would be unfair to assimilate these companies to overstaffed agencies commonly found in the irrigation sector, and reducing staff would only yield very limited gains.

In such a situation, companies can only control their expenses by deferring or leaving the maintenance works unpaid. Moreover, for patronage or political reasons, companies may employ more persons than the number fixed by decrees. Most of the maintenance work is done by the companies themselves, or by public enterprises under contract without real competition, which may result in increased costs.

Due to the emergence of local pumping stations, DEs now supply only about 50% of the area they served originally. Their incomes are based on the area supplied and have therefore significantly declined, but the amounts in electricity bills have also decreased because of the smaller area now serviced. Electricity, however, is only one part of the expenses, and labour or other fixed costs have not decreased, because the number of persons paid by companies remains the same. Moreover, the contract between companies and cooperatives is based on an estimate of the irrigated and drained areas, but companies are not able to accurately determine the effective area. Cooperatives tend to under-report this area as a way to reduce the fee paid to the companies, contributing to widening the gap between DE incomes and costs.

The evolution of the legal status of DEs has constituted a significant step in the restructuring of water management after de-collectivization. Compared with a fully centralized management, it allows a better specification of responsibilities. The attempt to oblige DEs and IDMCs to balance their budgets, however, was a failure, despite efforts by DEs to improve fee collection. Defaulting by DEs could, in principle, be dealt with by resorting to provincial economic courts but since the BHH Company was made a national company under MARD such a move could be blocked by the provinces which control the courts. The situation changed in 1999, after riots erupted in Thai Binh in response to taxation perceived as abusive. DEs do not have to present balanced budgets anymore; following decentralization, the provinces became fully responsible for all financial matters relating to the DEs, and the payment of drainage pumping services in 'abnormal' years was devolved to them, rather than being handled by the MARD/central government. With the reduction in central funding and continuing need for capital maintenance and covering community drainage liabilities – the provinces ended up with bigger commitments to subsidy than they had before, driven by central policy but with the responsibility devolved.

Synthesis

The organizational and financial framework of water control in the Red River delta presents a complex and confusing image. While most countries in Asia have decided to provide irrigation service under high levels of operational subsidy, the Government of Vietnam attempted to recover a significant proportion of operating costs from farmers both before and after economic liberalization in the 1990s. Because of the relatively high cost of service provision arising from extensive pumping for both irrigation and drainage, pressure to recover costs intensified as the rest of the economy liberalized, squeezing the DEs between their service providers (electricity) and an already highly taxed farming population. On the one hand, the liberalization of the economy meant that production costs would have to be covered by the producers themselves and, on the other, the struggle for national food security after more than 30 years of war and scarcity, restrained the state from levying the full cost of hydraulic operations from farmers alone. Drainage service benefits the non-agricultural rural and urban population too; the public-good nature of this service justifies that the state covers part of the expenditures of the IDMCs and DEs and that the farmers and the state (central and provincial) shoulder cost recovery for irrigation and drainage. The Vietnamese State has tried to combine two political goals by striking a balance between rural stability and a service-cost approach to irrigation and drainage.

With the decentralization policy of the 1990s, the organization of the water control in the Red River delta became more complex. From a management point of view, some legislative capacity was transferred to the provincial level. From a technical point of view, the increasing involvement of cooperatives in irrigation and in the development of local pumping stations led to the effective redistribution of responsibilities between the IDMCs, DEs and the cooperatives. The resulting multiplication of circulars and rules for regulation at the central, provincial and communal levels created some heterogeneity and inequity in farm taxation. The water fees paid by farmers may be different from one cooperative to another. The calculation of income and expenditures of the DEs and IDMCs varies according to the province but this heterogeneity stems more from local political decisions than from the variety of hydraulic conditions.

The study also showed the benefits that can be drawn from decentralized and autonomous pumping stations, as opposed to centralized large-scale ones. Agriculture in the Red River delta grew dramatically in intensity and productivity thanks to the development of local pumping stations. The gains in flexibility and responsiveness to water needs came at the cost of what might appear to be excess pump capacity, but these gains are significant enough to encourage the development of local supply, even if the costs per hectare tend to be somewhat higher because of diseconomies of scale. The constraints of collective action are also better accepted by farmers within the limits of villages and communes, which are historically and culturally meaningful. Economically unsound development of local pumping stations may also be encouraged when farmers are able to access public funds and do not pay directly for the investments. The share of these investments paid by the communes varies with time and place.[19] Public funds can be sourced through the provincial budget, but this is an obscure point in which personal networks of influence[20] and the influence of the District Party Committees also play a great role.

Water pricing in the Red River delta is primarily geared towards ensuring partial financial stability. The closed nature of the Red River polders indicates that saving in pumping costs translates into financial gains but not into water savings at the macro level (in addition, contracts between cum and companies are generally made on the basis of area and not of volume). In any event, localized water shortages are due to inadequate management and insufficient hydraulic conveyance capacity of secondary canals in the face of uncoordinated pumping operations, rather than to a lack of water resources at the polder level. Even if water is not scarce and water savings largely irrelevant, decreasing abstraction would mean lower energy costs. While local stations have incentives to reduce their own costs, it must be noted that service by the cum is paid based on the area and is independent of the volume effectively supplied. Water charge mechanisms, therefore, have no direct impact on how much water is pumped and on the energy bill (Table 7.12).

The analysis of cooperative financial data suggests that farmers cover between 70% and 85% of O&M costs, not considering depreciation costs which remain dependent on state and/or provincial subsidies. This is, by world standards, a rather substantial contribution to cost recovery. In addition, the expression of the fee in terms of kilograms of paddy has successfully solved the common problem of erosion by inflation, by indexing costs to the price of food.

[19]Communes can use different local taxes or state subsidies to support such investments. In the late 1980s, for example, they used subsidies for agricultural input that were made redundant by the liberalization policy.

[20]Such networks may be linked to kinship, the village of origin, batches at the university or in the army, etc.

Table 7.12. Main actors and their strategies.

Actor	Constraints	Strategies	Observations
Farmers	No control on what the fees are for.	Develop local pumping capacity in order to facilitate intensification and diversification of agriculture. Get water when they need it, no matter what the cost is. Revolt if tax burden is unbearable.	Fees are area-based. Some village cooperatives introduced fees, based on actual costs: farmers pay for what they get and try to avoid unnecessary supplies.
Cooperatives	Need to cover their electricity and O&M costs. Direct pressure of farmers to get water on time. Need to earn income for other needs, services and activities at commune level.	Partly default on the fee to DE. Use fees for other purposes. Under-report the supplied area in contracts and negotiate water informally with central pumping stations. Get local pumping stations to be autonomous in irrigation management, and get access to more funds.	Investment in local pumps is expensive but their operation is cheaper than for centralized pumps.
Cums	Need to follow DE regulation. No means of controlling cooperatives' practices.	Satisfy the demand of cooperatives in order to complement their low official wages.	Do not have to justify pumping hours to the DE. Innovation: some cums sign contracts based on effective water consumed by cooperatives.
DEs	Have to ensure irrigation and drainage; do not control revenues; no flexibility for staff hiring. They are far from users and face cum staff's private strategy, opposed to DE's interest. The fees recovered do not balance their expenditures. Must cover the costs of drainage service in flood years and claim back subsidies – often paid one year late.	Adjust claimed service areas to almost match recovery. Defer maintenance. Wait for subsidies for big maintenance works. Take bank loans to cover operating expenses, particularly for electricity payment.	Have to justify pumping hours to the WRS. Do not have to present balanced budgets (since the abandonment of the 1997 reforms in 1999). Recovery at 92%.
Bac Hung Hai IDMC	Increasing indebtedness due to high operational costs, underpayment of bulk fees by DEs and failure by provinces to meet their financial obligations. Recovery only 72%. Occasional but irregular subsidies from the government.	Defer maintenance. Wait for government subsidies. Pass the debt on to a central state agency – in this case, the Ministry of Agriculture and Rural Development.	Do not have to present balanced budgets. Strategies to use the Economic Courts to enforce DE and other provincial payments have faltered.

Continued

Table 7.12. *Continued*

Actor	Constraints	Strategies	Observations
Province	Supposed to pay for the shortfall of revenue in case of a special year. Needs to provide subsidies to DEs since decentralization is in process (1999)	WRS/DE subsidizes third crop on centralized systems but not for cooperative pumping stations. Try to get subsidies from central government. Provide provincial directives to adjust national decrees to provincial conditions and policy.	Responsibility of each province lessened because four provinces are represented on the BHH IDMC board.
State	Has control over BHH IDMC through the Ministry of Agriculture and Rural Development and is requested by provinces to help for investments and major works. Has to adapt its policy to ecent changes which were not planned (emergence of local pumping stations) and decentralization to the provinces. The present institutional framework does not fit the present organization of irrigation. To get funds from international donors (which means to include some transfer measures in water policy); to reform without giving up strong administrative control of farmers.	Do not fully compensate for financial deficit as a way to maintain pressure on IDMC and to conserve its financial resources. Do not ask for full recovery in order to keep a balance between financial con straints and social peace.	Fear of countryside social unrest, as expressed in the Thai Binh riots of 1999.

Part of the fees is dedicated to the satisfaction of their irrigation needs but the opacity of the management of cooperatives does not allow farmers to estimate the adequacy of their payment with regard to the real costs incurred by cooperatives. This opacity is also allowed by the wide diversity of situations regarding water control (irrigation and drainage at field level may be achieved by gravity or with a complex mix of pumping operations) combined with an institutional diversity (the operations can be ensured by the cooperative and/or the company), which makes the calculation of fees very complex. Moreover, it also points to the fact that cooperative managers are generally administrative cadres, sometimes pursuing agendas beyond the scope of irrigation itself, as do company officials and district and provincial politicians. Local political practices are inclined to heavy investments in hydraulic equipment and in other infrastructures, such as roads, which can create an unbearable burden to the farmers: the 1999 riots in the Thai Binh province were motivated by mismanagement of the fees, which were raised and used for building roads and for paying bribes or extra salaries to the local authorities instead of for irrigation services.

One could argue that the water fees paid by farmers are still low in the Red River delta, but that the official water fees, which are often increased by many 'unofficial' subsidiary fees and taxes managed with little transparency, do not encourage farmers to contribute to the cost recovery of irrigation and drainage activities and generates mistrust. Water charges are not a goal in themselves, and are not something new in a context where farmers have paid taxes to the central state for centuries, but they should be linked to a clear definition of responsibilities and to management accountability. Interestingly, new contractual agreements signed between some cum and cooperatives aim to base payments on the cooperatives' effective water consumption, and tend to reinforce downward accountability.

The same opacity prevails regarding the management of IDMCs and DEs. The companies lack the incentives to present balanced budgets, since provincial or state subsidies will finally cover the deficit, and may be inclined to favour the satisfaction of their internal needs to the detriment of the quality of service. The permanent debt regarding the water diversion fee due to the IDMC and the electricity cost can be seen as a deliberate management policy: an *upward* transfer of the financial burden directly to the province, for the electricity, and to the state via BHH IDMC for the water diversion and main repairs.[21] This informal strategy was confirmed during interviews conducted with company officials during our research. At the same time, the companies also transfer part of their costs *downward* to the cooperatives by eliciting unofficial payments to field staff aimed at ensuring diligence and timely service. For the state, letting the debt grow might be a better strategy than purely making up for the financial shortfall with subsidies, since it allows maintaining a degree of pressure on provinces and districts by making manifest their lack of financial rigour (regardless of whether they are responsible for this or not). In addition, the state has now shifted some of its financial burden on to the provinces by making them responsible for covering possible extra costs in abnormal years.

All this shows that both within the cooperative level and between the cooperative and state companies, issues of water pricing are embedded within social networks based on kinship and political connections. Financial interdependence must therefore be seen not only as a mere contractual relationship, whereby financial flows are defined by reciprocal accountability and managerial rationality, but also as a part of the wider social and political web marked by shifting individual strategies, asymmetries of information and of bargaining power, and varying access to higher political strata. Just like in the case of cost recovery in the National Irrigation Administration of the Philippines, the model of contractual and financial autonomy of irrigation agencies proves to be oversimplistic in that it largely overlooks local politics (Oorthuizen, 2003).

Despite these qualifications, the crucial point is to determine whether the financial imbalance is the result of poor management and significant improvements are possible, or whether real constraints such as rising electricity bills, straightjacketing official regulation on fees and shrinking service areas do not allow companies to fare much better. Reality borrows from both ends, although our analysis tends to lean towards the latter. On the one hand, the overall financing of irrigation and drainage gives way to complex financial flows between nested levels of power and responsibility (farmers/cooperatives/companies and provinces/state), and the lack of transparency suggests that the economic efficiency of the service provision decreases due to financial losses at several levels. On the other, the debts of companies can be seen as implicit state subsidies made necessary by the political decision to keep water charges under a certain level. Since the overall taxation of households was shown to be quite high, this concern might be a practical recognition that surplus extraction by the state cannot be increased (the agricultural tax was reduced in 1993 in order to reduce the tax burden (Small, 1996), before being cancelled in 2001), and an indication that farmers' contributions might, after all, exceed what they get from the state in return, a point that

[21]At least for Bac Hung Hai and Bac Nam Ha polders.

needs further investigation. The apparent low percentage of companies' income spent on staff salaries (around 10%) indicates that administrative reforms might yield fewer gains than expected, although the extra payments made by farmers also show that these real costs might be higher than indicated.

In other words, the shortfall in the budgets of cooperatives, DEs and the BHH IDMC reflects both the non-optimal management of these organizations and the insufficient levies imposed on users, and the degree of defaulting that these organizations allow themselves is also a reflection of their perception of how much the state may be willing to pay, to avoid the transaction and political costs of engaging in more drastic or coercive reforms. It is also a measure of the 'distance' between the centre and decentralized administrative units, and of how upper bureaucratic layers fail to exert full control upon lower ones. From another point of view, it defines the trade-off between social considerations (constitutive of the Vietnamese Party's legitimacy) and macroeconomic constraints. Any institutional reform must question the distribution and share of responsibility in decision making, and introduce higher transparency in financing. This, of course, is an issue that cannot be restricted to the water sector and pertains to the wider question of political change in Vietnam.

References

Bach Trung Hung, Tran Ngoc Hân, Nguyen Van Dung and Le Quéré, E. (1999) Evolution de la riziculture de 1991 à 1995. In: *Agriculture Familiale et Gestion des Ressources du Milieu dans le Basin du Fleuve Rouge*. INSA, GRET, PFR, Maison d'Edition de l'Agriculture, Hanoi, pp. 147–159.

Beresford, M. (1988) *Vietnam: Politics, Economics and Society*. Pinter, London, p. 242.

Bousquet, M. in collaboration with the team 'Gestion Sociale de l'Eau du Programme Fleuve Rouge' (1994) *La Gestion de l'eau à An Binh: Evolution Historique, Technique et Sociale*. GRET, Paris and CNEARC, Montpellier, France.

Chassigneux, E. (1912) L'irrigation dans le Delta du Tonkin. *Revue de Géographie Annuelle*, tome IV, fascicule I, Paris.

Cuc, Le Trong, Gillogly, K. and Rambo, A.T. (eds) (1993) *Too Many People, Too Little Land: The Human Ecology of a Wet-Rice-Growing Village in the Red River Delta of Viet Nam*. Program on Environment. Occasional Paper No.15. East–West Center, Honolulu.

Dang The Phong and Fontenelle, J.-P. (1995) Bilan hydrique à l'échelle de la parcelle et de la maille, et pratiques individuelles d'irrigation pour la campagne de printemps 1993, dans le delta du Fleuve Rouge, au Vietnam. In: *L'agriculture du Delta du Fleuve Rouge à l'heure des Réformes*. INSA, PFR, Maison d'Edition de l'Agriculture, Hanoi, pp. 255–284.

Dang The Phong and Fontenelle, J.-P. (1997) Le passage à la gestion locale de l'irrigation dans le delta du fleuve Rouge: cas du district de Nam Thanh. In: *Spécial Vietnam, Numéro commun Cahiers Agricultures* 15, 93–99.

Dao Thê Tuan (1998) La transition agraire au Vietnam comme changement d'institutions. In: *Développement et Transition vers l'économie de Marché*. Actualité Scientifique, Universités Francophones, AUPELF-UREF, Montreal, Canada, pp. 457–471.

Do Hai Dang (1999) Action collective et jeux de pouvoir dans la gestion locale de l'irrigation. Le cas du delta du Fleuve Rouge (Vietnam). Masters thesis, Fondation Universitaire Luxembourgeoise, Arlon, Belgium.

Dumont, R. (1935) *La culture du riz dans le delta du Tonkin*. Sociétés d'édition géographiques, maritimes et coloniales, Paris.

Fforde, A. and Sénèque, S. (1995) The economy and the countryside: The relevance of rural development policies. In: Kerkvliet, B.J. and Porter, D.J. (eds) *Vietnam's Rural Transformation*. Westview Press, Boulder, Colorado, pp. 97–138.

Fontenelle, J.-P. (1998) L'eau de l'Etat et l'eau des villages: l'exemple de l'hydraulique du delta du Fleuve Rouge. In: *Sociétés Rurales et Environnement*. Karthala, GRET et Regards, Paris, pp. 75–95.

Fontenelle, J.-P. (1999) *The Response of Farmers to Political Change: Decentralization of Irrigation in the Red River Delta, Vietnam*. Liquid Gold Series, Paper 5. Agricultural University, ILRI, Wageningen, The Netherlands.

Fontenelle, J.-P. and Tessier, O. (1997) L'appropriation paysanne de l'hydraulique agricole du delta du Fleuve Rouge: processus et limites. *Autrepart 3*. Editions de l'Aube, ORSTOM, Paris, pp. 25–43.

Gourou, P. (1936) *Les Paysans du Delta Tonkinois*. Publications de l'Ecole Française d'Extrême Orient, Les Editions d'art et d'histoire, Paris.

Hémery, D. and Brocheux, P. (1995) *La colonisation Ambigüe 1858–1954*. La Découverte, Paris.

Kerkvliet, B.J.T. (1995) Rural society and state relations. In: Kerkvliet, B.J.T. and Porter, D.J. (eds) *Vietnam's Rural Transformation*. ISEAS, Westview Press, Boulder, Colorado, pp. 65–96.

Kerkvliet, B.J.T. (1999) Accelerating cooperatives in rural Vietnam, 1955–1961. In: Dahm, B. and Houden, V.J. (eds) *Vietnamese Villages in Transition*. Department of South-east Asian Studies, Passau University, Passau, Germany, pp. 53–88.

Le Ba Thao (1997) *Vietnam the Country and Its Geographical Regions*. The Gioi Publishers, Hanoi.

Lê Duc Thinh and Fontenelle, J.-P. (1998) Systèmes de cultures, stratégies paysannes et environnement institutionnel dans le delta du Fleuve Rouge. In: *Proceedings of the 15th International Symposium: Rural Livelihoods, Empowerment and the Environment. Going beyond the Farm Boundary*. Association for Farming Systems Research-Extension, Pretoria, South Africa, pp. 754–762.

Lê Thanh Khoi (1978) *Socialisme et Développement au Viêt-nam*. Puf, Paris.

Mai Van Hai (1999) Participation des agriculteurs aux travaux d'irrigation depuis le contrat 10. In: *Agriculture Familiale et Gestion des Ressources du Milieu dans le Basin du Fleuve Rouge*. INSA, GRET, PFR, Maison d'Edition de l'Agriculture, Hanoi, pp. 25–38.

Nguyên Duc Truyên (1993) *Les Effets de la Politique Economique Vietnamienne sur les modes de vie Paysans dans la Région du delta du Fleuve Rouge*. Mémoire de DEA, EHESS, Paris.

Nguyen Thi Hong Loan (2000) *Politiques de Gestion Financière des CEOH – Situation Actuelle des Activités Financières des Compagnies du Système Hydraulique de Bac Hung Hai*. Deltas INCO-DC project, VASI, Hanoi.

Oorthuizen, J. (2003) *Water, Works, and Wages: The Everyday Politics of Irrigation Management Reform in the Philippines*. Wageningen University Water Resources Series. Orient Longman, Hyderabad, India.

Sakurai, Y. (1989) *Land, Water, Rice, and Men in Early Viêt-Nam: Agrarian Adaptation and Socio-Political Organization*. In: Taylor, K.W. (ed.) The Center for South-east Asian Studies, Kyoto University, Kyoto, Japan.

Small, L.E. (1996) Irrigation operation and maintenance in Vietnam under economic restructuring. *Irrigation and Drainage Systems* 10(2), 245–262.

Taillard, C. (1995) Le Viêt-nam, émergence d'un nouveau dragon. In: Antheaume, B., Bonnemaison, J., Bruneau, M. and Taillard, C. (eds) *Asie du Sud-Est Océanie*. Géographie Universelle, R. Brunet Dir., Belin-Reclus, Paris, pp. 188–211.

Tessier, O. and Fontenelle, J.-P. (2000) Pression démographique et contraintes politiques: La paysannerie du delta du Fleuve Rouge dans la tourmente du XXème siècle. In: Gubry, P. (ed.) *Population et Développement au Viêt-nam*. Karthala, Paris, pp. 495–527.

Yvon-Tran, F. (1994) Une résistible collectivisation; L'agriculture au nord Vietnam: 1954–1988, PhD thesis. Université de Paris VII, Paris.

8 Water Pricing in Haryana, India

P.J.G.J. Hellegers, C.J. Perry and J. Berkoff

Introduction

Haryana is one of India's major irrigating states, with approximately 2.9 Mha undersurface irrigation. Water is scarce and irrigation water demand exceeds available canal water supplies. The major challenge facing the responsible government agencies is to manage water scarcity so as to minimize long-term damage to agriculture, fresh aquifers and soils.

It is indisputable that underwatering is pervasive and that, as non-agricultural demands rise, irrigation supplies will come under increasing pressure. Besides water shortages, agriculture is threatened by rising water tables in the western zone (about 52% of the area) and by falling water tables in the eastern zone (about 48% of the area). These proportions do not fully accord with the distribution of saline and fresh groundwater and suggest that brackish groundwater is already used for irrigation, presumably mixed with surface water and/or rainfall. In 1997, about 0.42 Mha were affected by high water tables, with 0.25 Mha totally waterlogged (GOH, 1998). Another source gives some 0.19 Mha affected by salinity and 0.33 Mha by sodicity (Agarwal and Roest, 1996). Interventions that improve on-farm water management reduce canal seepage, and installing drainage systems could help address these problems.

Priority objectives therefore specify the following:

- Increase the productivity of water in the context of declining long-term availability.
- Control abstraction of fresh groundwater to avoid decline and salinization of aquifers.
- Manage saline aquifers so as to reduce or avoid waterlogging and soil salinization.
- Finance adequate operation and maintenance (O&M) expenditures along with justified capital improvements.

Volumetric water pricing is often mentioned to address these problems, but the role it can play in meeting the objectives in Haryana is not clear. The main aim of this chapter is therefore to study the potential role of pricing policy in meeting the above priority objectives. To achieve this aim, the way water is currently allocated will be described and insight will be provided into the price, costs and value of irrigation water in Haryana.

The structure of this chapter is as follows. First, the study area and *warabandi* system that allocates water to all irrigators in proportion to their landholdings are described. Next, the price, cost and value of water are studied. An analytical framework is applied to assess the value of production and contribution of water to that level of production. Then policy recommendations are made and finally conclusions are drawn.

Study Area and *Warabandi* System

Study area

Haryana is located on the Indo-Gangetic plain in north-west India with a climate that is arid to semi-arid. It has an area of 4.4 Mha of which 3.8 Mha is cultivable and 2.9 Mha irrigable (GOH, 2004). The population totals 21 million of which 70% is rural. GDP per head is $660 (32% above the national average) and has been rising in real terms at up to 3% per annum. Agriculture accounts for 31% of GDP and, along with Punjab, Haryana led India's Green Revolution. Grain yields are some 30–40% above the national average and, with just 1.4% of India's area, this small state provides 30% of the national procurement of wheat and 10% of its rice. Gross sown area in 2001–2002 was 6.3 Mha and net sown area 3.6 Mha, giving an overall cropping intensity of 177% and an intensity on irrigated land of about 190–195%. There are three primary sources of water: rainfall, surface water and groundwater.

Annual rainfall averages 545 mm, ranging from more than 1000 mm in the extreme north-east to less than 300 mm in the arid west. Rainfall also varies from year-to-year and from season-to-season. About 80–85% occurs in *kharif* (June to September), and most of the rest in *rabi* (October to February). Evapotranspiration averages about 1550 mm so that irrigation is a prerequisite for successful cropping most of the time over most of the state.

Surface water comes from the Sutlej via the Bhakra canal system and from the Yamuna via the Western Yamuna system. Sutlej and other Indus allocations are regulated by the Bhakra-Beas Management Board (BBMB), which was created under the 1966 Punjab Reorganization Act. This Act and subsequent agreements govern the state shares in the three rivers (Sutlej, Ravi, Beas) assigned to India by the 1960 Indus Basin Treaty. Haryana has yet to obtain its full share and disputes continue, in particular relating to construction of the Sutlej Yamuna Link (SYL) canal, which would allow access to water from the Ravi and Beas. Yamuna allocations are governed by the Tajewala Headworks Agreement of 1954 as modified by the Punjab Reorganisation Act and other agreements.

Groundwater is abundant on the alluvial Indo-Gangetic plain. Recharge in Haryana has risen greatly as a result of surface irrigation. Brackish groundwater underlies up to two-thirds of the state, an area characterized by poor natural drainage, rising water tables and secondary salinization. The balance one-third is underlain by fresh groundwater and is characterized by falling water tables since use exceeds recharge by a considerable margin. By now, there are some 600,000 tube wells that are predominantly privately owned. Well owners commonly sell water to their poorer neighbours after meeting their own needs.

The *Warabandi* System

The irrigation management system in Haryana, as in other states in north-west India and Pakistan, was formalized under the Northern India Canal & Irrigation Act of 1873 (Eastern Book Company, 1982), based in part on earlier Moghul and British practices. Canals are designed based on the 'regime theory' with the aim of distributing suspended silt over the land. Surface supply is intended to be protective (i.e. to spread water over a large area inter alia to guard against famine) rather than to be productive (i.e. to meet full water demands of a specified irrigable area to maximize yields) (Ministry of Irrigation, 1982; Malhotra, 1988; Jurriens *et al.*, 1996). Supply is thus well below potential demand and water is rationed in proportion to irrigable area. Although often referred to as the *warabandi* system (literally 'fixed turn' system), *warabandi* is just one component of a complete system with the following main features.

Water allowance

Water is allocated in proportion to land, and farmers are free to use their allocation as they wish. In other words, the cropping pattern is a response to a pattern of supply (crops to water) rather than supply being a response to a cropping pattern (water to crops).

Delivery capacity (duty) is low, being typically no more than 0.15–0.175 l/s/ha at the outlet or perhaps 0.17–0.20 l/s/ha at the head allowing for canal losses. If given continuously, this satisfies the theoretical crop water requirements of no more than 20–30% of the irrigable land in kharif and 35–45% in rabi.

Reservoir operations

Reservoir operations are the responsibility of the BBMB. Subject to the priority normally given to hydropower and other non-agricultural uses, water is delivered to each irrigation canal headwork in line with the shares of the respective states. The seasonal operational plan is updated at least every 3 weeks to reflect actual water conditions.

The main system

The conveyance and distribution system is managed by the Irrigation Department (ID). Main/branch canals are operated with variable flow in response to BBMB allocations and – to a limited extent – demand (see below). Distributaries and minors are either full ON or full OFF, with flow reduction limited at most to 10–15%. When main or branch canals run full (e.g. if river flow exceeds diversion capacity) lower channels also run full.

Distributaries operate in rotation such that the sum of discharges in ON channels equals branch canal discharge allowing for losses. Priorities shift every 8 days so that each distributary has an equal chance of being ON. This design has come to be known as the structured design, with the system structured at the head of the distributary (the point below which flows are proportional and canals run full) (Albinson and Perry, 2002).

Adjustable gates on the main/branch canals support variable flow management. ON/OFF gates at the head of each distributary or direct minor allow canal rotation. Below this point, the system is un-gated with proportional division at each junction point.

Correct discharges in ON canals are critical to successful operation. Levels are monitored twice daily at key points. If flow at the tail falls below the design, action is taken to increase supply and/or close channels to maintain full supply. Canals are closed annually for maintenance, notably to check offtakes and restore cross sections.

Distribution below the outlet

Outlet capacities are based on duty. If the design duty is 0.15 l/s/ha, the capacity of an outlet serving 200 ha is 30 l/s. To ensure that the stream size is manageable by the farmer (in the range 25–40 l/s), *chaks* (outlet commands) are generally limited to between 100–300 ha and typically serve some 50–100 farmers.

All outlets are un-gated and run full when the minor is ON. The full flow in the watercourse is allotted to each farmer in turn on a weekly (168 h) schedule. Turn length is based on farm size. If the *chak* size is 200 ha and duty 0.15 l/s/ha, the farmer receives 30 l/s for 0.84 h/ha of land that he owns. If the *chak* size is 250 ha, he receives 37.5 l/s for 0.67 h/ha. Some limited adjustment may be made to these times to account for losses in the watercourse.

The farmer obtains water at the same time each week (the clock keeps ticking). If there is water, he has the right to the full flow. If not, he loses his turn. Equity is ensured by the rotation of supply to distributaries and the flow in the watercourse – if there is one – is owned at all times by a known farmer. The schedule rotates through 12 h at the end of each crop year to ensure equity in night-time irrigation.

The schedule below the outlet is known as the *warabandi* schedule. Farmers can either arrange this schedule among themselves (*kutcha warabandi*) or request registration by the authorities (*pucca warabandi*). In Haryana, almost all schedules are registered. It is then an offence to take water out of turn. It is also an offence to exchange or sell turns though this occurs in practice. Farmers maintain the watercourse at their own expense.

Groundwater

Groundwater is unregulated and the landowner has the right to exploit any aquifer lying below the surface of his land. In fresh groundwater areas, this means that the individual farmer has no incentive to limit extractions since others may continue to pump; and in saline areas, the farmer has no incentive to install drainage facilities since this would have to serve the whole locality to be effective. These two examples of 'the tragedy of the commons' are critical to understanding groundwater management.

In its essence, this system has survived since its inception in 1886 despite developments that include: (i) independence and partition; (ii) population growth; (iii) falling farm size; (iv) the Green Revolution; (v) the massive growth of mechanized pumping; and (vi) expansion and diversification of an increasingly market-based economy. The system's relative simplicity, transparency and low-cost help explain its robustness (Horst, 1998). Other factors include canal rotation 'which makes it difficult for the farmers to interfere with the "automatic" distribution by the proportional outlet structures on the distributary' (Jurriens *et al.*, 1996), and lack of ambiguity in the *warabandi* schedule – the irrigation turn is in effect a property right in water and farmers tenaciously defend their turn. Rationing does not, of course, meet precise crop water requirements. In Sirsa Circle, for the actual cropping pattern and after allowing for rainfall 'canal supply exceeds requirements by 50 mm (500 m^3/ha) during the winter period and the late-summer shortage is 210 mm (2100 m^3/ha)' (Agarwal and Roest, 1996). In fresh groundwater areas, groundwater can compensate for shortages.

The system does not, of course, always perform as designed, and deliveries may be inequitable both between distributaries or minors and along watercourses (Jurriens *et al.*, 1996). Shortfalls in O&M funds, farmer interference (notably in the outlet) and other factors are all of concern, although farmer interference is more prevalent where farm size and rural power are inequitable or rainfall is higher (or topography and soils are more variable [Berkoff, 1990]). On the other hand, some modifications to system operations may even be beneficial (illegal exchange or sale of turns, main-system-flow adjustments in response to waterlogging).

The system has worked well relative to other systems in India. Both relative agricultural success and a priori arguments suggest that it is well adapted to local conditions (Berkoff, 1990). Up to the 1950s, western Haryana was notoriously vulnerable to famine, yet now the state provides an astonishing share of India's grain and 'is emerging very fast as one of the leading states in the field of horticulture (though horticulture occupies only about 5.2% of cultivable area)' (GOH, 2004). The key indicator is the contrast between potential crop intensity based only on surface irrigation (55–75%) and actual intensity (190–195%) utilizing all three water sources. This contrast is explained in part by underirrigation. However, the main reason is the combined use of rainfall, groundwater and sub-irrigation by capillary rise, all of which have been augmented by surface irrigation. Rainfall, which in terms of volume may be the largest source, is much less productive without irrigation; groundwater and capillary rise reflect surface water recharge; and brackish water causes less damage – whether from irrigation or sub-irrigation – if used conjunctively with surface water and/or rainfall. The original intention of the system designers may have been to provide protective irrigation but the unanticipated spread of mechanized pumping along with sub-irrigation has led to one of the most productive agricultural systems in India, with high yields and a cropping intensity that approaches 200%. The question is, however, for how long, as it leads to dropping of groundwater tables and salinization.

Crop selection in response to supply (crops to water) means that the farmer, rather than the scheme operator, is primarily responsible for planning. In effect, the farmer maximizes farm income subject to his assessment of risk. Water rather than land or labour is generally the scarce resource so: 'farmers underirrigate some crops in relation to full potential evapotranspirative demand, because reductions in yield may be proportionally less than reductions in water applied' (Perry and Narayanamurthy, 1998). With regard to risk, rainfall is unpredictable but free; surface water is predictable within limits but incurs a small additional cost; and groundwater is predictable but more expensive. Groundwater and sub-irrigation may also be unusable or damaging. Farmers thus divide their farm into distinct plots on which they plant crops with differing water needs, allocating water between plots in the light of rainfall with the aim of meeting their implicit objective function. Based on field evidence from the Bhakra command, Perry and Narayanamurthy (1998) conclude that: 'Farmers generally aim to maximise returns to the scarce resource, but due to the uncertainties involved guard against unacceptable risk by reducing the area planted and increasing seasonal water allocations per unit area where supplies are less certain.'

Farmers are intensely concerned with their own welfare and, though there are good farmers and bad farmers, there is little doubt that, in general, they are equipped to perform this planning exercise. But their perspective is limited to their own interests, and this leads to the tragedy of the commons as described above. In fresh groundwater areas, water tables fall and groundwater irrigation on the current scale is unsustainable over the longer term. In saline groundwater areas, water tables rise and agriculture is threatened in complex ways by waterlogging and secondary salinity (Agarwal and Roest, 1996). Any modification to the present management system must also take these externalities into account (section under Recommended Policy Instruments).

Price, Costs and the Value of Water

Price paid for canal water

Charges for surface irrigation are levied on a crop-area basis: that is, rates per hectare vary across crops and are charged according to the area irrigated. The ID records crop areas, excluding those that utilize only rainfall and/or groundwater. Areas irrigated from canals are reported to the Revenue Department, which collects what is due as part of Land Revenue. This is incorporated in the general budget and does not directly determine budget allocations for recurrent costs. The general aim is to cover O&M costs, an objective that is almost achieved by the device of assigning only about one-third of ID recurrent costs to irrigation, with the rest assigned to non-irrigation users who receive priority at times of scarcity.

There is no explicit volumetric charge, although crop area and type are a proxy for volume. Table 8.1 shows crop-based charges ($/ha) along with their volumetric equivalents ($/m^3). The average charge can also be estimated from the total revenue derived from irrigation water charges ($1.00 = Rs 47.00). In 1999–2000, the net area irrigated by canals was 1.44 Mha, generating revenues of Rs 210 million ($4.47 million) (GOH, 2004), equivalent to an average of Rs 145 or $3.1/ha. If total surface water deliveries were about 9.4 Bm^3, this implies an average delivery of 6500 m^3/ha and an average water charge of $0.0005/m^3. This is comparable to the estimates in Table 8.1.

Table 8.1. Haryana: water charges by crop and volumetric equivalents.

	Rice (6,800 m^3)		Wheat (4,500 m^3)		Sugarcane (10,000 m^3)	
	\$/ha	\$/m^3	\$/ha	\$/m^3	\$/ha	\$/m^3
Haryana 2000	3.2	0.0005	2.7	0.0006	4.3	0.0004
Haryana 1999	2.4	0.0004	1.9	0.0004	3.1	0.0003

Costs of water delivery

Surface water costs

Annual recurrent costs of delivering water within Haryana to all users during the period 1996–2000 averaged about \$18 million per year. Annual deliveries were about 14 Bm^3, resulting in an average O&M cost of about \$0.0013/m^3. This confirms that the Haryana system is low-cost which reflects the highly centralized system of management, the relatively small number of control structures, limited staffing requirements and farmer responsibility for O&M costs below the outlet. This makes no allowance for capital costs, which are very substantial. One-third of the total O&M costs is allocated to irrigation (i.e. about \$6 million/year). In 1996–2000, irrigation received an average volume of 12.9 Bm^3/year (92% of the total), implying a cost to irrigation of \$0.0005/m^3. This was less than 1/20 of the cost per cubic metre attributable to other users (\$0.0107/m^3, given average deliveries of 1.12 Bm^3 and a share in costs of \$12 million). In return, non-agricultural users receive a more continuous and predefined service as well as priority at times of scarcity.

The World Bank-funded Haryana Water Resources Consolidation Project (World Bank, 1994) placed emphasis on cost recovery, requiring, first, a clear definition of the costs of system operations; second, political decisions on how costs should be allocated; and third, that charges be raised to cover O&M expenses over 6 years. This process was important in clarifying the situation, raising charges and highlighting the extent to which the ID provides water services to other users (drinking water to villages, industrial supplies, supplies to power stations, water to Delhi and water to other government departments, such as mining, fisheries and forests). Irrigation charges are nevertheless a highly sensitive political issue. In many Indian states, poor cost recovery stems from a combination of both low charges and low rates of collection. In Haryana, however, though rates are low, collection is 90% or more, in part due to collection of water charges as part of Land Revenue. Shortfalls at times of crises (floods, droughts, pest attacks) are usually offset by collection of arrears in subsequent years.

Groundwater costs and charges

Tube well water is charged by well owners at anything between \$0.2–1.6/h or at a flat rate of \$7.0–15.0/delivery/ha irrigated. The wide range reflects not only differing pumping heads, but also the extent to which tube well owners seek to recoup capital investment, exploit their monopoly powers, etc. (see section under Recommended Policy Instruments). If each delivery amounts to about 1250 m^3, a flat rate of \$7.0–15.0 is equivalent to \$0.006–0.012/m^3. This compares to an average quoted fuel cost of about \$0.005/m^3. It also suggests that at the lower end of this range charges are largely confined to marginal costs (mainly fuel). Whatever is covered, it is equivalent to 10–20 times the cost of surface supplies. The ratio would no doubt be higher if electricity was charged at an unsubsidized rate.

Value of water

Net returns

Tables 8.2 and 8.3 summarize farm budget estimates for the Sirsa district in the western

Table 8.2. Sirsa district: farm budgets – surface water and groundwater.[a] (Based on information from World Bank, 1998 and Aggarwal *et al.*, 2001.)

Crop	Gross return ($/ha)	Cropped area (ha)	Gross return ($)	Farm costs: surface water			Farm costs: groundwater			Net farm return	
				Inputs ($)	Labour ($)	Water ($)	Inputs ($)	Labour ($)	Water ($)	Surface water ($)	Groundwater ($)
Kharif											
Rice	327	0.83	270	55	71	2	55	71	32	142	113
Cotton	406	0.60	263	56	23	1	56	23	16	183	169
Chickpea	161	0.34	55	11	23	1	11	23	5	20	16
Rabi											
Wheat	449	2.00	899	165	82	4	165	82	29	647	622
Mustard	443	0.70	295	10	21	1	10	21	14	263	250

[a]Year unspecified.

Table 8.3. Sirsa district: farm budgets – five farm types, rabi 2000/02 and kharif 2002/03.

	Cropped area (ha)	Gross farm return ($)	Farm costs			Net farm return ($)
			Inputs ($)	Labour ($)	Water ($)	
Farm type 1	16.6	14,380	5,673	1,570	317	6,820
Farm type 2	6.1	5,109	2,020	435	126	2,528
Farm type 3	7.6	5,510	2,162	460	106	2,782
Farm type 4	7.1	3,960	1,987	223	93	1,657
Farm type 5	7.3	3,974	2,070	423	119	1,361

Farm type 1: paddy–wheat belt, head of canal, normal soils, four farms covering a total of 9 ha.
Farm type 2: paddy–wheat belt, middle of canal, normal soils, four farms covering a total of 3.3 ha.
Farm type 3: cotton–wheat belt, head of canal, normal soils, eight farms covering a total of 5.1 ha.
Farm type 4: cotton–wheat belt, middle of canal, normal soils, four farms covering a total of 5.9 ha.
Farm type 5: cotton–wheat belt, tail of canal, problematic soils, four farms covering a total of 5.7 ha.

zone. Table 8.2 is based on information derived from Aggarwal *et al.* (2001) and the World Bank (1998). It assumes that cropping and water use remain the same irrespective of the source of water. This is a simplification since cropping patterns might be expected to adapt to the improved security of supply and, perhaps, to the higher costs and volumetric basis of groundwater. Table 8.3 gives comparable data without distinguishing between surface water and groundwater irrigation, based on a survey of 24 farms in rabi 2001/02 and kharif 2002/03 (Appendix). The farms were divided into five categories on the basis of location in the canal system and type of land.

Despite considerable differences between the two sets of data, the tables confirm that water represents only a small part of farm costs, even in the case of groundwater, and that the costs of other inputs (seeds, fertilizers, pesticides, etc.) and labour are substantially greater. In the case of the Sirsa scheme, for example, the average shares of inputs, labour and water are 78%, 18% and 4%, respectively. Subsidies on other inputs are now limited and their costs approximate to trade-equivalent levels. The labour market is also relatively competitive given seasonal migration from eastern India and, though wages may exceed the opportunity cost of labour, this is becoming less significant as the economy develops. The major distortion in farm costs relative to the economic optimum in respect of irrigation is, therefore, due to low water charges and electricity subsidies.

Apparent returns to water

Tables 8.4 and 8.5 show net returns per unit of water after allowing for all financial costs, including those of water, for the two sets of farm budget data provided in Tables 8.2 and 8.3, respectively. Net returns to water are about $0.04/m^3$.

Discussion of price, costs and value of water

Care must be taken in interpreting these data. Expressing net farm returns in terms of the net return per unit of water seems to suggest that the profit over and above financial costs is wholly attributable to water. However, net returns might be similarly attributed to fertilizer or some other input while this profit represents the farmer's return to land, capital and management after allowing for other costs. If water was to be charged at a rate that equalled apparent net returns per unit to water and returns to land, capital and management would sink to zero (or, in the case of family labour, be no more than the going wage rate), which is unrealistic. On the other hand, water is a major constraint to increased agricultural production and Tables 8.4 and 8.5 suggest an extreme upper limit to the returns to water.

Returns to water are 50–100 times the water charge ($0.0005/m^3$), implying that water charges would have to rise substan-

Table 8.4. Sirsa district: water use and net returns by crop. (Based on information from World Bank, 1998 and Aggarwal *et al.*, 2001.)

Crop	Water use per hectare (m³/ha)	Total water use per farm (m³)	Net returns per farm		Net returns per unit of water	
			Surface water ($)	Groundwater ($)	Surface water ($/m³)	Groundwater ($/m³)
Kharif						
Rice, paddy	6870	5700	142	113	0.025	0.020
Cotton	4835	2900	183	169	0.063	0.058
Chickpea	2355	800	20	16	0.025	0.020
Rabi						
Wheat	2450	4900	647	622	0.132	0.127
Mustard	3715	2600	263	250	0.101	0.096

Table 8.5. Sirsa district: water use and net returns by farm type, rabi 2001/02 and kharif 2002/03.

Farm type	Average water use (m³/ha)	Total water use (m³)	Net returns per farm ($)	Net returns per unit of water ($/m³)
Farm type 1	9,200	152,700	6,820	0.045
Farm type 2	9,310	56,800	2,528	0.045
Farm type 3	6,170	46,900	2,782	0.059
Farm type 4	5,745	40,800	1,657	0.041
Farm type 5	7,425	54,200	1,361	0.025

Note: See Table 8.3.

tially before they had any significant impact on net farm returns, assuming that the water charge can be made volumetric (see next section). As is to be expected, water use was greater in the paddy–wheat than in the cotton–wheat belt, and net returns per cubic metre – at least in the cotton–wheat belt – declined towards the tail and were lower in farms with problematic soils. Table 8.4 suggests that net returns per unit of groundwater on the same basis were 2–10 times greater than groundwater charges ($0.006–0.012/m³).

This means that surface water charges would have to rise very substantially before they have an impact on water use. In other words, water demand at current charge levels under the current system of rationing is almost wholly inelastic. In the case of groundwater, this is less self-evident. Water charges are higher – for the least profitable case, net returns per unit are just double the charge – but water use is discretionary.

Recommended Policy Instruments

The above discussion suggests that water charges have minimal impact on surface water use. The system delivers a rationed supply that is sufficient for a limited part of the irrigable area. Since charges are well below the value of water to the farmer, there is no reason for him to reject any of his share since water can almost always be profitably used to meet the needs of irrigated crops, supplement rainfed crops, moderate underwatering, save on pumping costs or leach salts from the land. It is only if land is waterlogged or flooded that the farmer has reason to reject his share and the ID then often closes higher canals so as to alleviate problems that typically go well beyond the individual farmer. Instances where water cannot be profitably used are thus few and excess water, in any case, may do no harm. Far from rejecting his turn, therefore, the farmer resolutely defends it.[1]

Considerations in groundwater are very different. Not only are the charges made by well owners (much) higher than for surface water but a decision whether or not to turn on a pump is discretionary and does not prejudice access to the resource at a later time. The amount of freshwater extracted is thus a function of demand and not of availability. In areas of conjunctive use, surface water is a relatively stable, if limited base, supply; rainfall is variable and uncertain but free; and groundwater can be fine-tuned to 'optimize' net returns after exploiting other sources. That fresh groundwater is overpumped reflects the pattern of financial incentives, with richer farmers better able to adjust to falling water tables than poorer farmers. If falling water tables adversely affect water quality, then the resource may be lost and this of course then becomes the decisive concern.

In other words, so long as fresh groundwater is freely available, groundwater is provided on a volumetric basis and the amount demanded broadly optimizes farmer net returns subject to anticipated farm-gate prices, input costs, cross-elasticities and numerous other factors. Groundwater use in an imperfect and variable way thus reflects farmer willingness to pay. If conditions change (expected rise in farm-gate prices, electricity subsidies are withdrawn, etc.), the outcome is different. Net farm returns over-and-above financial costs (including water costs) are the farmer's return to land, capital and management and cannot be attributed to water as such. That the groundwater charge is so variable reflects variable spatial, temporal and farm conditions and numerous market imperfections. Even if extractions were to be effectively regulated, for instance to account for the externality costs associated with overpumping and/or salinization, the market would adjust to the new conditions with the price determined by the property rights created rather than by the current conditions of open access.

Surface irrigation is thus supply-driven and consumption is largely unaffected by water charges, while groundwater irrigation – no matter how imperfect – is demand-driven and consumption is a function of alternative water sources (rainfall and surface supplies) and (imperfect) market incentives. Given this background, what is the potential role of pricing policy in meeting the above objectives? The discussion is in two parts: (i) policies that require restructuring of the infrastructure; and (ii) policies that can be implemented with the present infrastructure.

Policies requiring restructuring of the infrastructure

Volumetric charges are often advocated as a mechanism for reducing water use and increasing output per unit of water. They require an infrastructure that can provide differentiated water supply and measurement at the point of sale. In the case of Haryana, they would thus require that the supply-based surface system (including the *warabandi* schedule) be replaced by a demand-based system that allowed water to be delivered in response to willingness to pay. To be effective, demand at the point of sale would have to be elastic with respect to price. At the theoretical limit, the charge would be ideally set such that demand and supply are brought into balance. For surface water in the Haryana context, volumetric charges could be levied at three possible levels: head of the watercourse, head of the minor or distributary and the farm.

Irrespective of how far differentiated supply is taken down the system, water rates must be sufficiently high to elicit a response

[1]If the farmer cannot defend his turn – if rural power is distributed inequitably or law-and-order breaks down – then the system is weakened. Persistent theft by head-enders can also wear the tail-ender down even under normal circumstances. Moreover, if rainfall is higher and the design supplements rainfall in kharif over the full irrigable area, or conditions are more variable than in Haryana, then there will be more instances when the individual farmer will want to reject water and this again tends to undermine this management system (Berkoff, 1990).

if they are to impact on water use. The increase required is in itself politically and socially infeasible. But there are more fundamental objections to volumetric pricing to the farmer. The present system is stable, simple and cheap to operate and this has major advantages for large schemes in developing countries (Horst, 1998). Moreover, the system already provides powerful incentives limiting water use and maximizing output; surface water use per hectare is already low and by Indian standards productive although unsustainable. This is even so if groundwater is saline. Where it is fresh, applications at the margin are charged on a volumetric or quasi-volumetric basis from groundwater. Farmers operate in real time, adjusting groundwater use in response to rainfall, surface supplies and financial incentives. Quite apart from the costs and risks of restructuring the delivery system, it is hard to imagine that volumetric pricing could be more successful.

Levying a volumetric charge at the head of the watercourse, minor or distributary is a less clear-cut issue and in some circumstances there may be a case for creating water user associations (WUAs) and/or organizations operating at the distributary or minor level. If WUAs and/or autonomous agencies are to be financially viable, they may limit demand in response to even moderately enhanced charges and may be willing to sell allotted shares if a market develops at this level. Being closer to the farmer, they may also be in a position to influence on-farm use even without volumetric charges to the farmer. However, the rationale for this has more to do with cost recovery and effective O&M than with enhancing the productivity of water, and where the system is functioning relatively well, as in Haryana, the uncertainties and risks are almost certainly unacceptable.

Account must also be taken of falling water tables, waterlogging and salinity. Declining water tables raise costs and disadvantage poor farmers. More importantly, they can affect quality since deeper aquifers are more saline than shallow aquifers. Rationing of surface water, ceteris paribus, has slowed the process of waterlogging and salinity.

Policies within the present infrastructure

If full volumetric water pricing of surface supplies is ruled out, what potential is there for modifying the present water charge system to reflect quasi-volumetric considerations? Possibilities can again be considered at three levels: main system, watercourse and the farm.

Main system rotation is equitable in terms of irrigable area. Given the homogenous character of an alluvial plain and equitable holding size, this also has the merit of transparency. Even so, differences between sub-commands – notably between fresh and saline areas and also in terms of rainfall, cropping patterns and other factors – could be reflected in differential schedules (Narayanamurthy, 1985). To a limited extent this already happens since the ID closes canals where waterlogging or flooding is acute irrespective of 'equity' considerations. One option would be to devise rotations that provide reliable but lesser supplies to saline areas (to ensure security, minimize recharge and slow the *rise* in the water table); and less reliable but greater supplies to fresh areas (since they already have security, excess deliveries can, if necessary, be recaptured by pumping or, alternatively, may slow the *fall* in the water table). Another option would be to devise schedules to meet differential demands of the predominant cropping pattern, e.g. differentiating between paddy–wheat and cotton–wheat (Narayanamurthy, 1985). This has the potential for bias and would tend to erode transparency. New schedules need to be articulated in a straightforward way.

The distinguishing feature of distribution within the watercourse is the *warabandi* schedule. Farmers have strong incentives to defend their turns and this is a major strength of the system. Trading beyond the watercourse implies a fundamental restructuring of the delivery system (see above) but trading along a watercourse is quite possible and undoubtedly occurs despite being an offence. Losses in the watercourse result in more water being delivered at the head than at the tail so that sale of tail-ender turns to head-enders adds to the surface water available (and incidentally may well be a factor in the inequities recorded in watercourse studies) (Jurriens *et al.*, 1996).

Farmers in any case differ in their resources, skills and wants, which leads to trades that may increase total welfare. Allowing trades along the watercourse is a market mechanism that could, in principle, increase productivity although it impacts on patterns of groundwater recharge and runs the danger of weakening the traditional and accepted system.

Differential crop charges imply a quasi-volumetric element at the farm level. Increased differentials and penal rates for crops that utilize large amounts of water could, in principle, make this approach more effective. However, cropping patterns cannot always be changed – paddy may be the only feasible crop in higher rainfall and waterlogged areas – and political objections would still have to be faced. A more interesting suggestion is made in the Indo-Dutch report (Agarwal and Roest, 1996). If water charges were to be based on the authorized water delivered to the farm rather than on the measured crop areas, they conclude that irrigated areas in saline regions, presumably in kharif could increase from 50% to 85%. Much of the rain-fed part of the farm would be converted to partial irrigation and the annual rise in saline water tables might be slowed – recharge would decline due to underwatering and greater evapotranspiration. As a result, waterlogging problems 'can be postponed by 5 to 10 years'. Of course, farmers even now irrigate crops on that part of their farm that they claim is rain-fed and subsequently mislead or collude with ID staff. Moreover, the act of measuring areas – indeed the whole land revenue tax process – contributes much to conserving the delivery and land tenure systems. Nevertheless, this proposal might receive further consideration.

Conclusions

Surface irrigation water in Haryana is distributed in proportion to holding size irrespective of soil type, crops grown, groundwater conditions or climatological factors. The amount delivered is sufficient in itself for no more than 20–30% of the irrigable land in kharif and 35–45% in rabi, leading to widespread underirrigation. Surface supplies are supplemented by (variable) rainfall and, if water is fresh, by groundwater pumping, so that cropping intensities are much higher than would be possible based just on surface supplies. Nevertheless, water remains a constraint on agricultural output and this is likely to intensify as non-agricultural demands grow. Agricultural production is also threatened by rising water tables in saline groundwater areas and falling water tables in fresh groundwater areas.

Effective rationing of surface supplies provides powerful efficiency incentives in water use, both directly and as pumping responds to variable rainfall and regular surface deliveries. This has been reflected in a remarkable growth in agricultural production despite constrained surface supplies. Moreover, the combination of main system rotation and *warabandi* below the outlet has proven robust and has demonstrated important advantages in terms of equity, transparency, social acceptance and low transactions costs. A shift from an accepted supply-based system to a demand-based system and volumetric pricing would involve major reconstruction of the physical infrastructure and a fundamental reform of accepted institutions and practices. The increase in the level of water charge needed to have a significant impact on water use would almost certainly be politically and socially unacceptable although small annual increments in water charges would be politically more acceptable than intermittent large increases in water charges. Thus, while in principle it might lead to a more responsive irrigation system, it is inconceivable that this could justify the costs and risks involved in making such a change.

More modest reforms of the supply-based system might include revised main system schedules, greater differentiation in area-based water charges, or replacement of area-based water charges by charges based on the water delivered during a *warabandi* turn. Main system schedules could in principle be modified to respond to soil or cropping conditions, for instance to provide more reliable but less abundant supplies to saline areas and vice versa, or to respond to the predominant cropping pattern in different areas. Water charges are presently collected along with land revenue and are based on the area of each crop irrigated by canal water. Charges are low but collection is relatively efficient and makes a reasonable contribution to meeting recurrent costs. Rates could be increased and the levels for different crops further differentiated to encourage planting of water-efficient crops. Alternatively, crop-based charges could be replaced by a charge dependent on the authorized water delivered during a *warabandi* turn, leaving the farmer to decide how best to allocate water on his farm.

Any such reforms need to be introduced cautiously given the risks associated with many modifications of the current accepted system. They would also at best have a modest impact on the long-term problems of falling water tables in fresh groundwater areas and waterlogging and secondary salinity in saline areas. Regulation of groundwater use represents a formidable challenge given the large number of wells and well owners. In the absence of an effective regulatory system, water tables will continue to decline until this is limited by rising pumping costs or deteriorating water quality. Waterlogging in saline areas can at best be slowed by reforms of the type discussed above. The only ultimate long-term solution would be costly investments in drainage and reclamation programmes.

Appendix: Overview of Outcome of the Spreadsheets

The returns to water in the Sirsa district of Haryana State in India were studied (see Fig. 8.A.1), using data on 24 farms. Eight farmers

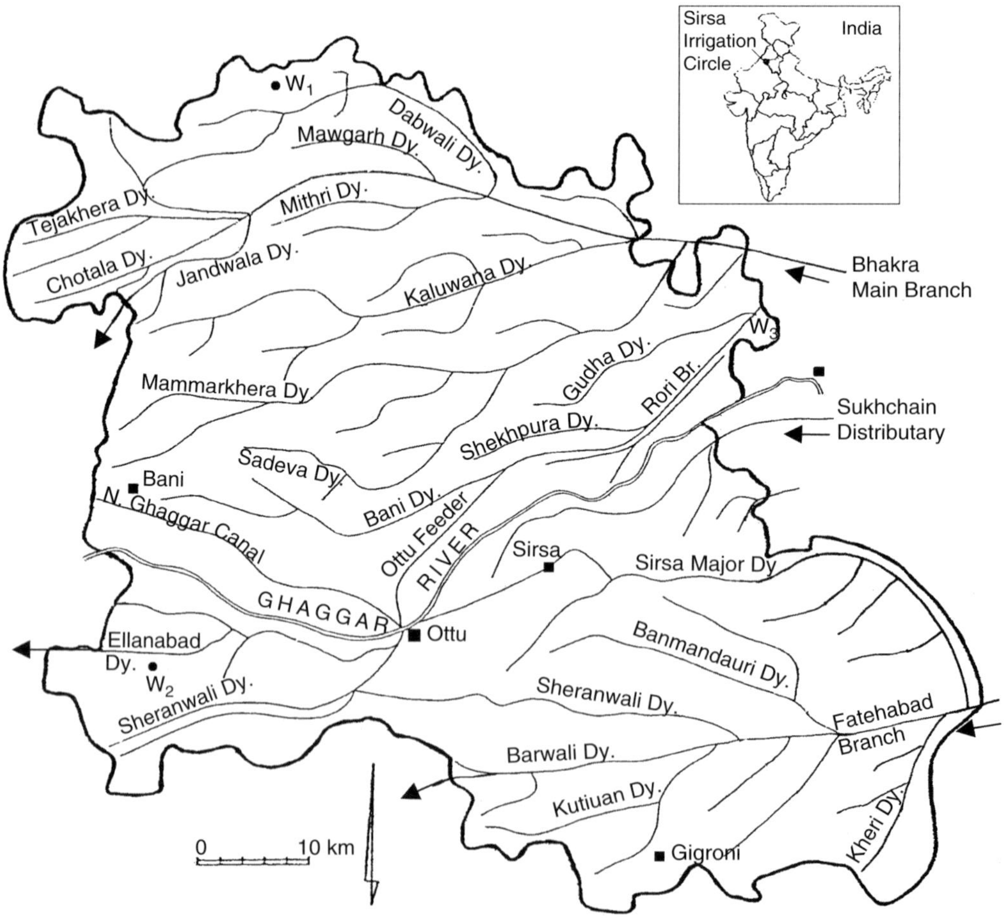

Fig. 8.A.1. Location of the Sirsa Irrigation Circle showing the canal network (Van Dam and Malik, 2003).

were selected from the Ottu Feeder in the paddy–wheat belt (four from Ram Pur Their and four from Sangatpura in eight Burji Ottu villages) along with 16 farmers from the Kasumbi distributary in the cotton–wheat belt in six villages (four from Fulkan, three from Kotli, two from Kanvar Pura, one from Ding, one from Kasumbi and five from Ban Mandori). These 24 farmers were divided into five farm categories on the basis of location in terms of the canal water source outlet and type of land. The data required for the AGWAT spreadsheets pertaining to rabi 2001/02 and kharif 2002/03 were collected from each respondent through personal interviews using structured questionnaires. The results are summarized in Tables 8.A.1–8.A.5. It is important to note that the data are based on an exceptional year, with very low canal water availability and rainfall.

Table 8.A.1. Farm type 1 (paddy–wheat belt, head of canal, normal soils, 4 farms; 9 ha).

				Farm costs				Water		
Crop	Gross return ($/ha)	Cropped area (ha)	Gross return ($)	Inputs ($)	Labour ($)	Water ($)	Net return ($)	Use (m^3/ha)	Total use (m^3)	Net return ($/$m^3$)
Kharif										
rice	894	7.88	7,046	3,256	1,155	233	2,403	13,782	108,600	0.020
Cotton	580	0.36	208	112	28	6	62	8,611	3,100	0.020
Rabi										
wheat	857	8.24	7,067	2,293	385	78	4,313	4,927	40,600	0.110
Mustard	655	0.09	59	13	2	1	43	3,333	300	0.150
Total	868	16.60	14,380	5,673	1,570	317	6,820	9,200	152,700	0.045

Table 8.A.2. Farm type 2 (paddy–wheat belt, middle of canal, normal soils, 4 farms; 3.3 ha).

				Farm costs				Water		
Crop	Gross return ($/ha)	Cropped area (ha)	Gross return ($)	Inputs ($)	Labour ($)	Water ($)	Net return ($)	Use (m^3/ha)	Total use (m^3)	Net return ($/$m^3$)
Kharif										
rice	880	3.04	2,675	1,188	366	96	1,023	13,816	41,800	0.030
Rabi										
wheat	801	3.04	2,434	831	69	29	1,505	4,934	15,000	0.100
Total	840	6.10	5,109	2,020	435	126	2,528	9,310	56,800	0.045

Table 8.A.3. Farm type 3 (cotton–wheat belt, head of canal, normal soils, 8 farms; 5.1 ha).

				Farm costs				Water		
Crop	Gross return ($/ha)	Cropped area (ha)	Gross return ($)	Inputs ($)	Labour ($)	Water ($)	Net return ($)	Use (m^3/ha)	Total use (m^3)	Net return ($/$m^3$)
Kharif										
rice	499	0.10	51	36	5	3	7	10,000	1,400	0.010
Cotton	792	2.96	2,342	918	228	62	1,134	9,460	27,700	0.040
Rabi										
wheat	772	3.37	2,598	974	198	32	1,393	4,154	13,600	0.100
Mustard	443	1.17	519	234	29	10	247	3,419	4,100	0.060
Total	725	7.60	5,510	2,162	460	106	2,782	6,170	46,900	0.059

Table 8.A.4. Farm type 4 (cotton–wheat belt, middle of canal, normal soils, 4 farms; 5.9 ha).

Crop	Gross return ($/ha)	Cropped area (ha)	Gross return ($)	Farm costs: Inputs ($)	Farm costs: Labour ($)	Farm costs: Water ($)	Net return ($)	Water: Use (m³/ha)	Water: Total use (m³)	Water: Net return ($/m³)
Kharif										
cotton	659	2.53	1,665	1,002	116	56	492	10,040	25,400	0.020
Guar	410	1.00	410	130	8	3	270	200	200	1.740
Rabi										
wheat	571	2.65	1,512	700	88	27	697	4,491	11,900	0.060
Mustard	423	0.88	373	156	12	7	198	3,750	3,300	0.060
Total	561	7.10	3,960	1,987	223	93	1,657	5,745	40,800	0.041

Table 8.A.5. Farm type 5 (cotton–wheat belt, tail of canal water, problematic soils, 4 farms; 5.7 ha).

Crop	Gross return ($/ha)	Cropped area (ha)	Gross return ($)	Farm costs: Inputs ($)	Farm costs: Labour ($)	Farm costs: Water ($)	Net return ($)	Water: Use (m³/ha)	Water: Total use (m³)	Water: Net return ($/m³)
Kharif										
cotton	644	3.12	2,009	1,085	256	81	586	11,795	36,800	0.020
Guar	513	0.79	407	112	10	2	283	253	200	1.800
Rabi										
wheat	476	3.12	1,484	828	152	34	470	5,192	16,200	0.030
Mustard	326	0.23	74	45	4	2	22	4,348	1,000	0.020
Total	533	7.30	3,974	2,070	423	119	1,361	7,425	54,200	0.025

Farms 1 and 2 experienced a shortage of family labour during the peak months of July (transplantation of paddy), October and November (due to harvesting of paddy, sowing of wheat crops and peaking of cotton crop on Farm 1). Both farms also experienced insufficient supply of canal water throughout the year, compensated for by groundwater pumped from tube wells. Highest net returns were found to be from mustard; net returns per cubic metre of water were smaller on Farm 2 than on Farm 1. Farms in the cotton–wheat belt experienced a shortage of canal water in the months of February, March, August, September and October. The cotton crop was more remunerative on Farm 3 than on Farms 4 and 5. The net returns were highest for Guar.

References

Agarwal, M.C. and Roest, C.J.W. (eds) (1996) *Towards Improved Water Management in Haryana State*. Final report of the Indo-Dutch Operational Research Project on Hydrological Studies. Chaudary Charan Singh Haryana Agricultural University, Hisar. International Institute for Land Reclamation and Improvement and DLO Winand Staring Centre for Integrated Land, Soil and Water Research, Wageningen, The Netherlands.

Aggarwal, P.K., Roetter, R.P., Kalra, N., Van Keulen, H., Hoanh, C.T. and Van Laar, H.H. (eds) (2001) *Land Use Analysis and Planning for Sustainable Food Security: with an Illustration for the State of Haryana*. Indian Agricultural Research Institute, India, New Delhi; IRRI, Los Baños, The Phillipines; Wageningen University and Research Centre, Wageningen, The Netherlands.

Albinson, B. and Perry, C.J. (2002) *Fundamentals of Smallholder Irrigation: The Structured System Concept.* IWMI Research Report No. 58. IWMI, Colombo.

Berkoff, D.J.W. (1990) *Irrigation Management on the Indo-Gangetic Plain.* Technical Paper No. 129. World Bank, Washington, DC.

Eastern Book Company (1982) *The North India Canal and Drainage Act, 1873* (along with State amendments and short notes). Lucknow, India.

GOH (Government of Haryana) (1998) *Management of Waterlogging and Salinity Problems in Haryana.* Master plan, prepared by high-level expert committee. Government of Haryana, Haryana, India.

GOH (2004) *Economic Survey of Haryana 2003–2004.* Planning Department, Government of Haryana, Haryana, India.

Horst, L. (1998) *The Dilemmas of Water Division. Consideration and Criteria for Irrigation System Design.* IWMI, Colombo.

Jurriens, M., Mollinga, P.P. and Wester, P. (1996) *Scarcity by Design; Protective Irrigation in India and Pakistan.* Liquid Gold Paper 1. International Institute for Land Reclamation and Improvement, Wageningen, The Netherlands.

Malhotra, S.P. (1998) *The Warabandi and Its Infrastructure.* Publication No. 157. Central Board of Irrigation and Power. New Delhi.

Ministry of Irrigation (1982) *Manual on Warabandi in Irrigation Projects.* Government of India, New Delhi.

Narayanamurthy, S.G. (1985) *Methods for Improving Canal Regulation in North-west India, Using the Example of the Bhakra Scheme.* World Bank, New Delhi.

Perry, C.J. and Narayanamurthy, S.G. (1998) *Farmer Response to Rationed and Uncertain Irrigation Supplies.* IWMI Research Report 24. IWMI, Colombo.

Van Dam, J.C. and Malik, R.S. (eds) (2003) Water productivity of irrigated crops in Sirsa district, India. Integration of remote sensing, crop and soil models and geographical information systems. WATPRO final report. Available at: http://library.wur.nl/way/catalogue/documents/WATPRO_final_report.pdf

World Bank (1994) *Haryana Water Resources Consolidation Project: staff Appraisal Report, Report No-12813-IN',* March 1994. The World Bank, Washington, DC.

World Bank (1998) *India Water Resources Management Sector Review – Initiating and Sustaining Water Sector Reforms.* Report No. 18356-IN. World Bank, Washington, DC.

9 The Energy–Irrigation Nexus in South Asia: Groundwater Conservation and Power Sector Viability

T. Shah, C. Scott, J. Berkoff, A. Kishore and A. Sharma

Introduction

Back in the 1950s, when energy use was considered synonymous with economic progress, state power utilities in India aggressively persuaded unwilling farmers to install electric tube wells. The chief ministers set ambitious connection targets and all manner of loans and concessions were made available to popularize tube well irrigation. The World Bank supported huge investments in rural electrification to promote groundwater use and agricultural growth, policies that appeared to be vindicated when the Green Revolution was found to follow the tube well revolution with a lag of 3–5 years. Repetto (1994) even asserted that 'the Green Revolution is more a tube well revolution than a wheat revolution'.

By the 1970s, the energy–irrigation nexus was a prominent feature of South Asia's agrarian boom, and groundwater irrigation had spread rapidly even within canal commands. The enthusiasm of the State Electricity Boards (SEBs) towards their agricultural customers however soon began to wane. The SEBs invariably charged their fees based on metered consumption but – as tube well numbers increased – metering and billing became an increasing burden. The costs of provision and maintenance of meters were perhaps the least of the SEB's worries. Farm power supply required an army of meter readers, and led to rampant meter tampering and power pilferage, underbilling and pervasive corruption. These high and rising transaction costs proved insupportable and, during the 1970s/1980s, state after state adopted a flat tariff linked to the horsepower (hp) rating. This eliminated the hassle and cost of metering in one go and, though still affording scope for malpractices such as under-reporting of the hp rating, this was much easier to control than pilferage under a metered tariff regime. In turn, however, as farm power emerged as a major driver of irrigated agriculture, chief ministers found electricity pricing to be a powerful vote-winner. Flat tariffs became 'sticky' and, unable to raise flat tariffs for years on end, yet still pressured to supply abundant farm power, the SEBs found their balance sheets turning red. The argument has thus turned full circle and the industry and its protagonists (e.g. the multilateral donors) have returned to the view that metering is a precondition for restoring the SEB's financial viability.

Support for metering is based essentially on the neoclassical economic theory that typically focuses on the 'transformation costs' of generating and distributing power, and the efficiency gains to be derived from economic pricing, while overlooking the 'transactions costs' incurred. In this chapter, our objective is to re-evaluate this debate from the perspective of the New Institutional Economics (North, 1997). We begin by assessing the scale of the energy–irrigation nexus in South Asia. The estimates quoted matter less than the broad conclusion that – by any measure – the nexus is far more important in South Asia than elsewhere in the world with the exception perhaps of North China. This is followed by a section describing what it would take to make a metered tariff regime work, the main comparison being with North China where such a regime does seem to work. Concluding that South Asia differs in too many ways to duplicate China's success, the rest of the chapter explores the potential for indirect management of the groundwater economy through the specific mechanism of electricity pricing and supply policies.

The central premise is that electricity pricing and supply in South Asia are closely linked with the policy goals of managing groundwater irrigation for efficiency, equity and sustainability. The chapter makes no claim that the solutions proposed would resolve all problems of aquifer management though it does suggest that they would complement measures in other subject areas. Nor does the chapter address broader environmental issues associated with sustainability. It takes as given the generally accepted view that rapidly falling groundwater tables can have deleterious effects on the rural economy and on the environment, and that pragmatic measures that moderate such declines are generally beneficial. A further premise is that the financial viability of the power utilities has been undermined by their farm power operations and that this can be attributed at least in part to the failure of the power and irrigation sectors to interact in an intelligent manner. Again, the problems of the utilities and their operations go well beyond the issues addressed in the chapter. But even if the solutions proposed are, in some sense, partial and second best, the chapter concludes that analysing the energy and groundwater economies as a nexus can help evolve joint strategies that would contribute significantly to the preservation of South Asia's groundwater resources while at the same time improving the viability of its power industry.

The Scale of the Energy–Irrigation Nexus in South Asia

South Asia in a world context

The energy–irrigation nexus focuses attention on a class of issues that is largely confined to South Asia and, to a lesser extent, North China (see below). Many other countries – e.g. the USA, Iran, Mexico – make intensive use of groundwater in agriculture (Fig. 9.1). However, in these countries this involves only a small proportion of their people; energy use by agriculture is a small proportion of total energy use; and the cost of energy use is only a small proportion of the total value added in farming. The opposite is the case over much of South Asia and North China (Table 9.1).

According to a World Bank estimate, groundwater irrigation contributes about 10% of India's GDP (World Bank and GOI, 1998) using 15–20% of the electricity generated. In contrast, in Mexico's Guanajuato province, heartland of its intensive groundwater-irrigated agriculture, a typical tube well is run by a 100–150 hp pump and operates for over 4000 h/year (Scott *et al.*, 2002). In India, Bangladesh and Nepal, the modal pump size is 6.5 hp and average hours of operation are around 400–500 h/year (Shah, 1993). In Iran, 365,000 tube wells lift 45 km^3 of groundwater/year (Hekmat, 2002); India uses 60 times more wells than Iran to extract three times as much groundwater.

Despite these differences, other countries can still find it difficult to enforce groundwater controls. In Mexico, the Commission National de Aqua (CNA) has struggled to establish and enforce a system of water rights. While this has helped to register most of its 90,000 tube well owners, Mexico still finds it impossible to limit

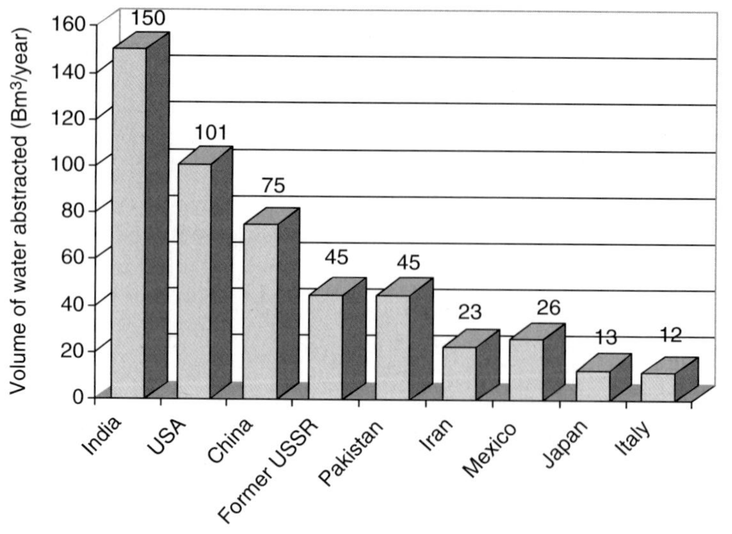

Fig. 9.1. Groundwater use in selected countries in the 1980s (MCM). (From Llamas *et al.*, 1992, p. 4.)

pumping to assigned quotas. Mexico has similarly been politically unable to remove substantial energy subsidies to agriculture or rein in groundwater depletion (Scott *et al.*, 2002). In Iran, when groundwater overdraft in the hinterland threatened water supply to cities, the government enforced a ban on many new groundwater structures, yet it is struggling to eliminate its annual groundwater overdraft of 5 km^3 (Hekmat, 2002). Even the USA has only found it possible to slow rather than stop the mining of the great Ogallala aquifer. If richer countries where groundwater irrigation is far less important cannot manage irrigators even in the face of serious environmental anomalies, how much less can it be expected of countries in South Asia where groundwater is relatively far more important and where it supports the livelihoods of millions of poor rural households?

Groundwater in South Asia

South Asia constitutes the largest user of groundwater in the world. Between them, India, Pakistan, Bangladesh and Nepal pump around 210 km^3/year, using some 21–23 million pump sets (13–14 million electric pumps and 8–9 million diesel pumps) (NSSO, 1999). If an average electric tube well (with pumping efficiency of, say, 25%)

Table 9.1. Dependence on groundwater in different countries. (From Hekmat, 2002, Iran; Mukherji and Shah, 2002, India; Scott *et al.*, 2002, Mexico; Shah *et al.*, 2003, China and Pakistan.)

Country	Annual groundwater use (km^3)	Groundwater structures (million)	Extraction/ structure (m^3/year)	Population dependent on groundwater (%)
Pakistan Punjab	45	0.5	90,000	60–65
India	150	21.3	7,900	55–60
China	75	3.5	21,500	22–25
Iran	29	0.5	58,000	12–18
Mexico	29	0.1	414,285	5–6
USA	100	0.2	500,000	<1–2

lifts water on average 30 m,[1] the electricity equivalent of energy used is around 69.6 billion kWh/year. At an assumed cost of Rs 2.5 ($0.05)/kWh, this implies a total cost of Rs 174 billion ($3.8 billion). We estimate the market value of the irrigation water produced is around Rs 450–550 billion ($9.8–12 billion) and its contribution to agricultural output at about Rs 1350–1650 billion ($29.3–35.9 billion).[2]

Growth in groundwater irrigation is relatively recent (Fig. 9.2). In India, gravity systems dominated until the 1970s but by the early 1990s groundwater had far surpassed surface irrigation in terms of area served and proportion of agricultural output (Debroy and Shah, 2003; Shah *et al.*, 2003). According to estimates of the Government of India (GOI), 60% of India's irrigated lands are now served by groundwater wells (GOI, 2001). Independent surveys suggest that the proportion may be as much as 75% if conjunctive use in command areas is included (Shah *et al.*, 2004b; NSSO 54th round). By now, pump irrigation in India accounts for 70–80% of the value of irrigated farm output; and rapid groundwater development is at the heart of the agrarian dynamism found in areas in Eastern India that had been stagnant for a long time. Furthermore, groundwater irrigation has helped make famines a matter of history: during 1963–1966, a small rainfall deficit left reservoirs empty and food production plummeted by 19%; during the 1987–1988 drought, the rainfall deficit was 19% but food production fell by only 2% thanks in large part to widespread groundwater irrigation (Sharma and Mehta, 2002).

In contrast to other countries, pump irrigation in South Asia also involves vast numbers of low-income households and a large proportion of the population. In 1999–2000, India's 81 million landowning families (http://labourbureau.nic.in/) had more than 20 million tube wells and pump sets among them, on average roughly one for every fourth landowning household. Moreover, a large proportion of non-owners are supplied through local fragmented groundwater markets (Shah, 1993). It is often argued that with 60 million tonnes of food stocks, India can now take a tough posture on groundwater abuse but this misses an important point. Quite apart from the practical difficulties of implementing such a policy, the contribution of groundwater to farm incomes and rural livelihoods is far more crucial than its contribution to food security, especially outside canal commands.[3] At the turn of the millennium, perhaps three-quarters of the rural population and over half of the total population of India, Pakistan, Bangladesh and Nepal depended for their livelihoods, directly or indirectly, on groundwater irrigation, many times larger than in Iran and Mexico. It is not surprising therefore that the energy–irrigation nexus has been at the centre of vote-bank politics in the region.

[1] Most groundwater irrigation in South Asia is based on open dug wells and shallow tube wells. Deep tube wells are less than 1% of all groundwater structures.

[2] The Centre for Monitoring Indian Economy estimates that electricity use in Indian agriculture in 2000–2001 was 84.7 billion kWh, much greater than our combined estimate of 69.6 billion kWh equivalent of the *total* energy use in agriculture for the four countries. However, these estimates for India include transmission and distribution (T&D) losses in non-farm sectors that are passed off as agricultural consumption (CMIE, 2003). Dhawan puts the value of the marginal product of power in agriculture at Rs 9.00/kWh ($0.20/kWh) in net terms and Rs 14/kWh ($0.30/kWh) in gross terms (Dhawan, 1999). We assume an average South Asian tube well uses 4 kWh/h, implying 17.5 billion h of pumping/year. At an average price of Rs 30/h (USc65/h), the market value of pump irrigation is Rs 522 billion ($11.34 billion). Those selling pump services typically claim a third of the crop. Based on this, we estimate the contribution to farm output as three times the market value of pump irrigation. An alternative approach assumes that a South Asian tube well produces Rs 25,000 ($543.48) worth of irrigation water/year contributing to Rs 75,000 ($1630) worth of crops. The World Bank asserts that groundwater contributes 10% of Indian GDP (World Bank and GOI, 1998). If so, our estimates are greatly understated.

[3] Dhawan (cited in Samra 2002), for instance, has asserted that in low rainfall regions of India, '[A] wholly [*groundwater*] irrigated acre of land becomes equivalent to 8 to 10 acres of dryland in terms of production and income.' (italics added).

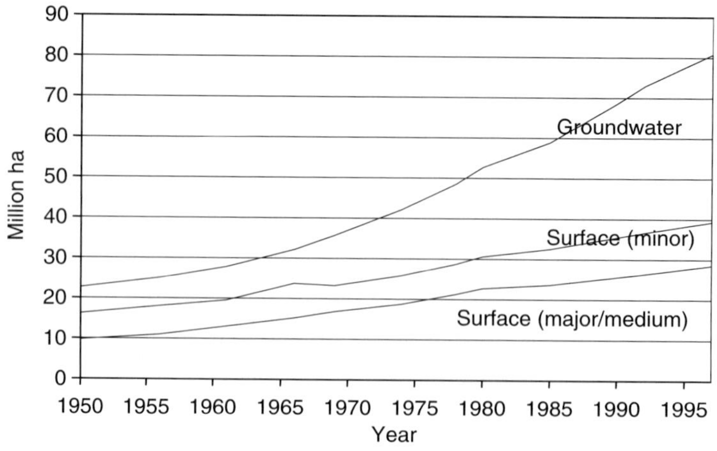

Fig. 9.2. India, irrigated area by source. (From Ministry of Water Resources, Government of India, 1999.)

Subregional patterns

Though groundwater is critical over much of South Asia, policy makers face conflicting challenges in different subregions. Particularly since 1970, agrarian growth has been sustained primarily by private pump investments. However, this has been highly uneven. In the groundwater-abundant Ganga–Brahmaputra–Meghna basin – home to 400 million of the world's rural poor – groundwater can have major livelihood and ecological benefits (Shah, 2001) but it is precisely here that economic development has been slow and halting. Eastern India is a classic example. After the eastern Indian states switched to a flat power tariff, the utilities found it difficult to maintain viability in the face of organized opposition to the raising of the flat tariff. As a result, the power utilities began to neglect the maintenance and repair of power infrastructure resulting, in turn, in a feeble rural power supply. Unable to irrigate their crops, farmers began en masse to replace their electric pumps by diesel pumps. Over a decade, the groundwater economy became more or less completely dieselized in large areas, including Bihar, eastern Uttar Pradesh and north Bengal. Figure 9.3 shows the electrical and diesel halves of India; in the western parts, groundwater irrigation is dominated by electric pumps but as we move east, diesel pumps become preponderant. The saving grace was that in these groundwater-abundant regions, small diesel pumps, though dirtier and costlier to operate, kept the economy going.

The issues in regions like north Gujarat, where groundwater is lifted from 200 to 300 m, are very different since such de-electrification could completely destroy the agricultural economy. In much of Pakistan, in the Indian Punjab, Haryana and neighbouring states, and in peninsular India, groundwater is being seriously overdeveloped to a stage that agriculture faces serious threats from resource depletion and degradation. The priority in these areas is to promote a constructive re-engagement of the power sector with agriculture and to find ways of managing groundwater use so as to make it socially and environmentally sustainable. It is in regard to these areas that this chapter is largely concerned.

In regulating groundwater use, the tools available to resource managers are few and inadequate, though the protection of the resource is proving far more complex and difficult than stimulating its initial development. The alternatives fall into two broad categories: (i) *direct management* through a system of metered tariffs and/or quotas; and (ii) *indirect management*, e.g. through the operations of the power market. These options are now considered in turn.

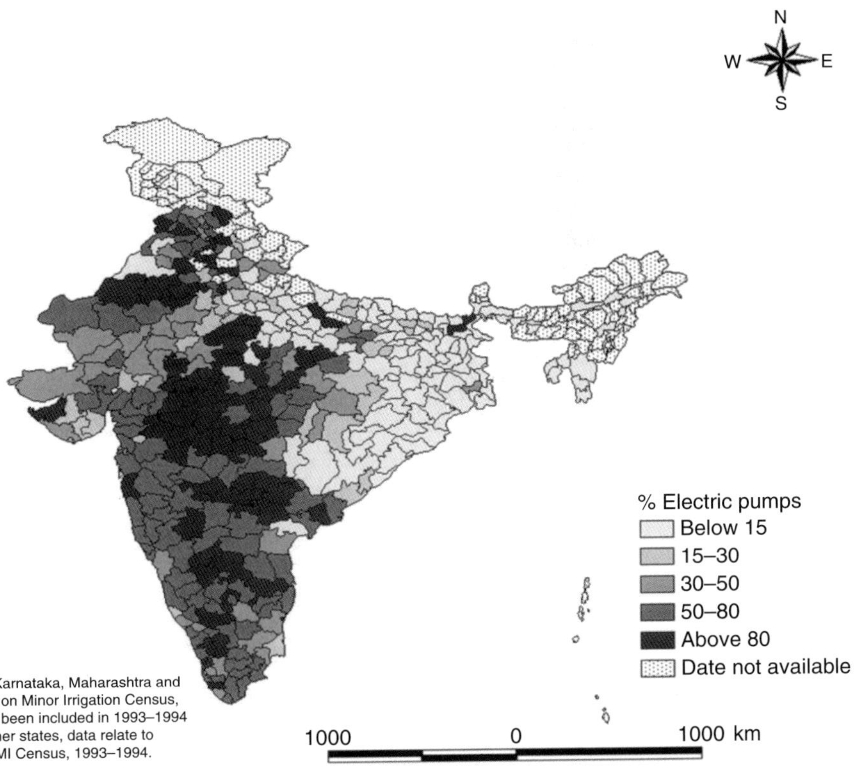

* Figures for Gujarat, Karnataka, Maharashtra and Tamil Nadu are based on Minor Irrigation Census, 1986 as they have not been included in 1993–1994 MI Census. For the other states, data relate to 1993–1994 based on MI Census, 1993–1994.

Fig. 9.3. Percentage of electricity operated groundwater structures to total mechanized groundwater structures.

Making a Metered Tariff Regime Work

Introduction

In India and elsewhere in South Asia there is a growing movement to revert to metered power supply. Despite widespread farmer opposition, the power industry believes that its fortunes will not change until agriculture is put back on a metered electricity tariff. Strong additional support is lent by those working in the groundwater sector where it is widely – and rightly – held that zero and flat power tariff produce strong perverse incentives for farmers to indulge in profligate and wasteful use of water and power because it reduces the marginal cost of water extraction to nearly zero. Annual losses to electricity boards on account of power subsidies to agriculture have been estimated at Rs 260 billion ($5.65 billion) in India, growing at an annual rate of 26%/year (Lim, 2001; Gulati, 2002). These estimates have, however, been widely contested, for instance it has been shown that SEBs have been classifying rising Transmission and Distribution (T&D) losses in domestic and industrial sectors as agricultural consumption since it is unmetered and so unverifiable.[4] But the fact

[4]Shah (2001) has analysed this aspect for Uttar Pradesh State Electricity Board and found agricultural power use to be 35% lower than claimed. Similarly, based on a World Bank study in Haryana, Kishore and Sharma (2002) report that actual agricultural power consumption was 27% less than reported, and the overall T&D losses were 47% while official claims made it 36.8%, making the SEB more efficient than it actually was. Power subsidy ostensibly meant for the agriculture sector but actually accruing to other sectors was estimated at Rs 5.50 billion/year ($0.12 billion/year) for Haryana alone.

remains that agricultural power supply under the existing regime is the prime cause of bankruptcy of SEBs in India.

Reflecting pressure from the power industry, GOI has prescribed that: (i) power on demand will be provided by 2012; (ii) all consumers will be metered in two phases, with phase I to cover metering of all 11 kVA feeders and High Tension consumers, and phase II to cover all consumers; and (iii) regular energy audits will be undertaken to assess T&D losses and eliminate power thefts within 2 years (Godbole, 2002). This is an ambitious agenda. Consistent with these policies, Central and State Electricity Regulatory Commissions have set deadlines for SEBs and state governments to make the transition to universal metering, and all new tube well connections now come with the option of a metered tariff with most states offering inducements to opt for metered connections. Support has also come from international agencies – notably the World Bank, USAID and ADB – which have begun to insist on metered power supply to agriculture as a key condition for financing new power projects.

Arguments for a metered tariff regime are several. First, metering is considered essential for SEBs to manage their commercial losses; you cannot manage what you do not monitor and you cannot monitor what you do not measure. Second, once farm power is metered, SEBs cannot use agricultural consumption as a carpet under which they can sweep their T&D losses in other markets. Third, metering provides farmers with the correct signals concerning the real cost of power and water, and encourages them to economize on their use. Fourth, for reasons that are not entirely clear, it is often suggested that a metered tariff would be less amenable to political manipulation than a flat tariff regime and easier to raise as the cost of supplying power rises. Finally, it is widely argued that a flat tariff is inequitable towards small landowners and to irrigators in regions with limited availability of groundwater. The logic in support of a metered tariff is thus obvious and unexceptionable. The problem is to make a metered tariff regime work as broadly envisaged. For this, three things seem essential:

- The metering and collection agent must have the requisite authority to deal with deviant behaviour among users.
- The agent should be subject to a tight control system so that he can neither behave arbitrarily with consumers nor form an unholy collusion with them.
- The agent must have proper incentives to enforce a metered tariff regime.

Under agrarian conditions that in many ways are comparable with those in South Asia, these three conditions appear to obtain in North China where a metered tariff regime works reasonably well (Shah, 2003; Shah *et al.*, 2004a). How is this possible? And if it works in North China why not in South Asia?

Why is metering effective in North China?

The Chinese electricity supply industry operates on two principles: (i) total cost recovery in generation, transmission and distribution at each level with some minor cross-subsidization across user groups and areas; and (ii) each user pays in proportion to his use. In contrast to much of India, tariffs thus reflect relative costs and agricultural use, which often attract the highest charge per unit, followed by household users and then industries. The operation and maintenance (O&M) of local power infrastructure is the responsibility of local units – the Village Committee at village level, the Township Electricity Bureau at township level and the County Electricity Bureau at county level. Responsibilities for collecting electricity charges are assigned to ensure that the power used at each level is paid in full at that level. At village level, the sum of power use for any given period recorded at individual meters has to tally with the power supply recorded at the transformer. The unit or person charged with fee collection pays the Township Electricity Bureau for power use at the transformer after allowing for

10% to account for normal losses. If the power supply infrastructure is old and worn out, line losses below the transformer make this difficult. With this supposition turning out to be true, an Electricity Network Reform program was undertaken by the National Government to modernize and rehabilitate rural power infrastructure. Where this was done, line losses fell sharply[5] and among the nine villages Shah visited in three counties of Henan and Hebei in early 2002, none of the Village Electricians interviewed had a problem tallying transformer records with the sum of the consumption recorded by individual users given the line-loss allowance of 10%.

An important reason why this institutional arrangement works is the strong local authority structures: the electrician is feared because he is backed by the Village Committee and powerful Party Leader; and the new service orientation is designed partly to project the electrician as the friend of the people. The Committee and Leader can also keep flagrantly arbitrary behaviour in check. The hypothesis that with better quality power and support service, farmers will be willing to pay a high price for power is exemplified in Henan where farmers pay a higher electricity rate compared not only to most categories of users in India and Pakistan (Yuan 0.7/kWh or US$0.0875/kWh, Rs 4.03/kWh) but also to the diesel price at Yuan 2.1/l. The village electrician in Henan and Hebei receives a fairly modest reward of Yuan 200/month, equivalent to half the value of wheat produced on a mu (or 1/30th of the value of output on 1 ha of land). For this modest wage, he undertakes to make good to the Township Electricity Bureau line and commercial losses in excess of 10% of the power consumption recorded on the transformers. If he can manage to keep losses to less than 10%, he can keep 40% of the value of power saved.

All in all, the Chinese have a working solution to a problem that has befuddled South Asia for nearly two decades. Following Deng Xiaoping who famously asserted that 'it does not matter whether the cat is black or white, as long as it catches mice', the Chinese have built an incentive-compatible system that delivers quickly rather than wasting time on rural electricity cooperatives and Village *Vidyut Sanghas* (Electricity Associations) being tried in India and Bangladesh (see below). Given the Chinese method of collecting metered electricity charges, it is well-nigh impossible for the power industry to lose money in distribution since losses are firmly passed on downstream from one level to the level below.

Why cannot a metering regime work in South Asia?

If South Asia is to revert to a metered tariff, the Chinese offer a good model. But there are two initial problems. First, agricultural productivity in China is much higher than in most of South Asia and even with power charged at full cost, pumping constitutes a relatively small proportion of the gross value of output. In South Asia, irrigation costs of this order (Rs 2100–8600/ha or $46–197) would make groundwater irrigation unviable except in parts of Punjab and Haryana. Second, while the South Asian power industry can perhaps approximate to the Chinese incentive system, it cannot repli-

[5]The village electrician's reward system encourages him to exert pressures to cut line losses. In the Dong Wang Nu village in the Ci county, the village committee's single large transformer which served both domestic and agricultural connections caused heavy line losses at 22–25%. Once the Network Reform Program began, he pressurized the Village Committee to sell the old transformer to the Township Electricity Bureau and raise Y10,000 (partly by collecting a levy of Y25 per family and partly by a contribution from the Village Development Fund) to get two new transformers, one for domestic connections and the other for pumps. Since then, power losses have fallen to a permissible level of 12% here (Shah *et al.*, 2004a).

cate the Chinese authority system at village level. The absence of an effective local authority that can guard the farmers from arbitrary behaviour of the metering agent or protect the latter from non-compliance by users may create unforeseen complications in adapting the Chinese model by South Asia. These costs soar in a 'soft state' in which an average user expects to get away even if caught.[6,7] An important reason why metering works reasonably well in China is that it is a 'hard state': an average user fears the village electrician whose informal power and authority border on the absolute in his domain. Two issues in South Asia are thus critical:

- The relentless opposition from farmers to metering;
- The problems that forced the SEBs to switch to a flat tariff during the 1970s in the first place.

Moves towards metered power consumption have met with unprecedented farmer opposition and there are few takers for metered connections; instead, the demand for free power has gathered momentum.[8] Opposition to a metered tariff is in part due to an assumed threat to the subsidy contained in the existing flat tariff. In addition, farmers find the flat tariff transparent and simple to understand; it spares them the tyranny of the meter readers; they fear that, once metered, all manner of new charges will be added under different names; and they raise the issue of equity – if canal irrigators receive irrigation at subsidized flat rates in public schemes, why not provide the same terms to groundwater irrigators?

The extent of farmer resistance is evident in the repeated failure of SEBs to entice farmers to accept metering even at subsidized rates ranging from Rs 0.20/kWh to Rs 0.70/kWh (US$0.004–0.013/kWh) compared to an actual cost from Rs 2.50/kWh to Rs 3.80/kWh (US$0.05–0.08/kWh). In 2002, Batra and Singh (2003) interviewed well owners in Punjab, Haryana and western Uttar Pradesh. They noted that an average well owner would spend Rs 2530 ($55) and Rs 6805/year ($148/year) *less* on their total power bill in Punjab and Haryana, respectively if they accepted metering at prevailing rates of Rs 0.50/kWh (US$0.011/kWh) and Rs 0.65/kWh (US$0.014/kWh). Even so, they would not accept metering. In effect, this

[6]Transaction costs of charge collection will be high even under a flat tariff regime if farmers think they can get away with non-payment. Throughout India and Pakistan, replacing nameplates of electric motors on tube wells has emerged as a growth industry under the flat tariff. In Haryana, a World Bank study has recently estimated that the actual connected agricultural load was 74% higher than that shown by the official utility records (Kishore and Sharma, 2002).

[7]There are exceptions in South Asia, notably in the urban sector. Private electricity companies that supply power in cities like Ahmedabad and Surat also instill the fear of God in users by regularly meting out exemplary penalties, often in an arbitrary manner. The Ahmedabad Electricity Company's inspection squads, for example, are set steep targets for penalty collection for pilferage. To meet these targets, they have to catch real or imagined power thieves; their victims pay the fine because going to courts would take years to redress their grievances while they stay without power. Although these stories paint a sordid picture, the company would find it difficult to keep its commercial losses to acceptable levels if its customers were not repeatedly reminded of their obligation to pay.

[8]And farmers are getting away with it in many states. Electricity supply to agriculture became a major issue in India's 2004 parliamentary and state elections. Chief Ministers like Chandrababu Naidu of Andhra Pradesh, Narendra Modi of Gujarat and Jayalalitha of Tamilnadu suffered major electoral reverses arguably on account of farmer opposition to their stand on electricity supply to agriculture. The new Chief Minister of Andhra Pradesh announced free power to farmers the day after he assumed office; and Jayalalitha, who had abolished free power in Tamilnadu, restored it soon after the results of election. Gujarat's Narendra Modi softened his hard stand on farm power supply; and in Maharashtra, Shiv Sena chief Bal Thakre announced his promise to provide free power to farmers should his party come to power.

is the price they are willing to pay to avoid the hassle and costs of metering.[9]

India has a long history of electricity cooperatives in an attempt to improve accountability and improve performance in the sector, originally under a metered regime (Gulati and Narayanan, 2003, p. 129). However, despite 50 years of effort to make these work, including with donor support, they have not succeeded.[10] The 50-year-old Pravara electricity cooperative in Maharashtra survives but only by owing the SEB several billions of rupees in unpaid past dues (Godbole, 2002). Recent experiments with new metering solutions include that of Indian Grameen Services, an NGO which organized Transformer User Associations in Hoshangabad district of Madhya Pradesh; the idea was that the SEB would set up a dedicated plant if farmers paid unpaid dues and agreed to a metered tariff. However, before the 2004 elections, the chief minister 'waived' past dues and the Hoshangabad association disintegrated, its members disillusioned. Orissa organized similar Village *Vidyut Sangha*'s (Electricity User Associations); while these are now defunct, Orissa has achieved modest success in improving metered charge collection by using local entrepreneurs as billing and collection agents. However, less than 5% of rural load in Orissa is agricultural, and this approach may be much more difficult in, for instance, Gujarat where agriculture may account for 50–80% of the total rural load.

It is too early to learn lessons from these experiments though there is a prima facie case that a direct approach to incentives on the Chinese model might be preferable. What is clear is that the old system of metering and billing – under which the SEBs employed an army of unionized meter readers – just will not work.[11] If the logistical difficulty and transaction costs of metering prior to 1975 were so high that a flat tariff seemed the only way of containing them, how much more so is this now that there are ten times as many electric tube wells? Even with far fewer connections, a 1985 study in Uttar Pradesh and Maharashtra by the Rural Electrification Corporation estimated that the cost of metering rural power was 26% and 16%, respectively, of the total revenue of the SEB from the farm sector (Shah, 1993). And this estimate included only direct costs, e.g. the cost of the meter and maintaining it, of the power consumed by the meter, of reading the meter, and of billing and collecting. These costs are not insignificant[12]; but of much greater relevance is the cost of contain-

---

[9]According to Batra and Singh (2003), farmers resist metering 'because of the prevalence of irregularities in the SEBs.' Complaints of frequent meter burning (which costs the farmer Rs 1000 per meter burnt or $22), false billing, uncertainty in the bill amount etc. were quoted. They suggest farmers also resist metering because of the two part tariff (energy charge and rental for meter) system offered as an alternative to flat tariff. They are reluctant to pay the minimum bill (rental charge), which they have to pay even if they do not use the pump in a given month. In Gujarat which had metered tariff until 1987, an important source of opposition to metering is the arbitrariness of meter readers and the power they had come to wield over them; in many villages, farmers had organized for the sole purpose of resisting the tyranny of the meter reader. In some areas, this became so serious that meter readers were declared persona non grata; even today, electricity board field staff seldom go to the villages except in fairly large groups, and often with police escort.

[10]Thus, Madhav Godbole notes, 'But if co-operatives are to be a serious and viable option [for power distribution], our present thinking on the subject will have to be seriously reassessed. As compared to the success stories of electricity co-operatives [in USA, Thailand and Bangladesh], ours have been dismal failures' (Godbole, 2002, p. 2197).

[11]A 1997 consumer survey revealed that 53% of power consumers had to bribe electricity staff for services supposed to be free; 68% said that grievance redressal was poor or worse than poor; 76% found staff attitudes poor or worse; 53% found repair and fault services poor or worse; 42% said they had to make 6–12 calls just to register a complaint; 57% knew of power thefts in their neighbourhoods; 35% complained of excess billing; 76% complained of inconvenience in paying their bills (Rao, 2002).

[12]A recent World Bank Study estimated that the cost of metering all farm power connections in the small State of Haryana would amount to $30 million (Rs 1380 million) in capital investment and $2.2 million/year (Rs 101.2/year) in operating costs (Kishore and Sharma, 2002). The Maharashtra Electricity Tariff Commission estimated the capital cost of metering the state's farm connections at Rs 11.50 ($0.25) billion (Godbole, 2002).

ing pilferage, of tampering with meters, of under-reading and underbilling by meter readers in cohort with farmers over vast areas.[13]

Most SEBs find it difficult to manage a metered power supply even in the industrial and domestic sectors. In Uttar Pradesh, 40% of low tension (LT) consumers are metered but only 11% are billed on metered use; the rest are billed based on a minimum charge or an average of past months of metered use (Kishore and Sharma, 2002). In Orissa, under far-reaching power sector reforms, private distribution companies have brought all users under a metered tariff regime. However, 100% collection of amounts billed has worked only for industry; in the domestic and farm sectors collection as a proportion of billing declined from 90.5% in 1995/1996 to 74.6% in 1999/2000 (Panda, 2002). All in all, the power sector's aggressive advocacy of a metered tariff regime in agriculture is based, in our view, on an excessively low estimation of the transaction costs involved.

From a *Degenerate* Flat Tariff to a *Rational* Flat Tariff Regime

Introduction

The preoccupation of water and power sector professionals in aggressively advocating reversion to a metered tariff regime – and of farmers to frustrate their design – is, in our view, detracting from the discussion of pragmatic approaches that have the potential for promoting a better-managed, groundwater-based agrarian economy in coexistence with a viable electricity sector. In other words, if direct management is impractical in South Asia what are the options for *indirect* management? One option is indirect management based on carefully designed electricity supply and pricing policies and the adoption of an 'intelligent' flat tariff regime.

The major advantage of the rational flat tariff would be in putting a brake on groundwater depletion in western and peninsular India. Growing evidence suggests that water demand in agriculture is inelastic to pumping costs within a large range. While a metered charge without subsidy can make power utilities viable, it may not help much to cut water use and encourage water-saving agriculture. If anything, the evidence suggests that farmers respond more strongly to scarcity of these resources than to their price. Pockets of India where drip irrigation is spreading rapidly – such as Aurangabad in Maharashtra, Maikaal in Madhya Pradesh, Kolar in Karnataka and Coimbatore in Tamilnadu – are all regions where water and/or power is scarce rather than costly. A rational flat tariff with intelligent power supply rationing to the farm sector holds the promise of minimizing wasteful use of both resources and of encouraging technical change towards water and power saving. Such a strategy might reduce annual groundwater extraction in western and peninsular India by as much as 12–21 km^3/year and reduce power use by 4–6 billion kWh, valued at Rs 10–15 billion/year ($0.22–0.33 billion/year).

A flat tariff is often written-off as inefficient, wasteful, irrational and distortionary besides being inequitable. In South Asia, this has indeed proved to be the case. It was the change to a flat tariff that encouraged political leaders to indulge in populist whims such as doing away with the farm power tariff altogether (as in Punjab and Tamil Nadu) or pegging it at low levels regardless of the true cost of power supply. Such examples have led to the general perception that flat tariffs have been responsible for ruining the electricity industry and for causing groundwater depletion in many parts of South Asia. But, in our view, the flat tariff regime has been wrongly maligned since, as applied in South Asia, it is a *degenerate* version of what

[13] Rao and Govindarajan (2003) lay particular emphasis on geographic dispersion and remoteness of farm consumers in raising transaction costs of metering and billing: 'To illustrate, a rural area of the size of Bhubaneshwar, *the capital of Orissa state*, will have approximately 4000 consumers. Bhubaneshwar has 96,000. *The former* will have a collection potential of Rs 0.7 million/month ($15,217/month); for Bhubaneshwar, it is Rs 22.0 million/month ($0.48 million/month).'

might otherwise be a rational pricing regime. A zero tariff is not rational; nor is a flat tariff without proactive rationing and supply management.

Marginal cost pricing is far from universal in other sectors

To most analysts, a flat tariff violates the marginal cost principle that advocates parity between the price charged and the marginal cost of supply. Yet, businesses commonly price their products or services in ways that violate the marginal cost principle but make overall business sense. For instance, flat rates may be charged to stimulate use so as to justify the incremental cost of providing a service. In the early days of rural electrification, SEBs charged a flat-cum-pro-rata tariff to achieve two ends: to ensure that each tube well used at least the power to justify its investment in laying cable and poles; and the flat component of the tariff encouraged users to achieve this level. India's telephone department still provides the first 250 calls for a flat charge even though all calls are metered, the idea being to encourage telephone use to a level that justifies the incremental cost of providing the service.

But the most important justification for a flat tariff regime is to save on the transaction costs of doing business. Organizations hire employees on a piece rate when their work is easy to measure; but flat rate compensation is prevalent worldwide since it is not easy to measure the marginal value of an employee's output on a daily basis. Urban public transport systems offer passes to commuters at attractive flat rates in part because commuters offer a stable business and equally because it reduces queues at ticket windows, and the cost of ticketing and collecting fares daily. Cable operators in India still charge a flat tariff for a bunch of television channels rather than charging for each channel separately because the latter would substantially increase their transaction costs. A few years ago, the Indian Income Tax Department offered businesses in the informal sector to pay a flat income tax of Rs 1400/year ($30.4/year) rather than launching a nationwide campaign to bring millions of small businesses within its tax net because the transaction costs of doing that would have been far higher than the revenue realized. A major reason municipal taxes are levied on a flat rate is the transaction cost of charging citizens based on the value they place at the margin on the municipal services.

Are all these businesses that charge for their products or services on a flat rate destined to make losses? No. They often make money *because* they charge a flat rate. Many private goods share this one feature with public goods like municipal services and defence: the high transaction cost of charging a differential price to different customers based on their use as well as the value they place on the product or service. So they recover their costs through a flat rate and remain viable through deft supply management. Canal irrigation is a classic example. Volumetric supply has long been advocated but nowhere in South Asia is volumetric water pricing practised in canal irrigation given the prohibitive costs of collecting volumetric charges (Perry, 1996, 2001). This is due to such factors as: (i) the large number of potential small farmers; (ii) the difficulty of excluding defaulting farmers; and (iii) the propensity for farmers to frustrate sellers' effort. While volumetric pricing of canal irrigation may be possible in, say, South African irrigation systems where a branch canal serving some 5000 ha might have 10–50 white commercial farmers, an Indian system serving the same area might contain 6000–8000 farmers (Shah *et al.*, 2002). The only way of making canal irrigation systems viable in the Indian situation is to raise the flat rate per hectare to a level that ensures overall viability.

Supply restriction is inherent to *rational* flat rate pricing; by the same token, flat rate pricing and on-demand service are incompatible in most situations. In that sense, consumption-linked pricing and flat rate pricing represent two different business philosophies; in the first, the supplier will strive to 'delight the customer' as it were, by providing on-demand service without quantity or quality restrictions of any kind[14]; in the latter, the customer has to adapt to the supplier's constraints in terms of the overall quantum available and the manner in which it is supplied. In the case of buffet meals, restaurants give customers a good deal but save on waiting costs, which are a substantial element in the economics of a restaurant. In the Indian *thali* system, where one gets a buffet-type meal served on one's table, the downside is that one cannot have a leisurely meal since the restaurant aims to maximize the number of customers served during a fixed working period and in a limited space. Thus, there is always a price for the value businesses offer their customers through products and services offered on a flat tariff; but that does not mean that the seller or the buyer is any the worse for flat rate pricing.

The flat tariff in irrigation

The reason that the flat rate tariff, as currently practised for pump irrigation in South Asia, is degenerate – and the power industry is in the red – is that the power utilities have failed to manage a rationed power supply. Under the flat tariff system as practised, most SEBs try to maintain farm power supply at 8–15 h/day throughout the year. This is comparable to maintaining a surface canal at full supply every day of the year. Raising a flat tariff to a level that covers the cost of this service is politically untenable.[15] A domestic consumer may assess a good quality service as power of uniform voltage and frequency supplied 24 h a day, 365 days a year. But the irrigators' idea of good quality service is power of uniform voltage and frequency when their crops face critical moisture stress. Ideally, the business objective of a power utility should be to supply the best-quality service consistent with the flat tariff pegged at a given level. With intelligent management of power supply, it should be possible to satisfy irrigation power demand by ensuring a supply of 18–20 h a day for 40–50 key moisture-stress days, with some power available at other times.[16] The pattern of farming demand differs in significant ways from that of domestic and industrial customers. It is this that provides the main opportunities for 'value improvement,' that is, 'meeting or

[14]On-demand power supply is the norm in most developed electricity systems and on-demand irrigation also typifies most groundwater systems worldwide. In contrast, fully on-demand surface irrigation is only found in a very few fully reticulated systems backed by adequate water supplies. Under the vast majority of conditions, balancing water supply and demand in surface irrigation requires quota limitations of some sort.

[15]In Madhya Pradesh, the latest state to announce power pricing reforms, the Chief Minister announced a sixfold hike in flat tariff. No sooner was the announcement made than there was a realignment within the ruling party, and cabinet ministers began clamouring for a leadership change. Subhash Yadav, the Deputy Chief Minister, lamented in an interview with India Today: 'A farmer who produces 10 t of wheat earns Rs 60,000 ($1304.35) and he is expected to pay Rs 55,000 ($1195.65) to the electricity board. What will he feed his children with and why should he vote for the Congress?' (India Today, 2002, p. 32). The farmers stopped paying even the revised flat charges and just before the May 2004 assembly elections, the Chief Minister waived all past electricity dues. Even so, he could not save his seat. His Congress government, until now eulogized for a progressive development-oriented stance, was trounced at the polls. Analysts attributed his defeat to the government's failure on three fronts: *Bijli, Pani* and *Sadak* (electricity, irrigation and roads).

[16]No doubt there will always be a few farmers who might demand a very different schedule to that of the predominant farming pattern in a specific area. These will typically be entrepreneurial farmers growing high-return, specialized crops. Options for these farmers include on-farm storage, duplicate diesel pumps, market solutions, etc. Even so, some activities at the margin may be precluded. But in a country as vast as India, conditions somewhere will be suitable for meeting such specialized demands and, given the other advantages associated with the proposed 'rational flat tariff' system, this is likely to be a minor issue.

exceeding customer expectations while removing unnecessary cost' (Berk and Berk, 1995).

Groundwater irrigators are envious of farmers in canal irrigation projects since they pay so little for their water. But a typical canal irrigator may get surface water no more than 10–15 times in a year and often he would be happy to get water six times in a year. In the new Sardar Sarovar project in Gujarat, the policy is to provide farmers a total of 53 cm depth of water in 5–6 instalments. For an irrigation well with a modest output of 25 m^3/h, this would mean the ability to pump for 212 h/ha. In terms of water availability, an electric pump owner with 3 ha of irrigable land would be at par with a farmer with 3 ha in the Narmada command if he gets 636 h of power in a year and would be considerably better off if the 636 h of power comes when he needs the water most. When Gujarat commits to year-round supply of 8 h/day of farm power, in effect it offers tube well owners water entitlements that are, in theory, 14 times larger than the water entitlements that the Sardar Sarovar project offers to farmers in its command area.[17] Under a metered tariff, this may not matter since tube well owners would use power only when the value generated exceeds the marginal cost of pumping. But under a flat tariff, they would have a strong incentive to use some of these 'excess water entitlements' for low marginal value uses just because it costs them nothing on the margin to pump groundwater.

A rational flat tariff, if well managed, can confer two main benefits. *First*, it may curtail wasteful use of groundwater. If farm power supply outside the main irrigation seasons is restricted to 2–3 h/day, it will encourage farmers to build small on-farm storage tanks for meeting multiple uses of water. Using a progressive flat tariff – by charging higher rates per connected hp as the pump size increases – would provide an additional incentive to purchase and use smaller-capacity pumps to irrigate smaller areas, e.g. in regions where resource depletion is rampant. Above all, a restricted but predictable water supply would encourage water-saving irrigation techniques more effectively than raising the marginal cost of irrigation. *Second*, given the quality of power T&D infrastructure in rural India, restricting the period of time when the farm power system is 'ON' may by itself result in significant reduction in technical and commercial losses of power. The parallel with water supply systems is clear. In a 1999 paper, for example, Briscoe (1999) wrote that throughout the Indian subcontinent, unaccounted-for-water as a proportion of supply is so high 'that losses are "controlled" by having water in the distribution system only a couple of hours a day, and by keeping pressures low. In Madras, for example, if the supply was to increase from current levels (about 2 h of supply a day at 2 m of pressure) to a reasonable level (say, 12 h a day at 10 m of pressure) leaks would account for about 900 million litres per day, which is about three times the current supply in the city!' Much the same logic works in farm power, with the additional caveat that the T&D system for farm connections is far more extensive than the urban water supply system.

Making 'Rational Flat Tariff and Intelligent Power Supply Management' Work

The preconditions for successful rationing

We believe that transforming the present *degenerate* flat power tariff into a rational tariff regime will be easier and more beneficial in the short run in many parts of South Asia than trying to overcome farmer resistance to metering. We also believe that doing so can significantly cut the losses of power utilities from their agricultural operations.

[17]At a rate of 25 m^3/h, a tube well can pump 73,000 m^3 of water if it is operated whenever power supply is on. At the water entitlement of 5300 m^3/ha prescribed in the Narmada project, this amount of water can irrigate 13.77 ha of land.

Four preconditions seem both important and feasible:

- *Separating agricultural and non-agricultural power supply.* The first precondition for successful rationing is to separate agricultural from non-agricultural power supply to rural settlements. The most common way this is done now is to keep 2-phase power on for 24 h so that domestic and (most) non-agricultural uses are not affected and ration the 3-phase power necessary to run irrigation pump sets. This is working but only partially. Farmer response in states like Gujarat is rampant use of phase-splitting capacitors with which they can run pumps even on 2-phase power. There are technological ways to avoid this. For instance, the 11 kV line could be adapted to shut off as soon as the load increases beyond a predetermined level. The costs of such infrastructural modifications could be significant and their feasibility varies. A pragmatic approach is therefore essential. Nevertheless, many SEBs have already begun separating the feeders supplying farm and non-farm rural consumers. For instance, Gujarat has embarked on an ambitious program (*Jyotirgram Yojana*) to lay parallel power supply lines for agricultural users in 16,000 villages at an estimated cost of Rs 9 billion ($196 million). In Andhra Pradesh, the separation of domestic and agricultural feeders is 70% complete (Raghu, 2004). This would ensure that industrial users in the rural areas who need uninterrupted 3-phase power supply and domestic users remain unaffected from rationing of power supplies for agricultural consumers. Another complementary infrastructural investment is to install meters to monitor power use so that power budgeting can be implemented effectively. For this, meters at transformer and feeder levels will be required. Many states have already installed meters at feeder level.
- *Gradual and regular increase in flat power tariff.* Flat tariffs have tended to remain 'sticky'; in most states, they have not been changed for 10–15 years while the cost of generating and distributing power has soared. We surmise that raising the flat tariff at one go to close this gap between revenue and cost per kWh would be too drastic an increase. However, as has been proposed by the Electricity Regulatory Commission in Gujarat, farmers would be able to cope with a regular 10–15% annual increase in the flat tariff far more easily than a 350% increase at one go.
- *Explicit subsidy.* If we are to judge the value of a subsidy to a large mass of people by the scale of popular opposition to curtailing it, there is little doubt that, among the plethora of subsidies that governments in India provide, the power subsidy is one of the most valued. Indeed, a decision by a ruling party to curtail the power subsidy is the biggest weapon that opposition parties use to bring down a government. So it is unlikely that political leaders will want to do away with power subsidies completely no matter what the power industry and donors would like. However, the problem with the power subsidy in the current degenerate flat tariff is its indeterminacy. Chief ministers issue diktats to SEBs about the number of hours of power per day to be supplied to farmers; that done, the actual subsidy availed of by the farmers is in effect left to them to usurp. Instead, governments should tell the power utility the amount of power subsidy it can make available at the start of each year; and the power utility should then decide the amount of farm power the flat tariff and the government subsidy can buy.
- *Off-peak power.* In estimating losses from farm power supply, protagonists of power sector reform systematically overestimate the real opportunity cost

of power supplied to the farmers. For instance, the cost of supplying power to the domestic sector – including generation, transmission and distribution – is often taken as the opportunity cost of power to agriculture, which is clearly wrong since a large part of the high transaction costs of distributing power to the domestic sector is saved in power supply to agriculture under a flat tariff. Moreover, under current conditions, a large part of the power supplied to the farm sector is off-peak load power. Indeed, but for agriculture, the power utilities would be hard-pressed to dispose of this power.[18] It is true that irrigation demands are also seasonal, and that this will become more transparent under an 'intelligent' tariff regime. However, more than half of the power supplied to the farm sector is at night and – despite probable farmer reluctance to accept – this proportion could increase further. The important point here is that, in computing the power the prevailing flat tariff and pre-specified subsidy can buy, the utilities should use a lower opportunity cost of the off-peak supply to the extent it is applicable.

In summary, there is substantial scope for cutting costs and improving service. The existing policy in many states of maintaining power supply to the farm sector at a constant rate during pre-specified hours is irrational and the prime reason for wasteful use of power and water.[19] Figure 9.4 provides a notional indication of the extent of this waste. Ideally, power supply to the farm sector should be so scheduled as to reflect the pumping behaviour of a modal group of farmers in a given region when subject to a metered power tariff at full cost. While this might not meet the needs of *all* farmers, it would be good enough. Of course, it may be difficult to simulate behaviour for farmers subject to a flat tariff. In many states there are a few new tube wells whose owners pay for power on a metered basis but they are charged so low a rate that they behave much like farmers who pay a flat tariff. Another method would be to compare electricity use before and after a flat tariff to gauge the extent of overutilization of

[18]The cost of power supply has three components: Energy Costs, Fixed Generation Costs and T&D Costs. The first two account for about 60–80% of the total cost to serve. The energy cost, which is variable, depends on the length of time of power consumption but fixed generation costs depend on how much a farmer consumes at peak load. T&D costs depend on where the consumer is connected in the system. Since the contribution of agricultural power consumption to peak load is often very little, the opportunity cost of power supply to agriculture is lower than the overall average cost of supply. Moreover, agricultural consumption, most of it off-peak helps smoothen the load curve for the whole system and saves the back-up cost which is high for coal-based plants and insignificant for hydropower plants.

[19]In Tamilnadu, where farm power supply is free, 14 h of 3-phase power – 6 h during day and 8 h during night – is supplied throughout the year. In Andhra Pradesh, 9 h of 3-phase power supply is guaranteed, 6 h during the day and 3 h during the night (Palanisami and Kumar, 2002); this was recently reduced to 7 h when the new government announced free power. This implies that, in theory, a tube well in Tamilnadu can run for over 5000 h/year and in Andhra Pradesh for 3200 h. If the real cost of power is taken to be Rs 2.5/kWh (USc5.4/kWh), depending on how conscientious he is, a Tamilnadu farmer operating a 10 hp tube well can avail of a power subsidy ranging from Rs 0–93,750/($0–2038)/year; and an Andhra Pradesh farmer, Rs 0–60,000)/year ($0–1304/year). The stories one hears of farmers installing automatic switches that turn on the tube wells whenever power supply starts suggest that a large proportion of farmers are overusing in using power and water. Palanisami and Kumar (2002) mention that many borewell owners lift water during the night to fill an open well using an automatic switch and then lift water during the day from the open well to irrigate their fields! True, they would not indulge in such waste if they had to pay a metered rate at Rs 2.5 (USc5.4)/kWh; but they would also not do this if they got only 3–4 h of good quality power at convenient hours on a pre-announced schedule.

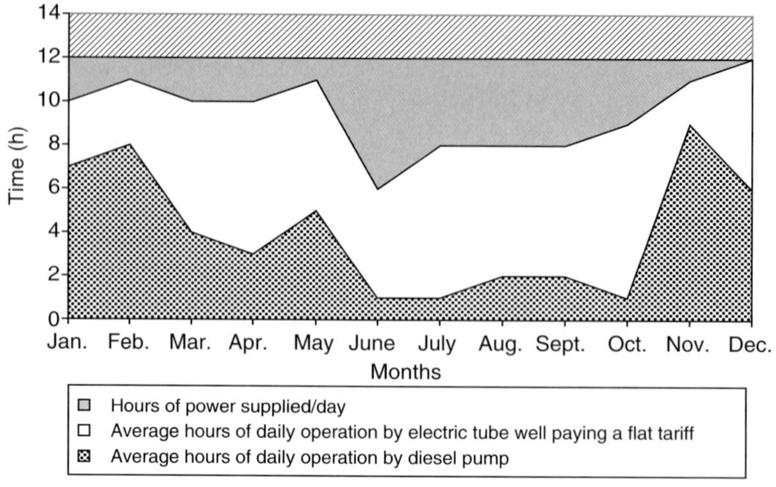

Fig. 9.4. Minimizing waste of power and water through supply management.
Note: This is a schematic diagram. The numbers are indicative and not based on actual field data.

power and water attributable to a flat tariff.[20]

However, it is the pumping behaviour of diesel pump owners, subject to the full marginal cost of energy, that might provide the best indicator. Several studies have shown that diesel tube wells are operated for half or less the time of electric tube wells that pay a flat tariff (Mukherji and Shah, 2002).[21] Batra and Singh (2003) interviewed 188 farmers in Punjab, Haryana and central Uttar Pradesh to explore if pumping behaviour of diesel and electric owners of water extraction mechanisms (WEM) differed significantly. They found no significant differences in Punjab and Haryana[22] but their results for Central UP suggested that diesel pumps are used when irrigation is needed and electric pumps when electricity is available. Very likely, a good deal of the excess water pumped by farmers owning both electric and diesel pumps is wasted in the sense that its marginal value product falls short of the scarcity value of water and power together. Figures 9.5 and 9.6 present the central premise: the excess of pumping by electric over diesel tube wells is indicative of the waste of water and power encouraged by the zero marginal cost of pumping under the present degenerate flat tariff regime. Mukherji and Shah (2002) present results from a survey of 2234 tube well irrigators across India and Bangladesh in late 2002.

[20]An extreme case is Tamilnadu where electricity consumption per tube well shot up from 2583 kWh/year under metered tariff in the early 1980s to 4546 kWh in 1997–1998. However, this jump would represent three components: (i) increased consumption due to degenerate flat tariff; (ii) increased consumption because of the increased average lift caused by resource depletion; and (iii) T&D losses in other segments that are wrongly assigned to agriculture. Palanisami (2001) estimated that 32% of the increased power use was explained by additional pumping and 68% by increased lift. However, he made no effort to estimate the (iii), which we suspect is quite large.

[21]We recognize that comparing hours of operation is not the same as comparing the quantity of water extracted. But, in understanding the economic behaviour of tube well owners, comparing hours is more meaningful than comparing water produced. In any case, ceteris paribus for the same hours of pumping, an electric pump produces more water due to its higher efficiency.

[22]Punjab and Haryana have much more productive agriculture compared to other parts of India with the cost of irrigation being just 8–10% of the gross value of produce. This might explain why the pumping pattern is inelastic to the energy cost. However, this is just a hypothesis and needs to be further confirmed.

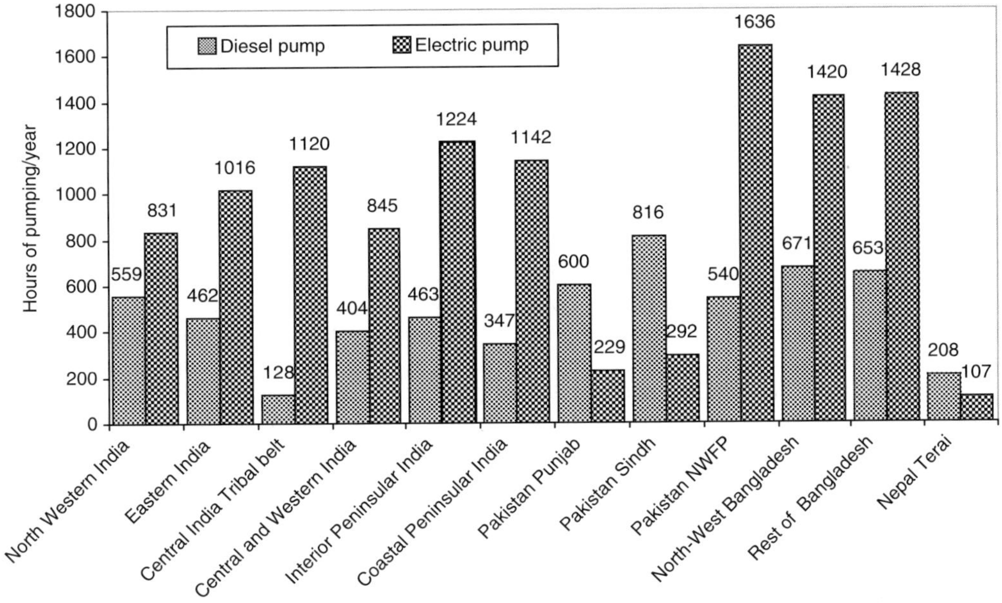

Fig. 9.5. Flat electricity tariff induce farmers to pump more.

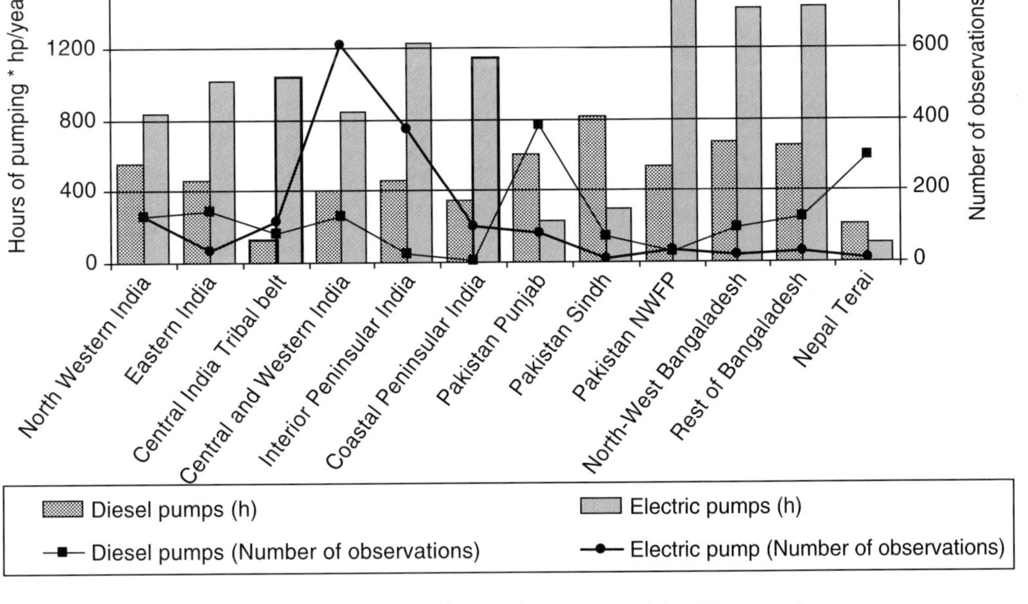

Fig. 9.6. Impact of flat tariff on average annual hours of pumping weighted by pump horsepower.

Figure 9.5 shows that electric tube well owners subject to a flat tariff invariably operate their pumps for much longer time compared to diesel pump owners who face a steep marginal energy cost. Since it can be argued that diesel pumps, on average, have a larger capacity than electric pumps we also compare pumping hours weighted by hp ratings. Figure 9.6 shows that hp-hours pumped by flat-tariff paying electric pumps are also significantly higher than those pumped by diesel pumps everywhere. The survey suggests that the difference in annual pumpage is some 40–150%; some of this excess pumping no doubt results in additional output but much of it very likely does not and, to this extent, is a social waste that needs to be eliminated.[23]

If, based on an analysis of the level and pattern of pumping by diesel pump owners, a power utility can shave off potential excess pumping by fine-tuning power supply schedule around the year, a flat tariff can become both viable and help eliminate 'waste.' The average number of hours for which diesel pumps operate is 500–600/year. At 600 h of annual operation, an electric tube well would use 450 kWh of power/hp; if all the power used is off-peak load commanding, say, 25% discount on a generation cost of Rs 2.5/kWh (US$0.05/kWh), then farm power supply by the power utility would break-even at a flat tariff at Rs 844/hp/year ($18.3/hp/year) as against Rs 500/hp/year ($10.9/hp/year) in force in Gujarat since 1989. Gujarat is committed to raising the flat tariff eventually to Rs 2100/hp/year ($45.65/hp/year) at the instance of the Gujarat Electricity Regulatory Commission. If it does so, farmers might well topple the government. A more viable and practical course would be to raise the flat tariff in steps to, say, Rs 900 ($19.6) at first and then to Rs 1200 ($26.09), and to restrict annual supply of farm power to 1000–1200 h compared to 3000–3500 h/year as at present. A 5 hp pump lifting 25 m^3 of water/h over a head of 15 m can produce 30,000 m^3 of water/year in 1200 h of tube well operation, sufficient to meet the needs of most small farmers in the region.

[23]It is probable that the real savings in power are proportionately greater than the real savings in water since a part of the excess water pumped returns to the aquifer. This can be a significant factor, especially where irrigation depends on shallow groundwater circulation.

Alternative Approaches to Rationing

The strongest evidence in support of our argument for *intelligent* rationing of farm power is that, for more than a decade, most SEBs in India have *already* rationed power to farmers in some way. For instance, Andhra Pradesh, where the new government announced free power, also announced that farm power supply would henceforth be restricted to 7 h daily. Nobody – farmers included – considers 24 h uninterrupted power supply to agriculture to be feasible or defensible under the flat tariff regime in force. Negotiations between farmer groups and governments almost everywhere in India are carried out in terms of the minimum hours of daily power supply the government can guarantee; and this can be termed the current default.

The current default is perhaps the least intelligent way of rationing power supply to agriculture because it fails to achieve a good 'fit' between the schedule of power supply and farmers' desired irrigation schedules. It leaves farmers frustrated on days when their crops need to be watered most and leads to wasteful use of power and groundwater when the need is least. From where the SEBs' present power rationing practices stand today, they only have to gain by achieving a better fit between power supply schedules and farmers' irrigation schedules. Farmers keep demanding that the 'constant hours/day' be raised because the default system does not provide enough power when they need it most. There are a number of ways of rationing that would raise farmer satisfaction and control power subsidies so that (i) it reduces farmers' uncertainty about the timing of power; (ii) it achieves a better fit between power supply schedules and irrigation schedules; or (iii) both. We suggest below a few illustrative alternative approaches that need to be considered and tried out with a view to increasing farmer acceptance and containing the subsidies provided as well as the wastage of power and water (Fig. 9.7).

- *Agronomic scheduling*. Ideally, SEBs should aim to achieve the 'best fit' by matching power supply schedules with irrigation needs of farmers to the extent this is feasible within the context of their overall operations. Under this approach, the power utility would constantly study: (i) irrigation behaviour of farmers in regions and subregions by monitoring cropping patterns, cropping cycles and rainfall events; (ii) matches power supply schedules to meet irrigation needs; and (iii) minimizes supply in off-peak irrigation periods. The advantages of such a system are that farmers would be happier, the total power supply to agriculture can be reduced, power and water waste would be minimized, and the level of subsidy availed is within SEB control. The key disadvantage of this approach is that it is highly management-intensive and, therefore, difficult to operationalize.
- *Demand-based scheduling*. In this approach, feeder-level farmer committees or other representational bodies of farmers assume the responsibility of ascertaining members' requirements of power, and provide a power supply schedule to the utility for a fixed number of allowable hours for each season. This is a modified version of agronomic scheduling in which the power utility's research and monitoring task is assumed by feeder committees. This may make it easier to generate demand schedules but more difficult to serve it. Moreover, the organizational challenge this approach poses is also formidable.
- *Canal-based scheduling*. Tube well irrigators outside canal commands justify demands for power subsidies by comparing their lot with canal irrigators who get cheap canal irrigation without

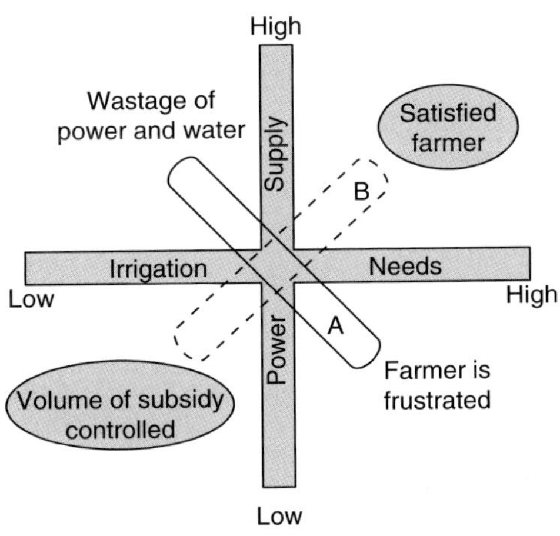

A: Mismatch between power supply and irrigation needs; existing system in which the farmer is frustrated

B: A win–win scenario: power supply is good and reliable, when the irrigation needs are high (satisfied farmer), and low power supply when irrigation needs are low (volume of subsidy is controlled)

Fig. 9.7. Improving farmer satisfaction and controlling electricity subsidies through intelligent management of farm power supply.

any capital investment of their own. However, under the present degenerate flat tariff, tube well irrigators often have the best of both the worlds. At 10h of power supply/day, an Andhra Pradesh tube well irrigator could in theory use 300–500 m^3 of water every day of the year. In contrast, under some of the best canal commands, farmers get irrigation for 10–15 times in an entire year. Under this approach, power rationing aims to remove the inequity between tube well and canal irrigators by scheduling power supply to mimic the irrigation schedule of a bench-marked public irrigation system. And although this will impose constraints on tube well irrigators, it can drastically reduce power subsidies from current levels. For that very same reason, it will face stiff resistance from tube well irrigating farmers.

- *Zonal roster.* An approach to rationing that is simpler to administer is to divide the state into say seven zones, each zone assigned a fixed day of the week when it gets 20h of uninterrupted, quality power throughout the year; on the rest of the days, it gets 2h. This is somewhat like a weekly turn in the *warabandi* system in canal irrigation systems in Indian and Pakistan Punjab. The advantages of this approach are that: (i) it is easy to administer; (ii) the agricultural load for the state as a whole remains constant, so it becomes easy to manage for SEB; also (iii) level of subsidies is controlled; and (iv) power supply to each zone is predictable so that farmers can plan their irrigation easily. Disadvantages are that: (i) farmers in deep water table areas or areas with poor aquifers (as Saurashtra in Gujarat) would be unhappy since they must pump for longer to obtain the same supply; and (ii) zonal rostering would not mimic seasonal fluctuations in irrigation demand as well as in agronomic rationing.
- *Adjusted zonal roaster.* The zonal roaster can help farmers plan their cropping pattern and irrigation schedules by reducing uncertainty in power supply but it does not do much to improve the 'fit' between irrigation need and power supply across seasons. In most of India, for instance, following the same zonal roaster in different seasons makes little sense. Modifying the zonal roaster system so that power supply offered is higher in winter and summer than in the monsoon season would improve the seasonal fit as well as reduce uncertainty.

Any approach must necessarily be consistent with the characteristics of the power operations in the particular subregion concerned. Systems analysis of power operations will thus be a critical step in evaluating feasible alternatives. The issues concerned go beyond the scope of this chapter but, clearly, choices will need to be flexible in the light of ongoing experience.

It will not always be possible to meet the precise needs of all farmers and a period of adjustment and experimentation may be necessary before the final arrangements are implemented. Power utilities in South Asia have never had the necessary understanding of irrigation requirements that this implies, which is a major reason for the constant hiatus between them and the agriculture sector. One reason is that SEBs employ only engineers (Rao, 2002). This important aspect has been overlooked in the power sector reforms under way in many Indian states, which focus on the institutional architecture of unbundling power operations. Distributing power to agriculture in South Asia is a very different activity to supplying urban and industrial demands and there is a real danger that private distribution companies will exclude agriculture as being 'too difficult and costly to serve,' as Orissa's experience is already showing.[24] Perhaps, the most appropriate course would be to promote a separate distribution com-

[24]The Orissa Electricity Regulatory Commission has already opened the gate for the power utility to ask agriculture to fend for itself, when it decided that 'any expansion of the grid which is not commercially viable, would not be taken into account in calculating the capital base of the company. In future, unless government gives grants for rural electrification, the projects will not be taken up through tariff route' (Panda 2002).

pany for serving the agriculture sector with specialized competence and skill base; and predetermined government subsidies to the farming sector should be directed to the agricultural distribution companies.[25]

Supporting intelligent management

Which of the above approaches should be adopted is thus a pragmatic decision in the light of local conditions. Farmers will no doubt resist rationing of power supply and any reforms will need to be introduced sensitively in association with farmer and political representatives and flexibly in response to ongoing experience and results. Moreover, farmer resistance can be reduced if reforms are accompanied by such measures as:

- *Enhancing predictability and certainty.* More than the total quantum of power delivered, in our assessment, power suppliers can help the farmers by announcing an annual schedule of power supply adapted broadly to match the demand pattern of the majority of farmers. Once announced, the utility must then stick to the schedule so that farmers can be certain about power availability.
- *Improving supply quality.* Whenever power is supplied, it should be at full voltage and frequency, minimizing the damage to motors and downtime of transformers due to voltage fluctuations.
- *Better matching of supply with peak periods of moisture stress.* Most canal irrigators in South Asia manage with only 3–4 canal water releases in a season. There are probably 2 weeks during the monsoonal season in a normal year and perhaps 5–6 weeks during winter when the average farmer experiences great nervousness about moisture stress to his crops. If the power utility can take care of these periods, 80–90% of farmers' power and water needs would be met. This might not, for instance, help sugarcane growers in Maharashtra, Gujarat and Tamilnadu; but then they are the large part of the power utilities' problems.
- *Better upkeep of farm power supply infrastructure.* Intelligent power supply management to agriculture will inevitably be a tricky business. If rationing is done by an arbitrary increase in power cuts and the neglect of rural power infrastructure, it might result in disastrous consequences as it did in East India. As described above, the saving grace was that in these groundwater-abundant regions, small diesel pumps, though dirtier and costlier to operate, kept the economy going. Where groundwater is lifted from 200 to 300 m, such de-electrification could destroy the agricultural economy.

Conclusions

We have argued in this report that neither a switch to a metered tariff regime at this juncture nor the raising of the flat tariff fourfold as, for instance proposed in Gujarat, is likely to be successful in South Asia and would in all probability backfire. Metering is highly unlikely to improve the fortunes of power utilities that have found no smarter ways than in the 1970s of dealing with the high transaction costs of metered farm power supply, which led to a flat tariff regime in the first place. However, if agriculturally dynamic states like Punjab and Haryana – where non-farm uses of 3-phase power supply are extensive and growing in the villages and where productive farmers can afford higher costs of better quality power supply – want to experiment with metered power supply, they would be well advised to create microentrepreneurs to retail power, to meter individual power consumption and collect revenue as in China rather than experiment with electricity cooperatives. It should, however, be borne in mind

[25]T.L. Sankar argues for the need to set up separate supply companies for farmers and rural poor that will access cheap power from hydroelectric and depreciated thermal plants and be subsidized as necessary directly by governments (Rao, 2002, p. 3435).

that the largest and most difficult problem lies in containing user efforts to frustrate the metered tariff regime, by pilfering power, illegal connections, tampering with meters and so on. While abuse remains possible in respect of a flat rate tariff, the opportunities are quite fewer. The ongoing experiments on privatization of electricity retailing in Orissa may produce useful lessons on whether metering-cum-billing agents can drastically and sustainably reduce the cost of metered power supply in a situation where tube well owners account for a significant proportion of electricity use.

Contrary to popular understanding, a rational flat tariff can be an elegant and efficient regime, which requires a complex set of skills and an understanding of agriculture and irrigation in different regions. A rational flat tariff and intelligent power supply management in fact could achieve much that a metered tariff regime is designed to achieve at much lower real cost and a much greater likelihood of success. The flat tariff will undoubtedly have to be raised, but the schema we have set out could cut power utility losses from farm power supply substantially. Total hours of power supplied to farmers during a year will have to be reduced but the aim would be to provide farmers with good quality power at times of moisture stress when they need irrigation most. Power supply to agriculture will need to be metered at feeder and transformer levels as a basis for power scheduling and 'intelligent' management but the transaction costs of a metered charge at farm level would be saved. If concurrently the utilities begin treating farmers as customers, the adversarial relationship between them could in time turn benign. Moreover, a rational flat tariff would tend to maintain water markets as buyers' markets albeit less than under the present degenerate flat tariffs (for detailed arguments see Shah, 1993). A rational flat tariff – under which power rationing is more defensible than under a metered tariff – would allow an effective check on total use of power and water. Restricting the total hours of operation supply would curtail technical and commercial losses by SEBs and reduce power subsidies while a rational flat tariff has the potential for significantly curtailing groundwater depletion by minimizing wasteful resource use. In most instances, proportionately more power is likely to be saved than water due to the prevalence of return flows, but which of these two benefits is more valuable will depend critically on the context. Together, however, they have the potential for making a very substantial contribution to improving economic performance and strengthening resource sustainability.

References

Batra, S. and Singh, A. (2003) Evolving proactive power supply regime for agricultural power supply, Anand, India: IWMI-Tata Water Policy Program (Internal report available on request).

Berk, J. and Berk, S. (1995) *Total Quality Management*. Excel Books, New Delhi.

Briscoe, J. (1999) The financing of hydropower, irrigation and water supply infrastructure in developing countries. *Water Resources Development* 15(4), 459–491.

CMIE (Centre for Monitoring Indian Economy) (2003) *Economic Intelligence Service*. Centre for Monitoring Indian Economy, Mumbai, India, May.

Debroy, A. and Shah, T. (2003) Socio-ecology of groundwater irrigation in India. In: Llamas, R. and Custodio, E. (eds) *Groundwater Intensive Use: Challenges and Opportunities*. The Swets and Zetlinger Publishing, The Netherlands.

Dhawan, B.D. (1999) *Studies in Indian Irrigation*. Commonwealth Publishers, New Delhi.

Godbole, M. (2002) Electricity regulatory commissions: the jury is still out. *Economic and Political Weekly* XXXVII (23), 2195–2201.

GOI (Government of India), Ministry of Water Resources (1999) *Integrated Water Resources Development: A Plan for Action. Report of the National Commission for Integrated Water Resources Development Plan*, Vol. I. Ministry of Water Resources, New Delhi, September.

GOI (2001) *Report on Census of Minor Irrigation Schemes, 1993–1994*. Ministry of Water Resources, Minor Irrigation Division, Government of India, New Delhi.

Gulati, A.S. (2002) Energy implications of groundwater irrigation in Punjab. Paper for the IWMI-ICAR-Colombo Plan sponsored Policy Dialogue on 'Forward-Thinking Policies for Groundwater Management: Energy, Water Resources, and Economic Approaches' organized at India International Centre, New Delhi during 2–6 September 2002.

Gulati, A. and Narayanan, S. (2003) *Subsidy Syndrome in Indian Agriculture*. Oxford University Press, New Delhi.

Hekmat, A. (2002) Overexploitation of groundwater in Iran: Need for an integrated water policy. Paper for the IWMI-ICAR-Colombo Plan sponsored Policy Dialogue on 'Forward-Thinking Policies for Groundwater Management: Energy, Water Resources, and Economic Approaches' organized at India International Centre, New Delhi during 2–6 September 2002.

India Today (2002) Running for cover: Demand for a tribal chief minister and a proposed hike in power tariffs pose a serious challenge to Digvijay Singh's leadership, XXVII(47), 19–25 November 2002, p. 32.

Kishore, A. and Sharma, A. (2002) Use of electricity in agriculture: An overview. Paper for the IWMI-ICAR-Colombo Plan-sponsored Policy Dialogue on 'Forward-Thinking Policies for Groundwater Management: Energy, Water Resources, and Economic Approaches' organized at India International Centre, New Delhi during 2–6 September 2002.

Lim, E.R. (2001) Presentation made in Conference on 'Power Distribution Reforms', 12–13 October, New Delhi.

Llamas, R., Back, W. and Margat, J. (1992) Groundwater use: Equilibrium between social benefits and potential environmental costs. *Applied Hydrogeology* 1(2), 3–14.

Mukherji, A. and Shah, T. (2002) Groundwater socio-ecology of South Asia: An overview of issues and evidence. Paper presented in Conference on 'Intensive Use of Groundwater: Opportunities and Implications,' Valencia, Spain. 25–28 November.

North, D. (1997) *The Contribution of the New Institutional Economics to an Understanding of the Transition Problem*. United Nations University, Wider Annual Lectures 1.

NSSO (National Sample Survey Organization). (1999) *Cultivation Practices in India*. Government of India. National Sample Survey Organization, New Delhi. 54th round, January–June.

Palanisami, K. (2001) Techno-economic feasibility of groundwater exploitation in Tamilnadu. Presented in ICAR-IWMI Policy Dialogue on Groundwater Management, 6–7 November, 2001 at CSSRI, Karnal, India.

Palanisami, K. and Kumar, D.S. (2002) Power pricing, groundwater extraction, use and management: Comparison of Andhra Pradesh and Tamilnadu states. Paper for the Dialogue on 'Forward-Thinking Policies for Groundwater Management: Energy, Water Resources, and Economic Approaches', India International Centre, New Delhi, 2–6 September, 2002.

Panda, H. (2002) Power sector reform in Orissa and its impact on lift irrigation; an assessment and lessons. Paper for the IWMI-ICAR-Colombo Plan-sponsored Policy Dialogue on 'Forward-Thinking Policies for Groundwater Management: Energy, Water Resources, and Economic Approaches' organized at India International Centre, New Delhi during 2–6 September 2002.

Perry, C.J. (1996) *Alternative Approaches to Cost Sharing for Water Service to Agriculture in Egypt*. Research Report 2. International Irrigation Management Institute, Colombo.

Perry, C.J. (2001) *Charging for Irrigation Water: The Issues and Options with a Case Study from Iran*. Research Report 52. International Water Management Institute, Colombo.

Raghu, K. (2004) People's Monitoring Group, Andhra Pradesh, Power sector reforms in Andhra Pradesh. Presentation made in National Workshop on Electricity Act 2003 for NGOs, Pune, India, 26–28 July.

Rao, S.L. (2002) The political economy of power. *Economic and Political Weekly* xxxvii (17), 3433–3445. 17 August.

Rao, D.N. and Govindarajan, S. (2003) *Community Intervention in Rural Power Distribution*. Water Policy Highlight No 14. IWMI-Tata Water Policy Program, Anand, India.

Repetto, R. (1994) *The 'Second India' Revisited: Population, Poverty and Environmental Stress over Two Decades*. World Resources Institute, Washington, DC.

Samra, J.S. (2002) Impact of groundwater management and energy policies on food security of India. Paper for the IWMI-ICAR-Colombo Plan-sponsored Policy Dialogue on 'Forward-Thinking Policies for Groundwater Management: Energy, Water Resources, and Economic Approaches' organized at India International Centre, New Delhi during 2–6 September 2002.

Scott, C., Shah, T. and Buechler, S. (2002) Energy pricing and supply for groundwater demand management: Lessons from Mexico. Paper for the IWMI-ICAR-Colombo Plan-sponsored Policy Dialogue on 'Forward-Thinking Policies for Groundwater Management: Energy, Water Resources, and Economic Approaches' organized at India International Centre, New Delhi during 2–6 September 2002.

Shah, T. (1993) *Groundwater Markets and Irrigation Development. Political Economy and Practical Policy.* Oxford University Press, New Delhi.

Shah, T. (2001) *Wells and Welfare in the Ganga Basin: Public Policy and Private Initiative in Eastern Uttar Pradesh, India.* Research Report 54. International Water Management Institute, Colombo.

Shah, T. (2003) Governing the groundwater economy: Comparative analysis of nationl institutions and policies in South Asia, China and Mexico. *Water Perspectives* 1(1). March 2003.

Shah, T., Koppen, B.V., Merrey, D., Lange, M.D. and Samad, M. (2002) *Institutional Alternatives in African Smallholder Irrigation: Lessons from International Experience with Irrigation Management Transfer.* Research Report 60. International Water Management Institute, Colombo.

Shah, T., DebRoy, A., Qureshi, A.S. and Wang, J. (2003) Sustaining Asia's groundwater boom: An overview of issues and evidence. *Natural Resources Forum* 27, 130–141.

Shah, T., Giordano, M. and Wang, J. (2004a) Water institutions in a dynamic economy: What is China doing differently from India. *Economic and Political Weekly* xxxix (31), 3452–3461, 31 July–6 August.

Shah, T., Singh, O.P. and Mukherji, A. (2004b) Groundwater irrigation and South Asian agriculture: Empirical analyses from a large-scale survey of India, Pakistan, Nepal Terai and Bangladesh. Draft discussion paper presented at 3rd IWMI-Tata Annual Partners' Meet on Water and Welfare: Critical Issues in India's Water Future, Institute of Rural Management, Anand 17–19 February, 2004.

Sharma, S.K. and Mehta, M. (2002) *Groundwater Development Scenario: Management Issues and Options in India.* Paper for the IWMI-ICAR-Colombo Plan-sponsored Policy Dialogue on 'Forward-Thinking Policies for Groundwater Management: Energy, Water Resources, and Economic Approaches' organized at India International Centre, New Delhi during 2–6 September 2002.

World Bank and GOI (1998) *India-Water Resources Management Sector Review: Groundwater Regulation and Management Report.* World Bank, Washington, DC; Government of India, New Delhi.

10 Wells and Canals in Jordan: Can Pricing Policies Regulate Irrigation Water Use?

J.-P. Venot, F. Molle and Y. Hassan

Introduction

The Hashemite Kingdom of Jordan is one of the countries with the scarcest water resources in the world. Due to both physical water scarcity and a high demographic growth during most of the second half of the 20th century it has been estimated that the per capita endowment of renewable blue water (i.e. surface runoff and groundwater recharge) is now only 163 m^3/year, while the average domestic consumption is 94 l/capita/day nationwide (THKJ, 2004).

Most agricultural activities are concentrated in the Lower Jordan River Basin (LJRB) (Fig.10.2), a region of prime importance for the country: it includes 83% of the total population, most of the main industries, and 80% of irrigated agriculture of the country. It is endowed with 80% of the country's water resources and withdrawals in the basin total 75% of those at the national level (Courcier *et al.*, 2005). The bulk of irrigated agriculture is located in two contrasting environments: the Jordan valley, where a public scheme supplies approximately 23,000 ha; and the highlands, which include two groundwater basins of major importance, the Amman-Zarqa and the Yarmouk basins[1] (Fig.10.2), where most of the private tube well-based irrigation that has developed over 14,000 ha in the last 30 years is located.

The main water allocation problems in the LJRB are schematized in Fig. 10.1. Amman receives water from the Jordan valley, local aquifers, and from southern outer basins. To meet its growing water demand, there is a need to: (i) improve inflow from the Yarmouk river (dam); (ii) transfer more water from the valley to Amman (and hence reduce agricultural use, although treated wastewater [TWW] is sent back to the valley); (iii) reduce abstraction from aquifers by highland agriculture in order to preserve water quality, avoid overdraft and reallocate water to cities; and (iv) rely on (costly) imports from southern basins as little as possible.

In the early 1990s, Jordan's officials took the measure of the coming water crisis and policies underwent a paradigm shift from supply augmentation towards demand management. The World Bank and other development agencies were influential in calling for an agenda that

[1]The Amman-Zarqa and Yarmouk groundwater basins are roughly coterminous with the river basins bearing the same names.

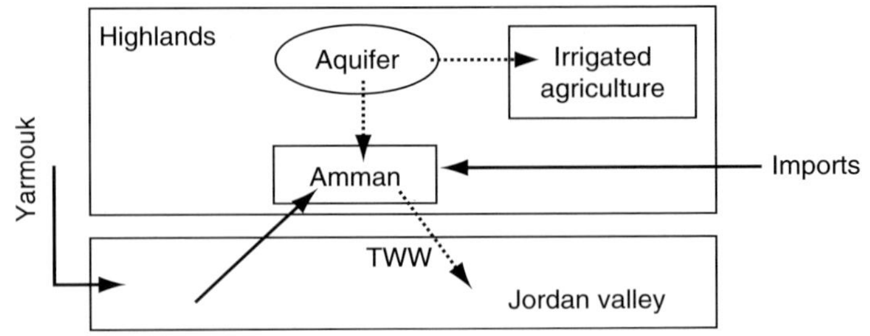

Fig. 10.1. Schematic representation of main water flows in the LJRB.

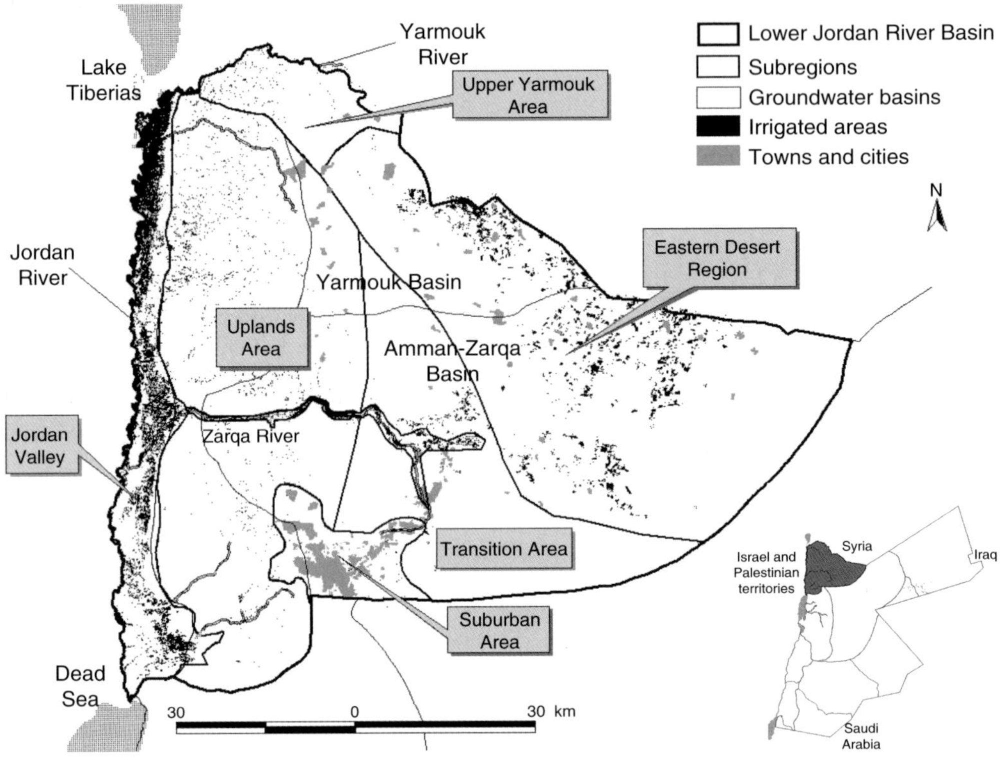

Fig. 10.2. Agricultural zoning of the LJRB (Adapted from Venot, 2004; unpublished land use classification from the Ministry of Water and Irrigation [MWI] and the Deutsche Gesellshaft für Technische Zusammenarbeit [GTZ].)

would include demand management measures and economic instruments to encourage efficient water use, transfer water to non-agricultural higher-value uses and reduce groundwater overdraft (Pitman, 2004). Pricing of irrigation water was chosen as an instrument to reduce demand for water (World Bank, 2003).

In the highlands, development of groundwater resources had been 'exacerbated by relaxed controls on drilling operations, and the near absence of controls on licensed abstraction rates' (THKJ and MWI, 1997b, 1998a). High rates of abstraction (up to 215% of the mean annual recharge in the Amman-Zarqa basin) prompted the government to design a new water strategy in 1997. Pricing policies were deemed to assist in controlling groundwater abstraction (with the ambitious task of taking abstraction rates 'close to the annual recharge by the year 2005'[2]) and to elicit shifts towards higher-value crops.

In the Jordan valley, more expensive water was expected to bring about efficiency improvements and a switch to less water-intensive crops, thus releasing water for Amman (World Bank, 2003). It would also assist in recovering state expenditures in public irrigation schemes: 'The water price shall at least cover the cost of operation and maintenance (O&M) and, subject to some other economic constraints, it should also recover part of the capital cost of the irrigation water projects. The ultimate objective shall be full cost recovery subject to economic, social and political constraints' (THKJ and MWI, 1997a, 1998b, 2004c; JRVIP, 2001a).

These reforms were to be embedded in the 1994 agriculture sector structural adjustment loan (ASAL) jointly funded by the World Bank and the German KfW and designed with the prime objective 'to support a transition to an optimal use of water and land resources' (ASAL, 1994; World Bank, 2003) and to tackle key problems of the sector: 'the lack of a national water policy, competing sector institutions, and insufficient attention to demand management'. Implementation of these policies proved to be problematic, since part of the government denounced the difficulties that increased agricultural water tariffs would cause and argued that administrative allocation together with efficiency improvement would be more efficient in saving water. Two hot debates arose.[3]

Regarding the highlands, the Groundwater Control By-law No. 85, passed in 2002 and further amended in 2004, was designed to regulate groundwater abstraction through the establishment of a block tariff system, with charging of water use over a threshold of 150,000 m^3/year/well. Regarding the valley, a block tariff system associated with crop-based quotas had been in place for some time and the debate revolved around possible increases in water charges. This chapter examines the rationale, the potential and the current impact of these water pricing policies in these two environments, and attempts to answer the following questions:

- What will be the likely impacts of the application of the by-law in the highlands?
- What will be the financial impact of increasing water prices in the valley, so as to cover O&M or capital costs?
- What is the likelihood of success of such policies in terms of water saving and raising economic efficiency, and what alternatives are available to meet these objectives?

In both the highlands and the valley, a typology of farming systems was established with the intent to discriminate the impact of policies on different types of farms and to assess what could be farmers' adjustments and responses in each case. Regional data aggregation then provided a wider picture of the water savings to be achieved, and of the financial impact on both farmers and the state. These results are developed in the final section, which discusses the disjuncture between expected and actual or estimated outcomes, points to commonalities and discrepancies between the two regions, and identifies measures which can improve the regulation of the water sector in Jordan.

Farming Systems in the Two Study Areas

Context

With the outflow of the Jordan river from Lake Tiberias virtually blocked by Israel, the lower Jordan river chiefly receives the water from its main tributary, the Yarmouk river. Several temporary streams of lesser importance named 'side-wadis', as well as the larger Zarqa

[2]This target was revised in 2004 and shifted to 2020 (Pitman, 2004).

[3]Opposition to higher water tariffs led to the occupation of the Parliament floor and further intervention by His Majesty the King (Pitman, 2004).

river, also incise the two mountainous banks and feed the valley: the valley is a 115 km long fertile plain located 300 m below sea level and where irrigation schemes have been built.

The highlands are composed of a mountain range running alongside the Jordan valley and of a desert plateau extending easterly to Syria and Iraq. While rain-fed cereals are grown near the mountains, precipitations become scarcer more to the east where only nomadic Bedouin livestock farming can be found, with a few localized plots of groundwater-based irrigated agriculture. The eastern desert region overlaps the Amman-Zarqa and the Yarmouk groundwater basins (cf. Fig. 10.2).

Irrigation is traditional in Jordan along the side-wadi valleys and on their alluvial fans spread in the Jordan valley itself, or wherever springs are available. Large-scale public irrigation dates back to the establishment of the Jordan Valley Authority (JVA) and to the construction, between 1958 and 1966, of the main 69 km long concrete canal – the King Abdullah Canal (KAC) – which parallels the river on its eastern bank. In 1962, a land reform led to the formation of thousands of small intensive farms (3.5 ha on average), and the settlement of numerous families, including Palestinian refugees (Khouri, 1981; van Aken, 2004). During the same period, several governmental projects aiming at settling Bedouins were implemented in the highlands and later gave way to a modern market-oriented agriculture developed by small to medium entrepreneurial farmers supplying growing cities and exporting their surplus around the Middle East (Elmusa, 1994; Nachbaur, 2004; Venot, 2004).

The heyday of irrigated agriculture was observed in the 1980s and early 1990s. In the Jordan valley, irrigation facilities were expanded and improved by the government, and modern irrigation and cropping techniques (greenhouses, drip irrigation, plastic mulch, fertilizer, new varieties, etc.), together with cheap labour from Egypt, became widely available. In the highlands, energy costs decreased and well-drilling techniques improved while land was cheap, fertile and not prone to diseases. During this period, agricultural revenues increased tenfold for vegetables and more than doubled for fruits: irrigated agriculture in Jordan enjoyed a boom in production and economic profitability that was described by Elmusa (1994) as the 'Super Green Revolution'.

With the growing competition from surrounding countries in the 1990s (Turkey, Lebanon and Syria) and the loss of the Gulf export market, the profitability of Jordanian agriculture decreased, strongly affecting farmers' revenue (Fitch, 2001; Jabarin, 2001) and taking the sector's contribution to the country's GDP down to 3.6%. Freshwater is increasingly transferred from irrigated agriculture (in the valley) to urban uses (in the highlands), affecting the agriculture sector which receives ever-decreasing quantities of water and becomes more vulnerable to droughts (Courcier *et al.*, 2005). In exchange, agriculture in the southern part of the valley is increasingly supplied with treated wastewater (McCornick *et al.*, 2001, 2002; THKJ *et al.*, 2002; JICA, 2004; THKJ and MWI, 2004b).

This chapter focuses on two main regions of the LJRB: (i) the eastern desert area (the only region of the LJRB highlands to be concerned by the by-law); and (ii) the northern and middle directorates of the Jordan valley (where JVA management rules apply). The total irrigated area in the eastern desert region totals 11,835 ha; 50% of this area is planted with olive trees, 34% with stone fruit trees (peach and nectarine trees essentially) and 16% with vegetables. In the northern and middle directorates of the Jordan valley, the irrigated area totals 19,345 ha, with 43% of vegetables, 42% of citrus, and the remainder of banana and cereals.

Farming system characterization

Farming systems were analysed in order to identify the different types of farms found in the valley and in the highlands. Understanding the socio-economic processes occurring at this microscale will allow us to better foresee the adjustments and the strategies developed by farmers in a changing context and the impact of water pricing policies on farmers. By complementing this microlevel analysis with regional data (statistic data, satellite image analysis) we can assess the possible evolution of regional irrigated agriculture as a whole.

Extensive farm surveys were carried out in the highlands by USAID/ARD in 2000/2001 (Fitch, 2001), but economic analyses were based on cropping patterns. This makes it difficult to discriminate responses by type of farmer. In order to sketch out farming systems that combine typical cropping patterns with socio-economic characterization (profile of the farmer, land tenure, labour use, costs, etc.), 30 in-depth farm surveys were carried out during the spring of 2003. Farming systems were then modelled in economic terms based on crop budgets whose consistency with USAID/ARD data was checked. Likewise, the main farming systems in the Jordan valley were identified and their economics modelled based on 50 farm surveys carried out also during the spring of 2003, and on other studies (ARD and USAID, 2001b; JRVIP, 2001c).

The highland surveys led to the identification of three main categories of farming systems (Table 10.1; a detailed description can be found in Venot *et al.*, 2007). They include settled Bedouins who have taken up vegetable (and sometimes fruit tree) cultivation, and urban-based entrepreneurs involved in high-value fruit production and closely managing their farm, although they often reside in Amman. Both Bedouins and entrepreneurs sometimes also maintain olive orchards in parallel. Other absentee owners adopt more extensive agricultural systems (with open-field vegetables or olive trees) and employ a manager. The main differences between these farming systems are the degree of capital use and intensification, and the direct/indirect type of management.

Generally speaking, farming systems in the Jordan valley are more intensive than in the highlands: farms are smaller (3.5 ha on average against 20–25 ha in the highlands) and net benefit per hectare (for similar crops and/or farming systems) is generally higher. The survey identified five main categories of farming systems (Table 10.2). They include family farmers who either own or rent the land and grow vegetables in open fields; entrepreneurial farmers who adopt capital- and labour-intensive techniques like greenhouses with a high return on investments; citrus orchards cultivated in the north of the Jordan valley and managed either by the family who owns the land, or by absentee investors interested in the social rather than the economic value of their farm; highly profitable bananas grown in the extreme north of the valley; and, finally, some poorer farmers with more extensive vegetable cultivation, associated with small orchards.

Control of Groundwater Overabstraction in the Highlands

The problem of groundwater overdraft

Since the 1930s, when the first wells were dug in the Azraq oasis, to the present, groundwater abstraction in the highlands has increased to meet the needs of agriculture, industries and cities, although the part of agriculture has decreased in both absolute and relative terms in the last decade. According to the official figures of the MWI for 2004, total groundwater abstraction in the LJRB reached 248 Mm^3, of which about half was used in agriculture (THKJ, 2004). In the highlands, in the Amman-Zarqa and Yarmouk groundwater basins, local groundwater abstraction reached 215% and 125% of the annual recharge, respectively. Taking return flows from municipal/industrial and irrigation uses into account, the overall net depletion of these aquifers comes down to 159% and 98% of their annual recharge, respectively.

The resulting drawdown of the aquifer is paralleled with a decline in water quality (due to increasing salinity and use of fertilizers and pesticides) and it is feared that both domestic and agricultural uses could be jeopardized, and further costly investments in water treatment needed (ARD and USAID, 2001a; JICA, 2004). In addition to these salinity problems, aquifer overdraft incurs growing pumping costs to all users and the abandoning of some wells (Chebaane *et al.*, 2004).

Groundwater policies and by-law No. 85 of 2002

Faced with such problems the Government of Jordan has tried to reorient its water policy through the Water Strategy Policy of 1997.

Table 10.1. Profile of main farming systems (highlands, eastern desert region).

	Settled Bedouins			Stone fruit tree entrepreneurs		Absentee owners	
	Family vegetable farm	Mixed farm vegetables and olive trees	Family fruit tree farms	Intensive entrepreneurial farmer	Intensive absentee owner – stone fruit trees and olive trees	Prestige olive tree farm	Extensive open-field vegetables
Land tenure/ water access	Rent	Ownership	Ownership	Ownership	Ownership	Ownership	Ownership/ Rent
Farm area range (ha)	20–25	(20–25+10–15)	10–20	25–35	40–80	15–35	20–25
Net benefit (US$/ha/yr)	1,100	621	6,900	16,000	14,850	300	600
Net benefit (US$/farm/year)	24,750	21,750	103,500	480,000	891,000	7,500	13,500
Number of wells	1	1	1	1	2	1	1

Table 10.2. Profile of main farming systems (Jordan valley, northern and middle directorates).

			Citrus farms		Banana farms		Poor farmers
Farming systems	Open-field vegetable family farms	Entrepreneurial greenhouse farms	Family farms	Absentee owners' farms	Family farms	Entrepreneurial farms	Mixed farms
Land tenure	Rent/ Ownership	Rent/ Ownership	Ownership	Ownership	Ownership	Ownership/ Rent	Rent/ Sharecropping
Farm area range (ha)	3–6	6–10	3–6	1–20	1–5	1–5	1–3
Number of family workers	2–5	1–2	3–5	1	3–5	1–2	4–10
Net benefit (US$/ha/year)	3,800	7,500	1,250	400	7,000	12,500	1,050
Net benefit (US$/farm/year)	17,100	60,000	5,625	4,000	21,000	37,500	2,100

Table 10.3. Water prices according to the volume abstracted in private agricultural wells. (From THKJ and MWI, 2002b, 2004a as mentioned in by-law No. 85 of 2002.)

Quantity of water pumped	Water prices in wells with former abstraction license – 2002 by-law	Water prices in wells with former abstraction license – 2004 amendment	Water prices in wells without former abstraction license
0 to 100,000 m^3	Free	Free	JD0.025/m^3 ($0.035)
101,000 to 150,000 m^3	Free	Free	JD0.030/m^3 ($0.042)
151,000 to 200,000 m^3	JD0.025/m^3 ($0.035)	JD0.005/m^3 ($0.007)	JD0.035/m^3 ($0.050)
More than 200,000 m^3	JD0.060 /m^3 ($0.085)	JD0.060/m^3 ($0.085)	JD0.070/m^3 ($0.098)

Several measures have been taken to decrease groundwater abstraction, including: (i) freezing of well-drilling authorizations in 1992; (ii) implementation of a tax of $0.35/$m^3$ for any water pumped and sold/used for industrial or aesthetic purposes (since 1994) as well as for domestic purposes (since 2002); (iii) a campaign to equip private wells with water meters; (iv) reduction of losses in urban networks; (v) promotion of less water-intensive/high-value crops; and finally (vi) promulgation of the groundwater by-law No. 85 of 2002 (Chebaane *et al.*, 2004). Government policies called for a massive reduction in abstractions by highland pumpers by 86 Mm^3/year until 2010, and by a further 36 Mm^3/year until 2020 (World Bank, 2001b). Water savings elicited by the new water charges were expected to reach about 40–50 Mm^3 over the next 3–5 years (Checchi and Devtech, 2003).

From 1962 to 1992[4] licenses to drill agricultural wells were granted by the government. Two-thirds of the licenses granted specified the maximum amount of water that each farmer could pump (most commonly 50,000 or 75,000 m^3/year, and sometimes 100,000 m^3/year after 1990; Fitch, 2001) but these limits were never enforced (THKJ and MWI, 1997b, 1998a). In 2002, the groundwater by-law introduced a system of quotas combined with taxation of any use exceeding the quota. However, instead of endorsing previous license quotas, the by-law allowed uncontrolled abstraction up to a limit of 150,000 m^3/year/well, a volume much larger than the limits mentioned in the licences. Rules for the taxation of the water pumped above this limit are detailed in Table 10.3.

It has been reported that farmer interest groups have got the authorities to cancel the former licenses against the acceptance of the principle of taxing volumes abstracted above a certain limit (Pitman, 2004): technical, institutional and political difficulties act as impediments to the effective implementation of the reforms.

In April 2004, the first bills, corresponding to water consumption between 1 April 2003 and 31 March 2004, were sent to farmers. Until November 2005, no employee of the MWI had been entrusted with the task of collecting fees. In these conditions farmers have not yet paid these bills.

Between May and August 2004, two amendments have modified the regulation: the first one is a lowering of the already low fees for the volumes abstracted in licensed wells between 150,000 and 200,000 m^3/year. Volumes will be charged at Jordanian dinar (JD) 0.005/m^3 instead of JD0.025/m^3 (cf. Table 10.3). The second amendment concerns abstraction from brackish aquifers: the higher the water salinity, the lower the fee; it will have an impact in the south of the Jordan valley and in the Azraq basin (east of the country) but not in the LJRB highlands.

Implementing the by-law is now possible since most of the wells are equipped with water meters (94% according to Al-Hadidi, 2002). However, several problems must be underlined. First of all, in 2001 only 61% of the meters were functioning properly (Fitch, 2001) and, although major replacement campaigns have been conducted, this problem is likely to recur.

[4]No drilling license has been delivered after 1992. However, the number of operating wells is continuously increasing as illustrated by the records of the Water Authority of Jordan for 2004. This may be due to the development of well metering.

Moreover, there is an important lack of material and human resources since controls are handled by only a few employees of the Water Authority of Jordan (WAJ). Another problem arises because meters are still not protected. Experience in the Jordan valley has shown that if water meters are not protected in a box closed with a padlock, they are likely to be broken or at least fiddled with (Courcier and Guérin, 2004). In the highlands, the risks of deterioration are reduced because the meter is paid for by the farmer but, on the other hand, tampering is quite easy and could become common.[5]

Financial impacts and expected adjustments in eastern desert's farming systems

Based on the description of farming systems presented earlier, this section explores the financial impact of the by-law on each type of farming system and how this impact could be mitigated by possible farmers' strategies.

Financial impacts of the by-law on farming systems

Table 10.4 summarizes financial impacts (before and after the 2004 amendment, Scenario A and Scenario B, respectively) on farms with licensed wells, assuming that actual withdrawals remain unchanged.[6]

Settled Bedouins with fruit tree farms and absentee owners with prestige olive trees will not be affected by the by-law since their current annual water consumption is less than 150,000 m^3/well. Fruit tree farmers will be very slightly affected by the by-law. Table 10.4 illustrates that the amendment considerably softened the financial impact of the by-law on settled Bedouins with vegetables or mixed farms and absentee owners with vegetables.[7]

To assess possible farmers' responses it is necessary to know what the present irrigation efficiency in the eastern desert is and to what extent the quantity of water supplied to crops matches their water requirements. Surveys have shown that orchards (especially olive trees)[8] are underirrigated with regard to full agronomic requirements: further water savings are thus unlikely. On the other hand, vegetable farmers abstract nearly 160% of the net crop water requirements, as evaluated by Fitch (2001). In this condition, the overall efficiency of water use in vegetable farms only reaches 62% and can be improved without affecting production. If we assume that on-farm irrigation efficiency can reach a maximum of 75%, vegetable farmers could decrease the amount they pump from 216,000 m^3 down to 179,760 m^3 while still meeting net crop water requirements.

The financial impacts at the farm level of four different scenarios are presented below: (A) the first scenario assumes a maximization of water savings by a decrease of water use down to 150,000 m^3/well/year (so that no fee needs to be paid), and a proportional reduction in the cultivated area (water use efficiency remains constant); (B) the second scenario assumes that farmers pay their water bills without changing their water consumption; (C) in the third scenario farmers increase irrigation efficiency up to 75% (still meeting crop water requirements) and reduce water abstraction; and (D) the fourth scenario is like Scenario C, but farmers do not reduce abstraction and use

[5]Anecdotal observations during our surveys showed that tampering and 'compromising' with WAJ employees did exist.

[6]Unlicensed wells in Jordan are mainly located near the Azraq oasis (east of the LJRB) and in the south of the Jordan Valley where they tap the brackish aquifer. For the sake of simplification, the following quantification assumes that all wells in the highlands of the LJRB have a license.

[7]For mixed farms, we have presented a case where farmers have only one well. In these conditions, impacts of the by-law are expected to be high. However, many of these farmers have two separate wells that they use indifferently to irrigate two different plots. In the latter situation, the by-law will not have any impact on them and no changes are expected to occur.

[8]Only 56% of olive-orchard requirements are met: this very low satisfaction (also observed by Hanson, 2000) illustrates their drought-tolerance quality and also their very low profitability. Deficit irrigation highlights that these orchards have a high social value but that their conventional economic profitability is not of prime importance to farmers. Farmer strategies do not boil down here to profit maximization.

Table 10.4. By-law impact on farm income in the eastern desert.

		Settled Bedouins			Stone fruit tree entrepreneurs		Absentee owners	
		Family vegetable farm	Mixed farm vegetables and olive trees	Family fruit tree farms	Intensive entrepreneurial farmer	Intensive absentee owner – stone fruit trees and olive trees	Prestige olive tree farm	Extensive open-field vegetables
Net benefit (US$/ha)		1,100	621	6,900	16,000	14,850	300	600
Water use (m^3/farm/year)		216,000	284,750	150,000	300,000	435,000	137,500	216,000
Actual abstraction costs	US$/ha	2,181	1513	1,373	1,310	1,310	432	1,875
	US$/farm	49,072	52,955	20,595	39,300	78,600	10,800	42,187
	% of current revenue	198	243	19.3	8.2	8.8	144	312
A – Extra water costs by-law	US$/ha	138	259	–	342	108	–	138
	US$/farm	3,110	9,050	–	10,250	6,475	–	3,110
B – Extra water costs by-law and amendment	US$/ha	76	217	–	295	61	–	76
	US$/farm	1,710	7,621	–	8,850	3,675	–	1,710
Revenue decrease	Scenario A	12.6	41.6	–	2.1	0.7	–	23.0
(% of current revenue)	Scenario B	6.9	35	–	1.8	0.4	–	12.7

Table 10.5. Financial impacts of the by-law (with amendment) on settled Bedouins farms and absentee owner vegetable farms according to the four response scenarios.

Farming system category		Settled Bedouins		Absentee owner
		Open-field vegetable family farm	Mixed farm vegetables and olive trees	Open-field farm
Scenario A	Volume abstracted (m³/well)	150,000	150,000	150,000
	Change in revenue – US$/ha and % of current revenue	– 341 (−31%)	– 264 (−43%)	– 186 (−31%)
Scenario B	Volume abstracted (m³/well)	216,000	284,750	216,000
	Change in revenue – US$/ha and % of current revenue	– 76 (−6.9%)	– 217 (−35%)	– 76 (−12.7%)
Scenario C	Volume abstracted (m³/well)	179,760	218,760	179,760
	Change in revenue – US$/ha and % of current revenue	– 379 (−34%)	– 426 (−68%)	– 379 (−63%)
Scenario D	Volume abstracted (m³/well)	216,000	284,750	216,000
	Change in revenue – US$/ha and % of current revenue	+ 129 (+12%)	+ 35 (+5%)	+ 46 (+8%)

the water saved to increase the cultivated area. We hypothesize that irrigation efficiency can be improved up to a maximum of 75% through a better design of the farm network, the use of higher-quality emitters, better on-farm operations, and a better monitoring of soil water reserves that would allow fine-tuning of irrigation, thanks to the involvement of more specialized technicians. The cost of such changes can be estimated at about $370/ha/year (Courcier, 2006, personal communication [by e-mail 20 May 2006]).[9] Contrary to common assumptions that farmers can easily save substantial amounts of water by just being 'more careful', improvements demand better knowledge and material and thus have a cost, especially in a situation where microirrigation is already in use. Assessing such costs is a difficult task, and the willingness/ability of farmers to achieve these improvements will depend on these costs.

Adjustments to be observed in open-field vegetable and mixed farms

Table 10.5 summarizes the impacts of the four scenarios on extensive vegetable farms run by settled Bedouins or absentee owners.

[9]This cost can be broken down into: $90/ha/year of incremental wage and $280/ha/year for dripper lines as well as for primary and secondary pipes, filters and tensiometers. To increase efficiency above 75%, there is an additional need for skilled engineers as well as for computerized systems that would cost about $1400/ha/year, with an initial investment of $1100/farm (Courcier, 2006, personal communication [by e-mail 20 May 2006]). The incremental cost to increase efficiency up to 75% is lower than the extra revenue that the farmer would derive from expanding his field and using saved water, but other constraints can explain why a 'farmer-maximizer' has not yet increased his irrigation system efficiency. These include aversion to risk or to incremental labour and time to be spent on the farm, as well as a low investment capacity, especially in a situation where most Bedouin farmers are indebted (Cheebane *et al.*, 2004). The relative high costs (compared to farmers' revenue) of increasing efficiency above 75% make such an evolution unlikely.

For settled Bedouins with vegetables in open fields, reducing the land area until water abstraction is curtailed down to 150,000 m^3/well/year (Scenario A) entails a decrease in income of 31%. Paying the water fee (B) is a much better strategy (−6.9%), even though farmers already face water costs which are higher than their net income (cf. Table 10.4). Improving efficiency without increasing cropping area (C) entails a 34% decrease in farm revenue. If actual costs of improving efficiency are lower than $76/ha, (a rather low value compared with our estimate of $370/ha), then strategy C is cost-effective. Strategy D seems a better option with a 12% increase in farm revenue, due to the expansion of the irrigated area. Conclusions for absentee owners are similar: Scenario D is the best option but another possible strategy for well owners would be to rent out their wells to large entrepreneurial fruit tree farmers or to cities (cf. below). It is noteworthy that these conclusions would not have been significantly different with the pre-amendment price of water.

These results confirm the fact that technology costs are in general much higher than corresponding savings in the water bill, unless prices are taken at very high levels. In other words, even in the present case where water costs are very high, saving water is rarely cost-effective for farmers, and price incentives alone are unlikely to reverse this situation. However, in regions with abundant land, savings derived from improved irrigation efficiency can be used to expand the cropping area in a cost-effective way (Scenario D). Since, under conditions of high water costs, higher water costs deplete incomes, they may also trigger adoption of higher-value crops.

To avoid paying any water fee (A), settled Bedouins with mixed farms would have to decrease their current abstraction of 284,750 m^3/year by 47%, incurring a drop in income of 43% (the farmer would first abandon his olive orchard and then shrink its [more profitable] vegetable area). The average income is so low that paying the fees (B) would entail a 35% decrease in revenue (pre-amendment water prices would have sent a stronger signal but at the cost of more than half the current income). Strategy C would be even worse with an expected decrease in revenue of about 68%. Finally, as in the case of vegetables, improving efficiency and increasing the cropping area (D) would offset the financial loss due to the by-law and increase farmers' revenue by 5%.

Adjustments to be observed in entrepreneurial fruit tree farms

Intensive stone fruit tree entrepreneurs will be slightly affected by the by-law. In line with their large water abstraction, farmers will have to pay high water fees (between $3675 and $8850/farm according to the farming system; cf. Table 10.4). However, due to the high profitability of these farming systems, this increase in water prices will have a negligible impact on farmers' revenue (~2%).

In all likelihood, Scenario B will prevail, that is, farmers will squarely foot the bill. In systems where trees are underirrigated and efficiency already high, Scenarios C and D are very unlikely. Scenario A, however, might also be an option if there is a possibility for farmers to rent an additional nearby well: this new well would provide both the shortfall of water needed for the old orchard and additional water for expansion. The availability of large flat desert areas would make this option quite easy (although it is illegal because areas attached to a particular well are normally specified) and economic calculations show that such an expansion would be profitable, even with the cost of well renting (about $18,000/well). This rent is also higher than the total revenue generated at present by extensive open-field farms managed by absentee owners and would also make this option attractive to them. This could accentuate the current increase in stone fruit production by entrepreneurial farmers in the highlands. In such a case, there will not be any water savings but higher productivity will be achieved through the shift from vegetables to fruit trees.

Water savings at a regional scale

A land-use mapping carried out by the MWI and the GTZ based on two mosaics of

LandSat images dated August 1999 and May 2000 was used to estimate irrigated areas within the Amman-Zarqa and Yarmouk groundwater basins, giving a total of 14,460 ha with a breakdown between olive trees, fruit trees and vegetables. Based on these estimates of irrigated areas and on crop water use data, we can approximate groundwater abstraction in the Amman-Zarqa and the Yarmouk basins and compare these values with earlier estimates from other sources, and with annual recharge values given by THKJ (2004).

Results show that gross agricultural abstraction records of the MWI are 20% below other evaluations. The MWI may underestimate present agricultural abstraction, partly due to the difficulties attached to water metering mentioned above. In our estimate, gross abstraction rates are presently reaching 249% and 195% of the annual recharge in the Amman-Zarqa and Yarmouk basins (or 179% and 168% if return flows of irrigation and municipal/industrial uses are considered, i.e. net abstractions of 121 and 63 Mm^3/year). These estimates will be used as a baseline situation in the following sections to assess possible water savings in the two groundwater basins considered.

Information on the different classes of agricultural wells according to their yearly production in the two groundwater basins of Amman-Zarqa and Yarmouk shows that out of the 606 wells located in these two basins, only 182 yield more than 150,000 m^3/year and will thus be concerned by the by-law (MWI records for 2004). Discounting government wells producing more than 500,000 m^3/year, this figure drops down to 166 wells that represent 38% of water abstracted in these two basins. Finally, as shown above, since only settled Bedouins with vegetables or mixed farms and absentee owners with vegetables are likely to respond to the by-law, only 83 wells in the eastern desert (90% of these in the Amman-Zarqa basin) will eventually be affected by the by-law.

Regional water savings can be assessed based on the four scenarios considered earlier by aggregating responses expected for each type of farm. Table 10.6 shows that the maximum gross water savings to be expected in vegetable plots in the eastern deserts are about 5.5 Mm^3/year (90% of these in the Amman-Zarqa basin). These savings would be obtained if all vegetable farmers decreased their water application and irrigated area by one-third on average, while maintaining their actual water use efficiency (Scenario A). This would lead to high agricultural losses ($2.5 million, not shown). This response, however, is not the one that the incentives in place are likely to prompt.

In Scenario B, nothing is changed except for a transfer of $0.21 million from vegetable farmers to the state coffers, or a total of $0.84 million if payments of all farms are considered. Improving efficiency without increasing cropping area (Scenario C) would reduce abstracted volumes to around 179,760 m^3/well/year in vegetable farms. In such conditions, gross water savings would reach 3.0 Mm^3/year and the regional gross overdraft would be decreased by about 2.2%. The net abstraction would not be affected by this change.

Finally, Scenario D would lead to increasing the depleted fraction by about 2.3 Mm^3/year (as cropping area and efficiency increase, and return flows are reduced), which would defeat the objective of the by-law. Generally speaking, encouraging higher efficiency in conditions where land is not a constraint is counterproductive to the objective of reducing the depletion of water resources. The fact, however, that expanding cultivation by using saved water is – on paper – financially profitable but not observed strongly suggests that the real costs of increasing efficiency may be higher than what has been considered here.

In conclusion, we can say that the implementation of the by-law in its current form will not lead to significant water savings. Because of the threshold of 150,000 m^3 and the weight of the public wells, 72% of the wells in the Amman-Zarqa and Yarmouk basins will not be affected by the by-law (a threshold of 100,000 m^3 would take this proportion down to 53%). Olive orchards, for example, which represent 32% of the total agricultural water abstraction in the highlands and qualify as the

Table 10.6. Impact of the by-law on vegetable farms at the basin level (eastern desert zone).[a]

	Abstraction (m^3 per year and per well)	Gross water savings (Mm^3/year)		Net water savings (Mm^3/year)		Depleted fraction in vegetables (Mm^3/year)	Government revenue from vegetable farms (Million US$)	Overall government revenue (Million US$)
		Mm^3/year	% of current gross overdraft	Mm^3/year	% of current net overdraft			
Present situation	216,000	–	–	–	–	11.1	–	–
Scenario A: maximizing water savings	150,000	5.5	4.0	3.4	4.2	7.7	–	–
Scenario B: payment of water fees	216,000	–	–	–	–	11.1	0.21	0.84
Scenario C: improving efficiency area constant	179,760	3.0	2.2	–	–	11.1	0.012	0.63
Scenario D: improving efficiency increased area	216,000	–	–	–2.3	–2.9	13.4	0.21	0.84

[a]For Scenarios C and D, all calculations have been done considering an achievable irrigation efficiency of 75% (in vegetable farms). For Scenarios A and B we considered the present efficiency in vegetable farms (62%). System efficiencies in olive and other orchards have been considered homogeneous at 70% and 80%, respectively, in the four scenarios.

prime target of policies because of their low water productivity (WP) (WP = \$0.05/m^3) will not be affected. If we add to this the facts that high-value crops such as fruit trees (WP = \$1.1/m^3) will be financially little affected and that farmers' behaviour is unlikely to change, then the 83 wells concerned correspond to only 18% of the total water abstraction (16.1 and 1.8 Mm^3/year in the Amman-Zarqa and Yarmouk basins, respectively).

Vegetable and mixed farms are most vulnerable to hikes in water charges: this is because their income is so low that any additional production cost will depress them further. However, it is unlikely that such pressure would result in significant water savings, since improving efficiency would require investment in technology and qualified labour that are: (i) higher than gains resulting from a reduced water bill; and (ii) beyond the capacity of most of these farmers, many of whom are indebted.

Upper (optimistic) estimates of reduction in gross water abstraction (Scenario A for vegetable and mixed farms) point to a decrease by 4%, that is, 5.5 Mm^3/year, a drop in an ocean of overabstraction, and quite short of the 40–50 Mm^3 hoped for.[10] Revenue to the government is expected to vary between \$0.63 and \$0.84 million/year, not considering the costs of collection and enforcement.

With higher charges (like in the pre-amendment price table, for example), olive orchards and fruit tree farms would remain insulated but the pressure would be made to bear on the most vulnerable vegetable and mixed farms; with a lower threshold, olive orchards would be under pressure too. In all likelihood, few of these farms would be in a position to invest in order to achieve better efficiency (nor would economies in the water bill ever offset the costs of doing so). Affected farmers might just decrease their area and water abstraction (incurring a loss in their income) until they reach the threshold and avoid water charges.

But they might as well sell their water to neighbouring fruit farmers, rent out their wells (if they own them) and move out of agriculture. This would amount to a shift in production from vegetable farming and olive trees to higher-value fruit production, and would definitely raise the productivity of water, but: (i) benefits would accrue to wealthier entrepreneurs; (ii) this would defeat earlier social policies aimed at settling Bedouins by providing them opportunities in the agriculture sector (Chebaane *et al.*, 2004), unless they are able to find equivalent or better job opportunities; (iii) the amount of water used would not be radically altered; and (iv) water demand would become extremely inelastic because of the high crop return; worse, the shift to higher efficiency fruit (or other) production could have the perverse consequence of allowing expansion of orchards, with lower return flow to the aquifer, greater depletion of water, and thus worsening of the status of the aquifer.

Because of the large share of unaffected farmers and likely impacts in terms of crop shifts rather than of improvements in efficiency, a substantial drop in water abstraction can only be obtained through the diminution of either the cultivated area or the number of wells in use. As demonstrated above, negative incentives (reduced thresholds, higher tariffs, petrol taxation, stricter enforcement, etc.) cannot achieve this without displacing weaker farmers and strictly prohibiting the selling/renting out of wells, but recent political crises suggest that such extreme measures are unlikely to be accepted. Attendant positive incentives, such as buying-out of wells (a measure envisaged by the government and considered positively by 50% of farmers [Chebaane *et al.*, 2004]), compensation for the uprooting of olive trees in the eastern desert (Fitch, 2001) and substituting treated wastewater for groundwater (ARD and USAID, 2001b) are more promising. Additional measures include reduction of losses in urban networks, educational and public awareness programmes for water users, allowing transfer of water to neighbouring orchards and the possibility of renting out wells (which would offer financial compensation but would not contribute to conservation objectives [Chebaane *et al.*, 2004]). Last, the removal of petrol subsidies for well operation or higher taxation of water must be accompanied by measures that provide

[10] If abstraction of all private wells was to be reduced to 150,000 m^3/year, total gross water savings would reach 12.5 Mm^3/year.

alternatives to people moving out of low-value agriculture, such as subsidies or secure market opportunities to help viable farms to intensify production.

Water Pricing in The Jordan Valley

Water allocation

From the beginning of large-scale irrigation in the Jordan valley, in the 1960s, a crop-based system of water allocation by quota has been used to supply water to irrigated schemes. Volumetric pricing was also initiated in 1961, with a cost of fils1/m^3 (Hussein, 2002; one fils is equivalent to JD0.001 or $0.0014). The official quota system has undergone several changes since the 1960s and has been mainly used as a guideline, with adaptations according to circumstances and national priorities (THKJ and JVA, 1988, 2001). According to quotas defined in 1988 (THKJ and JVA, 1988), each plot of vegetable grown between mid-April and mid-December received 2mm of water/day (during the rest of the year water was allocated on demand). Citrus and bananas were supplied with 4 and 8mm/day, respectively, from the beginning of May to the end of October (and on demand during the rest of the year, when demand is low). Historical large landowners (mainly citrus owners) as well as entrepreneurial farmers growing bananas are the main beneficiaries of these quotas.

Bananas and citrus are highly water-consuming crops and were traditionally cultivated in the northern part of the Jordan valley (Khouri, 1981; Elmusa, 1994): their higher quotas have now been frozen resulting in the institutionalization of some inequity in the access to water in the Jordan valley. Only the plots planted with bananas before 1991 are eligible to a 'banana allotment'. In 2004, however, in contradiction to its policy to reduce demand, the JVA legalized citrus orchards planted between 1991 and 2001, granting them the citrus allotment instead of the vegetable allotment they were receiving before. All other areas receive the vegetable allotment if the farmer declares to the JVA that he is cultivating his plot.

The 1997–1999 period was marked by a severe drought which, in 1999, made ad hoc reductions in farm allotments necessary. While some areas had to be left fallow, it is not clear whether impacts on yields were observed, but these reduced quotas have been maintained ever since (except in the south of the valley, where treated wastewater is used). In 1999, vegetables and citrus were allocated 75% of their allocation while bananas received 85% of their quotas. Allocations were reduced by 25% in 2000 and 2003, and by 50% and 40% during the summer 2001 and 2002, respectively.

In 2004, the JVA proposed new quotas expected to better match supply and crop water requirements (THKJ and JVA, 2004). These recommendations are close to the reduced quotas of 1999. On a regional scale, changing from the previous allocation system (2, 4, 8mm/day) to the new recommended values yielded total water savings in the northern and middle directorates (where the rules apply) of about 20.2Mm^3/year (between April and November), which were reallocated to domestic use in Amman.

O&M costs recovery

Revenues from irrigation water have gradually increased with time, as water charges established at fils1/m^3 in 1961 later increased to fils3/m^3, then to fils6/m^3 in 1989, and to an average of fils15/m^3 in 1996 (GTZ, 1993; FORWARD, 1998; the planned increase up to fils25/m^3 has been delayed).

Revenues from charges covered one sixth of O&M costs during the 1988–1992 period (GTZ, 1993; Hussein, 2002), which meant a corresponding average annual subsidy of $3.4 million. In 1995, less than a quarter of O&M costs was recovered. Charges were then increased more than twofold and data for 1997 point to a rate of recovery of O&M costs of two-thirds, with an average charge of fils15/m^3 (against fils18/m^3 of O&M costs) and a rate of defaulting of 20% reducing actual revenues down to fils12/m^3 (FORWARD, 1998; World Bank, 2001b).

Calculations for 1988–1992 showed that fixed asset depreciation and financing costs were twice higher than O&M costs proper (total costs were thus three times higher than O&M costs) (GTZ, 1993). THJK (2004) indicated that the ratio of average

Table 10.7. Crop-based water costs according to three different levels of price increase.

Cost of water ($/ha/year)	Vegetables	Citrus	Bananas
Current water costs	67	138	350
A. O&M costs recovery-block tariff system	94	191	485
B. Total costs recovery (O&M + capital costs)	278	573	1454
C. 80% of water costs borne by farmers in the highlands	586	1172	1740

capital costs to O&M costs was 2.07 for the 1997–2002 period.

Based on the actual block tariff system (FORWARD, 2000; cf. Appendix) we have estimated average costs per m^3 and per year for each type of crop according to the recent JVA recommendations (see details in Venot *et al.*, 2007). Total water costs for the farmers are higher in banana plantations ($350/ha/year) than in citrus orchards ($138/ha/year). They are lowest in vegetable farms which consume less water ($67/ha/year). Differences in water charges for each crop are lower than previously, since uses have been capped. The main beneficiaries of this evolution are banana farmers whose consumption rarely reaches expensive tariff blocks. The new JVA recommendations lead to lower water use and consequently to a lower overall level of O&M cost recovery, with an average charge of about fils13/m^3.[11]

In line with these recent evolutions, despite substantial differences between sources, we will consider here that current charges cover 72% of O&M costs and that full costs are three times higher than O&M costs.[12]

Economic impacts and adjustments at the farm level

This section provides financial evaluations of a rise in water prices according to three different scenarios. First, we will consider that water prices will increase up to a level where O&M costs of the JVA are recovered; this is the main objective of water pricing policies in Jordan (FORWARD, 1998; THKJ and MWI, 1998c, 2002a; Salman, 2001; THKJ *et al.*, 2002; THKJ, 2004). Second, we will consider a water price increase allowing the recovery of total costs of irrigation in the Jordan valley (O&M and capital costs). In these two scenarios, we consider that the actual block tariff system is maintained (cf. Appendix). Finally, based on a recommendation of THKJ (2004),[13] we will assess the impact of a hypothetical increase of up to 80% of the present average cost of water borne by farmers in the highlands, that is, about $0.116/$m^3$ (Al-Hadidi, 2002). In this third scenario, water is charged at a flat rate regardless of the total water used in the farm. (In the three scenarios, the rate of bill recovery is assumed to be 100%.) Table 10.7 specifies water costs for each crop and scenario and Table 10.8 for each farming system.

In Scenarios A and B, water prices are multiplied by a factor of 1.4 and 4.1, respectively, regardless of the crop planted. In Scenario C, and because of the implementation of a flat charge, water prices are multiplied by 8.5 for vegetables and citrus and by 5 for bananas. Table 10.8 shows that extensive farming systems (citrus and mixed farms) would be most impacted since water costs represent an important percentage of total costs (in citrus farms) and because their income is very low. On the other hand, intensive systems (greenhouse farms, for example) are not responsive to such policies since water costs are negligible compared to

[11] The JVA's revenue has decreased in line with declining allotments from 1999 onwards. This may have prompted the proposal to establish a monthly flat charge of JD2 ($2.8) on each water bill.

[12] In fact, since 2005, O&M costs are totally covered by the sale of water from the Mujib Southern Carrier to the Dead Sea industries. This recent change is not considered here in order to keep conservative estimates.

[13] 'The water production cost from private wells borne by the farmers (at present about fils100/m^3) should be taken as a guideline for adjusting the water tariffs charged by the JVA (at present fils10–12/m^3). The tariff for 'public' water of the JVA should not be lower than 80% of the average cost of the water produced from private wells' (THKJ, 2004).

Table 10.8. Impact of different levels of water price increase on farming systems in the Jordan valley.

Farming systems		Open-field vegetable family farms	Entrepreneurial greenhouse farms	Citrus farms		Banana farms		Poor farmers
				Family farms	Absentee owners' farms	Family farms	Entrepreneurial farms	Mixed farms
Allocation type		Vegetables	Vegetables	Citrus	Citrus	Bananas	Bananas	Vegetables
Net income (US$/ha/year)		3800	7500	1250	400	7000	12,500	1050
Total costs (US$/ha/year)		8150	21,000	1550	1200	8200	8600	2400
Actual water costs (% of income)		1.8	<1	11	34.5	5	2.8	6.4
Actual water costs (% of total costs)		<1	<1	8.9	11.5	4.3	4.1	2.8
Revenue decrease (% of the actual income) according to three different water prices	A. O&M costs recovery	<1	<1	4.2	13.2	1.9	1.1	2.6
	B. Total costs recovery (O&M + capital costs)	5.5	2.8	34.8	negative revenue	15.8	8.8	20.1
	C. 80% of water costs borne by farmers in the highlands	13.6	6.9	82.7	negative revenue	19.8	11.1	49.4

input and labour costs, and they will remain so at any politically acceptable price level (Wolf *et al.*, 1996).

Scenario A would have a limited impact on most farming systems in the Jordan valley. Revenues in vegetable and banana farms would decrease by less than 1% and 2%, respectively. Poor farmers would also be slightly affected by the increase (2.6%). Finally, citrus farming systems would be the most affected: revenues would decrease by 4.2% to 13.2%. In the latter case, most absentee owners would probably retain their orchard because it is not central to their livelihood, or would adopt other trees.

In Scenario B, farmers' revenues would decline more substantially. Productive systems (vegetables in open fields or under greenhouses) would again be slightly affected (revenue is expected to decrease by about 2.8–5.5%). These farmers would probably cope with this loss or seek (limited) on-farm water savings through better management, in a bid to decrease overall water costs (see below). Mixed farms developed by poorer farmers would be substantially affected (−20.1%): some farmers might be driven out of agriculture, looking for jobs in other economic sectors, while their plots could be rented to and cultivated by more entrepreneurial farmers.[14] Profitability of banana orchards would be moderately affected (revenues decrease by 8.8–15.8%). Despite their high revenues, some farmers might shift to other very profitable orchards such as date palm trees that are less water-consuming, especially if import tariffs on banana are lowered. Finally, citrus farms would be greatly affected: profitability of family farms would decrease by one-third, while absentee owners' farms would no longer be profitable: citrus areas would be expected to decrease substantially with many small owners (shopkeepers, civil servants, retirees, old farmers, widows, etc.) renting out their land or shifting to higher-value trees, and only a small fraction of rich absentee owners retaining their orchards.

Finally, Scenario C would have a dramatic impact on the Jordan valley agriculture. As in the two previous scenarios, citrus orchards would hardly be profitable anymore and would basically disappear, with the same replacement options as above. In banana farms, a partial shift to date palm trees and generalization of drip irrigation systems might be observed. Mixed farm operators would see their profitability decrease by one-half and would tend to be replaced by more entrepreneurial farmers. In the end, profitability of vegetables planted in open fields or under greenhouses would decrease by nearly 13.6% or 6.9%. This third option is hardly imaginable politically and would disrupt the valley economy.

Are improvements in irrigation and economic efficiency possible at the regional scale?

Whether substantial water savings are possible is highly variable and depends on what the actual irrigation efficiency is and, if any low value is observed, on the causes of such a state of affairs.[15] Improvement of efficiency is hindered by several constraints, both technical and socio-economic.

[14]Since 2001, land market transactions have been allowed in the Jordan valley. Renting plots is also a widespread practice. As land pressure in this valley is very high, any plot left fallow by a farmer is expected to be taken up by another farmer with a more intensive management and higher profitability. The irrigated area in this valley is unlikely to decrease, whatever water prices are.

[15]Because of the high diversity of situations, available data on efficiency are rather inconsistent (Al-Zabet, 2002; World Bank, 2002; Petitguyot, 2003; etc.). This is due to the extreme complexity and variability of use efficiency, and to what is considered: which crop and what type of farm; the plot, pumping unit or the valley level; the water-short period or the whole year; which ET and Kc values; total or effective rainfall; special water requirements for specific operations such as 'solarization' and in occasional periods of deficit irrigation. All these factors combined explain why the literature is not fully consistent (Ghezawi and Dajani, 1995; World Bank, 2001a; World Bank, 2002; Shatanawi *et al.*, 2005; USAID, 2006; etc.). Our estimates of annual irrigation efficiencies give 64%, 62% and 82% for vegetables, citrus and bananas, respectively.

- First, farmers experience many technical problems at the farm level that come from drip irrigation systems which have been installed without technical guidance (in 70% of the cases), direct connection of old dripper lines to the JVA's pressurized network,[16] problems of filtration and clogging, etc. (Wolf *et al.*, 1996; Courcier and Guérin, 2004; Shatanawi *et al.*, 2005).
- Second, whether much water can be saved just by farmers being more 'careful' and with limited additional costs is doubtful in non-gravity irrigation. Experiments by USAID/JVA and MREA/JVA suggest that with precision irrigation it might be possible to save around 25% of water applied. This is easier to achieve in citrus farms irrigated by open microtubes. Achieving better irrigation efficiency requires computerized monitoring, use of tensiometers, improved filtration, frequent renewal of drippers, qualified staff, etc., and is therefore very costly. With the impossibility to expand cultivated land, the incentive for the farmer to achieve such gains is low, since corresponding costs are too high, regardless of the price of water. If we keep the estimates used for the highlands ($370/ha/year for achieving an efficiency of 75%, and an additional $1130/ha/year for reaching 85%) we can see that economies in the water bill will never come close to improvement costs, even for Scenario B.
- Only very high-tech and capitalized farmers linked to high-value markets demanding high quality of products can adopt and master such practices. It is important to note that, historically, drip irrigation was developed in the early 1980s as a technical response to the need to produce high-value products (along with the adoption of mulch, fertigation, labour-saving technology, control of doses, homogeneity and quality of products, etc.) rather than to a lack of water per se.[17]
- Third, farmers also experience many difficulties because of deficiencies in collective pressurized networks which result in a high heterogeneity of water distribution (with deficits observed in higher parts, sandy soils or at the end of the lines); rotations are difficult to establish; water theft, rent-seeking and tampering of equipment are pervasive (GTZ, 2004).
- Fourth, despite being conceived as a demand-based system, subject to the limitation of quotas, the actual mode of operation of the JVA and the uncontrolled nature of the inflow from the Yarmouk river do not ensure enough reliability in water provision (Courcier and Guérin, 2004). Overirrigation can also be considered as a safeguard against uncertainty in supply.
- Fifth, the system of monthly quotas defines a ceiling to the abstraction of pumping stations from the main canal (KAC): demand may be higher than the quota during a few critical periods in spring and autumn (Petitguyot, 2003), when no savings are possible. Conversely, efficiency is often lowest when supply exceeds demand, with no alternative use for water and therefore little rationale for saving water.
- Last, the desirability of further water savings is not fully established, as it is feared that lower salt lixiviation would raise salinity problems in the valley (McCornick *et al.*, 2001). (In the early 1990s, the JVA encouraged farmers to take water free of charge in the winter months for leaching purposes; Wolf *et al.*, 1996).

The idea that farmers are wasting water only because its price is relatively low is therefore simplistic and mistaken; so is its corol-

[16]Irrigation water is provided to farmers through several pressurized networks serving areas of approximately 400 ha and pumping stations which draw water directly from the KAC.

[17]After the conversion of the open channel irrigation networks to pressurized systems (completed in the mid-1990s), which caused the reduction of the flow at the farm turnout from 20 l/s to 6–9 l/s, most farmers were obliged to shift to localized irrigation.

lary that raising prices will necessarily improve efficiency. A World Bank (2003) report indeed acknowledges that '[I]t was anticipated that increased water tariffs [of 1995] would reduce agricultural water use. This did not happen.'

Higher water charges also deplete incomes and, at least for low-value crops, tend to motivate shifts towards higher-value crops (Pitman, 2004; THKJ, 2004). Economic data in Table 10.8 suggests that, prima facie and as far as revenue per hectare is concerned, farmers would have an interest in shifting to vegetables or to high-value trees. Several points must be emphasized:

- First, although citrus (low productivity) and banana (water-intensive) may appear as undesirable there is little incentive for farmers to shift to vegetables (or to rent out their land to vegetable farmers) since they would then lose their higher quota with little hope of getting it back if they ever would like to revert to trees.
- Second, even if water prices were increased to cover all costs (Scenario B), banana farming would remain highly profitable and the shift to date palm trees (or other trees) not warranted (non-elasticity).
- Third, citrus would be made less attractive but large areas are owned by absentee owners whose livelihoods do not depend on their agricultural activity. Their orchards are linked to social prestige and recreational use and are not driven by economic motives. These owners may not shift to a more intensive and time-consuming activity for the sake of preserving their secondary agricultural revenue.
- Citruses in family farms are more likely to be replaced by more profitable trees (mangoes, guava, grapes, dates), or by vegetables, sometimes with the land being rented out to entrepreneurs. Yet these farmers have chosen to develop relatively extensive systems for a reason (lack of skill, capital, or alternative activities; ageing of farm-holder, etc.) and it will be difficult for them to shift to riskier, more intensive, and time-/input-consuming crops, unless market opportunities are identified.
- Last, it is worth mentioning that overestimating the capacity or willingness of farmers to adopt new crops or technologies and pushing for much higher water charges (Scenarios B or C) might lead to farmers responding to higher water tariffs by tampering with or destroying meters, bribery or defaulting. Unrest and political intervention would also be likely reactions. Such outcomes are not attractive for the government, which has little incentive to antagonize supportive segments of the society if gains are not expected to be substantial (Richards, 1993).

In conclusion it can be stated that all these elements strongly limit the scope for pricing mechanisms to achieve improvements in both irrigation and economic efficiency. Gains are possible but their magnitude and realization depend on the type of farm, and they cannot be obtained without support, including technical assistance, predictable water supply, secure markets, and subsidies to shift to drip irrigation (where this has not yet happened) and, gradually, to precision irrigation. Several alternative options have been proposed, along the following lines:

- Flexibility of water supply at the farm level is obtained not only through exceptional requests but also by digging farm ponds to buffer irregular supply (Shatanawi *et al.*, 2005), by using water from side-wadis and, wherever possible, by pumping groundwater. Many farmers already have implemented these options.
- Effective freshwater savings in the Jordan valley may come from the generalization of the use of treated wastewater blended with freshwater in the north of the Jordan valley, as proposed by ARD and USAID (2001b) (see also JRVIP, 2001b; McCornick *et al.*, 2002; and KfW *et al.*, 2006).
- Significant water savings could be achieved through a better in-season distribution of water in the KAC. With the

completion of the Wehdah dam on the Yarmouk river, it will be possible to have a more flexible management of water allotments (JRVIP, 2001b; Courcier and Guérin, 2004). Monthly quotas could be transformed into yearly quotas, with farmers keeping the latitude to distribute water along the year according to their needs (Petitguyot, 2003).

- With a more controlled water regime, it might be possible to adopt bulk allocation and bulk charging procedures, whereby water user associations would be in charge of managing a yearly amount of water and recovering charges (JRVIP, 2001a). This, however, is hindered by extant cultural and social structures and would require significant institutional transformations and changes in the agency(JVA)–farmer relationship (van Aken, 2004).
- The banana area could be reduced by substantially raising the price of the higher tiers of the quota so that revenue would be reduced without affecting other crops; it could also be made less profitable by removing duties on imported bananas, in line with WTO rules. Such economic incentives could contribute to inducing a shift towards other trees, but the potential loss of high banana allotments is likely to hinder this shift if no positive incentives are available.
- The most efficient way to reduce diversions to the valley (and to free more water for Amman) would be to gradually reduce quotas – as observed since 1999 – in order to force adjustments (high-tech management, change in crops, etc.). Additionally, a bonus might be granted to those who accept to shift from a high quota to the vegetable quota; of course, this would be hard to justify in the face of the recent contradictory measure of recognizing more citrus allotments.

The last point concerns cost recovery objectives: the analysis indicated that the prime objective of financial autonomy of the JVA is within reach. Charges could be slightly raised to ensure revenue, while defaulting should be controlled by stricter enforcement. Raising prices to full O&M costs would not dramatically affect farmers. It must be noted however that the 'fiscal drain' argument commonly raised to justify increased cost recovery is hardly convincing since the present O&M subsidy to the JVA is worth less than 0.1% of state expenditures at $3.7 billion.

Despite higher coverage of state-borne O&M costs, water charges do not instil any virtuous circle towards improved management and maintenance on both the manager and the farmer sides (Small and Carruthers, 1991). There is a lack of positive incentive stemming from the fact that charges paid by farmers do not benefit the scheme, managers do not depend on these payments (which are sent to the Ministry of Finance), farmers control neither part of the revenue nor water deliveries, supply is uncertain, and allocation not transparent enough. Under such conditions water pricing merely boils down to a taxation instrument. Bulk charging at the pumping station level and transferring responsibility for charging farmers individually to water user associations might be a way forward.

It is unlikely that raising fees much beyond O&M cost recovery can be tenable because of the limited effect on water use and the difficulty to justify charges higher than the JVA's expenditures, which would look like a transfer of wealth to the state. These factors and the fact that there is hardly any example of full cost recovery of public schemes in the world make Scenario B highly unlikely (not to mention Scenario C).

Discussion and conclusions

The results obtained in both the highlands and the valley have both similarities and discrepancies, and also bring out lessons that have wider validity.

**Limited effectiveness of increased prices in instilling higher efficiency*. Several modelling studies (Doppler *et al.*, 2002; Salman *et al.*, 2002; Shatanawi and Salman, 2002; Salman *et al.*, 2005 for the valley; Salman and Al-Karablieh, 2004 for the highlands) have shown that demand is only responsive to prices at levels which are in general not compatible with sustained farm incomes and equity. However, we have shown that the causes of efficiency losses are not all at the farm level and that further improvements require significant technological improvements which are costly and offset any gain derived from a reduced water bill (Pitman, 2004).

Consequently, the claim by the 2004 master plan (THKJ, 2004) that the full cost recovery for irrigation O&M pursued by the Ministry of Water and Irrigation will, among four objectives, 'increase conveyance system and on-farm water use efficiency' is not valid. From the correct assumption that 'low prices for irrigation water provide limited incentive to improve on-farm efficiencies' it is mistakenly inferred that raising prices will automatically improve on-farm efficiency and should therefore be 'a prime target for implementing improvements' (USAID, 2006). Despite evidence to the contrary, these claims are still pervasive among donors, development banks and some green NGOs (FOE, 2002). Removing public subsidies may have other virtues but should not be expected to bring about improvements in irrigation efficiency (or be justified by this).

**Intensifying agriculture: at what cost?* Consequently, the principal impact of higher charges would be to reduce the income of two categories of farmers: poor and often indebted farmers with more extensive agriculture, on the one hand, and absentee urban owners and rentiers with other income sources, on the other. Such a pressure would have a beneficial impact if these farmers were encouraged to adopt more intensive farming. One should note, however, that these higher-value cropping systems were already available to these farmers and there are good reasons why – despite their high return – they did not adopt them earlier. Farmers engaged in extensive agriculture lack capital to embrace such ventures, which incur considerable risk; rentiers lack the interest to burden themselves with intensive management and value their farm for reasons other than their profitability. Intensification must be driven by market opportunities and not forced by circumstances which would drag de-capitalized farmers into risky ventures with a high probability of going bankrupt. It is doubtful whether the benefits of pushing the more vulnerable farmers out of business would be higher than the social costs incurred.

Most countries are confronted with this necessity of balancing family farming and agrobusiness, and social stability and economic efficiency (the case of Spain in Arrojo, 2001; Berbel *et al.*, 2005). As a rule, state policies include investments/subsidies to allow modernization of family farms in order to better compete with highly capitalized operators.

**High-value crops: for which market?* The move towards a more intensive and higher-value agriculture is critically dependent on the availability of a market for it. With growing competition from other countries in the Middle East it is not easy to identify crops with a good return: farmers are neither immune to drops in prices following a too widespread adoption of promising crops nor all ready for, or capable of, handling the complexity of certain productions. Palm trees, for example, are salt-resistant and dates (so far) fetch high prices, but they have several drawbacks which make them largely unfit for small extensive farmers: they do not produce during a period of 5 years, post-harvest operations are difficult to master, and only high-quality products find their way to the best market niches.

**The politics of water management and policy.* The negotiations around the by-law and the amendment, carried out with a fair degree of participation of stakeholders (Chebaane *et al.*, 2004), showed that agricultural interests retain significant political and bargaining power; the government is unwilling to alienate the support of Bedouin tribes or part of the Palestinian population, and to prompt claims from Islamist radicals that Islamic law is violated (Richards, 1993). The teeth of the by-law were removed through the implicit abolition of former abstraction limits (which were lower than the 150,000 m^3 threshold adopted) and through the recent amendment which abated the already low water fees. Some groups of influential farmers, with strong political linkages and opposed to a control of water abstraction, have tried to stop the process and have managed to slow it down thanks to support in the parliament.

The fact that illegal citrus orchards in the valley have recently been regularized – quite contradictory to policy objectives – also suggests that the populations concerned have enough political clout to counter the reduction of quotas. All this confirms that water pricing schemes largely reflect the political economy of a country and that political counterweights are often raised when prices depress incomes. This does not mean that reforms are not desirable or should not be attempted; but this cautions us against simple-minded decisions and forces decision makers to weigh benefits against all costs.

******Improving allocation of water resources.* With such a minimal expected impact of price increases on efficiency, the objective of reducing demand to sustainable levels in the highlands and to volumes lower than current diversions in the valley through pricing measures is clearly unattainable and must be dismissed, in line with Berkoff (1994), who recognized 'that it is inconceivable that [charges] would be high enough to balance supply and demand'. Under such circumstances, the higher-level objective of regulating intersectoral allocation through prices, expressed in the ASAL despite considerable doubt from experts (Pitman, 2004), is quixotic.[18]

**State and donors: conflicting viewpoints.* Opposition to pricing by most quarters in the government is based on three considerations (Pitman, 2004): (i) social concerns and the view that farmers' access to groundwater is already too costly; (ii) the view that administrative allocation of surface water and technical/institutional improvements in management are more efficient and equitable than pricing in achieving sound management; and (iii) the understanding that alternative markets must be ensured before pushing farmers to abandon lower-value crops. With some caveats this study tends to confirm these misgivings.

Pitman (2004) notes that the 'social-welfare dimension of water was the largest divergence of views between the Bank and government over the agricultural sector' and critically soured relationships. A possible source of misunderstanding is that affected people include both poor farmers and rentiers, and that the former might be used to unduly shelter the latter from adverse policy measures.

**Safety nets.* Policy makers' misgivings may be well founded if one judges from experience in other domains where planned safety nets have been neglected, equity impaired and social objectives defeated. For example, the elimination of all direct subsidies to owners of small livestock herds over the period 1995–1997 has proven to be very effective in reducing herd sizes by 25% to 50%, overgrazing, and thus rangeland degradation and desertification. However, an official evaluation found that 'the poorest group – nomadic pastoralists – in the driest areas have fared worst as they do not have the income to buy even subsidized concentrates. All farmers monitored, with the exception of the medium-sized agro-pastoral farmers in the wettest areas in 1997/1998, had negative profits since 1996' (Pitman, 2004). Earlier consensus that attendant measures would be needed seems to have been later forgotten (Richards, 1993).

This suggests that too little attention is given to safety nets and the assumption that people can be reabsorbed by the labour market without much hardship is often not valid. Clearly, linkages to the macroeco-

[18] The claim by the World Bank (2003) that 'the partial tariff increase [in the valley] satisfied an immediate objective of maximizing transfer of water to the highlands' has no basis since this transfer is a bureaucratic decision completely independent of prices.

nomic framework must be strengthened if social objectives are to be fulfilled.

**From negative to positive incentives.* Negative incentives through prices that deplete incomes or force costly/risky adjustments generally raise considerable opposition which may express itself through political channels or in the streets. Such (stick) measures must be accompanied with positive incentives (carrot) (Al-Weshah, 2000). Positive incentives include a bonus for uprooting olive trees in the highlands or for accepting vegetable allotments in the valley (or tree allotments for banana growers), attractive buyout schemes of wells in the highlands, aid or crop insurance schemes for farmers tempted to diversify, etc. The government's refusal to raise prices before treated wastewater or market opportunities are available also indicates the fear of negative impacts in the absence of clear alternative opportunities and 'pull' factors.

**Enforcement and monitoring.* It is clear in both situations that individual metering is extremely demanding and hard to administrate. The percentage of broken meters both in the highlands and in the valley is likely to rise again after replacement campaigns. If fees significantly affect the economic situation of farms they will also probably trigger defaulting, tampering or destruction of meters, social unrest and political stress at unprecedented levels, and corruption or collusion between officials and farmers (GTZ, 2004). This does not mean that metering should not be attempted but reminds us of the costs involved and of the possibility that other approaches could be adapted more (e. g. charges based on crop and area in the highlands, or defined and recovered at the level of the pumping station in the valley).

**Quotas and regulation.* As shown from other situations where scarcity is high and volumetric control possible (Iran, Tunisia, Morocco, south of France, Italy, Spain, etc.), quotas are invariably selected as the main regulation instrument. This is because quotas are generally transparent, equitable, easy to understand, and effective in reducing demand without impacting incomes. Their implementation on wells, however, requires a major enforcement capacity. Their main drawback is their limited capacity to adjust to changes in demand. The present case provides such an example, where inefficiencies arise from the disincentive they generate for citrus and banana growers to adopt less water-intensive crops. A careful downward adjustment of quotas, as implemented since 1999, is, however, effective in skimming off the 'slack'.

Although the two situations show many commonalities, the comparison also evidenced a few meaningful discrepancies. The first difference is the possibility offered to highlanders to expand their plots. This allows them to capitalize on possible water savings and to increase cultivated areas (and benefits) in proportion. Since they may benefit directly from their financial or managerial efforts it is more interesting for them to improve efficiency than in the valley, where the sole reduction in the water bill (sometimes complemented by gains in yields) offers a limited incentive, while benefits go to Amman in the form of increased supply. Second, quotas in the highlands are merely thresholds which can be exceeded at limited cost, while those in the valley are rigid and cap diversions (although informal arrangements may offer some way out). Third, water supply in the highland is very reliable because it depends on individual wells and compact networks; in contrast, allocation and distribution in the valley are much more complex both technically (regulation of the KAC, rotation between farmers within pressurized networks, etc.) and socially (practices are embedded in complex social and political contexts). This difference explains why efficiencies are higher in the highlands (with the additional benefit that return flows tend to return to the aquifer while in the valley they mostly go to a sink: the Dead Sea). In sum, water management is technically simpler in the highlands but enforcement and control are problematic, while the opposite is true in the valley, where quotas are effective in controlling water use but management is heterogeneous and a uniform efficiency hard to achieve.

APPENDIX. Current and proposed irrigation water tariff structure in the Jordan valley. (From FORWARD, 2000.)

Water quality	Usage block (m³/month/3.5 ha maximum)	Irrigation tariff (per 1000 m³)	
		Current	Proposed
Freshwater	0–2500	JD8 ($11.5)	JD15 ($21.6)
	2501–3500	JD12 ($17.3)	JD30 ($43.2)
	3501–4500	JD20 ($28.8)	JD45 ($64.8)
	Over 4500	JD35 ($50.4)	JD55 ($79.2)
Low-quality water	0–2500	JD8 ($11.5)	JD8 ($11.5)
(freshwater mixed with	2501–3500	JD12 ($17.3)	JD12 ($17.3)
treated effluents or	3501–4500	JD20 ($28.8)	JD20 ($28.8)
highly saline water)	Over 4500	JD35 ($50.4)	JD35 ($50.4)

In conclusion, we can observe that there is pervasive overenthusiasm about what can be achieved through pricing policies, and that policy objectives are often listed without due attention to the contradictions they entail and the trade-offs they imply. Expectations of the ASAL, for example, were high but the goals of economic efficiency, equity and environmental sustainability central to the definition of Integrated Water Resource Management are not easily reconciled. In both, the highlands and the valley, substantial increases in volumetric charges would not elicit major water savings but would further depress the income from low-value or extensive crops. A shift towards high-value crops would not only raise water productivity but also entail a transfer of wealth to the government and to wealthier entrepreneurs, an evolution which is so far not considered desirable or politically palatable by Jordanian decision makers. It is therefore essential that negative incentives be accompanied by positive measures offering attractive alternatives (market options, subsidies for modernization, technical advice, etc.) and exit options with compensation.

Acknowledgements

This research has been supported by the Comprehensive Assessment of Water Management in Agriculture, the International Water Management Institute, the Ministry of Foreign Affairs, Government of France and the Institut de Recherche pour le Développement, France.

References

Al-Hadidi, K. (2002) *Valuation of Irrigation Water – The Case of Jordan*. Water Demand Management Forum on Water Valuation, 25–27 June 2002, Beirut, Lebanon. International Development Research Centre, Cairo.

Al-Weshah, R.A. (2000) Optimal use of irrigation water in the Jordan valley: A case study. *Water Resources Management* 14, 327–338.

Al-Zabet, T.G. (2002) Integrated agricultural and water management in the Jordan valley. In: Özay, M. and Biçak, H.A. (eds) *Modern and Traditional Irrigation Technologies in the Eastern Mediterranean*. International Development Research Centre, Ottawa.

ARD (Associate in Rural Development) Inc. and USAID (2001a) *Groundwater Management Action Plan, Amman Zarqa Basin Highlands*. USAID, Amman, p. 63.

ARD and USAID (2001b) *Plan for Managing Water Reuse in the Amman-Zarqa Basin & Jordan Valley*. Water Reuse Component Working Paper. Water Policy Support, Ministry of Water and Irrigation, Amman.

Arrojo, P. (2001) The impact of irrigation water pricing policy in Spain. In: Pricing water. Economics, environment and society. *Conference Proceedings, Sintra*. European Commission, Brussels, pp. 177–184.

ASAL-JORDAN – Agricultural Sector Adjustment Loan (1994) Issues in water pricing. Working Paper (Draft). Amman.

Berbel, J., Gutiérrez, C. and Viaggi, D. (2005) Summary and conclusions. In: Berbel, J. and Gutiérrez, C. (eds) *WADI – The Sustainability of European Irrigated Agriculture under Water Framework Directive and Agenda 2000*. European Commission, Joint Research Centre, Institute of Prospective Technological Studies. Seville, Spain, pp. 173–200.

Berkoff, J. (1994) Issues in Water Pricing. Working Paper. Jordan Agricultural Sector Adjustment Loan. The World Bank, Washington, DC.

Chebaane, M., El-Naser, H., Fitch, J., Hijazi, A. and Jabbarin, A. (2004) Participatory groundwater management in Jordan: Development and analysis of options. Groundwater: From development to management. *Hydrogeology Journal* 12 (1), 14–32, Springer.

Checchi & Company Consulting and Devtech Systems (2003) *Evaluation of USAID/Jordan's Strategic Objective 2: Improved Water Resources Management*. Final report Prepared for United States Agency for International Development (USAID).

Courcier, R. and Guérin, C. (2004) *Irrigation Optimization in the Jordan Valley: Main Lessons Learnt (2000–2004)*. French Regional Mission for Agriculture and Water, Amman.

Courcier, R., Venot, J.P. and Molle, F. (2005) *Historical Transformations of the Lower Jordan River Basin (in Jordan): Changes in Water Use And Projections (1950–2025)*. Comprehensive Assessment Research Report 9. Comprehensive Assessment Secretariat. Colombo.

Doppler, W., Salman, A., Al-Karablieh, E.K. and Wolf, H. (2002) The impact of water price strategies on irrigation water allocation under risk – The case of Jordan valley. *Agricultural Water Management* 55(3), 171–182.

Elmusa, S. (1994) *A Harvest of Technology: The Super-green Revolution in the Jordan Valley*. Centre for Contemporary Arab Studies, Georgetown University, Washington, DC.

Fitch, J.B. (2001) *Curtailment of Groundwater Use for Irrigated Agriculture in the Amman-Zarqa Basin. Uplands: An Economic Analysis*. For ARD-USAID, p. 38 + Appendixes. USAID, Amman.

FOE (Friends of the Earth) (2002) Jordan NGO shadow report: Nongovernmental report on the way to sustainability. Presented at The World Summit for Sustainable Development. Friends of the Earth, Middle East. www.foeme.org

FORWARD (Fostering Resolution of Water Disputes Project) (1998) *The JVA Cost/Tariff Model, Operation & Maintenance and Capital Costs Report,* 1998. Amman.

FORWARD (2000) *Assessment of Water Quality Variations in the Jordan Valley*. Amman.

Ghezawi, A. and Dajani, H. (1995) *Jordan's Water Sector Facts Manual*. Royal Scientific Society, Amman.

GTZ (Deutsche Gesellshaft für Technische Zusammenarbeit) (1993) *Study for the Recovery of O&M Costs of Irrigation Water in Jordan*. Report to the Jordan Valley Authority. Amman.

GTZ (2004) *Natural Resources and Governance: Incentives for Sustainable Resource Use*. Manual. Berlin.

Hanson, B.R. (2000) *Technical Report: Development of Irrigation Advisory Service Program in the Highlands Area, Groundwater Management Component*. MWI/ARD Jordan WRPS Task Order, Amman.

Hussein, I.A.J. (2002) Water planning in Jordan: Future scenarios. *Water International* 17(4), 468–475.

Jabarin, A. (2001) *Curtailment of Groundwater Use for Irrigated Agriculture in the Amman-Zarqa Basin Uplands: A Socio-Economic Analysis*. MWI and USAID-WRPS, Amman.

JICA (Japan International Cooperation Agency) (2004) The study on water resources management in the Hashemite Kingdom of Jordan. Draft Report Vol. I & II + Annexes. Prepared for the Ministry of Water and Irrigation, Amman. Yachiyo Engineering Co. Ltd, Tokyo.

JRVIP (Jordan Valley Improvement Project) (2001a) *Jordan Valley Improvement Project Phase A – Report 1: Institutional Reform*. The Hashemite Kingdom of Jordan, MWI, JVA, Amman. February 2001.

JRVIP (2001b) *Jordan Valley Improvement Project Phase A – Report 2: Water Resources*. The Hashemite Kingdom of Jordan, MWI, JVA, Amman. February 2001.

JRVIP (2001c) *Jordan Valley Improvement Project Phase A – Report 5: Agricultural Economics*. The Hashemite Kingdom of Jordan, MWI, JVA, Amman. February 2001.

KfW, GITEC, AHT Group AG, CEC and Ministry of Water (2006) *Feasibility Study for the Reuse of Treated Wastewater in Irrigated Agriculture in the Northern Jordan Valley – RRT2W – Workshop N 4 – Benefits and Risks of Treated Wastewater in Irrigated Agriculture*. AMRA Crown Plaza, Amman. 19–20 February 2006.

Khouri, R.G. (1981) *The Jordan Valley: Life and Society Below Sea Level*. Longman, London and New York.

McCornick, P., Haddadin, M. and Sabella, R. (2001) *Water Reuse Options in the Jordan Valley*. Water reuse component, Water Policy Support Activity. USAID, Amman.

McCornick, P.G., Taha, S.S.E. and El-Nasser, H. (2002) *Planning for Reclaimed Water in the Amman-Zarqa Basin and Jordan Valley*. American Society of Civil Engineering – Environmental and Water Resources Institute, Conference and Symposium on Droughts and Floods. Roanoke, Virginia.

Nachbaur, J.W. (2004) *The Jordan River Basin in Jordan: Impacts of Support for Irrigation and Rural Development.* MREA Working Paper. French Regional Mission for Water and Agriculture, Amman, 50 pp.

Petitguyot, T. (2003) *Water Consumptions and Available Resources: How to Improve JVA Allocation System in 41 North Scheme?* ENGREF, MREA Working Paper. French Regional Mission for Water and Agriculture, Amman.

Pitman, G.T.K. (2004) *Jordan: An Evaluation of Bank Assistance for Water Development and Management. A Country Assistance Evaluation.* World Bank, Washington, DC.

Richards, A. (1993) *Bananas and Bedouins: Political Economy Issues in Agricultural Sector Reform in Jordan.* USAID, Near East Bureau. Amman.

Salman, A. (2001) *Jordan Valley Improvement Project (JRVIP): Economic and Financial Assessment.* The Hashemite Kingdom of Jordan, MWI, JVA, Amman. February 2001.

Salman, A.Z. and Al-Karablieh, E. (2004) Measuring the willingness of farmers to pay for groundwater in the highlands areas of Jordan. *Agricultural Water Management* 68(1), 61–76.

Salman, A., Raddad, K., Shatanawi, M. and Al-Qudah, H. (2002) The economics of groundwater use in agriculture under different water prices and supply regimes in the upland area of Jordan. In: Hamdy, A., Lacirignola, C. and Lamaddalena, N. (eds) *Water Valuation and Cost Recovery Mechanisms in the Developing Countries of the Mediterranean Region.* Bari: CIHEAM-IAMB, 2002. pp. 177–190 *(Options Méditerranéennes: Série A. Séminaires Méditerranéens; n. 49).*

Salman, A.Z., El-Karablieh, E.K. and Fisher, F.M. (2005) An inter-seasonal water allocation system (SAWAS). *Agricultural Systems* 68(2001), 232–252.

Shatanawi, M. and Salman, A. (2002) Impact of full cost recovery of irrigation water on the farming economics in the Jordan valley. In: Hamdy, A., Lacirignola, C. and Lamaddalena, N. (eds) *Water Valuation and Cost Recovery Mechanisms in the Developing Countries of the Mediterranean Region.* CIHEAM-IAMB, Bari. pp. 105–121. (*Options Méditerranéennes: Série A. Séminaires Méditerranéens; n. 49).*

Shatanawi, M., Fardous, A., Mazahrhi, N. and Duqqah, M. (2005) Irrigation system performance in Jordan. *Options Méditerranéennes, Serie B* 52, 123–132.

Small, L.E. and Carruthers, I. (1991) *Farmer-Financed Irrigation: The Economics of Reform.* Cambridge University Press, Cambridge.

THKJ (The Hashemite Kingdom of Jordan) (2004) *National Water Master Plan.* German Agency for Technical Cooperation, Jordan.

THKJ and JVA (Jordan Valley Authority) (1988) *Jordan Valley Development Law No. 19 of 1988.* Amman.

THKJ and JVA (2001) *Amendment to the Jordan Valley Development Law No. 19 of 1988.* Amman.

THKJ and JVA (2004) Regulation of irrigation water usage and its monitoring for 2003. Regulation issued according to the rule (24-E) of the Jordan Valley Development law No. 19 of 1988 further adjusted by the By-law number 30 for year 2001.

THKJ and MWI (Ministry of Water and Irrigation) (1997a) *Water Utility Policy.* Amman, p. 14.

THKJ and MWI (1997b) *Jordan's Water Strategy and Policies.* Amman, p. 7.

THKJ and MWI (1998a) *Groundwater Management Policy.* Amman, p. 7.

THKJ and MWI (1998b) *Irrigation Water Policy.* Amman, p. 8.

THKJ and MWI (1998c) *Study on Strategic Aspects of Irrigation Water Management in the Jordan Valley.* JVA, GTZ, Amman.

THKJ and MWI (2002a) *Strategic Issues Facing the Jordan Valley Authority.* Amman.

THKJ and MWI (2002b) *Underground Water Control By-law. By-law No. 85 of 2002.* Amman.

THKJ and MWI (2004a) *Amendment to the Underground Water Control By-law. By-law No. 85 of 2002.* Jordan.

THKJ and MWI (2004b) *Water Sector Economics.* Amman.

THKJ and MWI (2004c) *Tariff Systems in the Water Sector.* Amman.

THKJ, MWI and JVA (2002) *2003–2008 Strategic Plan.* Ministry of Water and Irrigation, Amman.

USAID (United States Agency for International Development) (2006) *Jordan Water Strategy.* WATERIQC. Integrated water and coastal resources management. http://www.wateriqc.com/

Van Aken, M. (2004) *Social and Cultural Aspects of Current and Future Governance for the Management of Water Resources in the Jordan River Valley.* French Regional Mission for Water and Agriculture. Amman.

Venot, J.-P. (2004) *Changes in Water Management and Irrigated Agriculture in the Lower Jordan River Basin in Jordan – A Technical Review of Irrigated Farming Systems: Present Situation and Impacts of the Water Management Changes in Prospect.* Main Report-Technical Approach. Working Paper INA P-G, MREA, IWMI. French Regional Mission for Water and Agriculture. Amman.

Venot, J.P., Molle, F. and Hassan, Y. (2007) *Irrigated Agriculture, Water Pricing and Water Savings in the Lower Jordan River Basin (in Jordan)*. Comprehensive Assessment Research Report. Comprehensive Assessment Secretariat. Colombo. Forthcoming.

Wolf, G., Gleason, J.E. and Hagan, R.E. (1996) Conversion to drip irrigation: Water savings, facts or fallacy, lessons from the Jordan valley. In: Greydanus and Anderson (eds) Irrigation Water Conservation, Opportunities and Limitations. *Proceedings Water Management Seminar, October 1995. USCID, Sacramento.*

World Bank (2001a) *Jordan – A Pioneer in Water Sector Reform*. Hashemite Kingdom of Jordan Update Fourth Quarter 2001. World Bank, Washington, DC, pp. 11–15.

World Bank (2001b) *The Hashemite Kingdom of Jordan: Water Sector Review Update*. Main Report. World Bank, Washington, DC.

World Bank (2002) *Report No. PID10046*. Project Name: Jordan-Valley Development. P. 7. Washington, DC.

World Bank (2003) *Performance Assessment Report: Jordan*. Agricultural Sector Adjustment (Loan No. 3817-JO) and Agricultural Sector Technical Support Project (Credit No. 3818-JO). World Bank, Washington, DC.

11 Water Pricing in Tadla, Morocco

P.J.G.J. Hellegers, C.J. Perry and T. Petitguyot

Introduction

In 2002, Morocco had a population of 29.6 million of which 43% lived in rural areas; about 35% of the population are farmers. Agriculture accounts for 16.1% of the GDP, and average per capita income was $1190 (World Bank, 2003). The total area of Morocco is 71 million ha (including 26 million ha of Sahara), of which only 9 million ha are utilized as the agricultural area (13%). The average annual rainfall is less than 300 mm, but is variable in time and space (50 mm in Saharan zones and 2000 mm in mountainous regions).

Morocco's climate makes rain-fed agriculture uncertain and of generally low productivity, especially in the southern areas where rainfall is highly variable and, on average, far less than potential evapotranspiration. Production from rain-fed arable land consequently varies widely. About 1.6 million ha can be potentially irrigated, and 1.2 million ha (75%) are currently irrigated, of which 55% is government-managed, 30% owned and managed by local communities and 15% (mostly irrigated with groundwater) privately developed (FAO, 2001).

Irrigated areas produce 45% of agricultural value added and 75% of agricultural exports (Ait Kadi, 2002). Irrigation currently accounts for 88% of water withdrawals (domestic and industrial use account for 8% and 4%, respectively). The average availability of water is just 1045 m^3/person/year and projected increases in population are expected to reduce this value to about 750 m^3/person/year by 2020 (El Yacoubie and Belghiti, 2002).

In 1990, the estimated national water balance showed an availability of 11 Bm^3 with demand at 10.9 Bm^3. The supply of water is expected to rise to 16.8 Bm^3 by 2020 (as a result of dam construction and the development of additional aquifers). Demand for water is expected to be higher at 17.6 Bm^3, with irrigation accounting for 4.8 Bm^3 (70%) of this increase (Ait Kadi, 2002). Although these values are estimates, they indicate that Morocco's currently developed resources are fully utilized.

An additional concern is the deteriorating water quality, with increasing amounts of water needed to flush and dilute pollution loads (particularly high salinity and sediment).

This chapter focuses on the Tadla region. In Tadla, because of the increasing deficit of surface water farmers use groundwater. Water tables are falling and the water is often highly saline, prompting concern over the sustainability of groundwater development.

Overall, the main factor constraining agricultural production is the availability of water. With scarcity of canal water and overexploitation of groundwater, a number of policy-relevant issues have emerged:

- Reducing overall water consumption in agriculture;
- Increasing the productivity of the water consumed;
- Balancing the supply of, and the demand for, groundwater;
- Avoiding soil and water salinization;
- Providing a sustainable water service through better maintenance and cost recovery.

The role that volumetric water pricing can play in addressing these issues in Tadla is not clear. The main aim of this chapter is to assess the potential role of the water pricing policy. To achieve this aim the way water is currently allocated will be described and insight will be provided into the price, costs and returns to irrigation water in Tadla.

First, the Tadla scheme is described. Next, the price, cost and returns to water are studied. An analytical framework is applied to assess the value of production and contribution of water to that production. Then the possible impact of policy options is described. Finally, conclusions are drawn.

The Tadla Scheme, Policies, Infrastructure and Institutions

The Tadla scheme

The Tadla region is a plain 70 km long and 40 km wide. The cultivated area covers 255,000 ha, including 137,500 ha of rain-fed land and 117,500 ha of irrigated land. The Tadla irrigation system is the oldest large-scale scheme in Morocco. First operated in 1929, it consists of two separate subnetworks of lined open canals which receive water by gravity from two dams. These subnetworks are:

- Beni Amir, right bank of river Oum Er-Rbia, 27,500 ha, irrigated from the Ahmed El Hansali dam (670 Mm^3);
- Beni Moussa, left bank, 69,600 ha, irrigated from the Bin el Ouidane dam (1.30 Bm^3).

According to the initial project design, Beni Amir needs 420 Mm^3 of water and Beni Moussa, 710 Mm^3. However, since the 1980s, considerably less water has been allocated to the scheme (see section under *Water allocation at regional level and within the scheme*). In 2003, only 150 Mm^3 were available for Beni Amir and 350 Mm^3 for Beni Moussa (36% and 49% of the original allocation, respectively). As a result of this deficit, private groundwater development is widespread.

Water in national policies

Irrigation-sector development

Government policy in the agriculture sector has favoured investments in irrigation since 1968, when King Hassan II decided that 1 million ha should be irrigated by the end of the 20th century (this is referred to as the 'million hectares' policy). These investments have accounted for more than 65% of the total public investments in agriculture since 1965 (Herzenni, 2001). The objectives of this investment policy and irrigation it has supported are:

- To improve self-sufficiency through a better coverage of basic food needs;
- To find an equilibrium in the 'trade balance' through the development of exports;
- To improve the living conditions of the rural population;
- To add value to agricultural products through the development of agro-industries.

Morocco has adopted an integrated approach to large-scale irrigation development. Nine modern large-scale irrigation schemes have been established; they are government planned and financed, and each is managed by a Regional Office for Agricultural Development (ORMVA). The basic philosophy is that 'to attain the desired objectives, it is not sufficient to construct irrigation infrastructure as rapidly as possible, the state must also create the conditions enabling development to take place.' A comprehensive framework for this policy is defined by a variety of laws grouped in the Code of Agricultural Investments of 1969. The code is regarded as a contract between the state and the country's farmers to improve the national economy through irrigation development (Ait Kadi, 2002):

- The state finances the dams, the irrigation network and necessary on-farm development. Through ORMVA, it provides credit, selected seeds, fertilizer, farm equipment, etc. Finally, it guarantees the prices of certain crops (mainly sugarbeet and sugarcane) through contracts.
- In turn, the farmer is obligated to farm his irrigated land in the national interest and to repay the state 40% of the investment costs and 100% of the O&M costs through a land-improvement tax and volumetric water charges.

Water allocation and management

The original concept of irrigation in Morocco assumed relatively plentiful water, managed at project level, and provided for controlled cropping patterns, so that irrigation schedules could be set in relation to a known crop demand determined in advance by the government. This practice was abandoned in the 1980s, and farmers are now free to choose their own cropping patterns – generally increasing the potential demand. In parallel with this liberalization, water availability has declined significantly to schemes such as Tadla, and Morocco has adopted a policy of basin-level allocation of water among competing uses. Water management at scheme level is now based primarily on a rationing system, with each farmer given an entitlement of water, which the farmer may use, but there is generally less water than the farmer would wish to receive. A national program launched in 1993 aimed to increase the size of existing irrigation schemes and encourage more efficient water use.

The Water Law

A major step in water policy was achieved through the Water Law that was passed in September 1995. This law establishes institutions and defines rules for the sustainable use of water resources. Seven financially autonomous River Basin Agencies were created as a result of this law. The Agencies prepare a management plan for all water resources in their basin and implement it, deliver authorizations for any use of the public domain, and are responsible for the quantitative and qualitative monitoring of the resources.

Irrigation infrastructure and water distribution in the Tadla scheme

Water allocation at regional level and within the scheme

Among the nine large-scale irrigation schemes, the annual planned average water use is 5100 m^3/ha, but varies between 3000 m^3/ha/year (Tafilalet and Ouarzazate) and 7100 m^3/ha/year. (Tadla) (Benjelloun Touimi, 2002).

The amount that can be delivered to farmers depends on the water allocated to the scheme; this is decided each year at the level of the River Basin Agency. The amount to be released is calculated according to the projected inflows and available reserves in the two upstream dams; the amount released may be adjusted during the year depending on the actual rainfall. This release is shared between Tadla and other downstream irrigation schemes. As a result of the chronic droughts, irrigation expansion and the demand from other schemes, the allocation to Tadla is substantially less than the amount initially designed. In 2001–2002, only 27% of what was initially designed (710 Mm^3) was delivered to the scheme (Fig. 11.1).

As a result, irrigation in Tadla faces a severe shortage of water and the distribution rules have been adapted to deal with a shortage situation. Now that demand largely exceeds supply, no demand-oriented management can be carried out, and water

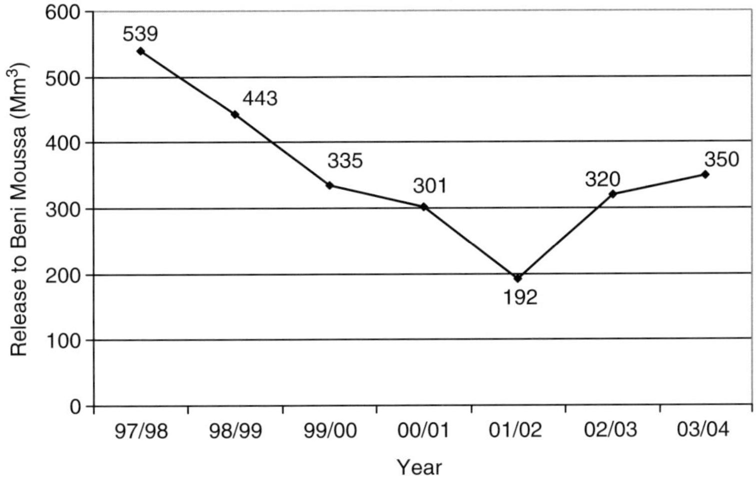

Fig. 11.1. Annual releases from the Bin El Ouidane to the Beni Moussa scheme. (From ORMVAT, 2004.)

Channel>>					
Farm 1 Plot A	Farm 1 Plot B	Farm 1 Plot C	Farm 1 Plot D	Farm 1 Plot E	Farm 1 Plot F
Farm 2 Plot A	Farm 2 Plot B	Farm 2 Plot C	Farm 2 Plot D	Farm 2 Plot E	Farm 2 Plot F
Farm 3 Plot A	Farm 3 Plot B	Farm 3 Plot C	Farm 3 Plot D	Farm 3 Plot E	Farm 3 Plot F
Farm 4 Plot A	Farm 4 Plot B	Farm 4 Plot C	Farm 4 Plot D	Farm 4 Plot E	Farm 4 Plot F
Farm 5 Plot A	Farm 5 Plot B	Farm 5 Plot C	Farm 5 Plot D	Farm 5 Plot E	Farm 5 Plot F

Fig. 11.2. Schematic of watercourse and farm plot layout (under which Trame B farms had the same cropping pattern).

allocation among farmers is based on a rationing system.

Water distribution

While the total seasonal allocation to the farmer is fixed, the schedule of delivery and the amount of water delivered at each water turn are based on the crops. In case of unexpected water scarcity during the season, priority of water delivery is given to specific crops. Farmers cannot transfer unconsumed water to another turn, so most of them take all the water they can get at each turn. Actual management therefore is effectively quota-based.

The infrastructure was designed for a specific situation, namely the irrigation of an obligatory cropping pattern at the farm level, with crops organized in homogeneous blocks served by a common watercourse. The system was logical when cropping patterns were enforced so that Plot A (Fig. 11.2: the tertiary channel is the bold line at the top; watercourses are indicated by the vertical double lines) for each farm was under the same crop and could be provided with a water delivery schedule suited to that crop (Cornish and Perry, 2003). Known as the Trame B model, this system simplified water scheduling and management because each watercourse was operated to serve a specific crop and its specific water requirements.

However, in the 1980s cropping patterns were liberalized to enable water to be distributed on a farm basis rather than on a crop basis, with the result that the 30-year-old design no longer corresponds to the current management situation. However, the ORMVA management still issues clear 'guidance' on feasible cropping plans prior to each season, based on the anticipated water availability per hectare, and the demand of individual crops (so that farmers opt for a larger area of less water – or a smaller area of more water-demanding crops).

Each farm has six plots, arranged horizontally. The left-most watercourse first serves Farm 1 Plot A, followed by Farm 2 Plot A, through to Farm 5 Plot A. Irrigation then continues to Farm 1 Plot B on to Farm 5 Plot B, through Tertiary 2, and so on. In any given irrigation turn, a farmer would have to come back as many as six times to irrigate his farm. This operating pattern is matched by the design of the infrastructure, which has division structures at each level to ensure accurate provision of the proper discharge to each area.

In recent years, a provisional allocation has been established at the beginning of the irrigation season (September) and farmers are informed about it. During the year, the actual volume delivered to a farmer is calculated by multiplying the number of hours of his turn by the flow rate (generally 30 l/s). This rationing provides a relatively transparent and equitable means of allocating water, ensuring that consumption of water is controlled.

In such a constrained system, the volumetric water fees paid by farmers (see section on *Price paid by farmers*) serve predominantly as a means of cost recovery.

Irrigation at scheme, farm and plot level

From the dam, water is conveyed by gravity through a system of concrete-lined channels, divided into a primary, secondary and tertiary levels (see Table 11.1). At the tertiary level, channels are suspended on pillars and can carry 120 l/s before branching off into 30 l/s earthen watercourse channels from which the farmers take water. At each branching point, there are *modules à masque* (stepwise or baffle distributors), which provide supplies to offtaking channels, relatively independent from the upstream water level in the parent canal.

Table 11.1. Cumulated length of lined channels in the Tadla irrigation scheme. (From ORMVAT, 2004.)

Channel type	Cumulated length (km)
Principal	200
Primary and secondary	360
Tertiary	1800
Total	2360

Field observations indicate that while individual modules can be adjusted to various flow rates (30, 60 or 90 l/s), most are fixed at a particular rate, ensuring consistent patterns of delivery. Since the water demand schedule for the various crops is different, the watercourses are arranged to run at right angles to the ownership boundaries so that each watercourse can be operated to serve the needs of a specific crop (Fig. 11.2).

The most frequently used irrigation technique at field level is the traditional *robta*. A plot is divided into several small basins, each one of about 10 m^2, irrigated via *seguias* (earthen watercourses) that convey water through the farms. The initial land-levelling has been gradually degraded as a result of the agricultural practices and the manual digging of the irrigation basins and watercourses in the fields.

The ORMVAT estimates irrigation efficiency (including internal conveyance) at farm level to be 50%; that is, only half of the delivered water is directly used by the crop. Taking distribution losses into account, the overall system efficiency is even lower, namely less than 45%. However, much of this wasted water is reused in the system: many drains are tapped through individual pumping, and the infiltrated water is the major inflow to the underlying aquifer, from which a large number of farmers pump water to complement surface supply. In a way, the fact that water tables are generally falling and large-scale waterlogging is not reported suggests that the estimated losses are already being fully exploited through local reuse. This issue is important, given the claim that reducing losses may improve availability only if 'losses' are not already being recaptured, although this particular reuse is accompanied with a decrease in water quality (Seckler, 1996; Cornish and Perry, 2003).

Groundwater use

All the latest studies (Hammani *et al.*, 2004) show that irrigation losses account for the major part of the inflow to the shallow aquifer. In the 1980s, severe and repeated droughts led many farmers to invest in pumping devices; they were encouraged to do so by the state, which provided subsidies and technical support. Nowadays, about 10,000 wells are used in the schemes and approximately 40% of the farms have wells.

Most of the pumps are powered by diesel engines, with an average discharge of 10–15 l/s. Farmers generally use groundwater to supplement surface water. As groundwater is generally more saline than surface water, conjunctive use at plot level may be necessary to avoid soil degradation and yield losses. However, farmers are much more concerned with quantity issues, and these medium-term risks are outweighed by the demand to increase the present supply (Petitguyot, 2003).

There has been a regular decline in the level of the shallow aquifer for 20 years now, and there is no regulation to control withdrawals of groundwater. As a result, many wells have dried up. Farmers who can afford new investments now deepen their wells or sink deeper tube wells (wells still represent 89% of the total but 25% are non-functional). Whereas shallow resources are of bad quality and may only be used for agriculture, deep aquifers are exploited by urban and industrial users, which will result in competition.

Institutions and governance

Water allocation in the river basin

According to the Water Law, River Basin Agencies (which are under the responsibility of the Ministry of Public Works) are in charge of developing and allocating water resources. Each year, the agencies and the basin's stakeholders agree on a programme for water allocation. Urban and industrial needs have priority over the agriculture sector. In the Tadla area, water use for electricity production has the lowest priority, and water for irrigation is released according to agricultural needs only.

Although farmers are represented on the agency board which sets up the annual programme, their influence is negligible (2 members out of 35 on the board) and only the ORMVA may interact with a significant power to negotiate agricultural allocation.

Organization of the ORMVAT

Morocco's nine major irrigation systems are operated by ORMVAs, which are semi-autonomous, regional public institutions under the responsibility of the Ministry of Agriculture. They are in charge of agricultural development (in both the irrigated sector and the surrounding rain-fed areas), including irrigation design, O&M and fee collection. About 1000 people work at the ORMVA in Beni Mellal, which is responsible for Tadla (400 on water management, 300 on extension and agricultural development, 300 on administrative tasks).

Pricing and cost recovery

The ORMVAs' financial resources come from fees paid by users, particularly irrigation water fees, and from state subsidies (investment subsidies and/or subsidies to balance operating budgets). An ORMVA accountant (who works for the Ministry of Finance) is responsible for supervising the collection of water fees. There are two forms of cost recovery:

- Recovery at source: This method applies to farmers who have production contracts with agro-industrial units, such as sugar mills. Here, the mill pays the ORMVA any water fee due, before paying the farmer for his crop.
- Direct payment: Farmers are individually invoiced every quarter using a cus-

tomer code, with invoices delivered by the *aiguadier* (ditch rider or water guard). Payment is due twice a year. Farmers incur penalties for late or non-payment (after 1 month, suspension of supply; after 2 months, an 8% increase in the amount due; after 1 year, there should be a court action). In reality, the issue of non-payment is strongly related to land status, as farmers who share the same undivided property receive only one invoice and face difficulties with respect to the division of the bill. Instead of court action, water supply pipes to many farmers in this situation are disconnected from the network. It is worth noting that in Tadla, the rigorous management of non-payers (they are quickly disconnected) means that the level of invoice payment is very high (more than 90%).

Between 1995 and 1998, a novel system was introduced for water accounting by farmers. In pilot areas, each farmer received a water consumption 'cheque book'. For each water turn, the farmer filled in a cheque for the ditch rider, and kept a copy for his own records. A part of the annual volume was allocated to each farmer for the season but the schedule of deliveries was variable, based on individual demand (within reason and subject to competing demands). The cheque book kept a running account of the total amount of water used. This approach was an innovative means of combining rationing with flexibility (the infrastructure allows for flexible delivery of the allocated quota), but proved difficult to manage during the severe drought of 1998. However, in 2002, this system was reintroduced, and is used as an incentive for farmers who use modern irrigation techniques and are able to irrigate a larger area per unit of water delivered.

Water user associations in the Tadla area

At the beginning of the 1990s, the government decided to develop participatory irrigation management, giving farmers a greater role in irrigation management. In Tadla, farmers showed little interest in the incentive offered to participate (i.e. a reduction in water fees), according to Papin (2003). Farmers also lack the historical experience in (organizing) irrigation that exists in other parts of the country. A law passed in 1990 provided a legal basis for establishing water user associations (WUAs), with responsibility for managing irrigation at the tertiary level. Tadla has 29 registered WUAs (11 in Beni Amir and 18 in Beni Moussa), representing 41% of farmers in an area covering 44,540 ha. However, most of these associations are not operational. A study carried out in 2001 (ENGREF, IAV Hassan II and CNEARC, Tadla, 2001, unpublished data) reported that only one WUA was active in Beni Moussa, and that this could be explained by a diversification of its activities to other sectors (road construction, basic education). The WUAs did not prove to be successful, and many farmers refused to pay the charges to finance a WUA. This could be seen as a compliment to the operation of the irrigation schemes by the ORMVA – the farmers found this satisfactory and did not see the need to add an additional layer of management.

Price, Costs and Returns to Water

Price paid by farmers

Surface water

Canal water fees are based on the Agricultural Investment Code of 1969 – a general law on agricultural water management, water pricing and service fee recovery. The Code provides a comprehensive cost recovery structure, including the full recovery of O&M costs (through water fees) and the partial (40%) recovery of capital costs (through the water fee), indexed over time to inflation. Water is charged on the basis of quantity received, which is metered in the case of pressurized systems and calculated on the basis of time and the nominal flow rate in the case of surface systems. Water fees can be increased, but the new fee must be approved by the Ministers of Public

Works, Agriculture, and Finance as well as by parliament.

Actual water charges in Morocco are relatively high by international standards and charged according to the volume of water delivered (although payment for at least 3000 m^3/ha is obligatory). In Tadla, the canal water fee in 2002 was \$0.02/m^3 (\$1.00 = MAD8.9). This was the lowest in Morocco because, unlike other areas, Tadla canal systems do not involve pump-lifts. In some regions, the rate is as high as \$0.062/m^3 (Ben Abderrazik, 2002). None the less, the canal water fee in Tadla has steadily increased over time, from \$0.005/m^3 in 1980, to \$0.01/m^3 in 1987–1988, and to \$0.015/m^3 in 1992, but this is, of course, also partly the result of inflation (El Yacoubie and Belghiti, 2002). In regions where pumping is a significant part of operational costs,[1] farmers do not pay the full O&M costs. Instead, these ORMVAs rely on an annual transfer of funds from the central government in order to meet operational expenses, and farmers are not charged for capital costs.

Groundwater

The pumping of groundwater from wells is a private undertaking of the farmers. Well owners pay the full cost of development and O&M. The energy cost can be estimated according to the discharge of the pumps (generally 15 l/s). Various sources indicate an average of \$0.03/m^3 (Papin, 2003; Petitguyot, 2003; Le Grusse *et al.*, 2004). The full cost of groundwater extraction (i.e. including energy costs, amortization and pump maintenance costs) is more difficult to estimate, as it depends on the actual utilization of the pump, the head and other parameters. According to the same sources, the total cost in Tadla is around \$0.06/m^3, with a high variability between farms. Some comments should be made about this value. First, it is not certain that farmers consider this total cost in their daily decisions whether to irrigate or not: investment costs are sunk costs and the marginal costs might be more relevant. Second, Le Grusse *et al.* (2004) note that many tube wells are shared by neighbouring farms, who may thus share the investment burden and reduce the total (and hence unit) cost. Also, compared to surface water, these costs integrate neither qualitative differences such as salinity (lower in canal water) nor an insurance value (groundwater protects farmers against network deficiencies in critical growing stages) that may greatly influence farmers' choice.

Groundwater is generally regarded by farmers as a supplementary resource to be used in the case of a deficit. The gap between the groundwater and surface water tariffs is not much wide and an increase in surface water tariffs might trigger the exploitation of groundwater.

Costs of water delivery

The costs incurred by the supplier in the provision of irrigation water services in Tadla are summarized in Tables 11.2 and 11.3. Annual O&M costs are \$11.5 million (for an area of 92,000 ha), which is \$125/ha/year. Annual total costs are \$13.5 million, which is \$147/ha/year including depreciation on capital. This relatively small difference between the O&M and the full costs is because Tadla is an old project – the first large irrigation project to be built in Morocco – and was (in current prices) therefore comparatively cheap at the time of construction. It requires, however, more maintenance. For a water delivery of 7400 m^3/ha, O&M costs are \$0.017/m^3 and full costs are \$0.02/m^3.

Official statistics indicate that current water charges cover more than the O&M costs (Table 11.4), which is consistent with the estimated farm payment for water (\$145–155/ha). If full water fee collection is achieved – i.e. if all users pay their bills in Tadla – more than 100% of the O&M expenditures are covered. The data indicate that system delivery losses (between diversion and delivery to farmers) are relatively low.

[1]Electricity is charged to ORMVAs at 20% below the commercial rate – around \$0.08/kWh.

Table 11.2. Annual O&M costs including labour (without capital depreciation) in $ million. (From ORMVAT, 2004.)

	Dept. of Irrigation and Drainage		Dept. of Construction		Dept. of Agriculture Development and Extension		Total	
	Amount	%	Amount	%	Amount	%	Amount	%
Direct costs	4.1	70	0.5	68	3.3	69	8.0	69
Indirect costs	1.8	30	0.2	32	1.5	31	3.5	31
Total costs	5.9	100	0.8	100	4.8	100	11.5	100

Table 11.3. Annual total costs (with capital depreciation) in $ million. (From ORMVAT, 2004.)

	Dept. of Irrigation and Drainage		Dept. of Construction		Dept. of Agriculture Development and Extension		Total	
	Amount	%	Amount	%	Amount	%	Amount	%
Direct costs	5.4	71	0.6	69	3.5	70	9.5	71
Indirect costs	2.2	29	0.3	31	1.5	30	4.0	29
Total costs	7.6	100	0.9	100	5.0	100	13.5	100

Table 11.4. ORMVAT budget and expenditures (in $). (From ORMVAT, 2004.)

Year	1996–1997	1997–1998	1998–1999
Provisional budget	11,954,205	10,493,814	10,218,438
Actual expenditure on operations	10,355,227	9,089,175	9,040,417
Pumping costs	161,250	135,361	138,854
Maintenance	58,182	100,206	86,563
Personnel	6,325,682	5,567,629	5,584,896
Costs of other operations	3,810,227	3,286,082	3,230,104
Actual expenditure on new investment	901,136	2,342,474	2,401,667
Total actual expenditure	11,256,364	11,431,649	11,442,083
Income (recovery from fees)	13,322,500	12,264,948	14,293,125

Returns to water

Agricultural production in Tadla consists predominantly of cereals (mainly wheat), sugarbeet, fodder and olive trees (Table 11.5).

A consistent analytical framework was applied to assess the returns to water for the typical cropping patterns observed for various farm sizes. The returns to water are calculated as the value of production, net of input costs, divided by the volume of irrigation

Table 11. 5. Area irrigated by crop in Tadla (5-year average 1998–2003). (From Petitguyot, 2003.)

Crop	Area (ha)	%
Cereals	46,000	39
Sugarbeet	13,800	12
Fodder	25,000	21
Olive	16,800	14
Citrus	8,100	7
Vegetables	8,500	7

Table 11.6. Summary data for Tadla. (From Hellegers and Perry, 2004.)

Farm	1	2	3
Size (ha)	4.8	6.0	7.7
Gross income ($/ha)	1453	1971	996
Net income before water charges ($/ha)	901	1470	612
Water charge if 100% canal ($/ha)[a]	156 (17)	145 (10)	145 (23)
Water charge if 100% well ($/ha)[a]	320 (35)	297 (20)	298 (49)

[a]Values within parentheses indicate % of net income.

water used. The appendix shows the results of a farm survey for three farms in Tadla, ranging in size from 4.8 to 7.7 ha. The first three tables show farm income assuming that irrigation is fully from canal water, while the last three tables show farm income assuming that irrigation is fully from groundwater. In fact, however, most farms use a mixture of sources. The exact mix could not be accurately assessed, so the calculations estimate the extreme cases.

The main crops grown on these farms include wheat (including seed multiplication), fodder crops (lucerne, berseem) and olives. The returns to wheat and broad bean are relatively high compared to the returns to lucerne, which may be explained by the relatively low price of lucerne, as it is often used as fodder for livestock. The appendix shows that the net return to water is about $0.10/m^3.

The key data for this study, summarized in Table 11.6, are gross income per hectare, net income (before water charges) and the proportion of net income (before deduction of water charges) accounted for by water charges. It is important to note that agricultural income given in Table 11.6 relates to crop production only.

These data indicate that farmers in Tadla spend a substantial proportion of their net income (10–23%) on canal irrigation services and even more (20–49%) if they irrigate entirely with groundwater. These results should be considered with care as they do not represent the high variability of production systems in Tadla. They are, however, consistent with other results found by Petitguyot (2003).

Discussion of price, costs and returns to water

The O&M cost of water delivered at the field in Tadla is $0.017/m^3, while the full cost is $0.02/m^3. The current volumetric canal water fee is high ($0.02/m^3) compared to other similar case studies and covers the O&M costs. The marginal cost of groundwater is $0.03/m^3. The costs of canal and groundwater are, however, less than the return to irrigation water ($0.1/m^3). As farmers spend a substantial proportion of their income on canal water, it is likely that current prices discourage wastage and give an incentive to concentrate usage on productive crops. It is, however, not likely that it will balance water supply and demand.

Possible Impact of Policy Options

Groundwater

As far as groundwater is concerned, the principle of state ownership of water has been in place since 1914. To stabilize groundwater extraction, the sustainable aquifer yield and the demand for groundwater need to be balanced. However, there are currently no defined entitlements for the use of groundwater. There is a restriction on the pumping of groundwater (i.e. no deeper than 40 m below the soil surface), although in practice this is rarely enforced and is therefore no effective policy instrument. The majority of farmers install wells without obtaining the required authorization.

An alternative policy instrument aimed at limiting groundwater extraction is currently being drawn up. Under the Water Law, the River Basin Agency is empowered to impose a tax on each volume of water extracted from individual wells ('consumer pays' principle). The administrative costs and technical complexity of charging for extraction on the basis of the number of pumping hours – as currently proposed by the government – would be high, and will not guarantee a reduction in usage (although the implied increase in the unit price of water will provide some incentive to reduce usage there is no assurance that sustainable supply and demand will be properly balanced). Given the problems with the enforcement of existing regulations on the installation and operation of pumps, it is not certain whether hours pumped will be easy to measure and used as an instrument for demand management. It is likely that bribery would increase and meters would be tampered with.

Canal water

The volumetric canal water charge will not reduce water consumption substantially as the level of the charge is only 20% of the returns to water. Rationing eventually governs demand. The present system of charging for canal water would not, in the absence of rationing, achieve a balance between supply and demand. A considerable increase in the price of water would be needed to balance the supply of, and the demand for, canal water. However, such an increase would lead to a significant fall in the returns to agriculture and increased migration to cities.

An additional threat posed by increasing canal water fees is that such an increase is likely to lead to the increased exploitation of groundwater. Moreover, although the recovery of charges is exceptionally high in Tadla, and further increases in canal water fees might reduce the rate of recovery, as has occurred in many other schemes in Morocco, where water fees have increased but not the total income of the water manager.

Further, a substantial increase in charges is likely to lead to a decrease in the rate of recovery, as suggested by El Gueddari (2002), who shows that in Morocco the rise in fees up to the O&M cost level has been paralleled by a decline of fee recovery from over 70% down to 55%. In Tadla, recovery is extremely high because of the strict application of the disconnection procedure in case of non-payment but a total of 8% of the farms are nevertheless reported to have been disconnected and only survive on groundwater (Petitguyot, 2003).

Rationing is therefore a more suitable instrument to govern demand and to foster the productive use of water.

A particular difficulty with volumetric water charges is that they do not ensure appropriate cost revenue levels for the scheme manager. In a dry year, there will be limited water to sell, and revenues will fall proportionately. In a year of high rainfall, demand for irrigation water will be limited, leading to revenue shortfalls. A two-part tariff (a fixed and a volumetric tariff) provides additional security of revenues to the manager.

In summary, Tadla has a technically sophisticated surface irrigation system capable of delivering differentiated irrigation schedules to individual farmers, but simple quota-based rationing is the basis for constraining demand. Volumetric water charges are only used to achieve cost recovery. It should be noted that to overcome scarcity of surface water, many farmers have invested in private tube wells, and that the unit price of this water is more than double that of the supplied surface water. Any increase in water tariff should be considered relative to the impact on this complementary resource, the use of which is not regulated.

Synthesis

The availability of water is, and will continue to be, the key factor constraining agricultural production in Tadla. Deteriorating water quality increases this concern. The scarcity of canal water and the significant exploitation of groundwater in dry years have led to the identification of several policy objectives, e.g. to reduce overall water consumption in agriculture, to increase the productivity of water, to balance the supply of, and demand for, groundwater, to avoid soil and water salinization and to provide a sustainable water service through better maintenance and cost recovery. The main aim of this chapter was to study the potential role of pricing policy in meeting these objectives.

The volumetric canal water fees currently charged in Tadla ($0.02/m^3) cover the O&M costs, but are only about one-fifth of the estimated return to water ($0.1/m^3). Such fees will not reduce water consumption, as supply is rationed through quotas at levels well under crop requirements and which preclude significant savings.

Balancing supply and demand through volumetric charges would require a very considerable increase in the price of water. This is not desirable for two reasons: an increase in the price of canal water would significantly reduce farm incomes, and such an increase could trigger an increase in the use of groundwater. Rationing, which is already used in Tadla, seems the most suitable instrument to govern demand for canal water, and has the additional benefits of low transaction costs, equity and transparency.

Under the current system in Morocco, the regional ORMVAT is responsible for the distribution and allocation of water from the principal canal down to individual farms, and for maintaining the system. The ORMVAT also collects water fees and plays a role in planning cropping patterns and providing agronomic advice.

Thus, Morocco is already using very suitable instruments – namely rationing and some volumetric charging – to govern the demand for canal water and to recover O&M costs. However, attention needs to be paid to policies to control groundwater use in an effective way.

Acknowledgements

The authors would like to thank Marcel Kuper and three anonymous reviewers for their helpful comments.

Appendix: Overview of the Outcome of the Spreadsheets

Table 11.A.1. Tadla farm budgets – canal irrigated. (From Hellegers and Perry, 2004.)

Farm 1 4.8 ha	Income per ha $	Crop area ha	Farm income $	Input costs $	Labour costs $	Water costs $	Net income $	Family labour use days	Family labour return $/day	Water use 000 m³	Gross water return $/m³	Net water return $/m³
Sugarbeet	1060	1.0	1058	443	165	155	295	59	5	7.8	0.1	0.1
Wheat	2680	0.8	2135	195	25	64	1851	22	83	3.2	0.7	0.6
Lucerne	1080	1.8	1944	1466	48	415	15	53	0	20.7	0.1	0.0
Broad bean	1645	0.6	987	60	6	43	878	27	33	2.2	0.5	0.4
Olive 1	920	0.6	552	68	86	52	345	21	16	2.6	0.2	0.2
Berseem	498	0.6	299	69	21	21	189	5	35	1.0	0.3	0.2
	Totals	5.4	6976	2301	352	750	3573	188	19	37.5	0.2	0.1

Cropping intensity = 112%; utilization of family labour = 78%; proportion of family labour in total used = 70%.

Farm 2 6 ha	Income per ha $	Crop area ha	Farm income $	Input costs $	Labour costs $	Water costs $	Net income $	Family labour use days	Family labour return $/day	Water use 000 m³	Gross water return $/m³	Net water return $/m³
Sugarbeet	1400	2.0	2772	1062	506	308	895	74	12	15.4	0.2	0.1
Wheat	2033	2.0	4025	421	147	160	3298	34	97	8.0	0.5	0.4
Broad bean	1410	1.0	1404	102	60	72	1170	47	25	3.6	0.4	0.3
Paprika	2600	1.0	2590	442	9	203	1936	3	565	10.1	0.3	0.2
Olive 2	1040	1.0	1036	83	178	127	648	26	25	6.4	0.2	0.1
	Totals	6.9	11,827	2110	900	870	7947	184	43	43.5	0.3	0.2

Cropping intensity = 116%; utilization of family labour = 77%; proportion of family labour in total used = 46%.

Farm 3 7.7 ha	Income per ha $	Crop area ha	Farm income $	Input costs $	Labour costs $	Water costs $	Net income $	Family labour use days	Family labour return $/day	Water use 000 m³	Gross water return $/m³	Net water return $/m³
Sugarbeet	600	1.5	901	672	4	234	(8)	118	(0)	11.7	0.1	0.0
Wheat	1550	3.4	5251	664	28	273	4286	84	51	13.7	0.4	0.3
Lucerne	480	2.6	1246	1543	14	598	(909)	245	(4)	29.9	0.0	0.0
Broad bean	1410	0.2	271	26	0	14	231	9	25	0.7	0.4	0.4
	Totals	7.7	7669	2905	46	1118	3600	456	8	55.9	0.1	0.1

Cropping intensity = 100%; utilization of family labour = 63%; proportion of family labour in total used = 97%.

Table 11.A.2. Tadla farm budgets – groundwater irrigated. (From Hellegers and Perry, 2004.)

Farm 1 4.8 ha	Income per ha $	Crop area ha	Farm income $	Input costs $	Labour costs $	Water costs $	Net income $	Family labour use days	Family labour return $/day	Water use 000 m³	Gross water return $/m³	Net water return $/m³
Sugarbeet	1060	1.0	1058	443	165	318	132	59	2	7.8	0.1	0.1
Wheat	2680	0.8	2135	195	25	132	1784	22	80	3.2	0.7	0.6
Lucerne	1080	1.8	1944	1466	48	850	(420)	53	(8)	20.7	0.1	0.0
Broad bean	1645	0.6	987	60	6	89	832	27	31	2.2	0.5	0.4
Olive 1	920	0.6	552	68	86	106	291	21	14	2.6	0.2	0.2
Berseem	498	0.6	299	69	21	43	167	5	31	1.0	0.3	0.2
	Totals	5.4	6976	2301	352	1538	2785	188	15	37.5	0.2	0.1

Cropping intensity = 112%; utilization of family labour = 78%; proportion of family labour in total used = 70%.

Farm 2 6 ha	Income per ha $	Crop area ha	Farm income $	Input costs $	Labour costs $	Water costs $	Net income $	Family labour use days	Family labour return $/day	Water use 000 m³	Gross water return $/m³	Net water return $/m³
Sugarbeet	1400	2.0	2772	1062	506	631	572	74	8	15.4	0.2	0.1
Wheat	2033	2.0	4025	421	147	327	3130	34	92	8.0	0.5	0.4
Broad bean	1410	1.0	1404	102	60	147	1095	47	23	3.6	0.4	0.3
Paprika	2600	1.0	2590	442	9	416	1723	3	503	10.1	0.3	0.2
Olive 2	1040	1.0	1036	83	178	261	514	26	20	6.4	0.2	0.1
	Totals	6.9	11827	2110	900	1783	7034	184	38	43.5	0.3	0.2

Cropping intensity = 116%; utilization of family labour = 77%; proportion of family labour in total used = 46%.

Farm 3 7.7 ha	Income per ha $	Crop area ha	Farm income $	Input costs $	Labour costs $	Water costs $	Net income $	Family labour use days	Family labour return $/day	Water use 000 m³	Gross water return $/m³	Net water return $/m³
Sugarbeet	600	1.5	901	672	4	479	(253)	118	(2)	11.7	0.1	0.0
Wheat	1550	3.4	5251	664	28	560	3999	84	48	13.7	0.4	0.3
Lucerne	480	2.6	1246	1543	14	1226	(1537)	245	(6)	29.9	0.0	0.0
Broad bean	1410	0.2	271	26	0	28	216	9	23	0.7	0.4	0.4
	Totals	7.7	7669	2905	46	2293	2425	456	5	55.9	0.1	0.1

Cropping intensity = 100%; utilization of family labour = 63%; proportion of family labour in total used = 97%.

References

Ait Kadi, M. (2002) Irrigation Water Pricing Policy in Morocco's Large-Scale Irrigation Projects. *Hommes, Terre & Eaux* 32(124), 25–33.

Ben Abderrazik, H. (2002) Pricing of irrigation water in Morocco. In: http://www.worldbank.org/

Benjelloun Touimi, M. (2002) Valorisation de l'eau d'irrigation dans les Grands Périmètres Irrigués au Maroc cas des Productions Végétales. In: http://www.worldbank.org/

Cornish, G.A. and Perry, C.J. (2003) *Water Charging in Irrigated Agriculture – Lessons from the Field.* Report OD 150. HR Wallingford, UK, 104 pp.

El Gueddari, A.B.S. (2002) *Système de Tarification de l'eau d'irrigation au Maroc: Principes et Evolution*. FAO Regional Office for the Near East, Cairo.

El Yacoubie, Z. and Belghiti, M. (2002) *Water Valuation, the Case of Morocco, Valuation of Irrigation Water in Morocco*. Ministry of Agriculture, Rural Development and Forestry, Rural Engineering Administration, Development and Irrigation Management Department, Beirut.

FAO (Food and Agriculture Organization of the United Nations) (2001) *Case Studies on Water Conservation in the Mediterranean Region*. IFTRID, Rome.

Hammani, A., Kuper, M., Bouarfa, S., Debbarh, A., Badraoui, M. and Bellouti, A. (2004) Evolution de l'utilisation conjointe des eaux de surface et des eaux souterraines dans le périmètre irrigué du Tadla (Maroc). INCO-Wademed Seminar, 19–21 April 2004, Rabat.

Hellegers, P.J.G.J. and Perry, C.J. (2004). *Water as an Economic Good in Irrigated Agriculture: Theory and Practice*. LEI-Report 3.04.12. Wageningen UR, Wageningen, The Netherlands.

Herzenni, A. (2001) Enjeux de la GPI (gestion participative de l'irrigation) au Maroc. Cas de la Grande Irrigation. Draft.

Le Grusse, Ph., Kuper, M., Hammani, A., Zemzam, S. and Bouarfa, S. (2004) Les strategies d'equipement en station de pompage des petites exploitations agricoles du Tadla. INCO-Wademed Seminar, 19–21 April 2004, Rabat.

ORMVAT (Offices Régionaux de Mise en Valeur Agricole du Tadla) (2004) Monographie du Périmètre du Tadla.

Papin, C. (2003) PIM in Tadla: Policy and field reality: A case study of the Beni Amir irrigation scheme, Tadla Region, Morocco. MSc thesis. Wageningen UR, Wageningen, The Netherlands.

Petitguyot, T. (2003) Agriculture irriguée et utilisations durables des ressources en eau souterraines et de surface – Une exploration micro-économique dans la plaine du Tadla, Maroc. Master thesis. ENGREF, Université Paris X.

Seckler, D. (1996) *The New Era of Water Resources Management*. Research Report 1. IIMI, Colombo, p. 17.

World Bank (2003) Morocco at a glance. Available at: http://www.worldbank.org/data/countrydata/aag/mar_aag.pdf

12 Water Pricing Policies and Recent Reforms in China: The Conflict between Conservation and Other Policy Goals*

B. Lohmar, Q. Huang, B. Lei and Z. Gao

Introduction

In response to growing awareness of water shortages and associated problems, China is debating and establishing a variety of policies to encourage water conservation. Among the policies under consideration is water price reform. Until recently, water in China was viewed as an easily accessible resource that could be harnessed to boost industrial and agricultural production. Thus, water prices were low, if they existed at all. While water shortages have been acknowledged since the early 1980s and local measures to conserve water were established in some areas, it was not until the late 1990s that China embarked on a concerted effort at the national level to promote water conservation and improve the overall productivity of China's relatively scarce water resources. In principle, the ultimate goal of water pricing is not only to generate funds to maintain and improve water delivery systems, but also to conserve water so that it can be allocated to areas where society places a higher value on it. Such areas may be the environment, other sectors of the economy or future water users. Devising mechanisms for doing so, however, is not easy and is rarely achieved. In many areas of the world, including China, water fees are often insufficient to fund operations and maintenance, let alone encourage conservation.

The realization of the need for stricter water conservation and higher water prices, however, comes at a time when one of the primary policy goals for China is increasing rural incomes. This creates a fundamental conflict facing water policy makers: encouraging water conservation while reducing the effect that higher water fees and lower water deliveries have on farm incomes. While this conflict has hindered attempts to increase water prices, it has caused China to search for creative ways to encourage conservation without adversely affecting incomes – in some cases even leading to an increase in incomes. However, there are other issues that reduce the effectiveness of increases in water fees to promote water conservation such as a general lack of farm-level volumetric measuring for surface water deliveries and rigid water delivery mechanisms that limit farmers' capacity to adapt to reduce the burden of higher water charges.

*The statements made in this chapter are those of the authors and do not reflect the views of the US Department of Agriculture or the China Ministry of Water Resources.

In this chapter we provide an overview and synthesis of China's irrigation water pricing policies. We review the history of China's agricultural water policies to provide a context and background for discussion of current policy issues. We also describe how agricultural water prices are determined, applied and collected. We conclude by discussing a series of issues that confound further reform and the effectiveness of pricing policies in promoting water conservation and farmers' capacity to adapt to higher water prices.

The findings and observations in this chapter are based on extensive fieldwork on irrigation management practices carried out by the authors over the past several years. Collectively, the authors have interviewed hundreds of farm households, village officials and local irrigation managers. These interviews mostly took place in the more water-scarce areas of northern China, but parts of southern China are also represented. In addition, China is a large and diverse country, and agricultural practices vary widely. Thus, we try to convey this variation in practices, but cannot always provide estimates of the extent to which any given practice is more, or less, common than the others.

Irrigation Policy in China: Background

Agriculture was critical to the development plans of China's new leaders in 1949. Agriculture not only provided employment for roughly 80% of the labour force in China at the time, but agricultural goods were also among the few products that China could export to earn hard currency to invest in industrial capital. Moreover, increasing agricultural production allowed China to provide inexpensive food to urban industrial workers and facilitate industrial development.

Because of the desire to develop quickly, policies adopted in the socialist period between 1949 and the late 1970s sought to harness China's water resources to boost agricultural and industrial production without regard to the opportunity costs of the water. During this period, irrigated area expanded rapidly as the command area for existing irrigation districts (IDs) was expanded and new districts were established. The People's Communes, or collective groups under the communes such as Production Brigades, organized much of the expansion of surface water infrastructure during this period. The projects were often short on capital and were completed by mobilizing large amounts of collectively managed labour. In addition to surface water IDs, local officials increasingly tapped groundwater aquifers with the adoption of electric pumps, particularly in areas of northern China such as the Hai river basin. The number of tube wells in China grew from 150,000 in 1965 to more than 2.3 million by the late 1970s (Shi, 2000). Because of the collective nature of the irrigation assets and the intent to boost agricultural production without regard to costs, the water channelled to agricultural fields (which were also farmed collectively during this period) was delivered without charging any fees. For groundwater, the communal entities that established the wells would pay the cost of electricity, but prices were also set below the actual cost of producing and delivering electricity.

In the late 1970s and early 1980s, China's agricultural policy underwent a major transformation with the adoption of policies together referred to as the Household Responsibility System (HRS). Instead of farming land collectively, local leaders allocated each farm household individual plots of land to farm by themselves. In return, farm households were obligated to (or 'responsible' for, hence the name) delivering a fixed quota of grain to the state-owned Grain Bureaus at a predetermined, but generally below-the-market, price. This system not only restored households as the primary production unit in agriculture but also ended the collective institutions that formerly built, managed and maintained much of China's irrigation infrastructure.

However, by restoring the household as the primary production unit and allowing them to earn profits on their production, households became responsible for purchasing their own agricultural inputs. Thus, water

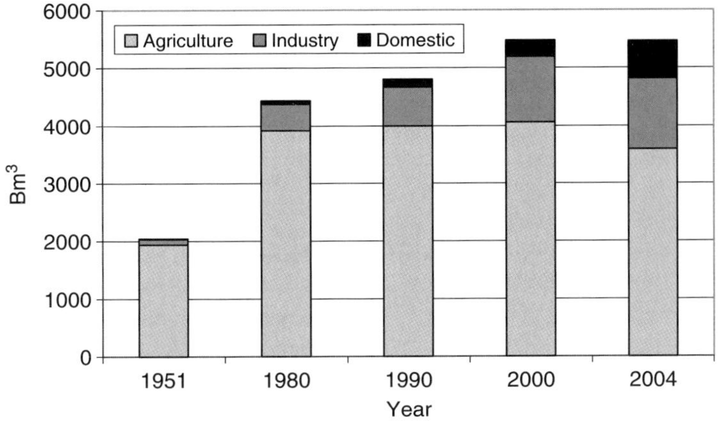

Fig. 12.1. Structural changes in China's water use, 1951–2004 (in Bm^3).

fees were introduced for households with irrigated land. The fees were set by the local Price Bureaus in accordance with guidelines set out by the national Price Bureau. While prices in water-scarce northern China were higher than in water-abundant southern China, these bureaucratically set prices still served to subsidize irrigation water (Lohmar *et al.*, 2003; Tsur *et al.*, 2004, Chapter 8). Irrigation water fees in this period were generally below the costs of recovery, storage and delivery of the water, not to mention the opportunity cost of the water in other uses. The operating costs of IDs are often high due to payroll and other obligations. Moreover, a lack of incentives to provide services (which, in part, arises from low water prices) generally led to poor irrigation delivery services and subsequent inadequate water fee remissions, exacerbating the problem of cost recovery. In some IDs, managers resorted to establishing the so-called multibusiness enterprises (such as fish farming or tourism enterprises) using ID assets to maintain payrolls (Lohmar *et al.*, 2003; Easter and Liu, 2005).

Despite the shortcomings of the system with regard to efficiency of pricing and cost recovery, it is important to note that a system of pricing with the intent to generate self-financing IDs was established in China after the reforms in the early 1980s. This system served as a base for subsequent reforms, and differs from some systems in other parts of Asia where irrigation is funded directly through government departments.

Before long, the policies that actively sought to harness water resources appeared to be reaching and surpassing availability constraints as signs of acute water shortages began to occur with increasing frequency. Water use growth doubled during the period 1949–1951, then doubled again between 1950 and 1980. After 1980, China's total water use growth slowed, and grew from around 4.4 Bm^3 to around 5.4 Bm^3 (Fig. 12.1). Moreover, water allocated to irrigation actually decreased from 1980 to 2004, primarily within the last 5 years, with water allocated to industry and domestic use driving the growth in water consumption over the period 1980–2004. Water in important river basins, notably the Yellow and Hai river basins, was increasingly used up entirely before reaching the ocean. In part because of the overexploitation of surface water, the exploitation of groundwater accelerated. By the 1990s, in some areas, the groundwater table was falling by over 2 m a year (Lohmar *et al.*, 2003).

To address the situation of tighter water supplies available for agriculture, the national government embarked on a series of water policy reforms to encourage water conservation in the 1990s. Since agriculture is still by far the largest user of water, these reforms generally targeted agriculture. Reforms were established at all levels of the water delivery system including river basin management reforms, ID management reforms, regulations on groundwater withdrawals, and village- and household-level incentives to adopt water-saving conveyance and delivery technologies (Lohmar *et al.*, 2003; Wang *et al.*, 2004).

Two factors that confound sound water resource policy making are a lack of clear jurisdictional control over water policy and a lack of sufficient property rights to water for users to benefit from using it more efficiently. In China, water is owned by the state, with the very limited exception of water in some local ponds and delivery systems owned by the local collectives that built them. The primary state agency charged with managing the state's water is the Ministry of Water Resources (MWR) and its provincial counterparts. However, several other agencies have some jurisdiction over water resources such as the Urban Construction Bureaus (accessing and delivering mostly groundwater for urban consumers), the Ministry of Land and Resources (measurement and evaluation of groundwater resources), the Price Bureaus (determining pricing guidelines) and others. The MWR administers the state's agricultural water resources through a system of IDs and withdrawal permits, which, in principle, cover all water diversions including those from groundwater. Each ID is given the right to withdraw a fixed amount of water from a surface source (river, reservoir or aquifer) and distribute this water to irrigators in their district. IDs are given substantial leeway to determine how to distribute the water allocated to them, but neither an ID nor the MWR has the right to determine water prices, which is done at the Price Bureaus. The disjointed nature of the right to determine price versus the right to determine delivery schedules results in poor incentives for the MWR to monitor withdrawals from surface sources into IDs, and to monitor local users' withdrawals from the irrigation system under the existing permit system. Groundwater is also owned by the state but, in practice, the villages that sit on top of the aquifers have de facto rights to the water. These are the most unencumbered water rights and, because of this, groundwater managers and users face stronger incentives to use water more efficiently, although the price is generally below its actual value in agriculture.[1]

How Water Is Priced in Agriculture

A variety of agencies, policies and local institutions affect agricultural water pricing in China. Water pricing differs by whether the water is diverted from a surface water system or pumped up from a groundwater aquifer. In addition, local policies and institutions also affect water pricing since many areas have established mechanisms to improve services, water fee collection or both.

Surface water

The most common form of irrigation water in China is surface water, which is generally less expensive than groundwater. As is well known, China has a history of surface water irrigation systems that dates back several thousand years. On top of this, China has greatly expanded surface water irrigation, and improved ageing systems in the years since 1949. Volumetric pricing for surface water at the village level began in the 1980s and prices have increased somewhat since that time, but to date volumetric pricing at below the village level is very uncommon. Prices for surface water vary substantially by locality, ranging from 0.01–0.05 Renminbi(RMB)/m^3 in the south to 0.05–0.15 RMB/m^3 in the north, and these have been rising over the past two decades and will likely continue to rise.[2] However,

[1]In cases where the groundwater table is very deep, the costs of pumping can be greater than its value when used to irrigate wheat. In these cases, farmers may forgo wheat production, rely on rainfall for wheat production, or invest in cash crop production in combination with a more efficient irrigation delivery system.

[2]One RMB is equal to about $0.12 at current exchange rates.

because farmers and sometimes village water managers can often neither choose when their irrigation water is delivered nor decide how much they receive, surface water management policies confound efforts by farmers or local leaders to respond and adapt to price changes. Moreover, additional pumping costs and voluntary labour requirements may increase the actual costs of the water as well (Webber *et al.*, 2006).

Surface water irrigation in China is delivered by IDs that vary significantly in size and management. Currently, there are 402 large IDs in China that have command areas exceeding 20,000 ha. Within larger districts, there is generally an array of smaller reservoirs and farm ponds to store water closer to irrigated areas. In addition, there are over 5000 'medium sized' IDs in China with command areas between 667 and 20,000 ha. Most of China's irrigated area, however, is serviced by the hundreds of thousands of small IDs, with a command area less than 667 ha and they service roughly half (55.3%) of China's irrigated land. These smaller IDs interact with local villages and are sometimes owned by a village or township. Larger IDs are generally segmented into smaller sub-districts that interact with the villages in their command area.

Traditionally, collection of surface water fees at the farm level is managed by the village, sometimes referred to as the collective. Every village has a person assigned as the primary water manager.[3] That person may be either a member of the village government itself or someone selected from outside the government to manage the irrigation deliveries and fee collection in return for a small stipend. Either way, the water manager is often in charge of informing farmers of water deliveries by the ID, particularly in more water-scarce areas in northern China. Sometimes, the irrigation manager interacts with the ID to arrange for timely deliveries as well.

Surface water prices in China are largely determined by bureaucratic decree rather than by any market mechanisms. The primary agency responsible for determining water prices is the Price Bureau, which sets national guidelines and is guided not only by the demand for water in the specific sector in question but also by general economic and political considerations. Water price guidelines established by the Price Bureaus differ according to the user of the water, with industrial users paying a higher price than agricultural water users.[4] Local Price Bureau offices then determine pricing guidelines for their local users, based on the national guidelines, and usually work with the relevant ministries (such as the provincial MWR office) in determining the local pricing guidelines.

Once water pricing guidelines are established and forwarded to the local ID, it determines how to price its deliveries. Local practices vary and in some areas the ID delivers water to villages at certain times throughout the irrigation season without carefully measuring the quantity. In return, the village is expected to remit to the ID a water fee assessed per mu of irrigated land.[5] In the past, this fee was often bundled with other fees and paid once or twice a year, and sometimes even paid in kind when farmers deliver their grain quota obligations. When this happens, farmers may not even know how much they pay for water because the water fee is bundled together with other payments (see Chapter 7, this volume, for a similar situation in Vietnam). However, since grain quotas are far less pervasive today than in the past, water fees are generally paid in cash. Recent policy initiatives also discourage the bundling of fees so that water fees are increasingly paid

[3]In larger villages, managers may interact with a group of subordinate water managers representing subgroups of households in the village.

[4]Within industry and within agriculture water prices can vary. Often water used as coolant for electricity generating plants is priced lower than water for other industrial uses, and sometimes even lower than for agriculture. Within agriculture, farmers growing cash crops may pay more for water than those growing grain.

[5]A mu is a Chinese unit of land area equivalent to 1/15th of a hectare.

independently of other local taxes and fees, which are being phased out (Gale *et al.*, 2005). When the water fees are collected independently, the village irrigation manager who coordinates deliveries typically collects the fee payments as well.

In addition to paying a per-mu fee for irrigated land, farm households also contribute labour to maintain and construct new irrigation infrastructure. This 'volunteer' labour contribution is a carry-over from the collective period when most rural infrastructure was constructed using teams of collectively managed labour under the communes and the communal subunits. Indeed, farmer labour during and after the collective period, coupled with investments made by collective contributions, largely built and maintained surface water infrastructure, giving farmers a sense of ownership and a natural stake in how these assets are managed.

IDs generally measure water flows at some point (usually at the branch or lateral canal) in the delivery process, and it is not uncommon that the volume of water delivered to a village is measured. In this case, it is possible to introduce some volumetric water charge at the village level. When irrigation deliveries are volumetric at the village level, many villages split the charges into two components: a fixed component generally intended to maintain the delivery system and a volumetric component intended to cover the operations and management costs of the ID. With the volumetric component, villages can find ways to reduce their water use and save money, giving them an incentive to conserve water. However, a limitation on this incentive is that it occurs primarily at the village level since even the village volumetric charge is typically divided into charges to individual households according to their irrigated land size rather than according to some measure of their water use. Thus, either the village leadership must initiate conservation practices or the individual farm households must organize to collectively establish conservation practices. Given the lack of incentives faced by the village leadership (they do not gain from reducing water charges) and the costs of organizing a collective effort, the incentives to establish water conservation practices under this system are not particularly strong.

Village payments for irrigation water do not go to the ID directly, but instead to the next higher level of the Water Resource Offices affiliated to the Ministry of Water Resources. IDs are under the authority of the Water Resource Office in the jurisdictional level that encompasses the entire command area of the ID. Smaller IDs may be under the authority of the township-level Water Resource Office, or the county-level office, and larger ones may be under the prefecture or even the provincial office. Payment made by farmers and collected by the village for irrigation water deliveries goes through these levels of bureaucracy before being remitted to the ID that delivered the water. In general, each level of bureaucracy will charge a fee for the service of handling these payments, further reducing the amount ultimately remitted to the ID. Moreover, the fees retained by the various levels of bureaucracy are typically tied to the amount collected, which in turn is often tied, to some degree, to the amount of water delivered to villages. This results in an incentive for local Water Resource Offices to maintain or increase water deliveries rather than an incentive to promote conservation. These water fee remittance practices may also serve to generate resistance to reform by local governments.

The system of pricing and fee payment outlined above leads to a number of inefficient practices and consequent problems. A primary problem is the low water price set under the Price Bureau guidelines. Low prices are a problem not only because they provide poor incentives to conserve water but also because the revenue received under these low prices are often insufficient to maintain delivery infrastructure (Lin, 2003). Because of low prices and poor infrastructure, IDs face few incentives to put energy into delivering the water in a timely manner. This leads to poor and untimely water deliveries that can reduce the value of water in agriculture (MWR, 2006). The poor delivery services then result in farmers refusing to pay their water fees, which exacerbates the problem of low-income generation for IDs.

To address these problems, China has begun to establish various types of irrigation management reforms. The goal of these management reforms is to improve water fee collection, water delivery services and ID fee remissions and, in some cases, to reduce water allocated to farmers. These reforms are most common in larger IDs in water-scarce areas of northern China, where the disconnection between irrigators and water delivery decisions and also directives to save water are greatest. Reforms take a variety of different arrangements, but generally they try to turn over management of local irrigation assets to individuals or groups that have a stronger incentive to provide services and collect fees. Moreover, the fees collected by these agents are remitted directly to the ID. For larger districts, they are remitted to companies established by the ID to manage deliveries in sectors of the district. These reforms can also be effective at reducing water applications when managers can earn income by reducing the water they deliver to irrigators.

Water user associations

The establishment of Water User Associations (WUAs) is a major movement to improve irrigation management in China. Originally promoted by the World Bank in the mid-1990s, in many respects, WUAs are similar to some of the progressive irrigation management institutions that already existed in China. Theoretically, a WUA is a farmer-based, participatory organization set up to manage the village's irrigation water. The idea is that farmers come together to elect a board to manage irrigation water issues such as fee collection, scheduling deliveries and negotiating volumetric pricing with the ID. In larger IDs, Water Supply Companies (WSCs) are established by the ID to sell volumetric water deliveries to the WUAs in their command area and also occasionally to other users. This management structure is very similar to the institutions in some IDs where the ID has established smaller-scale management groups to sell water volumetrically to local villages. In addition, under WUAs, water resource fee remission circumvents the local Water Resource Offices, increasing the actual amount received by the ID. By the early 2000s, more than 500 World Bank-sponsored WUAs and over 40 WSCs had been established in China, and roughly 1500 WUAs established outside World Bank-sponsored projects (Lin, 2003). The number of WUAs has almost certainly increased since Lin's study.

The motivation behind promoting WUAs is generally to improve services and fee remissions to the ID rather than promoting water conservation at the farm level. The idea is that farmers will have a greater stake in the system and, therefore, will be more willing to invest in water-saving conveyance infrastructure and remit water fees in return for improved irrigation services. Theoretically, it is only by way of increasing the potential for collective action, via regular WUA meetings, that the WUA may serve to promote water conservation. In addition, it is at the collective meetings that the irrigators discuss the amount of water they want to purchase from the WSC and how much it would cost. These discussions may also encourage farmers to reduce water use at the farm level and cut costs.

In practice, WUA management varies considerably and many, perhaps most, are not participatory. Indeed, the leadership of WUAs sometimes does not differ significantly from the village leadership. In a recent survey of WUAs, primarily in Ningxia province, Wang *et al.* (2004) found that the governing board for 70% of the WUAs surveyed was the village leadership itself, and of the 30% where the leadership appointed a manager, one half of the managers were former village leaders. Other researchers found similar close relationships between the village and the WUA leaderships (Gao and Li, 2002; Mollinga *et al.*, 2005). However, there are also indications that some WUAs allow for more participation via direct elections, etc. (Lin, 2003; Lohmar *et al.*, 2003; Easter and Liu, 2005). In a recent survey of village leaders throughout northern China, roughly half reported that WUA leadership was elected by villagers in areas with WUAs.[6]

[6]Unpublished results from the 2004 China Water Institutions and Management survey by the Center for Chinese Agricultural Policy, Chinese Academy of Sciences.

WUAs, whether participatory or not, may still be effective in improving water management, fee remission and promoting water conservation. Wang *et al.* (2004) argue that it is the incentives of the water manager appointed by the WUA that matter for a WUA to be successful. Under this criterion, managers appointed by the village government may effectively improve water management and fee remissions, and reduce water use when they have the right set of incentives to do so, regardless of whether they are selected by farmers or not. Initial evidence presented by the authors indicates that farmers' participation is not an important factor in a WUA's capacity to reduce water use and maintain yields. However, since nearly all the WUAs in the study are newly established, these results may be only short-term effects. Longer-term, and possibly more substantial, savings made via investments into infrastructure, system maintenance and agricultural practices, may be more likely to occur when farmers have a greater role in management decisions. In addition, some argue that since irrigation assets were largely built with farmers' volunteer labour and some investment from collective savings, farmers have earned a right to be involved in irrigation management decisions.

When WUA managers can claim profits generated by activities in implementing policies that achieve the goals of the WUA, the institution is most likely to be effective at achieving policy goals (Wang *et al.*, 2004). The main way WUA managers can do this is by working on the margins of the price of the volumetric deliveries and the per-mu fees paid by farmers. Essentially, WUA managers can allocate funds to line canals or other investments that reduce the water loss and improve water deliveries to fields. The amount of water saved (which can be as high as 40–50% from a lined canal) can be deducted from the planned amount of water purchased from the WSC and reduce the volumetric component of the fee payment.[7] Moreover, with increased effort in monitoring and supervision, water can be allocated to farmers more efficiently so that irrigators receive better irrigation services even though less water is being drawn from the system. Since the fees farmers pay are set in advance and generally fixed per-mu payments, if managers can reduce water deliveries but keep irrigators happy with timely service, the payments they make to the WSC for volumetric deliveries are reduced and water managers can then earn money. This system gives managers strong incentives to reduce water purchases from the WSC, yet maintain effective irrigation deliveries to the field to keep farmers satisfied so that they pay their irrigation fees. Often, as an inducement for farmers to accept more limited (but timelier) water deliveries, WUA managers will pass some savings on to the farmers by charging lower per-mu water fees. Thus, when effective, this management system can both reduce water withdrawals from the surface system and decrease farmer's water fees.

The establishment of WUAs, however, does not appear to have greatly improved the ID's ability to be financially self-sufficient. Lin (2003) notes that prices in Hunan and Hubei provinces where WUAs were established early are still well below the costs of deliveries and, in some cases, as low as 20% of costs. These shortfalls are due to the continuation of rigid pricing policies since the WUA reforms do not liberate local officials from pricing water outside the guidelines established by the Price Bureaus. The establishment of WUAs, however, does improve irrigation services and the timeliness of deliveries, and reduces conflicts between irrigators (Lin, 2003; Easter and Liu, 2005), which have increased farmers' willingness to pay water fees. But since these fees are still set at a low level, improvements in water fee submission do not seem to have improved the ability of the IDs to be financially self-sufficient. Some also argue that the lack of participation and the 'privatized' nature of WSCs and some WUAs cause them to forego needed investments in order to appear more solvent and pay bonuses to managers, an effect that could cause them to seek far more govern-

[7]These savings are not 'real water saving', a point addressed in the conclusion of this paper.

ment support in the future than they appear to be 'saving' at present (ICID, 2004).

Contracting canal management

Related to WUAs, canal contracting is an officially advocated reform in surface water management. It is similar to WUAs in that it turns irrigation management over from village officials to a specified manager, but instead of a whole village's irrigation infrastructure being turned over, just a lateral canal which may service only part of a village, is turned over to the manager. In addition, these managers are generally not appointed by the village leadership or selected by farmers but rather selected by some other process, sometimes via open bidding. The selection process generally stipulates a ceiling price for water that managers can charge, and often also a minimum investment that managers must put into the irrigation system to qualify for the right to manage the system. In some cases, managers can pocket the difference between the fees they collect, a stipulated return on their investment, and the volumetric-based payments they must make to the ID, although the terms of this arrangement vary.

Similar to managers in some WUAs, canal managers have incentives to improve management and reduce water allocations when they can earn money by doing so. Also as with WUAs, the incentive usually comes by reducing volumetric purchases from the ID or WSC while maintaining effective irrigation deliveries by improving conveyance infrastructure and management techniques. When canal managers reduce volumetric purchases of water from the ID, they may reduce farmers' water fees in order to pass some of the savings on to irrigators and maintain their support. While farmers have little control over how the manager implements these policies, widespread disapproval, particularly if it affects agricultural production, would likely bring about the intervention of higher-level officials. Improvement in service and conveyance also allows for less water to go into the irrigation system and, via reduced conveyance losses, still allows for sufficient water to reach the fields and keep farmers satisfied with their delivery services. Considering these factors, canal contractors can earn money by reducing water purchases and improving services and fee collection.

Concluding note

China's investments in surface water infrastructure over the last 50 years have provided many farmers with irrigation water and, through this, increased rural incomes and agricultural production. Recent reforms have also improved services and help assure that water revenues go to improving local management and infrastructure. However, management reforms vary greatly in their effectiveness and have yet to become widespread throughout China. Most surface water systems in China still suffer from a lack of volumetric pricing mechanisms, poor management incentives and bureaucratic fee remission practices. These practices result in poor services and inadequate fee remissions to the ID. Moreover, increasing competition for surface water resources, occasional water scarcity due to year-to-year variation in rainfall and various fiscal issues still plague surface water systems and exacerbate the problems of timely deliveries and services.

Groundwater

Groundwater prices which vary considerably more than surface water prices, are generally higher, and there are important quality differences as well. Because the Price Bureaus do not determine groundwater prices, the cost of groundwater deliveries can be as high as 1 RMB/m^3 in areas in the Hai river basin where wells can be more than 100 m deep.[8] Since groundwater prices vary with well depth, and well depth varies

[8]We use the term 'cost of deliveries' here because the water is generally free and other than paying for pumping and conveyance costs, only a small percentage of villages charge an additional 'water resource fee' on top of these costs.

Table 12.1. The cost of water, depth to water and water use per unit of land in Hebei province's groundwater-using communities, 2004. (From 2004 CWIM data, reported in Huang *et al.*, 2007.)

	Percentile of the cost of water	(1) Depth to water (m)	Average cost of (2) water (RMB/m³)	(3) Volume of water use per unit of land (m³/ ha)
	Wheat			
1	Average	31	0.24	4608
2	0–25%	14	0.08	6433
3	26–50%	21	0.20	5285
4	51–75%	52	0.30	2934
5	76–100%	53	0.56	2154
	Maize			
6	Average	34	0.24	2019
7	0–25%	20	0.06	2255
8	26–50%	34	0.16	2094
9	51–75%	57	0.26	1463
10	76–100%	68	0.52	1119

significantly, the variation in groundwater prices can be high even within smaller regions like the Hai river basin, where wells vary in depth from 10 to over 100 m. In Table 12.1, the implied pumping costs for groundwater in northern China range from 0.06 to 0.56 RMB/m^3, depending on well depth. In addition, groundwater may have a high saline content, especially the shallower groundwater tables. However, the water in deeper groundwater tables might be far colder than surface water and stress the crops when applied, particularly when the crop is in the seedling stage.[9]

Despite the higher prices and potential quality problems, groundwater is an increasingly important source of irrigation water in China, particularly in northern China where water is scarcer. In the past few decades, many areas in the Hai river basin have tapped into groundwater aquifers with diesel or electric pumps to irrigate their fields, allowing them to produce winter wheat and a second crop of maize, or sometimes cotton. Farmers also increasingly use groundwater as a buffer stock of irrigation water to be drawn when surface water is scarce and replenished (via seepage from surface irrigation) when surface water is available. Farmers often prefer to use surface water because it is cheaper and, sometimes, because of saline or temperature problems referenced above. Generally speaking, groundwater levels are continually falling in China. However, the groundwater levels in areas where groundwater is used conjunctively with surface water are falling at a slower rate than in areas without access to surface water.

Groundwater pricing policies vary from surface water pricing policies. One primary difference is that groundwater is increasingly priced volumetrically since the pumping necessary for groundwater makes it easy to measure the volume delivered (or at least extracted) and this volume is directly related to operating costs as well. Another difference is that there are no large IDs delivering groundwater. Groundwater resources are managed at the local level, most often at the village level. Groundwater systems are much smaller, often organized around a single well with just a few dozen households receiving water from that well. Thus, groundwater users have greater opportunities than surface water users to interact with the person managing the irrigation deliveries. In addition, proceeds collected by groundwater managers are, in general, remitted directly to the entity (the village committee or the well owner) that provides the services and infrastructure for the

[9] On several occasions, the authors have heard from farmers in China the complaint that cold groundwater stresses young crops.

groundwater and not used to support a larger bureaucracy as is the case with many surface water systems. Groundwater irrigation services and fee remission mechanisms are also improving as groundwater assets are increasingly managed by private interests rather than by collectives.

Collective groundwater services

As with surface water, rural collectives generally managed groundwater pumps and conveyance systems after the reforms in the late 1970s. Commonly, smaller collectives, known as village groups (*cunzhuang xiaozu*) and made up of 30–50 households (but sometimes as large as 100 or more households), work together to maintain the wells and infrastructure they inherited from the former communes to provide groundwater for members of the group. This arrangement works particularly well since village farmland is largely allocated to households via the village groups so that households belonging to the same group tend to have contiguous plots of farmland.

When groundwater irrigation assets are owned and managed collectively, prices paid by irrigators may not be based on water volume. Instead they are often collected on a per-mu basis much like fees collected for surface water. These fees are used to cover the costs of maintaining the well, paying for electricity and paying back any loans taken out for any additional investment. The person selected by the group to manage all this may also receive a small stipend from the fees collected. In many cases, the village itself rather than the village group manages groundwater pumping and delivery assets. Under these arrangements, they also generally collect water fees based on irrigated land; however, they may sell water volumetrically to village groups or to farmers themselves.

Non-collective groundwater services

Much like the case with surface water, some groundwater assets became inoperable due to the groundwater table falling below the depths of the wells or the well structure collapsing. This induced a need to re-establish wells, often deeper and more powerful, but without the communal institutions that first established them. The collective parties that became responsible for their maintenance had either unclear rights to the system or found it difficult to garner the fiscal resources necessary to rebuild the system. Thus, in the era after the HRS reforms, non-collective institutions rose to re-establish wells and groundwater irrigation in many regions where farmers had come to rely on this resource (Wang *et al.*, 2005b; Zhang *et al.*, 2005). These included private well owners and operators, as well as joint ventures, often with local governments as partners, in companies that supply groundwater to irrigators. This trend accelerated in the late 1990s as transfers from higher levels of government were reduced and villages had to find their own funding for local investments. According to data from field surveys reported by Wang *et al.* (2005a), private, rather than collective, interests established 80% of the new wells in the Hai river basin in the 1990s, an area that is particularly dependent on groundwater for irrigation.

With the increasing role of private funding for groundwater irrigation deliveries, water pricing practices became more rational and increasingly based on volumetric deliveries. For the most part, the newer, mostly privately financed, groundwater companies sell water volumetrically to irrigators. In addition to investments into the wells and pumps, managers also often invest in underground pipes to deliver water to spigots that may be a few hundred metres away from the well itself, thus reducing the amount of water lost in conveyance to the field. Generally, the fee charged to irrigators is based on the electricity used rather than on the volume of water, but electricity used is highly correlated with the volume of water pumped and the depth of the well. Importantly, the companies delivering groundwater for irrigation generally do not pay any resource extraction fees; the water is free to them once they pay the costs of accessing and delivering it to farmers' fields. To the extent that local governments invested in these ventures, they too have mostly turned to volumetric pricing to recover the investment. These companies possess no formal ownership rights to the water they sell to farmers since groundwater resources theoretically belong to the state.

Issues with Price and Irrigation Reform in China

The current reforms in agricultural water management and pricing in China are the beginnings of a movement away from the former policies that harnessed water as a cheap and readily available means to promote economic growth. However, the current reforms fall short of policies that actually price water to achieve efficient allocation. Instead, policy reforms serve primarily to trim the bureaucracy in irrigation water management, clarify incentives to improve services and streamline the path of fee remittances to improve the capacity for local IDs to be self-sufficient. Thus, there is still substantial capacity to improve upon the current reform efforts. A better understanding of a number of important relationships may help water pricing policies to be successful politically as well as economically.

Water price reform

Effective irrigation water price reform in China is hamstrung with the debate over how higher water prices might affect agricultural producers. Currently, improving the income and welfare of farmers is the number one policy goal of China's leaders. Several policies are geared toward achieving this goal, including the abolishment of age-old agricultural taxes, the introduction of direct subsidies for agricultural producers, and increased investments into rural health care and drinking water purification systems (Gale *et al.*, 2005). Irrigation water fees were once commonly bundled with the various taxes and fees that are now being abolished. Raising water fees is diametrically opposite to the policy of abolishing all the other fees farmers pay and serves to cancel out the effectiveness of the current policy. Indeed, some observers are concerned that, if water fees are increased, local officials will not lower the local taxes and fee payments but instead simply call them 'water fees'.

While there is concern over how water prices might affect rural livelihoods some also argue that price policy will not be effective at reducing water use in agriculture. Yang *et al.* (2003) argue that water demand is relatively inelastic so that raising prices would not serve to reduce water use but rather only to increase the revenues of the IDs. Estimates of the price elasticity of irrigation water vary, and certainly water use is more inelastic in the short term than over a longer period when farmers can adjust themselves by adopting conservation technologies or practices or changing cropping patterns altogether. However, estimated inelastic price responses are due, in large part, to a legacy of very low prices so that farmers face the inelastic portion of the water demand curve. Farmers' water deliveries are also often constrained so that even at the low prices they face, the quantities used do not represent the value of the marginal product of water.

If prices rise substantially, farmers will become far more responsive to price changes. Huang *et al.* (2007) and Liao *et al.* (2005) have found that farmers would reduce water applications significantly if the price of water is raised to a level that reflects the value of the marginal product of water. Both these studies used cross-sectional data, so they are likely to take longer-run adjustments into account. For example, from a survey of farm households in northern China, Huang *et al.* (2007) show that deeper water tables are associated with higher pumping costs and thus higher water costs (Table 12.1). For wheat, maize and cotton, farmers use significantly less water per hectare in areas where the pumping costs are higher than in areas where the water is less costly.

In addition, the share of production costs due to water charges varies significantly by crop and by water depth. Again,

Table 12.2. Cost of groundwater in total production cost.[a] (From 2004 China Water Institutions and Management.)

By crop: Cost share of inputs in total production cost (unit: %)

		(1) Water	(2) Fertilizer	(4) Other inputs[b]
1	Wheat	24	49.6	26.4
2	Maize	12.9	50.7	36.4
3	Cotton	10	37.3	52.7

By depth of groundwater

	Percentile of water depth	(1) Average water depth[a] (m)	(2) Total cost of water per hectare (RMB)[a]	(3) Share of water cost in total production cost[a] (%)
4	0–25%	6	471	13
5	26–50%	21	454	16
6	51–75%	58	622	21
7	76–100%	91	834	23

[a]The sample here is the plots that have detailed input information and have used groundwater in irrigation.
[b]Other inputs including labour and capital cost.

data from the survey of households in northern China carried out by Huang and the Centre for Chinese Agricultural Policy, Chinese Academy of Sciences indicate that, on average, groundwater applied to irrigated wheat is 24% of total production costs, while for cotton it is less than 10% because cotton uses less water and more labour, pesticides and other inputs (Table 12.2). For fruits and vegetables, the share of water in production costs is likely even less, but this is, in part, due to investments made into water-saving technologies for fruit and vegetable production. Given the low values and constrained water deliveries, Huang *et al.* (2007) show that doubling of water prices results in only an 8% fall in crop income. In addition, having begun a programme to subsidize farmers, China is well positioned to offer farmers a lump sum water conservation subsidy to replace the income loss when farmers adjust to higher water prices.

To understand the effect that water prices have on rural incomes, it is important to know that farm households in China today have sources of income other than agriculture. Roughly half the farm household income in China is from non-farm sources such as local wage labour, self-employment enterprises or migration.[10] In addition, of the remaining half that comes from agriculture, less than 20% is from growing grain, the rest from raising livestock, aquaculture and cash crops which bring a much higher return to water (NBS, 2005). Thus, a substantial increase in water prices may cause farmers to earn less producing grain, but it may also induce farmers to make investments and develop the marketing channels necessary for more lucrative cash crop or other agricultural operations. Indeed, Huang *et al.* (2007) found that as well depth rose from less than 10 to over 60 m, the sown area committed to non-grain crops rose from 15% to more than 30%. If higher prices induce conservation and the water is freed up to be reallocated to industrial use, and the increased water availability

[10]Migration in China is internal, primarily rural-to-urban migration and is largely temporary because migrants are not given permanent residential rights in urban areas. They typically return home at least once a year and often return to their native villages permanently after saving a sum of money working in urban areas. Since the migrants are still considered members of the rural household, their income is considered part of the rural household's income.

for industry increases industrial output and demand for labour, then this reallocation could serve to increase the fastest growing segment of farm household income: non-farm employment. There are, however, income distribution issues with increasing irrigation water prices since households most reliant on agricultural production generally have lower incomes than households with substantial off-farm income.

Farmers' response to scarcity

Given the recent reforms and price incentives in some areas, and the incentives offered at the local level, farmers are increasingly making choices that reflect increasing water scarcity (Liao *et al.*, 2005; Blanke *et al.*, 2006; Huang *et al.*, 2006a). As outlined above, such choices include reducing irrigation water applications to traditional and staple crops (intensive margin), and shifting into crops that bring higher returns to water (extensive margin). However, such behaviour is tempered by another important policy goal encouraged by local-level cadres: maintaining domestic grain production. With surface water, water conservation decisions are sometimes, perhaps often, made at levels above the farm household, such as wet–dry rice production, or reducing wheat irrigation deliveries from four to three times a year. Even with management reforms, water conservation decisions are often made at the WUA or canal manager level and farmers must act accordingly. With groundwater irrigation, however, increasingly lower water tables coupled with more private interests in the groundwater market serve to induce changes in agricultural practices on both the intensive and extensive margins that could affect the production of important grains and threaten self-sufficiency in these crops.

When water prices get high enough that it becomes unprofitable to irrigate staple grains or cotton, then farmers are faced with the choice of foregoing irrigation entirely or adapting to other crops that can bring a higher return to the water. Rising incomes and rapid urbanization in China have brought about a rise in the demand for fruits and vegetables, and these are often the crops farmers choose when faced with higher water prices. While these products tend to be more water-intensive than field crops, they are better suited to effective water-saving irrigation technologies such as greenhouses and drip irrigation systems, lowering the water withdrawal requirements to grow these crops.

The movement into higher-valued fruits and vegetables, however, does not come without risks. These risks, coupled with initial investments required to change crops, partly explain why farmers often do not move into cash crop production unless pushed by forces such as rising water costs. Markets for many cash crops tend to be fairly thin so if large numbers of producers decide to move into production of these crops, prices can drop dramatically. These swings can be even more pronounced with orchard crops where the lag-time between the decision to plant the orchard and production of the first crop can be several years. Fruit and vegetable crops are also more susceptible to spoilage than staple grains, increasing the risk of a loss unless expensive cold chain or other modern marketing infrastructure is in place. The fact that many of these crops require some, and sometimes substantial, initial investment increases farmers' exposure to these various risks.

Movement into higher-valued fruits and vegetables using water-saving irrigation technology may be constrained by growth in domestic demand for these products and problems in selling them on international markets. China has enormous production potential in these products and this potential likely outstrips the projected increases in domestic demand due to increasing incomes and urbanization. However, fruits and vegetables can often be problematic export commodities in that they are more likely to be subject to sanitary and phytosanitary (SPS) restrictions. This problem is even more acute in China due to the excessive use of pesticides by farmers in China, and the wide range of products it could potentially export (each product must go through SPS reviews in each importing country) (Huang and Li, 2005). Together, these policy and institutional constraints could serve to reduce farmers' capacity to increase the allocative efficiency of water by using it in the production of higher-valued crops.

Farmers' capacity to continue movement into high-value and labour-intensive crops may also run into conflict with other policy goals. China's long-standing insistence on maintaining near self-sufficiency in staple grains, particularly food grains like wheat, may serve to induce policies that discourage movement out of wheat production. Self-sufficiency of food grains conflicts with the goal of increasing rural incomes since wheat tends to be a low-income crop, and it is also threatened by increasing water scarcity since wheat production in water-scarce areas is almost entirely dependent on irrigation. But if large numbers of farmers move out of wheat production and into other crops that bring higher returns to scarce water and other inputs, China's wheat production may fall to levels below their self-sufficiency goals. Trade, in such a situation, would be beneficial and allow farmers to produce and export relatively labour-intensive crops and import relatively land-intensive wheat, while also reducing water withdrawals for agriculture. However, if self-sufficiency of wheat is threatened, China may establish policies that discourage movement out of wheat production and constrain farmers' capacity to adapt to water shortages.

Property rights

Policies that increase water prices will more effectively induce efficiency improvements and potentially benefit farmers and others in the rural economy if property rights to water are more clearly defined. Currently, the debate over the role of water property rights is as heated as the debate over water prices and there is no consensus over how to determine or allocate these rights (Liu, 2003;Huang *et al.*, 2006b; Jia and Duan, 2006). A variety of projects that examine ways to allocate water rights and promote water conservation are also being carried out in China, but there are institutional barriers that restrict the adoption of these practices such as state ownership of water resources and the lack of authority to transfer this ownership. In some projects, non-agricultural users (such as electric-power-generating companies) directly fund water-saving investments, such as canal lining, then maintain rights to the water these investments save (Xu, 2006). In other projects, officials are experimenting with policies that grant farmers the use rights to water at a low price and then allow them to sell these rights to other users or back to the water management authorities at a higher price (Jia and Duan, 2006).

While the debate over formal allocation of rights continues, norms and practices at the local level indicate that a set of de facto rights already exists and understanding these de facto rights will help determine how price changes might affect farm households. For example, withdrawal permits represent a partial right to water, but the rights are generally not sufficient to provide incentives to monitor and enforce the withdrawals. Villages and farmers have de facto rights to water as well, but these are very limited. In a recent survey of 130 farm households in China's Ningxia province, only 2% responded that they had the right to decide when to take (surface) water deliveries.[11] Thus, for some important rights such as the right to determine when to apply water to irrigation crops, someone (either the canal manager or, more likely, an entity above the canal manager that the households are unaware of) has the right to make that decision regardless of whether they have formal ownership rights to the water. Such a situation has implications for farmers' capacity to adjust to higher-priced or more restricted water deliveries. If farmers are limited in their ability to choose irrigation timing in order to increase the marginal return of more limited deliveries, or to apply it to different crops than the canal manager is taking into consideration, then farm households will bear a

[11]Unpublished survey results from a survey carried out by the Center for Chinese Agricultural Policy, Chinese Academy of Sciences.

greater burden from higher water prices than they would if they had more control over irrigation deliveries. Giving farmers more control over deliveries, however, can be done without granting them formal ownership rights to water.

Given the high transaction costs to measuring and monitoring water use at the farm level, and the desire for water management agencies to maintain control of such a valuable resource, local-level rights to water in surface water systems are few. In general, households only have the right to allocate water to their fields at times determined by others, and for which they pay a fixed fee. This arrangement limits their capacity to adapt by switching to crops that may use water more effectively but that require more timely and secure supply.

Real water savings

The changes in water management and pricing policies outlined in this chapter may reduce irrigation water applications but that does not mean they induce 'real' water savings. 'Real water savings' have different definitions but, in general, it refers to reducing non-recoverable water losses that result from excessive, non-essential evapotranspiration or water flows into the ocean or non-recoverable seepage. The Hai river basin in northern China, where water scarcity is most acute, is already a largely closed river basin in that very little water flows into the ocean; the Yellow river basin is effectively closed in that the flows in recent years have been due to policy decisions to maintain minimal flow through the estuary; and nearly all water beyond the minimum flow water is diverted to other uses. Thus, real water savings in these areas come primarily from reducing non-essential evapotranspiration, or ET. However, reducing ET is far more difficult, and in general, costly, than simply reducing withdrawals and field-level water deliveries (Kendy *et al.*, 2003).

Irrigation management reform efforts, investments into water-saving technologies and price-induced reductions in irrigation withdrawals all primarily serve to reduce water application rather than promote real water savings. Still, there are likely some real water savings that do come out of these policies by reducing evaporation of water from excessively irrigated fields and the adoption of some irrigation practices that reduce non-essential evapotranspiration in the process of delivering water to fields more effectively. This is an area that beckons more careful research and will play an important role in determining how effective policy measures are actually reducing water losses rather than reducing 'losses' that would otherwise be recovered and used elsewhere in the system.

Conclusion

Water pricing can be one of the most important policy tools for managing the demand for water (Dinar, 2000; Tsur *et al.*, 2004). The objective of water pricing is to signal to users the relative scarcity of water so as to provide them with incentives to save water. In addition, water prices serve to fund the diversion, storage and delivery systems that allow the water to be brought to the fields for irrigation uses. However, in China as in almost all other countries, water prices are set at such low levels that they do not reflect relative scarcity and are well below the value of water to agricultural users, making water pricing policy much less effective and inducing conservation. In China, it is only when groundwater is very deep that one observes high enough costs, coupled with volumetric pricing, and farmers respond to these costs by adopting conservation practices or switching to production of other crops. Water pricing policies may also have a significant effect on agricultural production and rural welfare, which are also important policy objectives in China. Therefore, understanding how farmers respond to changes in water prices and how these changes affect their livelihoods will help policy makers understand the impact that price reforms will have on rural incomes and agricultural production.

China is also burdened because irrigation water has been heavily subsidized in the past, and thus charging prices that reflect relative scarcity will come as a shock to farmers and will be difficult to promulgate. Water deliveries to agriculture (and other sectors) are often below the cost of deliveries and well below the value of water in agriculture and other sectors. The experiences in other countries reveal that transition from subsidizing irrigation deliveries to pricing water to the level that induces conservation is difficult in itself. This transition is often made more complex by other policy goals, namely to reduce farmers' overall fee payments and other locally assessed fiscal burdens. Moreover, China is undergoing a transition to a more market-oriented economy and rapid industrialization and development. Seeking to establish mechanisms to induce water conservation is, in part, due to these changes; yet it is made more complex by the rapidly changing environment, and the desire to do so while maintaining agricultural production; and reducing the negative effect such policies have on farm households is even more ambitious.

Given the legacy of inexpensive and available water for irrigation in China, there is substantial capacity to use water more efficiently both in agriculture and in other sectors. While price reforms to date have not established economically efficient prices, the benefits of further price reforms would be enhanced if complemented with policies that give farmers more decision-making power over how the water is used in agriculture. Current pricing projects, although confined to limited remote areas, may reveal mechanisms to advance reform in ways that do not conflict with other policies, including allocating water rights to farmers and allowing them to sell water to downstream users that will pay higher prices.

References

Blanke, A., Rozelle, S., Lohmar, B., Wang, J. and Huang, J. (2006) Water saving technology and saving water in China. *Agricultural Water Management* 87, 139–150.

Dinar, A. (ed.) (2000) *Political Economy of Water Pricing Reforms*. Oxford University Press, New York.

Easter, W. and Liu, Y. (2005) *Cost Recovery and Water Pricing for Irrigation and Drainage Projects*. Agriculture and Rural Development Discussion Paper No. 20. World Bank, Washington, DC.

Gale, F., Lohmar, B. and Tuan, F. (2005) China's new farm subsidies. International agricultural trade report. Economic Research Service, US Department of Agriculture, Washington, DC, http://www.ers.usda.gov/publications/WRS0501/

Gao, H. and Li, Z. (2002) *Irrigation Management Reform through WUA in Qingtongxia Irrigation System, Ningxia Yellow River Plain, China*. Working Paper. Irrigation and Drainage Development Centre, Ministry of Water Resources, Beijing.

Huang, J. and Li, N. (2005) Trade liberalization and Chinese agriculture, poverty and equality. *Nogye Jingji Wenti* (Issues in Agricultural Economy – in Chinese), No. 7, 2005.

Huang, Q., Dawe, D., Rozelle, S., Huang, J. and Wang, J. (2005) Irrigation, poverty and inequality in rural China. *The Australian Journal of Agricultural and Resource Economics* 49, 159–175.

Huang, Q., Rozelle, S., Lohmar, B., Huang, J. and Wang, J. (2006a) Irrigation, agricultural performance and poverty reduction in China. *Food Policy* 31(1), 30–52.

Huang, Q., Liu, W. and Wang, C. (2006b) Investigation report of water rights trade between Dongyang city and Yiwu city in Zhejiang province. Ministry of Water Resources (in Chinese), http://www.hwcc.com.cn/newdisplay/newsdispl;ayasp?Id=1821

Huang, Q., Rozelle, S., Howitt, R., Wang, J. and Huang, J. (2007) *Irrigation Water Pricing Policy in China*. Working Paper. Department of Applied Economics, University of Minnesota, St Paul, Minnesota.

ICID (International Congress on Irrigation and Drainage) (2004). *Irrigation and Drainage Services: Some Principles and Issues in Developing Countries*. ICID Position Paper. New Delhi.

Jia, Y. and Duan, J. (2006) Impact analysis of the pilot projects of water saving society at Zhanye City. *Water Resources Planning and Design*, No. 2 (in Chinese).

Kendy, E., Molden, D., Steenhuis, T., Liu, C. and Wang, J. (2003) *Policies Drain the North China Plain: Agricultural Policy and Groundwater Depletion in Luancheng County, 1949–2000*. Research Report 71. IWMI, Colombo.

Liao, Y., Gao, Z., Bao, Z., Huang, Q., Feng, G., Cai, J., Han, H. and Wu, W. (2005) China's water pricing reforms for irrigation: Effectiveness and impact. Draft Research Report. Irrigation and Drainage Department, China Institute of Water Resources and Hydropower Research and International Water Management Institute, Beijing.

Lin, Z. (2003) *Participatory Management by Farmers – Local Incentives for Self-financing Irrigation and Drainage Districts in China*. Environment and Social Development, East Asia and Pacific Region Discussion Paper. World Bank, Washington, DC.

Liu, B. (2003) Water rights in China. Paper prepared for the International Working Conference on Water Rights: Institutional Options for Improving Water Allocation, Hanoi, 12–15 February, 2003.

Lohmar, B., Wang, J., Rozelle, S., Huang, J. and Dawe, J. (2003) China's agricultural water policy reforms: Increasing investment, resolving conflicts, and revising incentives. *Agricultural Information Report*. Economic Research Service, US Department of Agriculture, Washington, DC.

MWR (Ministry of Water Resources) (2006) The problem and its effects existing in agricultural water price reform (in Chinese), http://gks.chinawater.net.cn/CWSNews_View.asp?CWSNewsID=24436

Mollinga, P., Hong, G. and Bhatia, A. (2005) Leadership and turnover: The contradictions of irrigation management reform in the Peoples' Republic of China. In: Shivakoti, G.P., Vemillion, D.L., Wai-Fung, L., Ostrom, E., Pradhan, U. and Yoder, R. (eds) *Asian Irrigation in Transition, Responding to Challenges*. Sage Publications India Pvt Ltd, New Delhi, pp. 310–345.

NBS (National Bureau of Statistics) (2005) *China Rural Statistical Yearbook*. National Bureau of Statistics (in Chinese), Beijing.

Shi, Y. (2000) Groundwater development in China. Paper for the Second World Water Forum, The Hague, The Netherlands, 17–22 March, 2000.

Tsur, Y., Roe, T., Doukkali, R. and Dinar, A. (2004) *Pricing Irrigation Water: Principles and Cases from Developing Countries*. Resources for the Future, Washington, DC.

Wang, J., Xu, Z., Huang, J. and Rozelle, S. (2004) *Incentives in Water Management Reform: Assessing the Effects on Water Use, Production and Poverty in the Yellow River Basin*. Working Paper. Centre for Chinese Agricultural Policy, Chinese Academy of Sciences, Beijing.

Wang, J., Huang, J. and Rozelle, S. (2005a) Evolution of tube well ownership and production in the North China Plain. *Australian Journal of Agricultural and Resource Economics* 49(2), 177–195.

Wang, J., Rozelle, S., Blanke, A., Huang, Q. and Huang, J. (2005b) The development, challenges and management of groundwater in rural China. In: Giordano, M. and Shah, T. (eds) *Groundwater in Developing World Agriculture: Past, Present and Options for a Sustainable Future*. IWMI, Colombo.

Webber, M., Barnett, J., Finlayson, B. and Wang, M. (2006) *Pricing China's Irrigation Water*. Working Paper. School of Anthropology, Geography and Environmental Studies, The University of Melbourne, Victoria, Australia.

Xu, W. (2006) Experience of spreading the construction of a water-saving society in Gansu province. Official speech, Ministry of Water Resources, online at www.gssl.cn/news/b/news.asp?newsid=1179.

Yang, H., Zhang, X. and Zehnder, A. (2003) Water scarcity, pricing mechanism and institutional reform in Northern China irrigated agriculture. *Agricultural Water Management* 61(2), 143–161.

Zhang, L., Wang, J., Huang, J. and Rozelle, S. (2005) *Groundwater Entrepreneurs in China: Selling Water to Meet the Demand for Water*. Working Paper. Centre for Chinese Agricultural Policy, Institute for Geographic Sciences and Natural Resource Research, Chinese Academy of Sciences, Beijing.

13 Water Pricing and Irrigation: A Review of the European Experience

J. Berbel, J. Calatrava and A. Garrido

Scope and Objectives

This chapter reviews the European irrigation sector and its water pricing policies. The first section provides an overview of the irrigation sector in terms of surface water, economic importance and water usage. The second section reviews some of the outstanding issues that have called the attention of the European Union and the Organization for Economic Cooperation and Development (OECD, 1999a,b, 2002) with respect to pricing irrigation water in Europe and OECD countries.

The third section examines the water pricing policies that were in place in EU's various member states and accession countries prior to the promulgation of the Water Framework Directive (WFD). The fourth section offers a detailed description of the WFD, which is by far the most important landmark in the history of the EU's water policy. It has profound implications for the way in which irrigation water will be priced after year 2010. The fifth section studies the existing literature on the likely implications and effects of the application of the WFD in the European irrigation sector while the last section summarizes the main conclusions.

The Irrigation Sector in Europe and Current Trends

The importance of the European irrigated acreage

By world standards, Europe is a densely populated continent. Over the centuries, the river systems have been heavily modified to support early industrialization, urbanization and navigation. As a continent, Europe spans the territory from the north of the Artic Circle to the south of Parallel 38. European agriculture uses 44% of the EU territory and exhibits great variability both along north–south and west–east transects, as a result of geographic and climatic diversity, from the temperate climates of the north to the arid climates around the Mediterranean Sea. The importance of irrigation thus increases from north to south, being an indispensable input for agriculture in most of the arid and semi-arid environments. In Mediterranean countries, irrigated farming accounts for a large share of total water withdrawals (83% in Greece, 68% in Spain, 57% in Italy and 52% in Portugal), while it represents less than 10% in Northern European countries. At the same time, there is a wide variety in farming patterns, the crops grown and the contents of water laws across the countries of the community. In particular, there are very large differences in average farm size between countries.

As Table 13.1 shows, irrigated acreage represents a significant percentage of the land with annual and perennial crops (FAO, 1997). In addition to the Mediterranean countries of the EU member states (Spain, Italy, France, Cyprus, Greece and Portugal), some of the eastern European countries (Albania, Bulgaria, Romania and the Russian Federation) have large irrigated areas. The irrigated area in the EU has grown from about 6.5 Mha in 1961 to nearly 12 Mha in 1996.

Current problems and trends

According to the WFD's preamble, the trends in most European countries indicate that the water supplies to the population are threatened by human-induced pressures and that aquatic ecosystems are undergoing severe processes of quality deterioration. As we will see below, reversing these trends is the main objective of the WFD.

National water figures conceal widely diverse situations among European regions. Those suffering from scarcity and deteriorating situations tend to coincide with those in which irrigation is the major water user. Large investments in infrastructure, supply-side policies and disregard for integrated policies have brought water systems in many regions to unsustainable use patterns. Quality and quantity issues are intertwined. Large population densities of generally wealthy populations have encouraged investments to take advantage of climatic patterns across Europe, mostly in tourism and second-home residences and, on the other hand, in agriculture. This demand-driven process, illustrated by the growth of private groundwater irrigation, has also been promoted by the EU Common Agricultural Policy as well as by national governments.

The increase in water abstraction in recent decades has led to a reduction in river flows and a decrease in the level of groundwater, the problem of quantity being aggravated by increasing levels of pollution. Both factors (quantitative and qualitative) involve a loss of good-quality water which is incompatible with the stated environmental objectives of the EU. For instance, in Spain where 28% of the water abstracted for irrigation is groundwater, there are 51 hydrological units with problems of overexploitation which total 710.7 Mm^3/year (MOPTMA-MINER, 1994).

Occasional water shortages with large adverse social and economic consequences, added to water-quality problems all around Europe, ignited a debate, which eventually gave rise to the European policy initiative that materialized in the WFD.

Genesis of the WFD and CAP reforms

Protection of the environment has been a key theme in recent EU legislation. The Maastricht Treaty (1992) made specific reference to environmental protection, safeguarding human health and achieving sustainable development. This was specifically agreed in Article 130-R:

> Community policy on the environment shall aim at a high level of protection taking into account the diversity of situations in the various regions of the Community.

But even before the Treaty was signed, environmental policy objectives were apparent in various types of legal acts.

The development of the European legislation for water resources falls into three 'waves'. The first wave goes back to 1975, and focused on water-quality standards and on the protection of surface water allocated for drinking. The second wave started in 1991 and focused, for the first time, not only on setting acceptable water-quality standards but also on controlling emission levels as a means of achieving the desired standards. The new legislation included the Urban Wastewater Management Directive, the new Drinking Water Quality Directive,

Table 13.1. Water and irrigation: Basic figures of European countries. (Compiled from FAO's databases.)

Country and year	Arable and permanent crops, 2000, FAOSTAT (ha)	Average precipitation 1961–1990 (mm/year)	Total renewable water resources, AQUASTAT (km^3/year)	Total renewable water resources, AQUASTAT (m^3/capita)	Full/partial control irrigation, AQUASTAT (ha)	Irrigated/arable land (%)
Albania (98)	699,000	996	41.70	13,306	340,000	48.6
Austria (98)	1,470,000	1,110	77.70	9,616	4,000	0.3
Belarus (93)	6,257,000	618	58.00	5,694	131,000	2.1
Belgium–Luxembourg (98)	837,000	853	21.40	2,003	40,000	4.8
Bosnia and Herzegovina (98)	650,000	1,028	37.50	9,429	2,000	0.3
Bulgaria (98)	4,636,000	608	21.30	2,680	800,000	17.3
Croatia (98)	1,586,000	1,113	105.50	22,669	3,000	0.2
Cyprus (94)	143,000	498	0.78	995	39,545	27.7
Czech Republic (98)	3,318,000	677	13.15	1,280	154,000	4.6
Denmark (98)	2,289,000	703	6.00	1,128	447,000	19.5
Estonia (95)	1,134,000	626	12.81	9,195	4,000	0.4
Finland (98)	2,191,000	537	110.00	21,268	64,000	2.9
France (98)	19,582,000	867	203.70	3,439	2,600,000	13.3
Germany (98)	12,020,000	700	154.00	1,878	485,000	4.0
Greece (98)	3,854,000	652	74.25	6,998	1,431,000	37.1
Hungary (98)	4,803,000	589	104.00	10,433	230,000	4.8
Italy (98)	10,825,000	832	191.30	3,325	2,750,000	25.4
Latvia (95)	1,874,000	641	35.45	14,642	20,000	1.1
Lithuania (95)	2,992,000	656	24.90	6,737	7,000	0.2
Macedonia (98)	599,000	619	6.40	3,147	55,000	9.2
Moldova (94)	2,190,000	553	11.65	2,712	312,000	14.2
The Netherlands (98)	944,000	778	91.00	5,736	565,000	59.9
Norway (98)	883,000	1,120	382.00	85,478	127,000	14.4

Continued

Table 13.1. *Continued*

Country and year	Arable and permanent crops, 2000, FAOSTAT (ha)	Average precipitation 1961–1990 (mm/year)	Total renewable water resources, AQUASTAT (km^3/year)	Total renewable water resources, AQUASTAT (m^3/capita)	Full/partial control irrigation, AQUASTAT (ha)	Irrigated/arable land (%)
Poland (98)	14,330,000	600	61.60	1,596	100,000	0.7
Portugal (98)	2,705,000	855	68.70	6,859	650,000	24.0
Romania (98)	9,865,000	637	211.93	9,445	2,880,000	29.2
Russian Federation (90)	126,820,000	460	4,507.25	30,980	6,124,000	4.8
Slovakia (98)	1,576,000	824	50.10	9,279	183,000	11.6
Slovenia (98)	204,000	1,162	31.87	16,031	3,000	1.5
Spain (98)	18,217,000	636	111.50	2,794	3,370,000	18.5
Sweden (98)	2,706,000	624	174.00	19,679	115,000	4.2
Switzerland (98)	437,000	1,537	53.50	7,462	25,000	5.7
UK (98)	5,928,000	1,220	147.00	2,465	170,000	2.9
Yugoslavia (98)	3,736,000	795	208.50	19,759	57,000	1.5

the Nitrates Directive and the Directive for Integrated Pollution and Prevention Control. The third wave addressed water resources in a holistic manner and culminated in the WFD.

OECD (2004) reports that the EU-15 reduced nitrogen use from 69 to 58 kg/ha in the decade after 1985–1986, but that the use of pesticides and total agricultural water grew by 5% and 20%, respectively. This occurred despite the contraction of 4% of EU-15's agricultural land.

The evidence after 25 years of European legislation was that water resources were still deteriorating, with consequences for ecological systems and human populations. In this context, and given the numerous unresolved problems encountered during the implementation of previous community water directives, the European Council of Ministers asked for a reform of the Water Policy (the 'third wave' of water legislation). The European Parliament and Council adopted the Water Framework Directive in September 2000, which was published in December 2000.

The Directive was subject to a long process of negotiations that was marred by disagreements between the Council of Ministers and the European Parliament that threatened to prevent the Directive from ever being adopted. This controversy can be interpreted as the culmination of conflicting interests between different actors at the local, national and European levels (Kaika, 2003). Ultimately, however, as a result of these complex negotiations, the final text of Directive 60/2000 integrated EU environmental principles (such as the 'polluter pays' principle and a high level of environmental protection that were included in the Maastricht Treaty) within a single document agreed to by all EU members.

The history of the Common Agricultural Policy (CAP) has also been one of adaptation to internal and external forces on the agriculture sector. The internal forces result from the very different positions of the agriculture sectors in member countries, and the changes in those positions over time. Historically, the CAP included heavy subsidies for production as well as for export and import restrictions, and for indirect subsidies such as on energy and irrigation costs. Likewise, the solution to rural deprivation was seen to be to support agriculture. There has been a long series of modifications of policy goals and instruments in reaction to changing agricultural policies. An important share of water consumption goes to crops heavily subsidized by CAP (e.g. sugarbeet, cotton and cereals in Spain, Italy and Greece, or maize in France).

On 26 June 2003, EU ministers for agriculture adopted a fundamental reform of the CAP. The reform will completely change the way the EU supports its farm sector. The new CAP will be geared towards decoupled forms of support that farmers will receive irrespective of their production levels. These new 'single farm payments' (that came into force in 2005) are linked to respect for environmental, food safety, animal and plant health, and animal welfare standards, as well as to the requirement to keep all farmland in good agricultural and environmental condition ('cross-compliance'). The effects of this reform on water demand will be important in continental areas (growing non-Mediterranean crops), which are those mostly affected by the change in agricultural support, and less important in areas where fruits and vegetables are the primary irrigated crops.

Issues in Irrigation Water Pricing: Costs and Incentives

In this section, we discuss some of the most salient issues in water pricing and draw some policy lessons in light of the discussion of experience from individual European countries given in the annex.

Issues related to the definition and measurement of irrigation costs

Cost recovery

WFD supports the achievement of economic objectives, specifically cost recovery for water services, including environmental and resource cost within each of the three sectors: agriculture, industry, domestic. WFD bases the concept of cost recovery on the concept of 'water services', and the complete meaning of this sentence has been defined in detail in the WATECO guide (2003) that develops the concept of full cost recovery by stating two levels of recovery: 'financial' and 'environmental and resource' costs.

In a perfect competitive market, the prices fall out of the market through the interaction of the buyers and sellers, and the optimum allocation of water is automatically achieved. But with water, and specifically in relation to environmental uses, the prices have to be 'invented'. The problem is to establish a set of prices that results in achieving the optimum allocation of water. As shown in the Annex, the charges for irrigation in the EU countries, as in most other countries, have been inadequate to recover capital and operating costs. Other levels of recovery have been introduced largely in regard to the issue of allocating the water between competing uses, in particular, between human and environmental uses.

The concept of 'water services' is defined in monetary terms as the economic cost of maintaining infrastructures and supplying water. This analysis should be done for the agriculture, industry and domestic urban water sectors. Additionally, on top of these monetary costs, WFD requires the estimation of both environmental and resource costs, and the definition of a programme to recover them. Differences between 'environmental cost' and 'resource cost' are difficult to implement in the real world. It would have probably been more useful to separate 'monetary' (O&M, depreciation, financial) and 'non-monetary' (environmental and resource) costs.

Resource costs are the most difficult to quantify. Usual notions of resource costs associate them with opportunity costs that are equivalent to the economic value of the opportunities forgone when allocating the resource to a given user. When water markets exist, resource costs can be assimilated to the market price of water netted of the costs incurred when abstracting or moving the water to its final destiny.

The difficulties of separating cost items are related to the different definitions of 'full cost recovery' prices that each country appears to follow. Appropriate policy action should also recognize that an irrigator's water use may entail additional social costs. These social costs may or may not be included in the definition of 'full cost recovery' rates, but it would certainly be in the interests of society to identify them and attempt to reduce them. The following sections clarify these notions and provide cost evaluations found in the literature and recent reports. We will use the following typology for monetary costs: (i) private farmer costs; (ii) irrigation scheme costs; and (iii) public water authority costs.

Private (on-farm) costs

Private irrigation costs include those items for which the irrigator is entirely responsible, and that farmers generally pay as any other cultivation cost, such as maintenance, energy and labour. There are two main drivers for the increased area under precision irrigation (drip irrigation): the scarcity of water and the scarcity of labour, which make automated irrigation systems very attractive for farmers who face the rising cost of both inputs.

Irrigation district or scheme costs

Irrigation districts distribute surface water and, less frequently, groundwater to individual farmers, and the costs of running and maintaining infrastructure and associated facilities serving a clearly identified set of irrigators are in principle paid by farmers irrespective of the kind of ownership of the district's infrastructure. In practice, there is abundant evidence of better district cost recovery in private associations than in state-run or publicly owned water infrastructure (OECD, 1999a). Most schemes are managed by irrigation districts, which usually are non-profit associations with legal status.

In countries such as Italy, Spain, Turkey or Mexico irrigation districts are assigned an instrumental role in water policy implementation and water management. According to the Spanish Water Law, irrigation districts (about

6200 *Comunidades de Regantes* are registered, covering 2 Mha) must have their statutes and by-laws approved by the Basin Agencies and perform a number of key tasks in water management. For instance, they collect farmers' charges and levies charged by the Basin Agency and transfer the revenue to the latter. They have also approved procedures to solve conflicts among irrigators, organize irrigation turns and develop and co-finance rehabilitation projects.

User associations in Spain are mostly collective organizations, irrespective of whether they are served with public concessions (either surface water or groundwater) or from private groundwater rights. The French Associations Syndicales Autorisées (ASAs) have similar characteristics although their size is usually very small, while the Sociétés d'Aménagement Rural (SAR) are purely private organizations. In Italy, water user associations are association of landowners with public status (meaning that they are regulated by law and subject to government supervision); much the same occurs in German '*Wasseverbandae*'. In Bulgaria and Romania collective user associations were created in the last decade.

On the other hand, in countries like Austria or Greece water user associations are controlled by public authorities. Lastly, in other countries such as England, Sweden, Ireland or Denmark where irrigation is predominantly an activity of private individual farmers it is not common to find agricultural user associations in charge of managing irrigation (and drainage) schemes. In England, local drainage boards are fairly common, e.g. in the Fens of East Anglia and in the Somerset levels.

Running costs of irrigation districts are borne solely by the irrigators. However, in most countries, investment costs, either in new schemes or in modernization or rehabilitation projects, receive significant subsidies. Most large irrigation infrastructures across OECD countries, irrespective of when they became operative, have been built with public capital grants. New irrigation districts are projected to be developed in the next decade in Spain or Portugal, although in the case of Spain, new irrigation projects are now very limited and targeted to areas undergoing depopulation.

An example of this type of cost is illustrated in Fig. 13.1 that shows the 'internal district cost' (net of Water Agency Tariffs) according to the amount of water supplied, for a selected group of Guadalquivir irrigation districts (surface water). The average internal distribution cost (entirely covered by users) was €0.037/m^3 in the Guadalquivir basin in 2003.

The internal costs of an irrigation district may be shared on a volumetric or per-hectare basis, or even defined by a binomial tariff. Table 13.2 reports various irrigation district costs selected from a number of irrigation districts across EU countries.

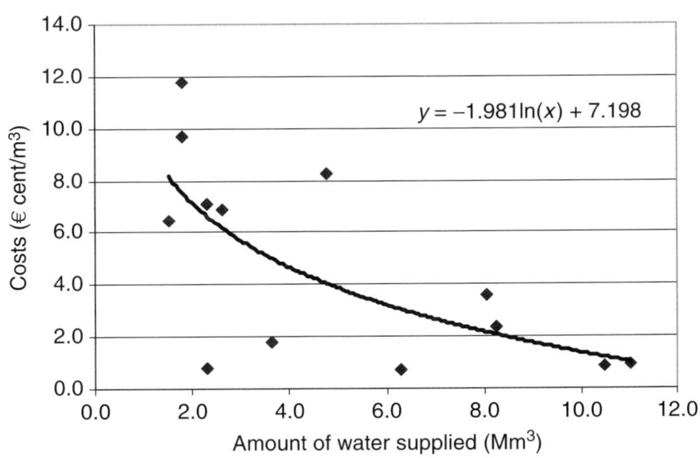

Fig. 13.1. District costs according to the amount of water supplied (Guadalquivir).

Table 13.2. Irrigation district costs.

Country/region	Districts	Origin of water	District's charges	Type of charges	Source
Spain (Valencia)	Traditional	S	€63.00–509.00/ha	Variable based on the no. of applications	García, 2002
	Traditional supported by State projects	S	€18.00–31.00/ha	Variable on the no.of hours of usage	García, 2002
	Traditional districts	S&G	€20.00–457.00/ha	Variable on no. of hours and no. of applications	García, 2002
	State Projects	S&G	€2.11–76.00/ha	Variable on no. of hours and no. of applications	García, 2002
	Private associations	S&G	€51.00–144.00/ha	Volumetric rate	'
Italy	Bonifica (Capitanata)	S	€10.54/ha + €0.061/m^3	Block rate pricing	Xiloyannis and Dichio, 2001
	Puglia	S	€63.00/irrigator + €120.00/ha	Flat rate	Xiloyannis and Dichio, 2001
Greece (Crete)	Large Irrigation Districts	G	€0.046–0.053/m^3	Volumetric rate	Chartzaloulakis and Angelakis, 2001
	Community projects Private projects	G G	€0.066–0.80/m^3 €0.15–0.23/m^3	Volumetric rate Volumetric rate	Chartzaloulakis and Angelakis, 2001
France	Charente	S	€7.30/ha + €0.0037/m^3	Volumetric rate	Montginoul and Rieu, 2001

Note: S: Surface; G: Groundwater.

Water authority costs

A critical methodological issue regarding cost recovery analysis is the definition of the financial costs of Water Authority services. The situation for full cost recovery may vary for each river basin. We illustrate difficulties with this definition with an example from the Guadalquivir basin in Spain, regulated with large dams and infrastructures. The use of historical values in cost recovery is generally accepted in most of the great public infrastructure projects. Specifically for water infrastructure, many dams are older than 50 years and have been theoretically fully paid for by users (farmers and urban users) even when there is a real positive salvage value and they are still in use.

We should remark that the use of 'marginal' or 'replacement cost' is not assumed in the WFD, and we may recall the water privatization in the UK in the 1990s, when the final value of assets was computed neither at 'historical value' (deemed too low a price) nor at 'replacement value' (deemed an excessive price) but estimated based on the 'present value of profit' or, in other words, the ability of buyers to pay. Economic theory defines the capital value of an asset as the present value of the future stream of profit and therefore neither historical value nor replacement value is relevant. In practice, the higher the amount for which the existing assets were sold by the government when the companies were privatized, the higher the charges for water and wastewater required to provide a commercial rate of return on those assets. In contrast, in the case of irrigation, Spain has been given permission by the EU Commission to use historical depreciation criteria for determining the extent of full cost recovery rates.

Generally, difficulties in defining financial costs also arise from how the costs of multi-purpose projects are distributed. In Spain, the sharing of costs between different uses is made by a 'stakeholder agreement' at basin level, considering the following variables:

- *Capital cost sharing:*
  - Flood control: The percentage of costs assumed to provide the public service of flood control may vary from 20% (most dams in Spain) to 70% of some special Mediterranean cases (e.g. Tous dam).
  - Urban (domestic and industry) versus irrigation: normally urban users have a different quality of service (daily, seasonal, yearly secured supply) versus irrigation that, in many cases, is residual use.
  - Energy (hydroelectric, refrigeration);
  - Environmental use.
- *Recovery of O&M*: Water agencies are multifunctional as they may not only control abstraction and pollution but, sometimes, also finance infrastructures to supply water. For example, the Guadalquivir Basin Authority recovers 75% of its O&M costs for public infrastructures through tariffs, but the remaining 25% is linked to the cost of environmental services (pollution and flood control, etc.).

Also the computation of financial costs should determine some technical parameters such as: (i) depreciation rate; (ii) salvage value of investment; and (iii) interest rate.

An example of water cost recovery estimation is described by Berbel (2005) who computes costs according to the current Spanish Water Law which states a 'cost recovery formula' defining a water tariff based upon computation of water agencies' O&M plus depreciation of water infrastructure (the depreciation rate is based upon historical costs without interest rate). When the cost definition and criteria of the Water Law are applied, we arrive at 99% of financial cost recovery. But when stricter accounting criteria are applied including faster depreciation and 5% interest rate, financial cost recovery is reduced to 71%. Finally, this percentage may be reduced if we compare the present average tariff in the Guadalquivir river (€0.0178/m^3) with the 'replacement cost' of €0.06/m^3 (full recovery rates for 'La Breña-2 dam' presently under construction).

On the other hand, still in Spain, in the eastern Valencia region, where the use of groundwater is very intense and predominates, the final cost of water for farmers ranges between €0.04/m^3 and €0.22/m^3 for surface water and groundwater, respectively, with an average of €0.115/m^3 (Carles *et al.*, 2001a,b; M. García, J. Carles and C. Sanchos, 2004, unpublished data).

To mention another example of cost recovery we may take the last large irrigation project associated with the 'Alqueva dam' (Portugal) with a present price of 1.8 cent/m^3 (not covering energy cost) but expected to reach 8 cent/m^3 in 2007 if full 'financial cost recovery' is to be achieved. This shows the importance of subsidies as the root of future water imbalances and environmental problems.

Water Authority costs include all cost items directly related to the supply of irrigation water, which are covered by water charges to users and by general taxpayers, with different degrees of cost distribution between the two groups. A common conclusion across countries (see Annex) is that irrigators have been, and still are, heavily subsidized.

Groundwater (on-farm) costs

Groundwater is the source of water for 20% to 100% of European irrigated farms, depending on the region and country. Irrigators using groundwater resources apparently pay all financial costs as they pay investment, maintenance and energy costs for pumping water because they are not supplied by any public scheme. Consequently, in most of the countries, users of groundwater do not pay any tariff to water authorities although some countries (France, the Netherlands, Denmark, the UK) charge a water abstraction fee. We will return to this concept in the next section.

Social and environmental costs

Non-monetary or social costs caused by irrigation are those inflicted on third parties or on the environment. In both cases, social costs originate from irrigators' use of valuable resources or from their polluting the resource base. The former is generally associated with the opportunity cost, and provides an indication of the value of the water allocated to alternative users. Irrigation can affect the environment through its direct impact upon water resources, soils, biodiversity and landscapes, as well as its secondary impacts that arise from the intensification of agricultural production through the transformation of rain-fed land into irrigated land (European Commission, 2000).

Recent work shows that social costs are far from negligible, and provides a solid basis for urgent policy action. The list of regions or basins where problems related to excessive irrigated water use have been identified would be very long. Generally, resource overdraft is caused by a water demand, both urban and agricultural, quite above the sustainable rate, where the cost paid by users is generally below financial (monetary) recovery cost for surface water and fully or partially paid for groundwater. In south-eastern Spain, where some trading of water occurs especially for fruit, vegetables and greenhouse production, water cost is only around 2% of total cultivation costs. This implies that water demand will inevitably tend to go beyond sustainable renewable use, indicating that the private cost of water does not reflect the scarcity of the resource.

Regarding pollution by nutrients, the main polluter in Europe is the agriculture sector, including rain-fed and livestock farming. Irrigated agriculture contributes to the increasing nitrate contamination, due to overfertilization. Examples of such direct effects have been found in the Adour-Garonne (France), in several Austrian regions such as the Marchfeld, the Pandofer plateau, and the Welser Heide and Eferding Becken areas, in a number of Spanish regions, mostly located along the Mediterranean coast and main river valleys, and in various nitrate vulnerable Greek zones such as Argolid, Kopas and the Thesaaly plain, where large irrigated areas are located (European Commission, 2000). Nevertheless, in most river basins the impact from livestock and rain-fed agriculture is higher than that from irrigation (e.g. in the Guadalquivir valley nitrate pressure generated by irrigated agriculture is around 22%, against 52% and 22% generated by rain-fed agriculture and livestock, respectively).

As a consequence of the above-mentioned evidence of irrigation pressure on the environment, countries such as France or the Netherlands try to 'internalize costs' by using an 'ecotax' on water use by irrigators. This ecotax on water abstraction (mostly groundwater) tries to internalize environmental and social costs, but the level of environmental cost recovery is quite low as seen from the first reports presented by the EU member states reporting on WFD implementation. The Spanish government is debating whether to charge an ecotax on all water use (both surface water and groundwater) to contribute to global integrated resource management at the basin level and meet the 2010 deadline set by the WFD for implementation of measures including water pricing. Provisional estimates for this ecotax (€1.00/1000 m^3) make it a 'political contribution of users' rather than an environmental cost recovery charge.

The use of water pricing incentives such as the ecotax is opposed by some authors as Martinez and Albiac (2004) who show that nitrogen pollution is most efficiently abated by targeting either the source or the emissions, and very inefficiently by imposing levies on used water. Nevertheless, most models of irrigation water demand predict a significant reduction when a water tariff is imposed (e.g. €0.06/m^3 in Aragon, Spain reduces water demand by 50%). This suggests that water and environmental policies must be closely linked and target the most pressing problems, be it water scarcity or nitrogen pollution. Still a further effort of empirical studies is required considering both short- and long-term farmer responses.

Numerous studies have shown that more efficient water use reduces agricultural pollution (Dinar and Letey, 1991; Weinberg *et al.*, 1993; Calatrava and Garrido, 2001). Yet, this does not imply that pollution control should be targeted with water pricing policies. Pollution control can be best performed within irrigation systems by providing precise water applications and monitoring.

Water use incentives

Incentives for conservation and efficient water use

According to the neoclassical definition of use externalities, most water problems in the European irrigation sector stem from situations where clear misalignments exist between farmers' private objectives and more general social objectives. The presence of divergences between private and social objectives is manifested by various trends. One is the widening of the divergence between farmers' low water marginal productivity in irrigated commodity production and the sum of the costs incurred by society for making the resources available to them (except for the case of high-value crops). Another is the confirmation that the water costs of competing users may be rising as a result of farmers' water use or polluting practices. Note that the manifestation of adverse incentives is perceived through time and not with snapshots. This implies that policy judgements should be preferably based on whether observed trends show improvements or are worsening.

A list of adverse incentives includes the following:

- *Per-hectare water charges*: Per-hectare charges (flat rate) are perhaps the most adverse incentive affecting irrigation across OECD countries. Very few irrigated districts relying on surface water have volumetric or other variable rate systems. The wide recognition of the need to change the tariff structures towards volumetric charges has not been accompanied by clear examples of policy implementation. To date, no rigorous evaluation has been made to measure the value of the efficiency losses resulting from the prevalence of flat rates. Montginoul and Rieu (2001) report that irrigators in Charente (France) are charged with two-part tariffs, but the fact that the variable rate is much lower than the marginal benefit of water use in the farms led the managers to impose water quotas in years of scarcity.

The comparison of water use levels of irrigators using surface water with those of farmers relying on groundwater may provide an indication of the effects of wrong signalling caused by flat rates. Hernández and Llamas (2001) show that groundwater users tend to use between 25% and 35% less than surface users. Yet, groundwater users with pressurized systems will obviously 'use' less than those on old surface water systems. In addition, return flows from upstream surface users may be used downstream, thus increasing the efficacy rates of surface water. All in all, a dollar value of such water use differences is difficult to come by, but may be equivalent to a lower bound of €15.00–35.00/ha, with the most conservative assumptions.

As will be argued below, numerous obstacles hinder progress in replacing flat rates with volumetric rates. Among them is the fact that it may not be efficient to do so, under a broad range of realistic situations. Work done by Tsur and Dinar (1997) illustrates how the efficiency gains may not justify the costs of restructuring tariffs. Chakravorty and Roumasset (1991) and Hafi *et al.* (2001) show that volumetric charges would have wealth redistributional effects in large districts with network losses. Another relevant obstacle is the lack of appropriate water-metering devices in many European irrigation districts.

Investment in irrigation technologies has ambiguous effects in general policy evaluations. Negative effects result from the fact that changes in technology may induce new crop patterns and increase total water consumption. García (2002) shows that drip irrigation technologies have been subsidized in the region of Valencia in Spain but, contrary to general belief, irrigators have not reduced application rates. Similar behaviour has been observed in the Guadalquivir river basin, in the sense that the adoption of drip irrigation has encouraged the planting of new crops (orchards, vegetables, etc.) that are more water-demanding than the preceding ones (Berbel, 2005).

- *Inelasticity of demand*: The number of studies showing that irrigators' water demand is highly inelastic in the short term, at least at low prices, is at odds with the fact that there are large differences in water consumption and application rates among irrigators and water districts. This means that differences are not governed by prices but by other factors. Response to price increases is not continuous as there is an optimum supply of water for each crop and the water production function implies that the optimum is not sensitive to price increases until a break-even point is surpassed, when a new crop is introduced or farmers simply go for rainfed crops. One would assume that if a set of irrigators seem to operate with low consumption rates, then another operating under similar conditions could be flexible enough to reduce its consumption. Whether it is a change of water price or a reallocation of water rights, the conclusion would be that the latter irrigators can and should reduce their consumption, following a relatively elastic water demand curve.

Before delving into this paradox, García (2002) suggests that water should be thought of as a productive input, whose demand elasticity depends on three factors: (i) the elasticity of substitution of water for other inputs; (ii) the price elasticity of demand for the good being produced; and (iii) the share of irrigators' total costs represented by water costs. The practical application of these principles is that water cannot be substituted by other inputs, except for large-water demanding crops like rice which are grown with little use of capital. Table 13.3 shows some examples in Europe of the relationship between water cost and total cost.

The assumptions embedded in this reasoning turn out to provide clues that may solve the paradox. If technology is fixed, water rights are not tradable and water allotments are fixed by the water authorities in the form of entitlements or quotas, then water demand is likely to be inelastic. Perhaps, looking at water use levels allowing for long-term adjustments, or looking at farms which do not rely on fixed allotments water demand would exhibit larger elasticities.

However, relaxing these three assumptions clears the puzzle, but opens two more.

Table 13.3. Water cost versus total cost. (From Berbel, 2005.)

Crop/system	Location	River/source	Output (€/ha)	Cost (cent/m³)	Water/ output (%)
Greenhouse	The Netherlands	Underground	120,000	15	0.8
Strawberry	Chanza (HU)	Guadiana	48,193	15	1.6
Greenhouse	Almeria	Mediterranea Andaluza	90,361	25	1.7
Maize	France	Several	6,000	10	3.3
Olive	Jaen	CH Guadalquivir	6,000	15	5.0
Cotton	Seville	CH Guadalquivir	4,000	8	12.0
Sugarbeet	Palencia	CH Duero	3,000	6	12.0
Wheat	Cordoba	CH Guadalquivir	1,506	8	10.6

Table 13.4. Water consumption differences among Valencian irrigation districts. (From García, 2002.)

(Type org, type rates) +, means more consumption –, means less consumption S, surface water G, groundwater		Type of district rates			
		Two-part based on no. of hours	Two-part based on no. of applications	Flat rates	Variable based on no. of hours
Type of organization	Traditional districts supported by state projects (S)	(–,–)	(–,+)	(–,+)	(–,–)
	Traditional districts (S&G)	(+,–)	(+,+)	(+,+)	(+,–)
	State projects (S&G)	(+,–)	(+,+)	(+,+)	(+,–)
	Private associations (G)	(+,–)	(+,+)	(+,+)	(+,–)

First, the adoption of water-saving technologies is generally found in districts whose water allotments, granted by basin agencies, have experienced a gradual decline. What is cause and effect is difficult to ascertain, because all that changes, namely, water consumption, allotments and technology adoption evolve simultaneously in response to administered scarcity; and it also shows that administered reallocation based on actual scarcity indeed begets adaptation rather than having to force this through prices at higher costs and to the detriment of equity.

The second one stems from the fact that in virtually all empirical attempts to measure water demand elasticities the districts studied do not face any opportunity cost resulting from their water consumption. This means that most demand analyses posit hypothetical price increases and then infer what would be the farmers' likely response using modelling techniques. Does this imply that water tradability or variable cost charges would suffice in reality to allow irrigators to exploit the efficiency gains found from cross-sectional studies? In other words, is the absence of variable prices responsible for the relatively inelastic demand found by numerous analysts? If the answer is positive, re-forming water charges may result in significant consumption decrease.

Recent work by García (2002), perhaps the most detailed analysis ever done in Spain to explain water use differences across all districts in the Valencia region, shows that water use variability is largely explained by three factors, namely, the type and institutional arrangement of districts, the origin of the used water and the type of pricing scheme. Yet, the analysis is carried out in a very innovative region, where tens of different crops can be grown. Table 13.4 summarizes the econometric results.

García Mollá's results suggest that traditional districts supported by state projects combined with 'two-part tariff systems' exhibit the lowest consumption levels. They also suggest that all districts using groundwater exclusively or in combination with surface water tend to consume more than those that rely exclusively on surface water resources, indicating perhaps unsustainable use. This result contradicts the conclusions of Hernández and Llamas (2001), and in our view shows that farmers seek maximum economic yields subject to the prevailing market, technological and institutional constraints. Under similar constraints, farmers' consumption will not be driven by the origin of their resources. Lastly, flat rates are directly associated with larger consumption, although causality is not properly established. García Mollá's work provides evidence that seriously disputes the results of Hernández and Llamas (2001), although this work did not consider Valencian districts. Perhaps the conflicting results can be reconciled by the fact that in Valencia surface water is very scarce and much less reliable than groundwater sources, whereas in Aragon or Andalusia surface water is generally abundant. The conflict of surface water versus groundwater is not so evident as there are many cases where surface water is available only 2–3 months during the year which needs to be supplemented by groundwater to allow for tree cultivation (which may need irrigation when surface water is not available), especially in Andalusia.

Bontemps *et al.* (2003) show that water demand in southern France is inelastic for low available volumes, and depends crucially on the weather conditions. Rieu (2005) shows that, although demand in Charente is elastic, local authorities have established quotas to avoid the negative effects on farm income. Overall, pricing policies in France seem to be driven primarily by the objective to ensure cost recovery and *agence*'s budget balance, although this is achieved by a great variety of pricing mechanisms (Rieu, 2005).

Dono and Severini (2001) add further evidence from southern Italy to the inelasticity hypothesis, and suggest that water demand turns increasingly inelastic as water charges increase, as the crops that may be able to pay higher prices are mainly high-value vegetables and fruits, which can support high water price increases.

Finally, Massarutto (2003) concludes that the demand inelasticity hypothesis should be framed in relation to the concept of 'exit price'. He claims that the effects on water demand are due to the fact that if water prices are below the exit threshold, they result in demand reductions caused by marginal adaptation of irrigation demand to price variations. Water demand elasticity is always very small, especially once the most obvious water-saving techniques have already been implemented. Above the exit price, water demand is brought to zero because farmers do not cover the input costs and they are better-off not using the water.

- *Users reallocation*: In a very authoritative essay, Brown (2000) documents the poor records of resource pricing to facilitate reallocation and more efficient use. Water reallocation either occurs because the government mandates it or (generally) because mechanisms are implemented to facilitate voluntary exchanges. At most, multi-layered policies, in which new pricing, lower allocations, rehabilitation projects and generous financing are included, can facilitate some trading. In Europe, water markets and liberalization are mostly understood as a process towards giving the private sector more pre-eminence in the areas of urban supply and wastewater treatment. It is only in Spain that there has been a serious attempt to provide for water right exchanges, which required a significant amendment to the water law, but which has so far been used very sparsely.

Other relevant incentives

- *Agricultural policies that promote water consuming crops*: Examples of crops, across the EU, with high water requirements supported by CAP programmes were numerous. Maize is considered a water demanding crop in temperate countries, and its EU growers were until 2003 entitled to a direct subsidy of €54.00/t of yield, which usually exceeds 10 t/ha. Since the CAP direct subsidies were defined to deliver equivalent levels of income support to all cereal, oilseeds and protein crops, they favour crops such as maize, rice, cotton or tobacco that demand much more water than oilseed crops such as sunflower or colza. With decoupling, this inconsistency has been eliminated, and farmers' use of water will not be driven by subsidy differences across crops.

Between 1973 and 1988, agricultural water use in France grew by 43%, largely due to generous public programmes which provided subsidies to farmers installing irrigation equipment, as well as guaranteeing generally low agricultural water prices. Most of the increase was used in maize production. This trend was reinforced after the 1992 CAP reform replaced production subsidies by per-hectare direct payments, as a result of the higher compensatory payments given to irrigated acreage than to non-irrigated acreage (Dubois de la Sablonière, 1997; Rainelli and Vermersch, 1998).

EU agricultural policy 'Agenda 2000' aimed at supporting a multifunctional, sustainable and competitive agriculture. It was based on the establishment of production-related direct aid payments and gave a prominent role to agri-environmental instruments to support farmers' income. In June 2003, the EU decided to replace, from 2006 onwards, most of the direct aid with a single farm payment scheme that is not linked to production. Beneficiaries will be obliged to accomplish certain environmental and food safety requirements.

Work done by Calatrava and Garrido (2001) shows that CAP's Agenda 2000 tended to increase irrigators' water demand in the Spanish region of Andalusia with respect to the pre-2000 situation, which was confirmed in the case of olive oils and vineyards. These authors show that the price support delivered to cotton producers in the region is largely responsible for the large benefit of water in the region. Pressure on water demand by farmers in the region has been on the rise, although recent changes in the Common Market Organization for cotton may have an inverse effect, as cotton support has been largely reduced. Many authors have established a connection between farm subsidies and irrigation water demand in Spain (Sumpsi *et al.*, 1998; Gómez-Limón *et al.*, 2002; Arriaza *et al.*, 2003; E. Iglesias, J.M. Sumpsi and M. Blanco, 2004, unpublished data). Their results show that, indeed, the elimination of farm subsidies has a larger impact on the farmers' welfare than the rise of water prices. When EU farm subsidies become completely decoupled from production in 2012, the economics of irrigation will be more guided by the relative productivity of crops and water accessibility than by relative farm subsidies granted to the crops.

Mejias *et al.* (2004) add further evidence to the water demand inelasticity hypothesis. In addition, they show that the EU policy based on full decoupling will likely reduce the income losses resulting from WFD's increased water tariffs, at least in Andalusia (Spain).

- *Subsidization of irrigation equipment*: Positive results come by increasing water productivity, which in turn would reduce the welfare windfall losses resulting from water price increases. Yet, Rainelli and Vermersch (1998) showed that one reason that explains the significant growth in French irrigated acreage was the subsidization of irrigation equipment, which reinforced the CAP incentives mentioned above (as with Spain, cited earlier).

The extent to which subsidization of irrigation equipment should be taken into account in water subsidization analysis is not clear. For one thing, a general belief is that these subsidies are redundant, as irrigators eventually invest in equipment with or without subsidies. Some of the reasons guiding their investment plans are labour cost reductions, lower input application costs through fertigation and upgrading product quality.

- *Inadequate cost recovery rates*: Low charges eventually translate into poorly maintained water infrastructures, which in turn reduce irrigators' competitiveness and 'capacity to pay'. Yet adequate cost recovery rates are not sufficient to ensure proper conservation of half-century-old irrigation districts. For instance, half of the Spanish irrigated acreage was built before 1960, when farms were small and poorly mechanized, and the country had embarked on reclamation projects. Since 2001, 95% of the budget devoted to irrigation in Spain is targeted to finance modernization projects, which have reached 1.3 Mha and a budget of €4 billion (Barbero, 2005). Beneficiary farmers must pay only 50% of the project's costs, for which they are granted preferential loans. But the process is becoming very costly, as projects have been refocused to include environmental, structural, technological and land planning/tenancy components. The gains are private in the form of more efficient and productive districts, as well as public in the form of water conservation and reduced pollution. By no means would farmers' full cost recovery rates suffice to finance such projects. Yet they are praised and uncontested.

The WFD and Economic Analysis

The WFD is an environmental norm rather than a general regulation instrument and its main objective is the sustainable use of water through the long-term protection of resources. Article 4 lays down the environmental objectives of the WFD. To avoid possible sources of conflict the WFD explicitly states that it aims to reach a more sustainable use of water resources which can protect or enhance regional development. Also, the WFD establishes derogation or dispensation mechanisms set by Article 4 ensuring that the environmental objectives can be challenged by other socio-economical considerations as long as these are transparent. Member states will define quality objectives ('good ecological status') but coordination is guaranteed through a calibration exercise to standardize norms for all Europe, producing a benchmarking of water quality and harmonizing the definitions of the good environmental status, in order to avoid different national standards for defining the 'good environmental status'.

With respect to the use of water for irrigation, the WFD mentions a number of important aspects, namely:

- River basin management, whereby water resources are managed at an integrated catchment level (including both surface and underground resources);
- Cost recovery for water services, whereby those who benefit from using water (as a resource or a sink for waste) pay for such services, including the environmental costs, which presently are associated with remediation costs;
- Participation of stakeholders in the planning and decision-making process;
- Protection of groundwater and wetlands.

It can be noted that there is very little reference to flood prevention (important to all member states) and drought management (essential to Mediterranean countries). A key concept about water price is that cost recovery refers to water services and not to the water resource itself. Specifically, Article 2, paragraph 38, defines water services as:

> All services which provide, for households, public institutions or any economic activity: (a) abstraction, impoundment, storage, treatment and distribution of surface water or groundwater, (b) waste-water collection and treatment facilities which subsequently discharge into surface water.

This definition has an enormous relevance for the correct interpretation of the principle of 'recovery of the costs of the water services' mentioned in WFD Article 9.

Many issues have created barriers to an early agreement on WFD (see Annex: Summary of European Countries' Experiences), but one of the most controversial was Article 9 in the first drafts of the proposal. This article originally obliged EU members to charge the full cost of water to users. The final agreement was much vaguer, establishing merely that EU members should try to recover all water service costs, including environmental costs, in accordance with the 'polluter pays' principle.

WFD Article 9 requires member states to take into account the principle of recovery of the costs of water services:

> Member States shall take account of the principle of recovery of the costs of water services, including environmental and resource costs, having regard to the economic analysis conducted according to Annex III, and in accordance in particular with the polluter pays principle.

At the same time, Article 5 mentioned in Annex III shows what could be entitled a synthesis of 'Theoretical Economic Analysis of Water Use' for WFD implementation.

Although the economic analysis of water use is an important element of the WFD, precise instructions about the methodology to carry out the economic analysis are not defined in the text. However, in the common strategy for the implementation of WFD, guidance is given by the WATECO reference committee that has developed a guide to illustrate the process of introducing economic analysis and cost recovery into the WFD implementation (*Economics and the environment: The implementation challenge of the water framework directive: a guidance document,* WATECO, 2003). In addition the European Commission has launched a series of pilot studies to illustrate the application of the WFD in a number of EU basins.

Estimating costs is an important aspect of the economic analysis, and the guide, WATECO (2003), summarizes this aspect as follows:

- Assessing whether the principle of *recovery of the costs* of water services is met;
- Conducting a *cost-effectiveness analysis* of alternative policy measures/projects;
- Assessing the costs of alternative management options in the *designation of 'heavily modified water body'*;
- Assessing the need for a derogation based on an economic appraisal of disproportionate costs (such as for the setting of *less stringent objectives or a time derogation).*

Note that the Directive defines costs as *economic costs*, which are the costs to society as a whole, as opposed to *financial costs*, which are the costs to particular economic agents. Catchments may ask for 'derogation' when the application of the WFD has disproportionate costs. However, derogations apply to the environmental goals and not to the application of cost recovery instruments. Finally, WFD considers the use of water pricing as an instrument among many others aimed at the final objective that is 'reaching a good ecological status', and should not be considered as an end in itself.[1] Nevertheless, there is a general consensus that the application of WFD should contribute to increasing levels of water price for both surface water and groundwater.

The problem with the definition of costs for cost recovery analysis in the WFD

The main document for this exercise in the WFD is the guide, WATECO (2003). The analysis of financial cost in the WFD has been already explained under the section on 'Water Use Incentives', and first reports done by member states in 2005 show that financial costs are closer to full recovery than expected a priori in most countries, according to the definitions given in the guide.

However, the definition of environmental and resource costs in WATECO is not very precise. Regarding environmental and

[1]More background is provided in COM 477 (European Commission, 2000a) on how water pricing could be used for cost recovery purposes. Many references are made to agricultural water use and to cost components. For further analysis of this concept, readers should refer to WATECO, 2003.

resource costs, rarely considered in current water tariffs, it is difficult to predict to what extent they will be considered, let alone whether accounting methods and potential increases of tariffs and levies will be defined. The European Commission has been rather conservative in requesting member states to add these costs to the rates. In this sense, it is difficult to predict the extent of positive environmental effects. As an estimation of non-monetary costs, an ecotax on abstraction is applied in the Netherlands, Switzerland and the UK as an attempt to internalize this cost.

There is lack of information on the environmental cost of water use, and this applies in particular to irrigation. Nevertheless, some European Water Agencies estimate the extra cost (over financial cost value) of the damage to the environment of water use in global terms (urban, industry, agriculture) at 20–25%; the application of economic instruments to agriculture, specifically to irrigated agriculture, should increase the financial pressure on farming.

An additional difficulty for assessing this cost is the pressure-impact evaluation (i.e. local water abstraction impact on global quality of water).

Finally, the methodology for evaluating the resource cost is not already clearly defined in Commission documents, and we believe that this may be included only after financial cost and environmental cost are fully recovered, which is not the case yet. But, apart from being unrealistic, does charging the resource cost make sense when financial cost recovery has not been reached anywhere, either in relation to surface water or groundwater services?

Likely effects of the WFD

As previously mentioned, the WFD aims to establish a framework for the management and protection of water on the basis of individual river basin districts. In that sense, economic instruments are only one of the possible policy measures to reach the 'good ecological status' that is the final objective of the Directive, including reaching a balance between abstraction and recharge of groundwater, with the aim to achieve water use sustainability. The WFD task force on groundwater will undoubtedly be discussing the definition of safe yields, sustainability and so on. Member states are expected, before 2010, to enforce water pricing policies which give an adequate incentive to improving the efficiency of water use, contributing to the environmental goals of the Directive.

Pricing for financial cost recovery and conservation

Virtually all analyses of the effects of price increases in irrigation predict that the agriculture sector would be severely hit by the strict application of the WFD, especially smaller and family farms. Studies consider tariff increases between €0.03/m^3 and €0.1/m^3, which are frequently below full cost recovery rates, and predict reductions in farm income ranging from 10% to 50% (Garrido and Calatrava, 2005).

Water pricing will have different impacts depending upon specific characteristics of each farming type. Berbel and Gutierrez (2004) found differences in the water demand curves for three regions in Spain (two) and Italy (one) (see Fig. 13.2 and Table 13.5). The Italian case, which was based on vegetable cultivation, shows a much lower level of water consumption and a much more rigid behaviour of the demand curve due to the high profitability of the crops cultivated. In the Foggia region (southern Italy), where excellent marketing channels for high-valued fruits and vegetables as well as drip technologies exist, there is almost no possibility of water saving. Furthermore, in the Italian case, increasing the price of water would have almost no effect in terms of diminishing water use, and would merely deflate farmers' incomes.

On the other hand, in the Duero valley (northern Spain), where irrigation is mostly based on sugarbeet, the impact of water price rise is that water demand collapses when price is above this crop's productivity and irrigation is abandoned. Climate in this region is very extreme with long and cold winters and hot summers, and irrigation is

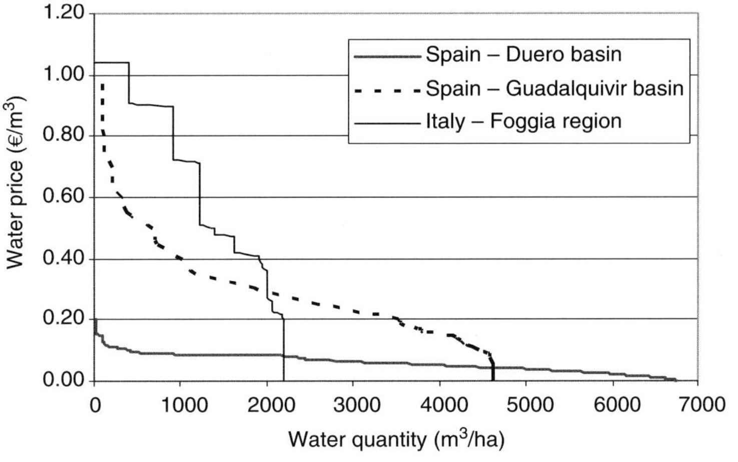

Fig. 13.2. Water consumption according to water price (three regions).

Table 13.5. Water demand characteristics. (From Berbel and Gutierrez, 2005.)

Duero (northern Spain)	Guadalquivir (southern Spain)	Foggia (Southeast Italy)
Demand disappears at €0.15/m³	Demand varies from €0/m³ to €1.00/m³	Demand varies from €0/cm to €1.00 /m³
Elastic demand	Inelastic up to €0.1. Then, elastic	Inelastic up to €0.23. Then, elastic
High response to water price	Low response to water price	Low response to water price

used to grow mainly low-value and heavily subsidized crops. The impact of the new EU single direct payment will likely include a drastic reduction of water demand because of crop shift.

The Guadalquivir case is somewhere in the middle, with some crops dependent on subsidies and others under market competition. In this area, water demand is approaching that of the Foggia case, as an increasing part of demand is already under drip irrigation (olive, citrus and other fruits, 44% of water consumption and 47% of area). As drip irrigation is linked to high-value crops (fruits and vegetables), water demand becomes more 'structural' and 'rigid', and the likely effect of water pricing is that the impact will go directly to decrease farmers' income, as significant water saving is already in effect.

These three cases show that the specificity of agricultural systems requires a detailed local analysis. Nevertheless, in general terms, price increase towards full cost recovery may be in the range from 20% to 400% over present levels, depending mainly upon two factors: the depreciation criteria adopted in infrastructural cost recovery and the inclusion of all subsidies in order to determine the cost recovery tariff. For less favoured areas, such as the Duero river (Spain), this increase will imply a substantial reduction in area irrigated, farm income and employment. On the other hand, high-value crops (Foggia, Italy) may bear price increases but with the consequence of a transfer of income from farmers to the Water Agency. In any case, the key factor for water saving will be not only the price increase itself but also the use of quantitative controls, so that flat rates (payment by area) are changed to volumetric or mixed rates (payment by quantity).

Regarding financial cost recovery and groundwater resources, in many European countries the most profitable agriculture (horticulture, fruit trees and greenhouses) is based on the use of groundwater. In the case of Spain, more than 20% of farmers, producing more than 50% of total irrigation value, and located in some of the less water-endowed areas, will be exempt from increases in water charges resulting from the WFD. Most of the irrigation in northern European countries (the UK, Holland) is also based on groundwater, but again water is used for crops with a high marginal value of water, as it increases mainly quality and not quantity (irrigated crops get significantly better prices, see Table 13.4) and water costs are below 1% of total costs. In other cases, such as France, groundwater is also the main source for irrigation, but in this case most of the water goes to maize (heavily subsidized by CAP). Most likely, the application of the WFD will not result in groundwater prices similar to those applied in surface water schemes.

We may consider that groundwater already recovers 100% of 'financial private cost', but the WFD implies that this source of water should contribute to environmental and resource costs as we will see in the next section, and before 2010, member states should define their plans to use price instruments, regardless of the source of water.

Other Policy Instruments Related to WFD

The set of policy instruments related to WFD implementation go beyond water pricing, as irrigated agriculture will also suffer from restrictions in the use of chemical inputs and possible ecotaxes on fertilizers or pesticides. Agriculture and livestock (both irrigated and rain-fed production) are responsible for water pollution by nitrates and phosphorus. Under the 33 priority substances proposed with the implementation of the WFD, heavy metals such as cadmium (linked to phosphorus in agriculture) and about 11 pesticides must be regulated. Consequently, measures for the adoption of Good Farming Practices will greatly influence irrigated agriculture in the near future. In this sense, we should recall that new irrigation techniques (e.g. drip-irrigated crops) may improve efficiency in the 'product output/fertilizer pressure' ratio as fertilization is in the water directly applied to plant, reducing losses.

Additionally, the future decoupling of farm subsidies, established by Council Regulation 1782/2003, is accompanied by a cross-compliance policy that conditions payments to the farmers achieving 'Good Agricultural and Environmental Conditions' in their parcels, and complying with several European Directives. Five of these relate to the environment, namely the Wild Birds Directive, the Groundwater Directive, the Sewage Sludge Directive, the Nitrates Directive and the Habitats Directive. Cross-compliance policy aims to speed up compliance with several European Directives that were not being adequately implemented by member states.

It is also very important to integrate the implementation of the WFD and the new CAP. First, the more choice farmers have in selecting the crops, the most efficient is water use, and the least income-reduction effects result from water conservation policies (Mejias *et al.*, 2004). Upon the reform of EU agricultural policy, several analyses have explored whether the incentives to use water would change as a result of more decoupled measures of farm income support. It is shown that more decoupled measures of support may make pricing policies more effective and less negative for farmers' benefits. Gómez Limón *et al.* (2002) show that agricultural and water policies may have conflicting objectives. Yet the trend towards more decoupled measures of support will likely ease the tension which, at least in the EU, has been found in many studies.

Beyond the existing possibilities in the Rural Development Programmes (RDPs), such as agri-environmental schemes, to reduce groundwater consumption or finance technology adoption, there is a need for a further consideration of compulsory water use practices in the Codes of Good Practices of the RDP and the cross-compliance scheme. Currently, issues such as restrictions on fertilizers and pesticides are included in these codes, but other issues such as drainage or irrigation technology adoption could also be included.

The costs to farmers derived from the compliance with other environmental EU Directives related to water (see WFD, Article 22) should therefore be added to the costs of complying with the WFD itself, adding further technical and economic constraints.

Indirect effects on agricultural labour are also to be considered. Irrigated agriculture is very important in rural areas of southern Europe and social impacts of water pricing are likely to be high and to raise local opposition. However, these should be compensated by increased demand for labour in other sectors (for instance, the irrigation technologies sector), and some kind of compensation scheme could be established for rural areas that will be seriously affected. A positive effect would be the reduction in the pressure put on public budgets, as expenses collected will increase, investment required for new infrastructures will be reduced and funds will be available for other projects.

Concluding Remarks

The EU WFD will profoundly change the basis for setting irrigation water pricing policies in the 25 EU member states. The implications of implementing WFD's Article 9 will depend on the evaluation of the costs of the water serviced to agriculture, and the proportion of costs that is eventually imposed on the irrigators' final charges. In most countries, irrigation water charges are lower than the financial cost recovery level and, generally, environmental cost is not considered. Some of the non-EU countries, like Romania, Bulgaria and Croatia, may soon develop similar policies to get ready for accession to the EU.

Most of the water pricing policies are related to surface water under public schemes, but the use of groundwater may account locally for 100% of irrigation. In the case of groundwater, financial costs of abstraction are fully recovered, but environmental costs are usually not included. But the WFD implies management of all water resources both surface and underground in order to reach 'good ecological status'. Most countries do not consider any form of ecotax for groundwater, or any kind of economic instrument in areas with local aquifers at risk of overexploitation.

Agricultural water tariffs are quite heterogeneous across countries, regions and even within regions. Tariff structures apply almost exclusively to surface water and they rarely reflect relative water scarcity, as they result from complex geographical, technical and institutional factors. Fixed per-hectare tariffs are predominant in southern European countries, mostly in districts supplied with surface water from publicly developed infrastructure, while volumetric charges prevail in northern countries.

The level of cost recovery is very low, and charges are in most cases far below urban or industrial ones. Noteworthy exceptions are the cases of the Netherlands, Sweden and the UK. Some countries (Switzerland and Croatia) have established discharge fees for agriculture, while others, such as Portugal and France, sometimes charge for drainage.

The WFD represents a unique world experience for a number of reasons. First, because it is a decisive step to make farmers responsible for the costs their use imposes on the water system and on the government's budget. Even if 'full cost recovery principles' are loosely applied on irrigation charges, and despite the fact that methods for accounting these costs may not be agreed upon by all member states, the gap between costs and charges will be transparent. Second, member states will need to justify on the grounds of cost and benefit analyses any dispensation to meet the WFD objectives. Thus, member states are accountable to the European Commission for setting full cost recovery rates and for taking into account the 'polluter pays' principle.

Yet, doubts exist on a number of issues before conclusions can be drawn about the effectiveness of this pricing policy. First, the

EU encompasses widely different irrigation sectors and economies, but policy objectives are inspired on fairly similar tenets. In some Mediterranean regions land planning and rural development are inextricably linked to the irrigation sector. The transition to full cost recovery prices will not be easy in many of these areas. Despite the initial reluctance shown by farm lobbies, many countries, including Spain, have submitted their 2004 economic reports to the EU Commission. The Spanish report, for instance, indicates that the rate of cost recovery in irrigation is slightly below 100% (Maestu, 2005).

Second, water-quality issues and more efficient allocation are still the most pressing problems in some of the water-stressed regions. If society is in need of more environmentally friendly and more frugal irrigation systems, it may pay to address other factors before squeezing farmers' income with higher charges. This is why many Mediterranean experts have coined the WFD as a 'Northern European' water policy.

Third, despite the above, the WFD will serve as a laboratory experiment conducted on a massive scale and over a large array of conditions. As the EU must set common rules (under the Common Implementation Strategy) they must be written, reported and disseminated to be ready for application in any corner of the EU. On the way to the 2010 deadline for the application of new water prices, the world may benefit from the EU experiences, positive and negative, as well.

Finally, we should consider that WFD is an innovative and ambitious norm as it is the first example of a significant use of economic instruments applied to natural resource management. We cannot quote any other significant example of natural resources (land, soil, etc.) subject to a similar treatment at the scale and socio-economic implications the WFD does.

Annex: Summary of European Countries' Experiences

In this annex we review the irrigation pricing policies that were in place in a selection of European countries before the WFD was passed in 2000. Table 13.6 attempts to summarize each country's main figures and water pricing schemes.

Belgium

Less than 5% of agricultural land in Belgium is irrigated. The agricultural sector in the Flanders region consumes on average 216 Mm^3/year of water, out of which 6.5% goes to agro-industry, 12.4% to livestock, 8.9% to greenhouses and 72% to open-air irrigation (Nys, 1998). Water management and pricing policies in Belgium fall under the responsibility of regional governments.

Agricultural water charges depend on the source of water: users linked to water pipes pay the same as households; users abstracting directly from groundwater sources pay (as from 1998) a levy on declared volumes; and users relying on surface water also pay a levy based on declared quantities.

Bulgaria

Agriculture consumes 13% of total water consumption in Bulgaria, that is, 1212 Mm^3 in 1999 (Kuobratova, 2001). Altogether 582,000 ha (65% of total agricultural area) are equipped for irrigation, although less than 10% of these are effectively irrigated. Agricultural water management is the responsibility of the Ministry of Agriculture and Forestry, 23 Irrigation Systems Companies (ISC) and 176 Irrigation Water User Associations (IWUAs) (OKO, 2001). Most irrigation water is supplied by the Irrigation Systems Companies, although the importance of collective irrigation is on the rise.

The 1999 Water Act establishes fees for both the use of water and the use of public water facilities, with exemptions for very low consumption and smaller farms. Irrigation water pricing depends on the source of water. Each ISC and IWUA establishes its own price structure. Water prices for IWUAs that manage state infrastructures are set at a lower level than for other agricultural users. IWUAs

Table 13.6. Irrigation water charges in several European countries.

Country	Pricing agency	Water supply fee/rate	Environmental water tax	Discharge levy/pollution tax	Subsidies	Reference
Belgium	Regional governments	Volumetric, depending on source Same as urban users	n.a.	n.a.	n.a.	OECD, 1997; Nys, 1998, personal communication
Bulgaria	Irrigation companies and irrigation districts	Water abstraction fee Water use fee: fixed (up to €5.00/ha) or volumetric (€0.007–0.075/m^3)	n.a.	n.a.	Part of O&M	OKO, 2001
Croatia	Government agencies	Volumetric, based on water quality Use fee: €0.01–0.04/m^3; Protection fee: €0.12 /m^3.	Yes	Yes	Heavily, for O&M maintenance	Ostojic and Luksic, 2001
Czech Republic	Government privatization process	n.a.	n.a.	n.a.	O&M, until privatized	Raskin *et al.*, 1996; OECD, 1999a,b
Denmark	Government	€0.55/m^3			Rate can be deduct from tax proceeds	OECD, 1997
France	Basin agencies	Binomial (average €0.08–0.390/m^3) Catchment and consumption components	n.a.	For livestock	Yes	Duchein, 1997; Montginoul, 1998; OECD, 2002
Germany	Länders	n.a.	Yes		Tax rebates	IISD, 1998
Greece	Governmental agencies	Volumetric in Crete: (€42.00–196.00/ha)	No	No	60% of total supply costs	Lekakis, 1998, personal communication
Hungary	Basin authorities and users associations	Basin abstraction fee and water fee: fixed (€5.00–36.00/ha) or volumetric (€0.004–0.034/m^3)	No	n.a.	Part of O&M	OKO, 2001

Continued

Table 13.6. *Continued*

Country	Pricing agency	Water supply fee/rate	Environmental water tax	Discharge levy/pollution tax	Subsidies	Reference
Italy	Public agencies	Concession fees and water rates (flat, binomial and increasing block rates) (see Table 13.2 for examples)	No	No	Part of capital costs	Destro, 1997; Xiloyannis and Dichio, 2001
The Netherlands	Water control boards	Abstraction tax €1.04/m^3	Yes	Yes	None	OECD, 1997
Portugal	Public and private suppliers	Two-tier rate: fixed: €12.00–211.00/ha; Vol: €0. 012/m^3	No	If applicable	O&M and part of capital	Castro, 1997; Bragança, 1998
Romania	Central government	€0.4/1000 m^3 for all regions	n.a.	n.a.	Part of supply costs Electricity costs	OKO, 2001
Slovakia	Basin authorities	Prices negotiated: maximum of €0.046/m^3 and average of €0.031/m^3 Regardless of use type	n.a.	n.a.	Part of supply and of irrigation costs	OKO, 2001
Spain	Basin authority and irrigation districts	Collected by district/users Fixed, volumetric or both Covers supply and district costs (see Table 13.2 for examples)	No	No	O&M and part of capital	MAPA, 2001
Sweden	n.a.	Private abstraction costs	Yes	n.a.	None	Bergvall, 1998, personal communication.
Switzerland	Regional agencies	Yes	n.a.	Yes	None. Total prices: €0.025–1.56/m^3	Siegrist, H., Switzerland, 1998, personal communication.
UK	Regions	Water abstraction fee: €0.08–0.023/m^3	Yes	Including supply fee	None	OECD, 1997; Knox and Weatherhead, 2003

managing their own infrastructures receive no price subsidies (Kuobratova, 2001).

There is a water abstraction fee and a water supply charge. The water supply charge can be either a fixed per-hectare one (up to €5.00/ha) or a volumetric one (€0.007–0.075/ m^3) or both. Depending on area and water source, irrigation water prices range between €0.01/m^3 and €0.09/m^3, while the average on farm costs (including irrigation operations) ranges between €0.13/m^3 and €0.18/m^3. Water tariffs cover part of operation and maintenance (O&M) costs and, in some cases, part of investment costs (OKO, 2001). Furthermore, the government has subsidized irrigation water prices for both surface water and groundwater in years of scarcity (Kuobratova, 2001).

Denmark

Irrigated farming in Denmark represents about 35% of all consumptive uses of water. Farmers using water for irrigation are subject to the 1994 'Green Tax Reform' that imposes a water rate of €0.55/m^3 of raw water. However, they are allowed to deduct this tax from their value-added tax proceeds. There is a major concern about pesticide pollution of groundwater and further environmental fees are likely to be imposed on irrigating farmers.

France

Basin authorities, where most users and stakeholders are represented, can exercise considerable scope in water planning and management and in setting water charges. Water charges in France have two components: a *basin* component (based on the average volume abstracted) and a *consumption* component (levied on the difference between abstractions and return flows). The criteria used to set charges vary substantially across basins, and mostly depend on characteristics such as the probability of drought, the type of user, capital costs, ownership and other basin characteristics (Duchein, 1997). Charges cover O&M and part of capital costs (Chohin *et al.*, 2003).

Farmers pay a binomial tariff, comprising both a fixed per hectare and a volumetric charge. Average water charges for irrigation range between €0.08/m^3 and €0.31/m^3 (Montginoul, 1997). In areas where per-hectare charges are paid, the average tariff is €106.00/ha (Chohin *et al.*, 2003). The ASA and the SAR charge average volumetric tariffs of €0.047–0.054/m^3 (Chohin *et al.*, 2003). Some SARs also have optional binomial tariffs (€40.00/ha and €0.07/m^3, or €25.00/ ha and €0.17/m^3). Others, like the SAR in the Languedoc Roussillon region (BRL), have pricing schemes that discourage use above certain thresholds.

In general, water charges across all irrigation units in France have been increasing over time, for three basic reasons. The first is the 1992 Water Code, which sought to broaden the revenue base for water supply companies in order to ensure their financial stability. Second, there has been a large increase in irrigated acreage across France, adding more pressure on several basins during summer or drought periods. Third, pollution is now considered as another 'use' of public waterways and water bodies, so that water authorities can sometimes justify charging 'resource-based' prices which can be added to other accounting and/or capital cost components. Farmers only pay pollution fees for water used in cattle production but not in crop production. Rieu's (2005) comprehensive review of water pricing policies in France shows that policies have been geared towards cost recovery objectives. Yet, there are large capital costs differences across basins and irrigated areas, creating a large range of capital costs recovery, between 15% and 60%. The pricing systems vary from 'average cost' to 'marginal cost', which are jointly used with systems of quotas.

Germany

German irrigated agriculture is not very extensive and general water policies tend to override agricultural policies. Water management is the responsibility of the Länders.

Traditionally, water prices have been based on extraction, treatment and transportation costs. Until several Länders started to establish 'water taxes' in 1988, water remained significantly undervalued with respect to other sectors (IISD, 1998). However, these water taxes deviate from the commonly accepted definition of water charges for two reasons. One is that they are generally levied only in cases where a permit or license is required. Since water metering in the agriculture sector is not common in Germany, the allotted volumes stated in licenses are far below actual abstractions carried out by licensees. The second reason is that the revenues collected from water taxes have often been used to compensate farmers for restrictions on fertilizer use in vulnerable areas. Furthermore, tax rebates (up to 90%) exist for those farmers who can provide evidence of being financially impaired by the tax. However, these rebates are conditional upon the implementation of water-saving strategies, and on using surface water instead of groundwater.

Greece

The relative contribution of agriculture to the GNP of Greece is one of the highest in Europe. Greece has about 1.33 Mha of irrigated land, which represents 38% of its total arable area and almost 10% of the country's total land surface. About 20% of the active population makes its living out of agriculture. Irrigated farming accounts for more than 80% of the nation's total water consumption. Irrigated acreage has increased by about 65% in the last 20 years, as a result of a strong political commitment to increase both agricultural production and farm incomes in rural areas. It is also the result of private initiatives, which currently represent about 60% of the total irrigated acreage, mostly equipped with sprinkler or drip technologies.

The remaining 40% of the total irrigated acreage (532,000 ha) is composed of cooperative irrigation projects jointly undertaken by the Local Land Improvement Boards (TOEV) and the National Land Improvement General Boards (GOEV). TOEVs manage water allocation, collect farmers' fees and manage collective facilities. GOEVs are semi-governmental organizations that finance works affecting more than one TOEV. Public projects are mainly equipped with modern irrigation technologies, although 41% of the irrigated area still uses gravity irrigation systems. The construction of irrigation projects comes under the responsibility of rural regional authorities assisting irrigation facilities aimed not only at economic objectives but also at environmental consumption and social objectives as well. Recently, a government-controlled experiment, the Organisation for the Development of Western Crete (OADYK), has begun to operate in Western Crete, providing water for drinking and irrigation purposes. It is a non-profit, self-financed organization.

The Greek water economy is presently approaching 'maturity', and there are few new opportunities to expand irrigation supplies. Irrigation water demand has been slowly increasing in the past decades with a tendency to reach stabilization (Margat, 2002). Public investments in reclamation projects have decreased about 32% since the 1970s. Although there are some ongoing initiatives which combine environmental objectives with better water and irrigation management, no significant effort has yet been made to make farmers pay for the important rehabilitation and maintenance costs which will be needed in the future. Both the challenging natural conditions of Greece and the relative economic importance of its agriculture sector are factors which explain the delay in implementing water pricing reforms in this sector. Of equal importance is the widely held perception in Greece that water supply projects are multipurpose facilities that contribute towards social progress, environmental conservation and protection.

According to Lekakis (1998), access to water resources has not yet been fully regulated, and the organization of the water management agencies and water suppliers is essentially governed by the civil code. This institutional framework, together with the remarkable hydrologic complexity of the country, explains why it is not possible to identify any common trends in Greek agricultural water pricing systems. Another factor which contributes to this heterogeneity is the fact that more than 40% of agricultural water demand is met by groundwater resources so that water fees are totally dependent on extraction costs, including fuel or electricity consumption. TOEVs set fees to cover administration, maintenance and operation costs of their collective facilities. On average, the revenues collected with these charges represented about 60% of TOEVs' total expenses in 1994, the rest being covered by the state. Lekakis (1998) also provides an estimated range of pumping costs of €42.00–196.00/ha. Charges paid to TOEVs cover only part of O&M and nothing of capital costs, while individual irrigators pay both of these (Chohin *et al.*, 2003).

Hungary

Hungarian agriculture consumes less than 10% of the total water consumption in the country, of which 92.5 Mm^3 are used for irrigation, 337 Mm^3 for fish farming (water supplied to fish ponds) and 125 Mm^3 for other uses (OKO, 2001, data for 1997). The irrigated area in Hungary was 108,400 ha (1998) although 264,300 ha were equipped for irrigation (4.3% of total agricultural area).

The 1995 Water Act establishes the need for public licenses for water use. Water management in Hungary depends on three Ministries: the Ministry of Agriculture and Rural Development; the Ministry of the Environment; and the Ministry of Transport and Water, which is responsible for the 12 existing District Water Authorities (OKO, 2001). Farmers in a particular area are grouped in Water Management Associations.

The water supply charge consists of a water abstraction fee (that depends on source type, quality of water and type of use) and a water price. The water abstraction fee is set by the government to finance its water management costs. The water price is freely set regionally by the water supplier. The water price for irrigation can be a fixed amount per hectare (between €5.00/ha and €36.00/ha, with higher values in modern districts) or a volumetric tariff (between €0.004/m^3 and €0.034/m^3). Fees account for 20% of farmers' water-related costs (OKO, 2001) and for 0.5–2% of the gross value product of crops produced (Strosser, 2003). Water tariffs cover part of O&M costs. However, in some regions it even covers all capital costs.

Italy

Irrigated agriculture accounts for 27% of agricultural land, 30% of farms and about 50% of total agricultural production. Around 60% of Italian agricultural exports are produced by irrigated agriculture (Leone, 1997; Bazzani *et al.*, 2003). Italy has unequally distributed water resources, abundant in the Po valley but scarce and unreliable in the South. Irrigated land is mostly located in the northern Po valley (about 2 Mha) and in the southern Capitanata region (about 450,000 ha). Farming in Italy represents about 61% of consumptive use of water, with irrigation estimated at 50% of withdrawals. Water demand for agriculture has been decreasing since 1970, although future water demand for irrigation is forecasted to stabilize around the present level of consumption (Massarutto, 2001; Margat, 2002).

The Land Reclamation Act (1933) converted all water bodies to the public domain, and set forth the principles which have guided the management of water resources in Italy ever since. The poorly maintained water distribution system in Italy relies mainly on 'Reclamation and Irrigation Boards' (RIBs) (*Consorzi di Bonifica e Irrigazione*) that are managed by associations of landowners, entities regulated by public law that control land reclamation and water distribution in a certain area. RIBs distribute about 90% of the water used for irrigation (ANBI, 1992, 1998). Consortia have self-financing capacity to foster rural development, as well as to build irrigation projects. The government provides funds to cover all project capital costs, while the Consortia are responsible for managing and maintaining these systems, and collecting charges from farmers.

The average water cost at the farm level is about €36.00/ha, but actual tariffs range from €2.00/ha to €355.00/ha. The tariff system is usually based on the running costs of servicing an area. It is only in a small part of the total irrigated area that water is measured and volumetrically priced. For instance, in the Romagna Occidentale Irrigation Board, 87% of the total area, served by open canals or non-metered pipe systems, pays per-hectare charges (€42.60/ha and €132.20/ha, respectively), while the remaining 17%, equipped with metered, pressurized distribution systems, pay €20.66/ha, plus a volumetric component (Bazzani *et al.*, 2005).

Italian farmers pay much less than other users. Charges cover only part of O&M costs and nothing of investment or depreciation costs (Chohin *et al.*, 2003). Massarutto (2003) reports a range of 70–100% O&M recovery rates in northern Italy and 20–100% in the South. In Sardinia, rates vary within each Consortia based on the type of water conveyance system, pressure, crops and irrigation technology, ranging from a flat rate of €51.00 for drip irrigators in Nurra Consortia to €392.00 for rice growers in Campidano di Oristano Consortia (Aiello *et al.*, 1997). Xiloyannis and Dichio (2001) find large water consumption differences for the same crops between a district in Bassilicata (a flat per-hectare rate) and another in Puglia that uses a block-rate system (a flat rate of €10.00/ha plus a variable rate of €0.09/m^3 (0–1300 m^3/ha), €0.056/m^3 (1300–2000 m^3/ha), €0.091/m^3 (2000–3000 m^3/ha) and €0.126/m^3 (for any unit exceeding 4000 m^3/ha).

The Netherlands

Irrigation accounts for 60% of total arable land in the Netherlands. Dutch agriculture uses 149 Mm^3 of tap water every year (25 Mm^3 in greenhouses, 38 Mm^3 in irrigated arable land and horticulture, and 86 Mm^3 in cattle farming). Water supply is the responsibility of a company wholly owned by the municipalities within its supply area. The water boards or *waterschappen* have responsibility for land drainage/flood defence and, in some provinces, for water-quality management. They work in close cooperation with the residents/landowners of their areas, who elect them.

The water boards' costs are fully covered by water users, including farmers who pay the full supply costs and, where appropriate, the full drainage costs as well (National Reference Centre for Agricultura, 1998, unpublished data). The agriculture sector contributes 27% of the total levies raised for quantitative water management. Unlike in most other countries, the Dutch agriculture sector contributes more revenue to water management costs than it is actually spent in its direct benefit, with a discrepancy of about 5%. The reason is that the main task of water boards is flood protection and land drainage. On average, water supply costs to agriculture amount to €1.04/m^3.

Farmers in the Netherlands are subject to a groundwater extraction tax, especially when they draw on tap water resources for cattle production. If they decide to extract groundwater directly themselves, a permit from the Central Government is required if pumping capacity exceeds 10 m^3/s or if the farmer uses more than 1 Mm^3/year, and the farmer has to pay the abstraction tax plus a small provincial tax. Most farmers install small pumping facilities, so they do not have to pay these taxes. Hellegers *et al.* (2001) conclude that the price of groundwater was inefficient and provided fewer incentives for the adoption of modern irrigation technology than a system that considers the cost of depletion and groundwater contamination in the price of groundwater.

Portugal

Portugal is relatively well endowed with water resources, although huge differences exist between the North and the South. Irrigated land constitutes about 60% of the nation's total water supply and 25% of the agricultural area.

The Portuguese Water Law combines public and private ownership of water resources. Unlike most countries, the state's role in promoting irrigation projects in Portugal has traditionally been quite limited. Purely public irrigation projects make up only 19–25% of the 650,000 ha of irrigated land, most of which are located in the southern regions, which make the role of public water pricing policies less important for national-level water management strategy. Traditionally, water abstractions have been allowed free of charge, provided that users do not generate significant levels of pollution. However, major institutional and legal progress has been recently made in terms of implementing water charges for public projects.

Agricultural water tariffs are levied by user associations in accordance with very complex mechanisms and formulae. The complexity arises because WUAs sometimes supply municipal water as well, property size affects the water charges, and charges are combined with drainage fees in projects that require drainage (Castro, 1997). Project beneficiaries are required to pay a yearly set charge called TEC (*Taxa de Exploração e Conservação*) which includes a selection of no more than three of the following components: (i) a fixed charge per reclaimed or ameliorated hectare of land (ranging from €14.00 to €211.00); (ii) a fixed charge per irrigated hectare (ranging from €24.00 to €114.00); (iii) a volumetric charge per cubic metre, if metering is possible (ranging from €0.008/m^3 to €0.021/m^3); (iv) a drainage fee, when drainage of excessive water is required (ranging from €15.00 to €62.00); and (v) a crop-based fee applicable for specific crops and projects (ranging from €13.00 to €68.00) (Bragança, 1998).

Although the capital cost charge element has never achieved its intended objective of full cost recovery, the Portuguese system has the peculiarity to compute its payable fees using different interest rates, with the rates varying with soil quality and the crops grown. For instance, Brangança (1998) reports significant water price differences paid by farmers in Sorraia: €0.01/m^3 for rice (17,200 m^3/ha) and maize (7200 m^3/ha), and €0.0131/m^3 for tomato (5400 m^3/ha). It is clear that the 'ability-to-pay' principle, combined with other agricultural policy objectives, underlies these price differentials. None the less, charges in Sorraia were gradually raised in the period 1991–1997, up to levels that exceed O&M costs.

Romania

The total agricultural area in Romania is 14.8 Mha, of which 9.8 Mha are arable and 3.1 Mha are developed and equipped for irrigation. Of these, only 440,000 ha were irrigated in 1998 because of abandonment and decay of facilities. Romanian agriculture consumes about 10% of water in the country, of which 284 Mm^3 are used for irrigation and 664 Mm^3 for fish farming (OKO, 2001, data for 1997).

The 1996 Romanian Water Law establishes the need for public permits for any water abstraction. The Ministry of Waters, Forests and Environmental Protection is mainly responsible for water management and protection, which are implemented by the 12 basin branches of the National Company '*Apele Romane*'. Irrigation user associations have been existing only since 1999. Water prices in Romania are set by the government for each type of water use, so that all farmers in the country pay €0.4/1000 m^3 of irrigation water used, and the government also covers all electricity costs (OKO, 2001). In those areas where irrigators' associations have developed they have set their own charges to cover their own supply costs.

Spain

Spain's irrigation practices go back to the times when the Muslims occupied the Iberian Peninsula, starting in the 8th century, and further developed Roman irrigation techniques. This explains why there is so much diversity across regions and even between neighbouring irrigation areas. Irrigation water demand in Spain has been slowly increasing in the past decades and is expected to continue growing with a tendency to level out (Margat, 2002). The era of Spanish modern water legislation began in 1985 with the Water Act that replaced the 1879 Water Act. The 1879 Act and the 1911 Irrigation and Land Reclamation Act jointly granted very generous economic conditions to irrigators who benefited from state water projects (Garrido and Calatrava, 2005). Spain's present charging systems are, in general, far from complying with the WFD. We analysed cost recovery for the Spanish case in the section devoted to the impact of the WFD.

About 70% of all Spanish irrigated area is serviced by communities of irrigators. In addition to administering the resources and infrastructures they share water among irrigators, and have a major role in water management both at the River Basin Authority (RBA) and district levels. They are active members in the governing and planning boards, and have permanent seats in the Basin Assembly of Users.

Farmers pay a 'regulation levy' and a 'water use tariff' to the RBA through the irrigation district, and an additional tariff to cover the costs of the irrigation district itself (called '*derrama*'). Irrigation districts that abstract their water directly and that do not use publicly developed infrastructures only pay the regulation levy plus their own pumping, transport and application costs. A fixed per-hectare tariff is applied in 82% of the Spanish irrigated area, while volumetric tariffs are applied in 13% of the irrigated area, mostly in those districts that are served with groundwater and/or that incur energy costs (MAPA, 2001). Binomial tariffs, which combine both a volumetric component, to cover variable costs such as energy or labour, with a fixed per-hectare charge, are applied in 5% of the irrigated area. Average tariffs paid for irrigation water in areas where water is supplied by RBAs is €0.02/m^3, except for the agricultural users served from the Tajo-Segura Transfer who pay about €0.09/m^3, while areas that use groundwater pay an average of €0.04–0.07/m^3, based on extraction and other O&M costs.

United Kingdom

Irrigation in the UK is all supplementary irrigation. There are 147,895 ha of irrigated land in the UK (Weatherhead and Danert, 2002) growing mainly potatoes and vegetables in the Anglian region (the drier eastern England, 50%), the Midlands (19%), Thames (10%) and other southern regions (9%). In some regions and seasons, irrigation may make up to 80% of abstractions. There is also some irrigation in Wales, Scotland and northern Ireland.

Although water is becoming increasingly scarce in the east of England, irrigation represents only 3% of all water diversions. However, water used for irrigation doubled in the period 1975–2000, with an underlying increase in water use for irrigation in the eastern counties of 3%/year in that period. In response to seasonal water shortages and restrictions on summer abstraction licenses, total water stored in on-farm reservoirs doubled from 33 Mm^3 to 64 Mm^3 in the 1984–1995 period (MAFF, 2000). In Scotland and northern Ireland, water resources are abundant, and farmers can take water from adjacent rivers simply by applying for permission, which is granted at no cost.

Irrigation based on river diversion is unsupported. Since the 1960s all abstractions in England require a license. Since 1997, there has been a succession of reviews and policy changes covering all aspects of water management in England and Wales, including the water abstraction licensing system, and the elimination of barriers to the trading of water licenses.

From 1993 onwards, each region is allowed to set charges to recover its water control costs (Rees, 1997). Farmers pay a fee when applying for a water abstraction license, as well as an annual charge that depends on the location, the return flow generated by each irrigation technology, water quality, and the season in which the abstraction is made. Prices vary from €0.008/m^3 in Yorkshire to €0.021/m^3 in Northumbria. A review of irrigation costs shows that average irrigation costs for large irrigated areas (greater than 50 ha) are about €0.43/m^3 for direct application in the field, rising to about €0.70/m^3 with clay-lined storage reservoirs, and over €0.90/m^3 with artificially lined reservoirs. Water costs are less than 7% of total costs. Thus, at current abstraction charges, summer direct abstraction is always cheaper than winter stored water. Summer charges would need to rise to about €0.27/m^3 for winter stored water to be a cheaper option. But additional summer water is not available in many situations. The average total costs using trickle systems range between €0.80/m^3 and €1.35/m^3 (Knox and Weatherhead, 2003).

References

Aiello, G.M., Cané, G., Fadda, G.A., Murrau, A. and Onanao, V. (1997) Caratteristiche dell' irrigazione collettiva in Sardegna. *Revita Irrigazione e Drenaggio* 44, 58–63.

ANBI (1992) *L'uso Irriguo Delle Acque*. Associazione Nazionale Bonifiche Irrigazioni, Roma.

ANBI (Associazione Nazionale Bonifiche Irrigazioni) (1998) *La Verità Su Un Territorio Bonificato: l'Italia*. Associazione Nazionale Bonifiche e Irrigazioni, Roma.

Arriaza, M., Gómez Limón, J.A. and Ruiz, P. (2003) Evaluación de alternativas de desacoplamiento total de ayudas COP: El caso de la agricultura de regadío del Valle del Guadalquivir. *Economía Agraria y Recursos Naturales* 6, 129–153.

Barbero, A. (2005) The Spanish national irrigation plan. OECD workshop on agriculture and water: sustainability, markets and policies, 14–18 November, 2005. Adelaide. Oral presentation.

Bazzani, G.M., D'Amore, A., Di Pasquale, S., Fabbri, A., Gallerani, V., Gomez Barbero, M. and Viaggi, D. (2003) La agricoltura de regadìo en Italia y el impacto de la directiva marco de aguas. In: FENACORE (ed.) *La Agricultura de Regadío Europea y La Directiva Marco de Aguas*. Federación Nacional de Comunidades de Regantes, Madrid, pp. 139–172

Bazzani, G.M., Pasquale, S.D., Gallerani, V., Morganti, S., Raggi, M. and Viaggi, D. (2005) The sustainability of irrigated agricultural systems under the Water Framework Directive: first results. *Environmental Modelling & Software* 20(2), 165–175.

Berbel, J. (2005) Análisis económico del agua en la Directiva Marco. Su aplicación a la Cuenca del Guadalquivir. Conferencia ISR, Córdoba, Spain, 28/abril/2005. Available at: http://www.isrcer.org/jornadas.asp

Berbel, J. and Gutierrez, C. (eds) (2005) *Sustainability of European Irrigated Agriculture under Water Framework Directive and Agenda 2000. EUR 21220*. Office for Official Publications of the European Communities, Luxembourg, Belgium.

Bontemps, C., Couture, S. and Favard, P. (2003) Estimation de la demande en eau d'irrigation sous incertitude. (Irrigation Water Demand Estimation. With English summary). *Economie Rurale* July–August 17–24.

Brown, G.M. (2000) Renewable natural resource management and use without markets. *Journal of Economic Literature* 38, 875–914.

Calatrava, J. and Garrido, A. (2001) Agricultural subsidies, water pricing and farmers' response: implications for water policy and CAP reform. In: Dosi, C. (ed.) *Agricultural Use of Groundwater: Towards Integration between Agricultural Policy and Water Resources Management*. Kluwer Academic Publishers, Dordrecht, The Netherlands, pp. 241–257.

Camacho, E. (2005) La Eficiencia de las Comunidades de Regantes. FERAGUA-C.H.G (en prensa).

Carles, J., García, M. and Avellá, L. (2001a) Aspectos económicos y sociales de la utilización de las aguas subterráneas en la Comunidad Valenciana. In: Hernández, N. and Llamas, M.R. (eds) *La Economía Del Agua Subterránea y Su Gestión Colectiva*. Fundación Marcelino Botín and Editorial Mundi-Prensa. Madrid, pp. 153–173

Carles, J., García, M. and Vega, V. (2001b) Gestión colectiva de las aguas subterráneas en la Comunidad Valenciana. In: Hernández, N. and Llamas, M.R. (eds) *La Economía Del Agua Subterránea y Su Gestión Colectiva*. Fundación Marcelino Botín and Editorial Mundi-Prensa. Madrid, pp. 291–321.

Castro, J. (1997) Portugal. In: Dinar, A. and Subramanian, A. (eds) *Water Pricing Experiences. An International Perspective*. World Bank Technical Paper 386. World Bank, Washington, DC, pp. 99–103.

Chakravorty, U. and Roumasset, J. (1991) Efficient spatial allocation of irrigation water. *American Journal of Agricultural Economics* 73, 165–173.

Chartzoulakis, K.S. and Angelakis, A.N. (2001) *Water Resources Management in Crete (Greece) with Emphasis in Agricultural Use. Transnational Workshop on Managing Water in Agriculture through Pricing: Research Issues and Lessons Learned*. CNR-ISPAIM, Ercolano, Italy, 24–26 May.

Chohin, A., Rieu, T. and Montginoul, M. (2003) Water policy reforms: pricing water, cost recovery, water demand and impact on agriculture. Lessons from the Mediterranean experience. Water pricing seminar, Barcelona, 30 June–2 July.

Destro, S. (1997) Italy. In: Dinar, A. and Subramanian, A. (eds) *Water Pricing Experiences. An International Perspective*. World Bank Technical Paper 386. World Bank, Washington, DC, pp. 64–68.
Dinar, A. and Letey, J. (1991) Agricultural water marketing, allocative efficiency and drainage reduction. *Journal of Environmental Economics and Management* 20, 210–223.
Dono, G. and Severini, S. (2001) *The Agenda 2000 CAP Reform and Its Impact on Irrigation Water Use: A Regional Programming Model for a Central Horticultural Area. Transnational Workshop on Managing Water in Agriculture through Pricing: Research Issues and Lessons Learned.* CNR-ISPAIM, Ercolano, Italy, 24–26 May.
Dubois De La Sablonière, F. (1997) *France: An Incentive Policy for Sustainable Management of Irrigation Water in the Loire-Brittany Basin*. OECD Workshop on the Sustainable Management of Water in Agriculture: Issues and Policies. Directorate for Food, Agriculture and Fisheries and Environment Directorate, Athens, 3–6 November.
Duchein, A. (1997) *France: Partnership between the Agricultural Community and the Basin Agencies*. OECD Workshop on the Sustainable Management of Water in Agriculture: Issues and Policies. Directorate for Food, Agriculture and Fisheries and Environment Directorate, Athens, 3–6 November.
European Commission (2000) *Pricing Policies for Enhancing the Sustainability of Water Resources*. Communication from the Commission to the Council, the European Parliament, and the Economic and Social Committee (COM (2000) 477 final). European Commission, Brussels.
FAO (Food and Agriculture Organization of the United Nations) (1997) *Yearbook*. FAO, Rome.
García, M. (2000) Análisis de la influencia de los costes en el consumo de agua en la agricultura valenciana. Caracterización de las entidades asociativas para riego. Tesis doctoral. Departamento de Economía y Ciencias Sociales. Universidad Politécnica de Valencia, Valencia, Spain.
García, M. (2001) Análisis de la influencia de los costes en el consumo de agua en la agricultura valenciana. Caracterización de las entidades asociativas para riego. Tesis doctoral. Departamento de Economía y Ciencias Sociales. Universidad Politécnica de Valencia, Valencia, Spain.
García, M.M., Carles, J. and Sanchís, C. (2004) Caracteristicas institucionales y territoriales y su influencia en el costo del agua como input en la agricultura. VII Congreso Nacional de Medio Ambiente. Madrid (Spain), 22–26 November. Unpublished.
Garrido, A. and Calatrava, J. (2005) Recent and future trends in water charging and water markets. In: Garrido, A. and Llamas, M.R. (eds) *Water Policy in Spain*. Resources for the Future, Washington, DC.
Gómez-Limón, J.A., Arriaza, M. and Berbel, J. (2002) Conflicting implementation of agricultural and water policies in irrigated areas in the EU. *Journal of Agricultural Economics* 53(2), 259–281.
Hafi, A., Klijn, N. and Kemp, A. (2001) Efficient pricing and allocation of irrigation water. A model of the Murrumbidgee irrigation area. ABARE Conference Paper 2001. 4, Canberra. Available at www.pc.gov.au/ research/ swp/rdia/
Hellegers, P., Zilberman, D. and van Ierland, E. (2001) Dynamics of agricultural groundwater extraction. *Ecological Economics* 37 (2001), 303–311.
Hernández, N. and Llamas, M.R. (eds) (2001) *La economía del agua subterránea y su gestión colectiva*. Fundación Marcelino Botín y Ediciones Mundi-Prensa, Madrid.
IISD (International Institute for Sustainable Development) (1998) Germany: water taxes. International Institute for Sustainable Development. http://iisd.iisd.ca/greenbud/germany.html.
Kaika, M. (2003) The water framework directive: a new directive for a changing social, political and economic European framework. *European Planning Studies* 1(3), 299–313.
Knox, J.W. and Weatherhead, E.K. (2003) *Trickle Irrigation in England and Wales*. Technical Report to Environment Agency. Water Resources R&D Project W6-070/TR. Environment Agency, Bristol, UK.
Kuobratova, M. (2001) Water, soil and landscape related issues and policies in Bulgaria. Seminar on Sustainable Agriculture in Central and Eastern European Countries: The Environmental Effects of Transition and Needs for Change. 10–16 September. Nitra, Slovakia.
Leone, G. (1997) Stato delle irrigazioni in Italia. *Rivista L'Acqua*, No. 5.
Maestu, J. (2005) Economic analysis of water use according to Article 5 of the WFD – Main results in Spain. Paper presented at the 6th International Conference of the European Society of Ecological Economics, Lisbon, 14–17 June.
MAFF (Ministry of Agriculture, Fisheries and Food) (2000) *Towards Sustainable Agriculture: A Pilot Set of Indicators*. Ministry of Agriculture, Fisheries and Food Publication PB 4583. HMSO, London.
MAPA (Ministerio de Agricultura, Pesca y Alimentación) (2001) *Plan Nacional de Regadíos. Horizonte 2008*. Ministerio de Agricultura, Pesca y Alimentación, Madrid.
Margat, J. (2002) *Tendances Contemporaines et Perspectives d'Evolution des Demandes en Eau dans les Pays Méditerranéens, Plan Bleu*. Sophia, Antipolis, France.

Martínez, Y. and Albiac, J. (2004) Agricultural pollution control under Spanish and European environmental policies. *Water Resources Research* 40(10).

Massarutto, A. (ed.) (2001) Water pricing, the common agricultural policy and irrigation water use. Draft Report for European Commission. Udine, Italy.

Massarutto, A. (2003) Water pricing and irrigation water demand: efficiency vs. sustainability. *European Environment* 13/2003, 100–119.

Mejias, P., Varela-Ortega, C. and Flichman, G. (2004) Integrating agricultural policies and water policies under water supply and climate uncertainty. *Water Resources Research* 40(7).

Montginoul, M. (1998) France. In: Dinar, A. and Subramanian, A. (eds) *Water Pricing Experiences. An International Perspective*. World Bank Technical Paper 386. World Bank (February 1998), Washington, DC, pp. 46–53.

Montiginoul, M. and Rieu, T. (2001) Irrigation water pricing reforms and implementing procedures: experience acquired in Charente and in Morocco. In: Pricing water: economics, environment and society. *Conference Proceedings Sintra, 6 and 7 September 1999*. European Commission.

MOPTMA-MINER (Ministerio de Obras Publicas, Transporte y Medio Ambiente) (1994) Libro Blanco de las Aguas Subterráneas. Madrid. Pub. Ministerio de Obras Públicas, Transportes y Medio Ambiente.

OECD (1987) *Tarification des Services Relatifs à l'Eau*. OECD, Paris.

OECD (Organization for the Economic Development and Cooperation) (1997) *Water Subsidies and the Environment*. OECD/GD(97)220, Paris.

OECD (1999a) *Agricultural Water Pricing in OECD Countries*. Document ENV/EPOC/GEEI(98)10/Final. OECD, Paris.

OECD (1999b) *The Price of Water. Trends in OECD Countries*. OECD, Paris, 173 pp.

OECD (2002) *Transition to Full-Cost Pricing of Irrigation Water for Agriculture in OECD Countries*. Document COM/ENV/EPOC/AGR/CA(2001)62/Final. OECD, Paris.

OECD (2004) *Agriculture and the Environment: Lessons Learned from a Decade of OECD Work*. OECD, Paris.

OKO (2001) *Agricultural Water Management Policies in Bulgaria, Hungary, Romania and Slovakia*. Final Report. OKO Consulting, Budapest. http://www.rec.org

Ostojic, Z. and Luksic, M. (2001) *Water Pricing in Croatia, Current Practices and Trends*. The Regional Environmental Center for Central and Eastern Europe, Zagreb. Available at http://www.rec.org

Rainelli, P. and Vermersch, D. (1998) Irrigation in France: current situation and reasons for its development. Study submitted to OECD Environment Directorate.

Raskin, P.D., Hansen, E. and Margolis, R.M. (1996) Water and sustainability: global patterns and long-range problems. *Natural Resources Forum* 20(1), 1–17.

Rees, J. (1997) United Kingdom. In: Dinar, A. and Subramanian, A. (eds) *Water Pricing Experiences: An International Perspective.*. World Bank Technical Paper 386. World Bank, Washington, DC, pp. 139–143.

Rieu, T. (2005) *Water Pricing for Agriculture between Cost Recovery and Water Conservation: Where Do We Stand in France?* OECD Workshop on Agriculture and Water: Sustainability, Markets and Policies 14–18 November, 2005, Adelaide.

Strosser, P. (2003) *Water Pricing in Central Eastern Europe. The Example of Hungary*. World Bank Water Pricing Seminar. World Bank, Barcelona, Spain, 30 June–2 July. http://www.worldbank.org/html/fpd/water/Barcelonameeting.htm.

Sumpsi, J.M., Garrido, A., Blanco, M., Varela, C. and Iglesias, E. (1998) *Economía y Política e Gestión del Agua en la Agricultura*. MAPA y (ed.) Mundi-Prensa, Madrid, p. 351.

Tsur, Y. and Dinar, A. (1997) The relative efficiency and implementation costs of alternative methods for pricing irrigation water. *The World Bank Economic Review* 11(2), 243–262.

WATECO (2003) *Economics and the Environment. The Implementation Challenge of the Water Framework Directive*. Accompanying Documents to the Guidance. European Commission, Brussels.

Weatherhead, E.K. and Danert, K. (2002) *Survey of Irrigation of Outdoor Crops in 2001 – England*. Cranfield University, Bedford, UK.

Weinberg, M., Kling, C.L. and Wilen, J.E. (1993) Water markets and water quality. *American Journal of Agricultural Economics* 75, 278–291.

Xiloyannis, C. and Dichio, B. (2001) *How to Improve Water Efficiency in Fruit Trees Species in Southern Italy*. Transnational Workshop on Managing Water in Agriculture through Pricing: Research Issues and Lessons Learned. CNR-ISPAIM, Ercolano, Italy, 24–26 May.

14 Policy-driven Determinants of Irrigation Development and Environmental Sustainability: A Case Study in Spain

C. Varela-Ortega

Introduction: Water Use and Agricultural Policies

Objective and contents of the chapter

This chapter analyses the role that water and agricultural policies play in the evolution of irrigated agriculture and water use and, as a consequence, on the conservation of aquatic ecosystems. Using an illustrative case study from central Spain, the chapter focuses on the joint impacts of the implementation of agricultural policies (the EU Common Agricultural Policy, or CAP) and water conservation policies (both European and national) on the development of irrigated agriculture, groundwater abstraction, and the conservation of depleted aquifers and associated wetlands. The chapter is divided into three sections. The first introductory section provides a general picture of how water policies and agricultural polices have determined to a great extent water consumption trends in the Mediterranean countries of the EU. A subsection illustrates how groundwater use for irrigation has been determined by the evolution of policy programmes in the region of Castilla-La Mancha of Spain's southern central plateau, introducing the case study. The second section examines the specific agricultural polices and water policies that have been successively applied in the area of study. Special attention is given to analysing the capacity of these polices to respond to the societal needs of socio-economic development and ecosystem conservation as well as to the comparative cost-effectiveness of the different public policy programmes. A subsection compares the impact of these policies with the alternative mechanism of water pricing. The third section includes some concluding reflections.

Evolution of water use and irrigated agriculture

The evolution of irrigated agriculture in the Mediterranean countries as in other countries worldwide has been determined by policies that relied to a great extent on technical solutions for water supply enhancement. Publicly funded large water infrastructures resulted in water deliveries at subsidized costs, increasing the burden on the public budget and leading to environmental damage (Rosegrant *et al.*, 2002; Benoit and Comeau, 2005).

In contrast to the one-sided water supply paradigm of the past, public authorities in many countries in the world are now confronted with the challenge of elaborating demand-side, integrated, and cost-effective water management policies. These policies will have to be designed and enforced to address the dual aims of achieving a more efficient use of water among sectors and social groups while ensuring the sustainability of water resources. The increasing incorporation of economic, social and institutional aspects, as well as public participation and the involvement of stakeholders, has proven to be effective for integrated water management and hence for food production, protection of water ecosystems and overall socio-economic development (Bromley, 2000; Rosegrant *et al.*, 2002; Margat, 2004; Benoit and Comeau, 2005). The recently enacted EU Water Framework Directive (WFD), which is mandatory for all member states, is an example of new integrated water management policies (EU, 2000).

In the EU, agricultural policies have affected water consumption in irrigated agriculture, most acutely in the arid and semi-arid regions of southern Europe that extend along the Mediterranean littoral and its hinterland. During the 1980s and 1990s, the CAP encouraged expansion of irrigation in response to production-based subsidies with contradictory effects in many irrigated areas. On the one hand, irrigation expansion led to unquestionable socio-economic benefits to the rural areas concerned but, on the other, it generated negative externalities with clear detrimental consequences to aquatic ecosystems (Baldock *et al.*, 2000; Varela-Ortega *et al.*, 2002).

Over time, the CAP evolved with the aim of promoting a more balanced integration of the agriculture and environmental sectors by incorporating environmental objectives into agricultural policy programmes. The first initiative was the McSharry reform of 1992, which added to the CAP specific environmental programmes governed by explicit regulations. The subsequent reform of Agenda 2000 gave a new impulse to introducing agri-environmental instruments into the CAP regime by making access to production-related direct payments conditional upon compliance with certain environmental standards. This new system of cross-compliance became mandatory for all member states under the Luxembourg reform of 2003, which promotes a multifunctional sustainable agriculture with direct payments for specific programmes substituted by a single farm payment fully decoupled from crop production.

The effect of the new CAP regime on irrigated agriculture (the implementation of which started in 2005) remains uncertain though several studies have underlined the potential of the new instruments for achieving compatibility between agricultural production and water resources conservation (Petersen and Shaw, 2000; Varela-Ortega *et al.*, 2002; Brouwer *et al.*, 2003). In particular, it can be expected that in many areas in Spain and in other member states, the decoupled single farm payment (SFP) will induce a land use shift away from highly productive and heavily water-consuming crops (such as maize). As the SFP was calculated as the annual average of the total payments received by a given farm during a 3-year reference period (2000, 2001, 2002), these crops are losing their financial comparative advantage, since they no longer benefit from the high production-related subsidies of the previous CAP programmes. Moreover, the new CAP requires the application of cross-compliance schemes that protect the environment and natural resources. These also can be expected to have a substantial impact on irrigated crops and water use.

Agricultural policies are not, however, the only policies that affect irrigated agriculture. In Spain as elsewhere in the EU, the reformed CAP is being implemented in parallel with the WFD, which calls for the adoption of water pricing instruments that incorporate the principle of full cost recovery of water services. If rigorously implemented, the WFD could well call into question the viability of a substantial proportion of irrigated farms in some areas of Spain (certainly in less fertile regions) (Berbel and Gutiérrez, 2004; Gomez-Limón and Riesgo, 2004; Mejías *et al.*, 2004; Varela-Ortega *et al.*, 2006b; Garrido and Calatrava, 2007). How these two ongoing policies will interact in the varied regions of Spain, how they will affect water use, irrigated agriculture, land use patterns, the conservation of natural resources, and the socio-economic development of rural areas, are still being investigated and constitute a major concern for public authorities.

Irrigation development and groundwater use in Spain: a policy-driven response

Groundwater is a strategic source of water in arid and semi-arid regions that face uneven distribution of rainfall and recurrent drought spells, such as the Mediterranean region. The use of groundwater for irrigated agriculture has expanded in recent decades relative to the use of surface water due to its accessibility to many private irrigators, the low cost of the associated irrigation infrastructure, high farming profitability, and lower vulnerability to climatic vagaries. New technologies for well drilling, pump installation and improved knowledge of hydrology have allowed an increasing number of independent private irrigators to resort to groundwater for farming in a 'silent revolution' (Llamas and Martinez-Santos, 2006). As a result, irrigation expansion has induced important socio-economic developments in rural areas due not only to the increase in direct farming activity but also to the indirect effects of secondary irrigation-related activities. Irrigation development and the resulting increase in groundwater abstractions have, in turn, however, caused overexploitation of aquifers and the progressive degradation of associated wetland ecosystems of high ecological value.

Depletion of aquifers by intensive irrigation has occurred in several regions of great environmental value in Spain. A remarkable example can be found in the western part of the region of La Mancha, on the southern central plateau. In this area, past CAP programmes encouraged irrigation expansion with positive social effects, including an increase in farm incomes, the creation of employment opportunities, the development of irrigation-related firms, population stability and overall socio-economic development (Martinez Vega *et al.*, 1995). On the other hand, the CAP programme has led to the overexploitation of the western La Mancha aquifer and to the subsequent degradation of the associated wetland ecosystem of the nearby national park 'Tablas de Daimiel' (Rosell and Viladomiu, 1997; Varela-Ortega and Sumpsi, 1999). This policy contradiction is depicted in Fig. 14.1 and illustrates how agricultural policies and environmental policies need to have common and coherent objectives. With the aim of remedying this ecological impact, a special agri-environmental programme (AEP) was launched in 1993 under the CAP environmental regulation of 1992.

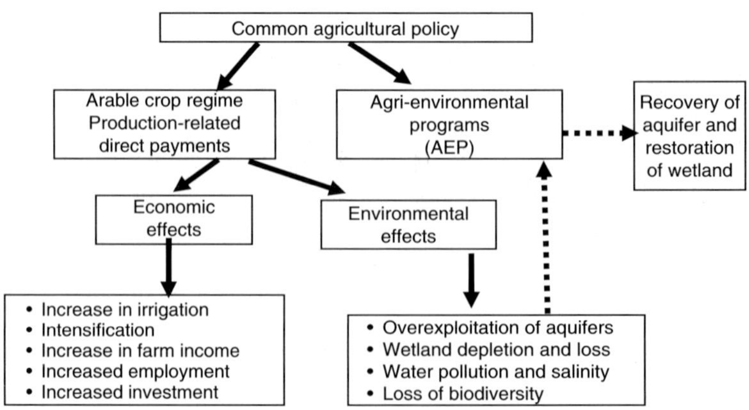

Fig. 14.1. A policy contradiction in the CAP agricultural and environmental programmes.

Irrigation Development and Environmental Sustainability: A Case Study in Spain

The unresolved controversy: groundwater irrigation or wetland conservation?

The wetland known as 'Tablas de Daimiel' in the western La Mancha region is unique and one of the most peculiar geomorphologic formations of the Spanish territory. The last example in Europe of a continental ecosystem known as a 'fluvial table', covering an area of about 2000 ha, this extraordinary wetland was formed by the overflow of the neighbouring rivers (Guadiana and Cigüela), its formation being favoured by the flat surrounding terrain and the high water table of the western La Mancha aquifer. The wetland is a unique habitat for the conservation of European and North African aquatic birds, with large populations of nesting and hibernating waterfowl and numerous species of aquatic flora and fauna. As a result, the wetland has attracted national as well as international recognition and has been registered under a number of national and international agreements, being made a UNESCO Biosphere reserve in 1981, a RAMSAR site in 1982 (Ramsar, 2006), a Special Protection Birds Area under the EU Birds Directive, and a Natura 2000 site under the Habitats Directive (Baldock *et al.*, 2000; MIMAM, 2006).

Over time, this fragile ecosystem has been progressively degraded as a result of excessive groundwater abstraction from the western La Mancha aquifer (Llamas *et al.*, 2001; CHG, 2006). The central aquifer covers an area of about 5000 km^2 and it had a surplus water balance up to the mid-1970s, before irrigation started to expand in the region. The expansion of irrigation has a clear policy-driven component. Figure 14.2 shows the evolution of water abstraction and irrigated area from 1985 to 2005. It also

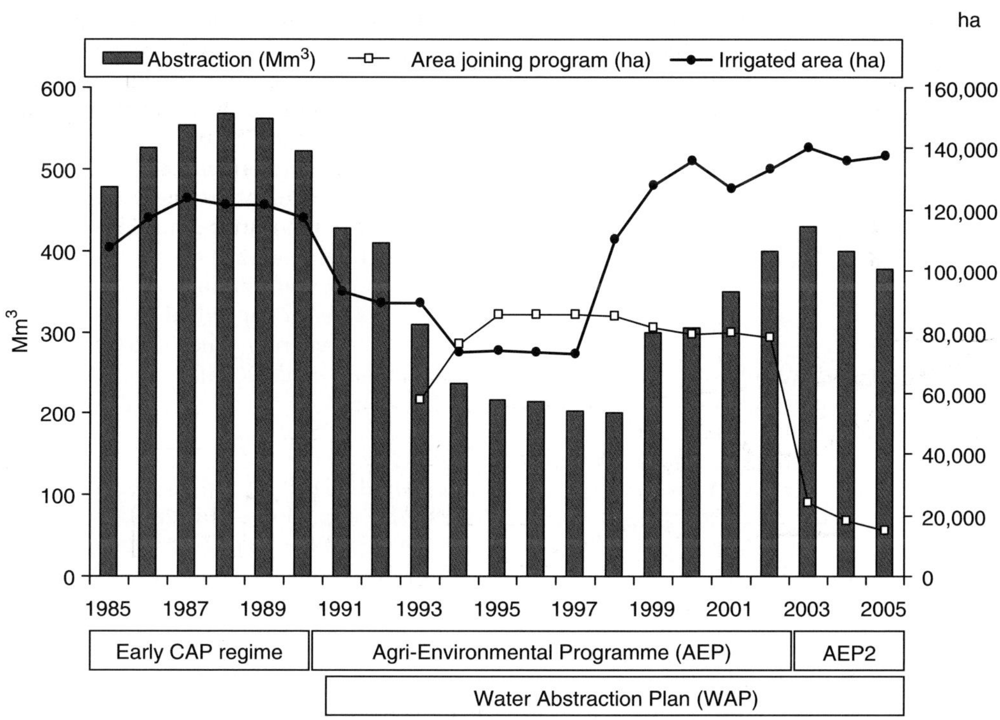

Fig. 14.2. Water abstractions, total irrigated surface and surface joining the AEP in the western La Mancha aquifer: 1985–2005. (From CHD (Confederación Hidrográfica del Guadiana) (2006) – JCC-LM (Junta de Comunidades de Castilla-La Mancha) (2006).)

shows the corresponding policy programmes that were applied during this period.

Following Spain's integration into the EC in 1986, the trend in irrigation expansion was reinforced. From the mid-1980s into the 1990s the intensity of well drilling and water abstraction by private irrigators increased considerably in response to the CAP subsidies. In the early 1990s, annual water abstractions rose to more than 500 Mm^3, greatly exceeding the natural recharge rate of the aquifer estimated at 230 Mm^3/year (CHG, 2006). As a consequence, return flows diminished considerably, the water table lowered and the aquifer was officially declared overexploited in 1991 (MOPTMA-CHG, 1995). The groundwater also suffered from salinization problems and contamination, while eutrophication of surface water produced changes in vegetation, peat fires, and a generalized decline of flooded lands that had devastating impacts on the local flora and fauna. Furthermore, the profitability of irrigated farming simultaneously diminished due to both the decrease in water availability and rising costs of deeper well drilling (Iglesias, 2001; Varela-Ortega *et al.*, 2002).

Two policies in the upper Guadiana basin: one objective and two instruments

The national policy

The official declaration that the western La Mancha aquifer was an 'overexploited aquifer' came about in 1991 and the River Basin Authority adopted a specific regulation that imposed a strict Water Abstraction Plan (WAP) (CHG, 2006) with the aim of restoring the overexploited aquifer. This regulation imposed strict water abstraction quotas on licensed wells and prevented the drilling of new ones. Maximum permitted water volumes were established according to farm size and crop type and, on average, the maximum allowable volume was set at 2000 m^3/ha, well below the preceding average water entitlement of around 4200 m^3/ha. Quotas were modified on an annual basis depending on climatic and demand conditions.[1]

Since the enactment of the 1985 Spanish Water Act, all water resources have been in the public domain, and irrigators have usufructuary water rights through administrative concessions granted by the Water Authority. Reflecting public ownership of the water, the WAP was defined by a water quota instrument and the farmers were not granted any compensation for the income foregone as a result of these compulsory measures. The water quotas were controlled either directly by water meters installed on-farm or – in most cases – indirectly by the crops grown by each individual farmer, making policy enforcement and control a difficult and costly exercise (MOPTMA-CHG, 1995). Moreover, the drastic reductions in the allowable quotas led to considerable social unrest and to free-riding behaviour in the form of illegal drilling of wells and excessive abstraction. This behaviour is common to other areas in the world where subterranean water is the major source of water for irrigation farming (Provencher and Burt, 1994; Shah *et al.*, 2000; Varela-Ortega and Sagardoy, 2003; Schuyt, 2005; Llamas and Martinez-Santos, 2006). Farmers opposed the cropping restrictions and water use limitations, given the lucrative price and subsidy incentives provided under the CAP. In sum, this water conservation policy faced major implementation difficulties and high transaction costs, as is typical of other similar cases of environmental policies (Whitby *et al.*, 1998; McCann *et al.*, 2005).

The EU policy

Following the CAP reform of 1992, a special 5-year AEP was adopted for the area in 1993 with the objective of recovering the wetlands of the National Park by reducing water abstraction from the aquifer. This programme proceeded in parallel with the national WAP but

[1]For 2006, the established permitted water volumes were 2640 m^3/ha for farms under 30 ha, 2000 m^3/ha for farms between 30 and 80 ha, and 1200 m^3/ha for farms above 80 ha (vineyards were granted a special entitlement of 1000 m^3/ha) (CHG, 2006).

was voluntary and had a social component that granted income compensation payments to irrigators in return for reductions in their water use. The initial 5-year programme was extended for another 5 years (1993–1997 to 1998–2002). Three levels of water use reductions were established, namely a 50%, 70% and 100% reduction in the irrigators' original (fixed) water entitlements (not subject to WAP annual adjustments). These reductions corresponded to three levels of income compensation payments, respectively (see Table 14.1). Thus, the policy instrument used for attaining the policy objective was a combination of voluntary water quotas and an income compensation scheme.

The overwhelming majority of farmers in the area joined the first AEP. By 1997 close to 90% of the total 120,000 ha of irrigated lands came under this programme and annual water abstractions were reduced by 60% or about 300 Mm^3, greatly exceeding the programme's objectives, which had targeted a reduction of 255–270 Mm^3 per annum (JCC-LM, 1999). While it was estimated that the water use restrictions of the compulsory WAP induced an average farm income loss of around €200–250/ha (MAPA-JCC-LM, 1992; Rosell and Viladomiu, 1997), the AEP with its income compensation scheme greatly reduced the social distress created by the WAP and encouraged farmers to shift to less water-demanding crops and to adopt water-efficient technologies (Rosell and Viladomiu, 1997; Iglesias, 2001). This water-saving behaviour was reinforced by the nationwide 5-year drought that lasted from 1991 to 1995. Due to the higher resilience of groundwater, the impact of the drought was much less severe in this area than in the lower part of the Guadiana basin where surface water irrigation is predominant (Llamas and Martinez-Santos, 2006). The programme had a much larger impact than foreseen and was able to achieve its environmental and socio-economic objectives (Rosell and Viladomiu, 1997; Iglesias, 2001). Its main drawback was its high cost in terms of public funds so that the cost-effectiveness of the policy was increasingly questioned (Varela-Ortega and Sumpsi, 1999).

The coupling of national and EU policies

The AEP was modified in 2003, reinforcing the environmental objectives promoted by the new CAP reform (also enacted in 2003). For this second phase, only the 50% and 100% water reduction levels were considered and the level of payments was based on farm size, with larger farms receiving a lower payment. Furthermore, water use volumes under the second phase were to be calculated not based on initial entitlements but on the water volumes established annually under the WAP, which reduced the permitted volumes even further. The second

Table 14.1. Evolution of the EU AEP. (From JCC-LM, 2006.)

EU AEP of western La Mancha aquifer Income Compensation Payments €/ha					
Level of reduction in water consumption (%)	First Phase (1993–2002) AEP1 Payments are independent of farm size			Second Phase (2003–2007) AEP2 Payments are modulated according to farm size	
	1993	1997	2001	2006	
50	156	164	179	1–40 ha	209
				40–80 ha	125
				>80 ha	63
70	258	271	296		
100	360	379	414	1–40 ha	518
				40–80 ha	311
				>80 ha	155

phase thus coupled, for the first time, the EU and National Policies within a common framework aiming to reduce water consumption in the agriculture sector, restore the aquifer and conserve associated wetlands. Since the new water quotas of the AEP2 were calculated as 50% or 100% of the WAP permitted volumes, they were substantially lower than the water quotas of the first phase of the EU programme, and thus the income compensation payments offered barely covered the income loss: the programme was hence abandoned by the majority of farmers. The total area where farmers participated in the programme was no more than 15,000 ha in 2005, as compared to close to 90,000 ha in 1997, and the total water use reduction was considerably lower than in the previous programme. Table 14.1 shows the evolution of the AEP during its two phases.

Besides seeking to control public expenditures, the merging of the two water policies (the Spanish WAP and the EU AEP) also reflected the EU WFD, enacted in 2000. As the first EU initiative designed to promote a comprehensive basin-based integrated water policy, this directive requires all EU member states to achieve 'good ecological status' of all watercourses by 2015. This meant that the River Basin Authority was required to strengthen the control of water abstractions and illegal drillings, so as inter alia to limit water abstraction by the agriculture sector to the maximum permitted total annual volume (200 Mm^3) compatible with the aquifer's natural recharge. For this reason, a Special Plan for the Upper Guadiana basin was recently presented with strict water consumption limitations for the irrigation sector, along with a socio-economic restructuring plan and the strengthening of public participation procedures (CHG, 2006).

The policy matrix given in Table 14.2 summarizes the agricultural and water policies that affect the study region. In the matrix, policies have been characterized by their objectives, instruments, and environmental and societal effects, including private (e.g. farmers' income) as well as public effects (e.g. enforcement and cost-effectiveness). The agricultural policies include the McSharry reform of the CAP in 1992 and the recent 2003 reform. The water policies are divided into two blocks: the first block includes the policies specific to the area of study, that is, the (national) WAP and the two phases of the subsequent EU AEP. The second block includes the general water policy (i.e. the WFD) that affects all regions of the EU.

The matrix underlines the interactions between agricultural and water policies by showing how the water quota of the first phase of the AEP is linked to the initial water endowments that prevailed prior to the last CAP reform in 2003. The matrix also shows how the quota instrument of the second phase of the AEP is linked to the (national) WAP, emphasizing the recent coupling of the national and EU policies.

Public Policies for Cost-effective and Sustainable Groundwater Management

In this section we present the methodology and results of the recent research EU project (NEWATER) conducted in the study area with the objective of analysing the respective environmental and socio-economic effects of the application of agricultural and water conservation policies.

The basic characteristics of the methodology are, first, the elaboration of a knowledge-base supported by considerable fieldwork and stakeholder consultation and, second, a farm-based non-linear static mathematical programming model of constrained optimization. The model describes the behaviour of representative farmers confronted by different policy scenarios. Following previous work in the area of study (Varela-Ortega *et al.*, 1998, 2002) the model incorporates new risk parameters and maximizes a utility function subject to technical, economic and policy constraints. The utility function is defined by a gross margin and a risk vector that takes into account climate as well as market prices variability. Activities are defined by a given cropping area and associated production

Table 14.2. Policy Matrix for Agricultural and Water Policies.

Policy		Policy objectives	Policy instruments	Environmental effects	Societal effects
Agricultural Policies	CAP McSharry Reform, 1992 Agenda 2000	• Farm income stability • Environmental protection	• Direct payments tied to production Original water Allotments	• High water consumption • Increase in irrigated surface • Overexploitation of aquifer • Loss of wetlands	• Increase in farm income • Increase in employment • Socio-economic development of rural areas
	CAP Luxembourg Reform, 2003 2005...	• Competitive agriculture • Farm income stability • Environmental sustainability • Rural development	• Single farm payment decoupled from production • Cross-compliance schemes: payments subject to compliance with environmental regulations	(expected) • Lower water use • Increase in extensive productions and low-water-demanding crops • Higher environmental sustainability	(expected) • Decrease in farm income • Larger application of rural development programmes
Specific Water Policies	Spanish WAP 1991...	• Reduction of water consumption • Aquifer stability • Wetland recovery	Water Quotas Established for aquifer recovery are lower than original allotments • Compulsory	• Less effective than expected • Lower water consumption • Use of modern irrigation techniques • Increase in low-water-demanding crops	• Farm income loss • Social unrest • Farmers opposition and litigations • Low implementation rate • High enforcement costs
	EU AEP1 1993–2002	• Independence of EU and National Policy • Independence of programme from the Spanish WAP • Reduction of water	Water Quotas Fixed as a 50%, 70% and 100% reduction from original water allotments	• Lower water consumption • Use of modern irrigation techniques • Increase in low-water-demanding crops • Recovery of the aquifer	• Large adoption rate by farmers • Compensation payments are sufficient • Farm income gain • Employment increase

Continued

Table 14.2. *Continued*

Policy		Policy objectives	Policy instruments	Environmental effects	Societal effects
		consumption and wetland recovery	+ Income compensation payments • Voluntary	• Partial restoration of wetlands	• Social stability • Low enforcement costs • High public cost • Low cost-effectiveness
	EU AEP2 2003. …	Dependence of EU and National Policies • Modulation of the programme and being tied to the Spanish WAP • Reducing water consumption and promoting wetland recovery	Water Quotas Fixed as a 50% and 100% reduction from quotas of the Spanish WAP + Income compensation payments	• Lower water consumption • Use of modern irrigation techniques • Increase in low-water-demanding crops • No recovery of the aquifer due to low implementation of the programme • Recharge of wetlands by water transfer from the Tajo basin	• Low adoption by farmers as water allotments are already low in the WAP upon which the new quotas are fixed • Compensation payments are not sufficient to attract farmers • Farm income loss • High public cost and low total water
General Water Policies	EU WFD 2000–2015	• Good ecological status of all water courses • Sustainable use of water resource • Integrated water management • Cost recovery of water services	• River Basin Organization as management unit • Planning and integrated management of all water resources • Economic instruments: Water pricing and application of the PPP • Development of programme measures in all basins	• Amelioration of the ecological conditions of watercourses • Lower water use in some areas • Increase in water use efficiency • Protection and recovery of wetlands	• Transparency and public participation • Accountability and cost-effectiveness assessment of policy measures • May reduce the economic viability of certain irrigated farms in southern EU

technique, irrigation method and soil type. The problem-solving instrument used is GAMS (General Algebraic Modelling System). The technical coefficients and parameters of the model were obtained from fieldwork carried out during 2006 in the study area, consisting of surveys and interviews with farmers, irrigation community representatives, technical experts, river basin managers, and regional government officials. The model was duly calibrated and validated, using the risk aversion coefficient as the calibration parameter and comparing results with data on crop distribution, land and labour in the study area.

The study area was represented by a set of four statistically representative farms that characterize the variety of production systems and farms types in the area. These representative farms correspond to the Irrigation Community of Daimiel that covers around 20,000 ha of irrigated lands and have 1450 affiliated members. The typology of representative farms is shown in Table 14.3.

Policy options

For comparative purposes, policy options have been selected for two reference years (2001 and 2006). All policies have been explained in the previous section and are summarized in the policy matrix (Table 14.2).

For year 2001 (based on results of previous research, Varela-Ortega *et al.*, 2002), two policy alternatives have been selected: (i) the CAP Agenda 2000 measures (reference policy), that include direct payments (a yield-based differentiated hectare premium which is higher for irrigated lands than for rain-fed lands); and (ii) the AEP that was in place in 2001 which includes water reduction quotas and an income compensation scheme.

For year 2006 (based on the model explained above) we have the 2003 CAP reform applied in conjunction with a water conservation policy chosen from amongst three options: (i) the WAP; (ii) the AEP2 with 50% water consumption reduction; and (iii) the AEP2 with 100% water consumption reduction. The WAP is mandatory and the two AEP are optional. The 2003 CAP reform is defined by a partial 75% decoupling scheme, the modality chosen by Spain, and the 4% modulation of subsidies.

The aggregate results of the policy analysis of 2001 are summarized in Table 14.4 (AEP1 70% was the modality chosen by the great majority of farmers) and are based on results of previous work (Varela-Ortega *et al.*, 2002), while the weighted average aggregate results of the policy analysis for 2006 (current policy options) are shown in Table 14.5.

Table 14.3. Farm Typology for the Irrigation Community of Daimiel in the Region of Castilla-La Mancha (2006). (From Field work analysis (2006) updated from Sumpsi *et al.*, 1998 (crop distributions are approximate).)

	F-1	F-2	F-3	F-4
Area (ha)	8	24	30	70
Soil quality	Low	High	Medium	Medium and low
Cropping pattern	Vine (100%)	Winter cereals (30%) Maize (5%) Horticulture (30%) Melon (20%) Set-aside (15%)	Winter cereals (25%) Maize (5%) Melon (25%) Vine (30%) Set-aside (15%)	Winter cereals (58%) Maize (2%) Horticulture and melon (30%) Set-aside (10%)
Coverage (% of area)	22	19	28	31

Table 14.4. Results of Policy Analysis (2001). (From Own elaboration based on Varela-Ortega *et al.*, 2002).

Aggregate results		Policy option: Reference agenda 2000	Policy option: AEP1 (70% reduction)
Farm income	Total	655	698
(€/ha)	%	100	107
Water consumption	Total	3776	1500
(m^3/ha)	%	100	40
Public expenditure	Total	212	386
(€/ha)	%	100	182

Table 14.5. Results of Policy Analysis (2006).

Aggregate results (2006)		Policy option: Reference policy CAP ref. with partial decoupling	WAP (mandatory)	AEP2 50% reduction	AEP2 100% reduction
Farm income	Total	944	765	676	584
(€/ha)	%	100	81	72	62
Water consumption	Total	3285	2495	1247	0
(m^3/ha)	%	100	76	38	0
Public expenditure	Total	100	82	339	630
(€/ha)	%	100	82	338	628
Water shadow price	Total	0.033	0.058	0.137	0.678
(€/m^3)	%	100	177	221	2058
Water costs (€/ha)	Total	201	154	79	0
	%	100	76	39	0
Water costs (€/m^3)	Total	0.06	0.062	0.063	0
	%	100	101	103	0
Water productivity (€/m^3)	Total	0.29	0.31	0.54	0
(average)	%	100	104	184	0
Income compensation of AEP (€/m^3)	Total			0.159	0.197
Crop distribution (%)	Rain-fed	12.4	34.3	54.6	100
	Irrigated	78.6	65.7	45.4	0
Labour (man-day/ha)	Hired	26.8	20.9	11.3	0.1
	Total	39.7	27.8	16.1	4.0

Discussion on the results is presented as follows:

On water consumption

The results for 2001 (Table 14.5) show that water use reduction under the first phase of the agro-environmental programme (AEP1 – Table 14.2) more than achieved the original AEP's objectives, reaching about 1500 m^3/ha. This was below the target of 2000 m^3/ha, as most of the farmers joined at the 70% reduction level (with water consumption on average reduced by 60%). However, as discussed above, from 2003 onwards, and with the adoption of AEP2 (Table 14.2), the average water consumption in the reference policy was smaller than in 2001 (3285 m³/ha instead of 3776 m³/ha) and the WAP reduced it even further (to 2495 m³/ha, on average) with the purpose of restoring the aquifer. For the 50% reduction level, the AEP2 resulted in a reduction down to only 1247 m³/ha on average, clearly insufficient for most crop requirements (Table 14.5).

Extrapolating these results to the overall aquifer (see Fig. 14.5), AEP1 was joined by a majority of farmers and affected around 90,000 ha, resulting in an estimated total reduction in water abstractions of 250 Mm³. But under AEP2, fewer farmers joined the programme which extended to only 15,000 ha, while the total volume saved in the aquifer was 35 Mm³ only.

On cropping patterns

Figure 14.3 shows the aggregate results for two CAP scenarios: Agenda 2000 (yield-based payments) and the recently applied CAP reform with decoupled payments (75% partial decoupling scheme). The water quantities that appear on the graph's *x*-axis correspond to the water allotments of the water scenarios selected (see Table 14.5). Extensive irrigation denotes crops that use low water quantities, such as barley and wheat and intensive irrigation denotes crops that use large water volumes, such as maize or sugarbeet. Results show that the newly applied decoupled CAP policy induces a shift away from water-intensive crops, such as maize, which loses its high direct subsidies. In the new CAP, rain-fed agriculture appears even in the reference scenario (3285 m³/ha) while in the former CAP Agenda 2000, rain-fed agriculture appears only under the AEP 50% reduction (1247 m³/ha). On the other hand, the cultivation of horticultural crops increases under the new policy across all water scenarios due to their higher profitability and their technical suitability to water-efficient irrigation technologies such as sprinkler and drip irrigation.

On-farm income

The new AEP2 (2006) results in a clear reduction in farmers' income despite the compensation payments that are granted to the farmers that voluntarily engage in this programme. For the 50% and 100% reduction alternatives, income is reduced by 30% and 40%, respectively. In contrast, the AEP1

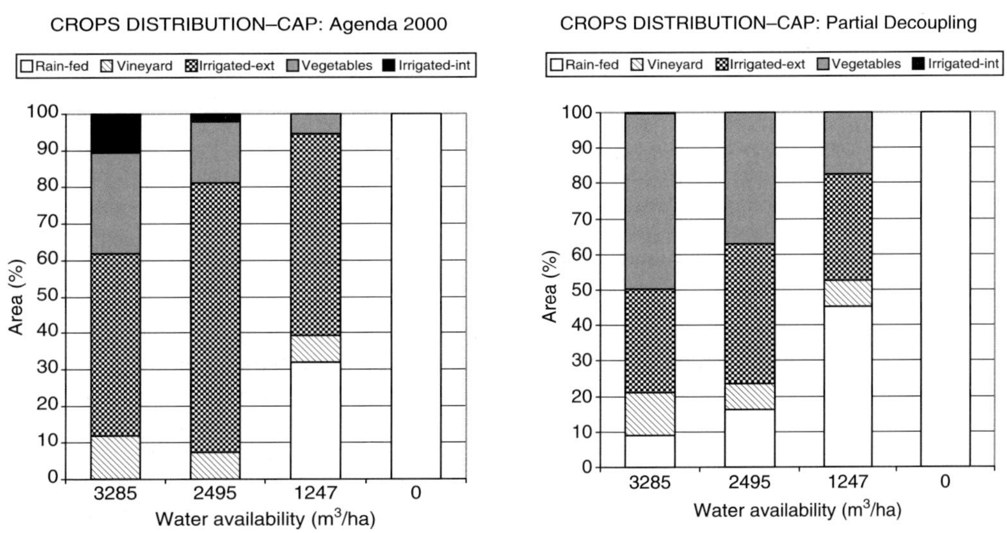

Fig. 14.3. Crop Distribution by Water Scenarios and Agricultural Policy Programmes. (From CAP, 2006.)

produced an increase of 6% in the income received by the farmers (Table 14.4). The reason is that, on the one hand, water allotments under AEP2 are calculated based on the WAP and thus amount to an average maximum permitted level of 1247 m^3/ha, lower than in the AEP1. On the other hand, income compensation payments in the previous programme were attractive enough for farmers to engage in the programme's 70% reduction level. Under the AEP2, income compensation is neither sufficient for the 50% reduction scheme nor for the 100% reduction level to make the programme attractive to the farmers.

On public expenditure

Both AEP1 and AEP2 are costly policies. In 2001, under AEP1, an average reduction of 60% relative to the original water allotment resulted in public expenditure almost doubling, rising by €386/ha. In 2006, under AEP2, public expenditure (including CAP payments) rose threefold to sixfold for the 50% and the 100% water reduction levels, respectively, corresponding to €339/ha and €630/ha, thus exceeding the impact on total farm income of this last option. The cost-effectiveness of these policies must therefore be questioned. Moreover, the direct costs (without the CAP payments) needed to reduce water use by one cubic meter are high under both options, amounting to €0.16 and €0.20 for the 50% and 100% reduction levels, respectively.

On water productivity

The average water values in all water scenarios are higher than the compensation payments, in unit terms, offered by the programme. Under the AEP2, for a 50% reduction level, average water productivity is €0.54/m^3, and the compensation offered to reduce consumption by half is €0.16/m^3. The same conclusion applies to the compensation offered under the alternative of abandoning irrigation altogether (€0.20/m^3). These results help explain the real situation in the area where the majority of the farmers are no longer willing to join the programme under this new stricter and less compensating scheme, as evidenced in the fieldwork survey and stakeholder interviews conducted in the zone (Varela-Ortega *et al.*, 2006a) and official data of the regional department of agriculture (JCC-LM, 2006).

Using average water values rather than marginal values for policy evaluation can, however, be ambiguous or even misleading as discussed extensively in the literature (Agudelo, 2001; Johansson *et al.*, 2002; Rogers *et al.*, 2002; Hanemann, 2006, among others). The reduction of water volumes under the AEP has been expressed in bulk volume terms as the compensation payment is equivalent for all units of water in the reduced allotment (€0.16/m^3 in the 1247/m^3 reduced allotment). However, the average value of water is not constant and increases as less water is supplied because farmers are likely to change their crops and technologies in response to water availability, as shown in the model results where cropping pattern changes according to the available water volumes and to the policy programmes. This can be shown in the results (Table 14.5) where average water value declines (from €0.54/m^3 to €0.29/m^3) as more water is delivered (from 1247/m^3/ha to 3285 /m^3/ha, respectively); thus the marginal value of water (shadow price of water in Table 14.5) is less than the average value.

The shadow prices of water thus increase as less water is supplied from €0.033/m^3, to €0.058/m^3, to €0.137/m^3 to a maximum of €0.678/m^3 as water allotments vary from 3285 m^3/ha, to 2495 m^3/ha, to 1247 m^3/ha and to 0, respectively. Similar results can be found for the region of Andalucia in Spain (Iglesias *et al.*, 2003). In our example, the results show that shadow price of water is greater (€0.678/m^3) than the compensation payment in unit terms (€0.197/m^3) for the first marginal unit of water. This result helps explain why the majority of farmers have proven unwilling to join the second phase of the AEP2.

The role of water pricing

Following the discussion of the previous sections, it is clear that water policies applied in the upper Guadiana basin have been ineffective in reducing water abstractions to a level compatible with the replenishment of the aquifer and hence the recovery of the wetlands. As the WAP is not fully enforced and the new AEP2 has been joined only by a small proportion of the irrigators, the quota instruments used in both programmes are not effective. In this situation it is interesting – for the purpose of policy analysis – to explore the potential effect of the application of an alternative instrument such as a water tariff structure.

The use of water tariffs has been discussed extensively in the literature as a major instrument for demand management policies and water conservation (Varela-Ortega *et al.*, 1998; Johansson *et al.*, 2002; Rogers *et al.*, 2002; Rosegrant *et al.*, 2002; Gomez-Limón and Riesgo, 2004; Garrido and Calatrava, 2007, among others). Water pricing policies can provide the farmers with the proper incentive to save water but, as water demand tends to be inelastic at low price ranges and institutional factors are determinant, volumetric pricing remains a controversial issue in many real-world examples and its wide application is still limited (see de Fraiture and Perry, Chapter 3, this volume; Molle and Berkoff, Chapter 2, this volume).

Subsequent research has been carried out by the author's research team in the area of study (Blanco, 2006), based on the same type of methodology and they have analyzed the effects of the application of simulated volumetric tariffs on irrigated farms. The results of this research can be used as a baseline for assessing the cost-effectiveness of the current policies applied in the area.

Two selected farms have been used for this analysis (E1 and E2) that correspond basically to the extensive large farm (F4) and the more intensive medium-size farm (F3) of Table 14.3. Table 14.6 shows the aggregate results of the application of increasing volumetric water tariffs on water demand, farm income, revenue collected by the water authority and public expenditure. Figures 14.4 and 14.5 show the water demand curves of the individual farms and the farm income variation when water tariffs are applied.

Water tariffs are applied once the current policy is in place (that is the WAP quota of 2049 m^3/ha) and we can see from the simulation results that water demand is reduced progressively and reaches an average level

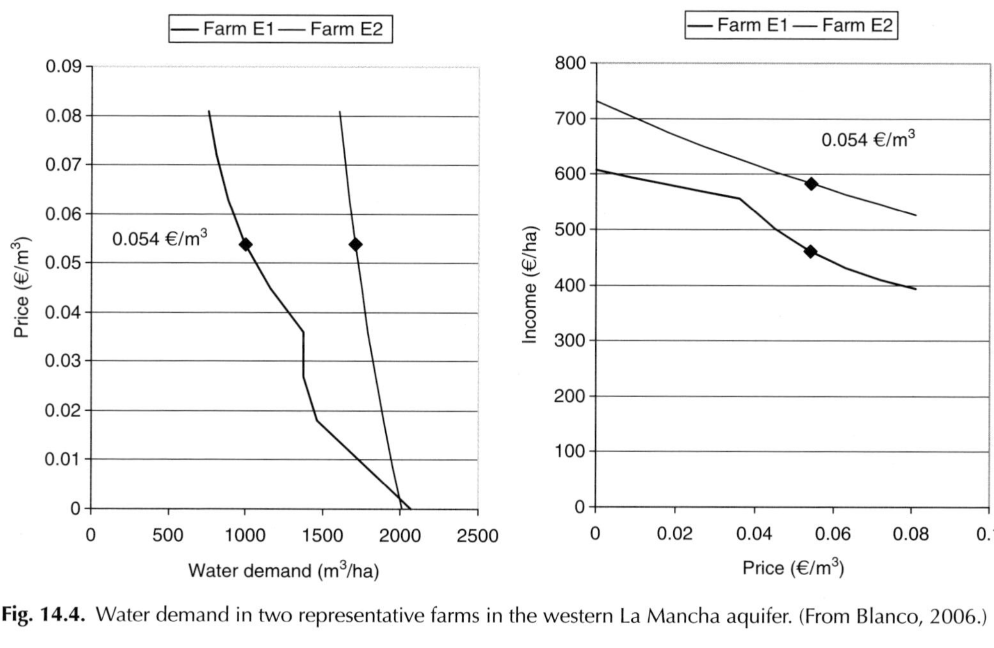

Fig. 14.4. Water demand in two representative farms in the western La Mancha aquifer. (From Blanco, 2006.)

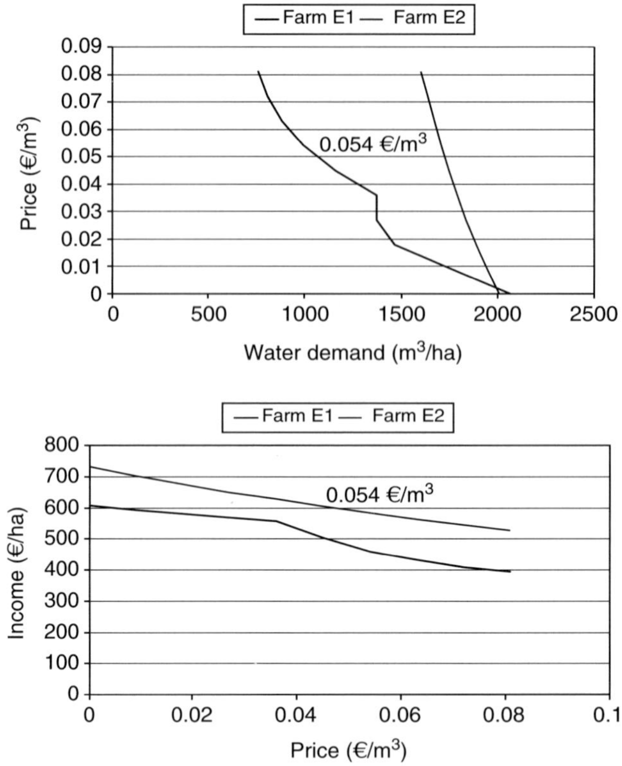

Fig. 14.5. Farm income variation in two representative farms in the western La Mancha aquifer. (From Blanco, 2006.)

Table 14.6. Effects of the Application of Volumetric Water Tariffs on Irrigated Farms in the Western La Mancha Aquifer. (From Own elaboration from Blanco, 2006.)

Water tariff (€/m³)	Water demand (m³/ha)	Farm income (€/ha)	Collected revenue (€/ha)	Government expenditure (€/ha)	Net public expenditure (€/ha)
0	2.049	646	0.0	115.4	115.4
0.009	1.822	627	16.4	97.5	81.1
0.018	1.596	610	28.7	79.4	50.7
0.027	1.518	594	41.0	74.3	33.3
0.036	1.503	578	54.1	74.6	20.5
0.045	1.342	534	60.4	78.2	17.8
0.054	1.215	499	65.6	80.9	15.3
0.063	1.127	472	71.0	82.9	11.9
0.072	1.064	452	76.6	84.2	7.6
0.081	1.018	435	82.4	64.7	−17.7

Note: Farm income figures for a zero water tariff are not exactly the same as for the WAP option in Table 14.5 due to slight differences in the farms considered but they are largely equivalent.

compatible with the natural recharge rate of the aquifer (1215 m^3/ha) at a water tariff of €0.054/m^3. This water consumption level is equivalent to the level attained by the AEP2 (of 50% reduction in water use) in Table 14.6. Figure 14.4 shows that water demand is more inelastic in the more intensive farm (E2) as higher productivity permits to absorb increased water use costs without drastically changing the cropping pattern towards less water-demanding crops or to rain-fed farming.

For this level of water tariff (€0.05/m^3), farm income is reduced by 23% (€147.6/ha) in the aggregate. However, aggregate results can be misleading. As shown in Fig. 14.5, in the more intensive farm E2, inelastic demand responses result in water use reductions of 15% that face higher income losses (about 21%), a result widely found in the literature. But the more extensive E1 farm responded to increasing water prices by shifting away from water-intensive horticulture (such as potato) to specialized low-water-demanding vegetables such as melon, a lucrative adapted crop in the area that is grown with drip irrigation. This explains why, in the aggregate, water use reductions are accompanied by a rather small income loss. In the case of the quota-based AEP2 income loss is barely 12% (€90/ha). However, public expenditure in the case of the application of water tariffs (that include only CAP subsidies) is, in fact, reduced by a small amount of €34/ha (from €115/ha to €81/ha) when prices rise to the desired target of €0.05/m^3 that recovers the aquifer. As water prices are administered prices, the revenue collected from the water fees by the water agency is public revenue and thus the overall net public expenditure is almost nil (apart that is from collection costs that have not been considered here). Conversely, AEP2 is an expensive policy, as pointed out in the previous section, and public costs rise more than fourfold to support this policy, reaching €339/ha, a substantially larger budget. This evidences the fact that agri-environmental polices that entail income compensation are not sustainable financially and their cost-effectiveness is indeed questionable.

Concluding reflections

- In general, water conservation polices that apply a strict quota system can achieve water use reductions and wetland recovery at low public costs. However, these policies are likely to be opposed strongly by the farmers, motivating costly litigation processes, a low uptake of the programmes and high enforcement costs to public authorities. Increasing the direct participation of stakeholders and stronger involvement in the decisions as well as social learning activities are strongly needed for the acceptance of this type of policies.
- Water conservation polices that include a quota system and an income compensation scheme (such as the AEPs applied in the area of study), can achieve the programmed water conservation target, provided that the compensation payment is attractive to the farmers. These policy programmes generally have a higher social acceptance and farmers' unrest can be avoided when compensation payments are sufficiently high to balance the income foregone by the farmers. However, these policies can be very costly, thus questioning the sound application of the policy. Moreover, such programmes conflict with the recently adopted EU WFD that, to ensure the good ecological status of all water bodies, requires the application of the polluter pays principle and a cost-effective evaluation of all programme measures (EU, 2000).
- It may occur that water quotas which entail compensation payments are too low (such as 50% from the permitted volumes) so that farm income can decrease if, for budgetary reasons, compensation payments are not sufficient to compensate for income loss. This will result in a limited adoption of the income compensation policy by the farmers and, in the aggregate, the policy may not meet the overall programmed water conservation targets.
- Water pricing policies can be effective instruments to induce water conservation strategies and are inexpensive

policies when compared with AEPs with an income compensation scheme. It seems likely that this kind of economic instrument could be effective for achieving the desired goals of reducing water extraction from the aquifer. For 50% water reduction levels, we may conclude that water pricing policies are more cost-effective than AEPs. However, even though volumetric pricing induces water use efficiency, it may produce distinctive effects across farm types. Due to the inelastic response of water demand to price changes in some farm types, a uniform water tariff may not achieve water conservation purposes in all areas. In addition, enforcing such water pricing schemes in private groundwater use has proved to be extremely difficult (see Venot *et al.*, Chapter 10, this volume).

- From an environmental perspective the application of a water pricing policy in this zone will be beneficial if reduction of irrigation in the area would achieve environmental objectives, but this policy would also entail economic and social costs to the area.
- Evaluating water productivity and water values needs careful attention. There is a tendency in the evaluation of water policies and projects to use average value estimates rather than marginal values, as marginal values require modelling estimates. A disparity between average and marginal values might be a crucial factor in misrepresenting the real value of water as, in most cases, average values are taken to be constant and hence overvalued (as argued by Hanemann (2006) in the case of the water transfer from the Ebro basin).
- Integrating agricultural polices and water polices is a key element for water conservation purposes. In fact, the new EU agricultural polices that incorporate, to a larger extent, environmental requirements, can play a major role in influencing water use trends and hence in meeting water conservation objectives. Water policies and agricultural policies should be designed and implemented in an integrated stakeholder-participatory manner, avoiding contradictions, finding synergies and integrating common objectives. However, the social context in which these policies will have to be implemented requires the selection of socially accepted instruments to balance the dual objective of protecting natural resources and maintaining farm-based livelihoods at tolerable social costs. This dual objective is best attained when strict water polices are combined with accompanying measures of rural development programmes and the establishment of water banks that permit a more flexible distribution of water allotments among farmers. This is the challenge facing the Spanish regional administration in charge of the application of both national and EU water policies in the area that we have studied. The requirements of the WFD to reach 'a good ecological status of all water bodies' in the EU with 'public transparency and participation' are providing incentives to the regional and national administrations to better enforce the water conservation policy. The new rural and social development program (*Plan Especial del Alto Guadiana*) being launched in this area is designed to diminish economic and social burdens. The design and enforcement of well-balanced polices are major tasks of policy makers in achieving successful water policies.

Acknowledgements

The author wishes to acknowledge the EU project NEWATER (New Approaches to Adaptive Water Management under Uncertainty, 2005–2009), and the Spanish Ministry of Education and Science (*Análisis de la gestión integrada del agua en la agricultura: aspectos socio-económicos, ambientales e institucionales*) for providing research funds for this study. The collaboration of the research assistants Irene Blanco, Gema Carmona and Paloma Esteve of the Polytechnic University of Madrid is also greatly acknowledged by the author, especially their valuable assistance during the fieldwork, data analysis and modelling.

References

Agudelo, J. (2001) *The Economic Valuation of Water: Principles and Methods*. Value of Water Research Report series N° 5. IHE Delft, The Netherlands.

Baldock, D., Dwyer, J., Sumpsi, J., Varela-Ortega, C., Caraveli, H., Einschütz, S. and Petersen, J.-E. (2000) *The Environmental Impacts of Irrigation in the European Union*. Report to the European Commission. European Commission, Brussels.

Benoit, G. and Comeau, A. (eds) (2005) *A Sustainable Future for the Mediterranean. The Blue Plan's Environment and Development Outlook*. Plan Bleu. Earthscan/James & James Publishers, London.

Berbel, J. and Gutiérrez, C. (eds) (2004) *Sustainability of European Irrigated Agriculture under Water Framework Directive and Agenda 2000*. Sixth Framework Program, European Commission, Córdoba, Spain.

Blanco, I. (2006) Impacts de politiques alternatives dans la gestion durable des eaux souterraines pour l'agriculture: le cas de l'aquifère de la Mancha Occidental (Bassin du Guadiana-Espagne). MSc thesis. Faculté de Sciences Economiques, Montpellier I, France.

Bromley, D.W. (2000) Property regimes and pricing regimes in water resources management. In: Dinar, A. (ed.) *The Political Economy of Water Pricing Reform*. The World Bank, Washington. Oxford University Press, New York, pp. 29–48.

Brouwer, F., Heinz, I. and Zabel, T. (eds) (2003) *Governance of Water-related Conflicts in Agriculture; New Directions in Agri-environmental and Water Policies in the EU*. Environmental and Policy Series, volume 37. Kluwer Academic Press, Dordrecht, The Netherlands.

CHG (Confederación Hidrográfica del Guadiana) (2006) *Plan Especial del Alto Guadiana*. Documento de Síntesis. CHG, Ciudad Real.

EU (European Union) (2000) *Water Framework Directive: Directive 2000/60/EC of The European Parliament and of The Council of 23 October 2000,* establishing a framework for Community action in the field of water policy. Official Journal of the European Communities (L 327) 22 December. Luxembourg, Belgium.

Garrido, A. and Calatrava, J. (2007) Recent and future trends in water charging and water markets. In: Garrido, A. and Llamas, M.R. (eds) *Water Policy in Spain. Issues in Water Resource Policy*. Resources for the Future Press, Washington, DC.

Gomez-Limón, J.A. and Riesgo, L. (2004) Irrigation water pricing: differential impacts on irrigated farms. *Agricultural Economics* 31, 47–66.

Hanemann, W.H. (2006) The economic conception of water. In: Rogers, P.P., Llamas, M.R. and Martinez-Cortina, L. (eds) *Water Crisis: Myth or Reality*. Marcelino Botin Water Forum 2004. Taylor & Francis, London, pp. 61–91.

Iglesias, E. (2001) Economía y gestión de las aguas subterráneas: El caso del acuífero Mancha Occidental. PhD dissertation. Polytechnic University of Madrid.

Iglesias, E., Garrido, A. and Gomez-Ramos, A. (2003) Evaluation of drought management in irrigated areas. *Agricultural Economics* 29, 211–229.

JCC-LM (Junta de Comunidades de Castilla-La Mancha) (1999) *Consejería de Agricultura y Medio Ambiente. Programa de Compensación de las Rentas Agrarias en las Unidades*. Hidrogeológicas de Mancha Occidental y Campo de Montiel. Comunidad de Castilla-La Mancha, Toledo, Spain.

JCC-LM (Junta de Comunidades de Castilla-La Mancha) (2006) *Consejería de Agricultura y Medio Ambiente*. Comunidad de Castilla-La Mancha, Toledo, Spain.

Johansson, R.C., Tsur, Y., Roe, T.L., Doukkali, R. and Dinar, A. (2002) Pricing irrigation water: a review of theory and practice. *Water Policy* 4, 173–199.

Llamas, M.R. and Martinez-Santos, P. (2006) Significance of the *Silent Revolution* of intensive groundwater use in world water policy. In: Rogers, P.P., Llamas, M.R. and Martinez-Cortina, L. (eds) *Water Crisis: Myth or Reality*. Marcelino Botin Water Forum 2004. Taylor & Francis, London, pp. 163–180.

Llamas, M.R., Fornés, J.M., Hernández-Mora, N. and Martínez Cortina, L. (2001) *Aguas Subterráneas: Retos y Oportunidades*. Fundación Marcelino Botín. Mundi-Prensa, Madrid.

MAPA-JCC-LM (Ministerio de Agricultura, Pesca y Alimentación – Junta de Comunidades de Castilla-La Mancha) (1992) *Plan Coordinado de Actuación en la Zona de la Mancha Occidental y Campo de Montiel*. Madrid (September).

Margat, J. (2004) *Atlas de L'eau Dans le Bassin Méditerranéen. CCGM/Plan Bleu/Unesco*. UNESCO, Paris.

Martinez Vega, J., Navalpotro, P., Cebrián, J.A. and Romero, R. (1995) Repercusiones sobre la sobre-explotación de acuíferos y de la PAC en la sustentabilidad de la agricultura manchega. *Estudios Geográficos* 56–19, 337–369.

McCann, L., Colby, B., Easter, K.W., Kasterine, A. and Kuperan, K.V. (2005) Transaction cost measurement for evaluating environmental policies. *Ecological Economics* 52, 527–542.

Mejías, P., Varela-Ortega, C. and Flichman, G. (2004) Integrating agricultural policies and water policies under water supply and climate uncertainty. *Water Resources Research* 40(7).

MIMAM (Ministerio de Medio Ambiente) (2006) Parques nacionales. Madrid (http://reddeparquesnacionales.mma.es).

MOPTMA-CHG (Ministerio de Obras Públicas, Transportes y Medio Ambiente – Confederación Hidrográfica del Guadiana) (1995). El problema hidráulico de la cuenca alta del Guadiana.

Petersen, J.E. and Shaw, K. (eds) (2000) Environmental Standards in Agriculture. *Proceedings of a Pan-European Conference on Meeting of Environmental Standards under Agenda 2000 (Madrid 5–7 October 2000).* IEEP (Institute for European Environmental Policy), London and WWF (World Wild Fund).

Provencher, B. and Burt, O. (1994) A private property rights regime for the commons: the case for groundwater. *American Journal of Agricultural Economics* 76, 875–888.

Ramsar (2006) www.ramsar.org

Rogers, P., de Silva, R. and Bhatia, R. (2002) Water is an economic good: How to use prices to promote equity, efficiency, and sustainability. *Water Policy* 4(1), 1–17.

Rosegrant, M., Cai, X. and Cline, S. (2002) *World Water and Food to 2025: Dealing with Scarcity*. International Food Policy Research Institute (IFPRI), Washington, DC.

Rosell, J. and Viladomiu, L. (1997) El programa de comoensación de rentas por reducción de regadíos en Mancha Occidental y Campo de Montiel. *Revista Española de Economía Agraria* 179, 331–350.

Schuyt, K.D. (2005) Economic consequences of wetland degradation for local populations in Africa. *Ecological Economics* 53, 177–190.

Shah, T., Molden, D., Sakthivadivel, R. and Seckler, D. (2000) The Global Groundwater Situation: Overview of Opportunities and Challenges. Research Report. International Water Management Institute, Colombo.

Sumpsi, J.M., Garrido, G., Blanco, M., Varela-Ortega, C. and Iglesias, E. (1998) *Economía y Política de Gestión del Agua en la Agricultura*. Mundi Prensa, Madrid.

Varela-Ortega, C. and Sumpsi, J.M. (1999) Assessment of cost-effectiveness of policy instruments for sustainable development in environmentally sensitive irrigation areas. Paper presented at the IXth Congress of the European Association of Agricultural Economists (EAAE), Warsaw, Poland.

Varela-Ortega, C. and Sagardoy, J.A. (2003) Analysis of irrigation water policies in Syria: current developments and future options. In: Fiorillo, C. and Vercueil, J. (eds) *Syrian Agriculture at the Crossroads*. FAO series APED (Agriculture Policy and Economic Development) n° 8. FAO, Rome, pp. 335–359.

Varela-Ortega, C., Sumpsi, J.M., Garrido, A., Blanco, M. and Iglesias, E. (1998) Water pricing policies, public decision making and farmers' response: implications for water policy. *Agricultural Economics* 19, 193–202.

Varela-Ortega, C., Sumpsi, J.M. and Blanco, M. (2002) Water availability in the Mediterranean region. In: Brouwer, F. and Van der Straaten, J. (eds) *Nature and Agriculture Policy in the European Union*. International Library of Ecological Economics. Edward Elgar, Cheltenham, UK, pp. 117–140.

Varela-Ortega, C., Blanco, I. and Esteve, P. (2006a) *Economic and Agronomic Aspects of Water Management in the Upper Guadiana Basin*. Stakeholder Meeting Report. Newater Project (Contract no 511179, Integrated Project in Priority 6.3 Global Change and Ecosystems) (www.newater.info). Research Directorate General, European Commission, Brussels.

Varela-Ortega, C., Simó, A. and Blanco, I. (2006b) *The Effects of Alternative Policy Scenarios on Multifunctionality: A Case Study of Spain*. Working Paper n° 15. ENARPRI (European Network of Agricultural Policy Research Institutes). CEPS (Center for Eurpean Policy Studies), Brussels.

Whitby, M., Saunders, C. and Ray, C. (1998) The full cost of stewardship policies. In: Dabbert, S., Dubgaard, A., Slangen, L. and Whitby, M. (eds). *The Economics of Landscape and Wildlife Conservation*. CAB International, Wallingford, UK, pp. 97–112.

Index

Note: Page references in *italic* refer to figures or tables in the text